ME/CFS – the Severely and Very Severely Affected

ME/CFS – the Severely and Very Severely Affected

Editors

Kenneth J. Friedman
Lucinda Bateman
Kenny Leo De Meirleir

Basel • Beijing • Wuhan • Barcelona • Belgrade • Novi Sad • Cluj • Manchester

Editors

Kenneth J. Friedman	Lucinda Bateman	Kenny Leo De Meirleir
Rowan University	Bateman Horne Center	Vrije Universiteit
Stratford, CT, USA	Salt Lake City, UT, USA	Brussels, Belgium

Editorial Office
MDPI
St. Alban-Anlage 66
4052 Basel, Switzerland

This is a reprint of articles from the Special Issue published online in the open access journal *Healthcare* (ISSN 2227-9032) (available at: https://www.mdpi.com/journal/healthcare/special_issues/me_cfs_issue).

For citation purposes, cite each article independently as indicated on the article page online and as indicated below:

Lastname, A.A.; Lastname, B.B. Article Title. *Journal Name* **Year**, *Volume Number*, Page Range.

ISBN 978-3-0365-9360-9 (Hbk)
ISBN 978-3-0365-9361-6 (PDF)
doi.org/10.3390/books978-3-0365-9361-6

Contents

About the Editors

Kenneth J. Friedman

Dr. Friedman received his Ph.D. from the State University of New York at Stony Brook in 1970. He trained under a NIH Post Doctoral Fellowship, followed by a NIH Staff Fellowship, before accepting a teaching and research position at the New Jersey Medical School. The inability of his daughter to obtain a diagnosis or treatment for her Chronic Fatigue Syndrome led Dr. Friedman to relinquish basic research and begin clinical research into ME/CFS and now PAPIS – Post Active Phase of Infection Syndromes. Dr. Friedman wrote the lead chapter in the New Jersey *Consensus Manual for the Primary Care and Management of ME/CFS*, and led the effort to create and contributed content to both the IACFS/ME's *Primer for Clinical Practitioners* and the International Writing Group's *Myalgic Encephalomyelitis/Chronic Fatigue Syndrome Diagnosis and Management in Young People: A Primer*. Dr. Friedman has guest-edited several ME/CFS medical journal special issues and has written chapters regarding ME/CFS for several major medical publishers.

Lucinda Bateman

Dr. Bateman attended the Johns Hopkins School of Medicine, completed her internal medicine residency at the University of Utah, and practiced as a general internist in Salt Lake City for 10 years before she started work in the Fatigue Consultation Clinic in 2000. In 2015, she founded the Bateman Horne Center (BHC), a 501(c)3 non-profit organization with a mission to improve the lives of people with myalgic encephalomyelitis/chronic fatigue syndrome (ME/CFS), fibromyalgia, and related conditions, through clinical care, education and research. She was an author of the 2015 Institute of Medicine publication recommending new evidence-based clinical diagnostic criteria for ME/CFS. The BHC was the clinical partner on two NIH-funded Collaborative Research Center (CRC) grants used to study the immunology, metabolomics and microbiome of ME/CFS. The BHC has also evaluated and treated 70-80 Long COVID patients in the clinic and has enrolled many in research.

Kenny Leo De Meirleir

Dr. De Meirleir received his M.D. in 1977 and his Ph.D. in 1985. His initial interests were in cardiology and cardiac rehabilitation, which led to his interest, research and publications in the field of ME/CFS. As of 2014, he has authored 681 publications with colleagues in Europe and the United States. Currently retired, he works 3 days per week diagnosing and treating patients with poorly explained chronic illness and serves as the director of Himmunitasvzw – a private institution in Belgium specializing in the treatment of ME/CFS, chronic Lyme disease and Long COVID.

Preface

ME/CFS – The Severely and Very Severely Affected

ME/CFS – The Severely and Very Severely Affected is the first themed, medical journal Special Issue created to specifically focus on the research and patient management strategies available to homebound (severely ill) and bedbound (very severely ill) ME/CFS patients. Here, it is available as a monograph. Of the scant resources that have been devoted to and developed for the totality of ME/CFS patients, the majority have been focused on mildly to moderately ill patients. These patients are able to present themselves in the offices of their healthcare providers or to a medical facility to receive care. But what of patients who are not ambulatory, or those for whom travel outside of a limited perimeter will result in an increased severity of disease, or be life threatening? How should care for these patients be provided? With an estimated 1.7 to 3.38 million Americans suffering with ME/CFS [1] over 65 million patients estimated worldwide,[2] and an estimated one in four of these patients being severely ill,[3] it is not appropriate to ignore the clinical needs of these patients nor neglect the need for both medical and clinical research on their exacerbated pathology. The conception and inception of *ME/CFS – The Severely and Very Severely Affected* preceded the COVID-19 pandemic, and more importantly, preceded the consequence of this pandemic: long-haul COVID or PASC. Nevertheless, in view of the similarities now being described between ME/CFS and long-haul COVID, it is reasonable to expect that a yet-to-be-undetermined number of long-haul COVID patients will become homebound and bedbound with an illness or illnesses not unlike severe and very severe ME/CFS. Thus, while it was not our intent to illuminate or forecast the long-term and perhaps life-long deficits produced by the COVID-19 virus, our Special Issue suggests a path forward in terms of research and providing care for those severely and very severely affected by long-haul COVID. *ME/CFS – The Severely and Very Severely Affected* offers methodological insights to researchers, patients, and clinicians alike, who will be confronted with a similar disease scenario; it has therefore taken on additional importance. In a monograph format, it serves as a resource for healthcare providers and patients alike.

Proposing and initiating *ME/CFS – The Severely and Very Severely Affected* defied conventional wisdom. Searching PubMed for papers using the terms [ME/CFS], or [Chronic Fatigue Syndrome], or [Myalgic Encephalomyelitis] and [very severely] ill retrieved no articles in the literature prior to our Special Issue. Searching PubMed for papers using the words [ME/CFS,] or [Chronic Fatigue Syndrome], or [Myalgic Encephalomyelitis] and [severely ill] yielded 12 papers in the literature prior to our Special Issue. With such scant literature on the subject and scant federal support for research on this group of patients, and with so few clinicians able to find a viable mechanism of supporting themselves while providing care for these patients, was it reasonable to attempt to establish a Special Issue devoted to these patients? In retrospect, the answer is a resounding "yes," as evidenced by the number of papers contained in this monograph. Herein are 25 articles, previously published in our peer-reviewed, PubMed indexed, journal Special Issue. The articles have been placed into categories and briefly described in the monograph's Summary.

This monograph contains two added audiovisual links (shown as the appendix): (1) Per author Whitney Dafoe's request, we provide a link and a QR code leading to to the audio recording of his manuscript. Thus, patients who are severely ill or otherwise unable to read his article may listen to it. (2) We are also providing a link to pre-existing videos that serve as an audio–visual summary of living life as a severely ill ME/CFS patient. Words alone cannot convey the impact of this illness. Videos convey not only the extent of debilitation, but also the emotional toll.

The Guest Editors wish to thank all contributing authors and the journal for having the courage to support our enterprise despite its uncertainties. With the anticipated population of long-haul COVID patients far exceeding the ME/CFS population worldwide, our hope is that this monograph, with its diversity of content addressing the severely and very severely ill, will serve as a guide to the path forward for patients, caregivers, governments, and the public at large.

1. Valdez AR, Hancock EE, Adebayo S, et al. Estimating Prevalence, Demographics, and Costs of ME/CFS Using Large Scale Medical Claims Data and Machine Learning. Front Pediatr. 2019;6:412. Published 2019 Jan 8. doi:10.3389/fped.2018.00412

2. Germain A, Barupal DK, Levine SM, Hanson MR. Comprehensive Circulatory Metabolomics in ME/CFS Reveals Disrupted Metabolism of Acyl Lipids and Steroids. Metabolites. 2020 Jan 14;10(1):34. doi: 10.3390/metabo10010034. PMID: 31947545; PMCID: PMC7023305.

3. Chang CJ, Hung LY, Kogelnik AM, et al. A Comprehensive Examination of Severely Ill ME/CFS Patients. Healthcare (Basel).

Additional Acknowledgements

- We thank all contributing authors and the editors of this journal for supporting the creation of the first monograph to singularly focus on severely and very severely ill ME/CFS patients.

- We thank the ME/CFS patient advocacy organizations that provided financial support for the initial publication of articles in our Special Issue of Healthcare and are reproduced in this monograph.

- We thank the journal's editorial staff for using fee-reducing mechanisms for some of the manuscripts published in our themed issue. The cost of manuscript publication did not prevent any manuscript from being published.

Kenneth J. Friedman, Lucinda Bateman, and Kenny Leo De Meirleir

Editors

Viewpoint

Extremely Severe ME/CFS—A Personal Account

Whitney Dafoe

Independent Researcher, 433 Kingsley Ave., Palo Alto, CA 94301, USA; whitney@whitneydafoe.com

Abstract: A personal account from an Extremely Severe Bedridden ME/CFS patient about the experience of living with extremely severe ME/CFS. Illness progression, medical history, description of various aspects of extremely severe ME/CFS and various essays on specific experiences are included.

Keywords: ME/CFS; extremely severe ME/CFS; severe ME/CFS; myalgic encephalomyelitis; chronic fatigue syndrome; personal account

Citation: Dafoe, W. Extremely Severe ME/CFS—A Personal Account. *Healthcare* **2021**, *9*, 504. https://doi.org/10.3390/healthcare9050504

Academic Editors: Christopher R. Cogle, Kenneth J. Friedman, Lucinda Bateman and Kenny Leo De Meirleir

Received: 25 February 2021
Accepted: 21 April 2021
Published: 27 April 2021

Publisher's Note: MDPI stays neutral with regard to jurisdictional claims in published maps and institutional affiliations.

1. Biography

Whitney Dafoe (see Figure 1) studied photography at Bennington College and The San Francisco Art Institute. His award-winning work in photography and film has been published and exhibited worldwide. Whitney first got ME/CFS at age 21, which made his education and photographic pursuits much more challenging and his ultimate goal of being a war and documentary photographer impossible. His condition worsened in 2009, going from mild to moderate ME/CFS. It then quickly deteriorated into severe ME/CFS in 2012. In 2014, it worsened again into extremely severe ME/CFS. In April of 2020, Whitney saw an improvement, back to severe ME/CFS, from the drug Abilify and, although still bedridden, is able to write for a limited time most days.

Figure 1. Whitney Dafoe Before Severe ME/CFS.

2. Introduction to ME/CFS

I have been struggling with health problems since 2004, when I was 21. Every time I traveled, my health seemed to plummet. However, I have always been inspired and dedicated and never thought I'd wind up where I am now. Therefore, I kept going, kept pushing myself to do everything I wanted to do. A trip to India in 2006 (See Figure 2) made the illness much worse. From 2009–2012, ME/CFS progressed to a moderate state. I started a wedding photography business in 2009 when I realized I could no longer hold a full-time job, thinking that it was a blessing in disguise because, once I got my health back, I would be making money doing something I loved. After a year, things were looking really good business wise, but it took me longer and longer to recover from the intense physical requirements of shooting a wedding. In 2010, when I couldn't recover in a week in order to shoot the next wedding, I decided I had to give it up and move back in with my parents, both heartbreaking decisions because of what they represented. For the next 2 years, I was bedridden much of the time, with my health and mobility slowly decreasing. In 2012 I was forced to rest in bed most of the day, saving up energy for little bits of projects, or working on some photographs for a half hour, or an hour on a good day.

Figure 2. Himalayas.

After seeing countless doctors and specialists in every area of medicine I could find for 8 years, since I was 21, having blood drawn over and over again and literally hundreds of tests done, I was finally diagnosed with Chronic Fatigue Syndrome/Myalgic Encephalomyelitis by Dr. Andy Kogelnik at the Open Medicine Institute in Mountain View CA. As you know, there is no cure.

The Symptoms of Chronic Fatigue Syndrome (CFS), or Myalgic Encephalomyelitis (ME), vary from patient to patient [1]. The most fundamental symptom is debilitating fatigue that worsens after physical or mental exertion. However, fatigue is much too mild a word. I like to compare the state I was in in 2012 to staying up for two nights in a row while fasting, then getting drunk. The state you would be in on the third day—hung over, not having slept or eaten in 3 days—is close, but still better than many ME/CFS patients feel every day. "Total body shut down" would be a better phrase, because you are at a point where your body physically does not have the energy to keep going.

3. From Moderate to Severe

ME/CFS began for me in 2004, when I was 21 [2]. I was in a mild state for 5 years, with my main symptom being lightheadedness that worsened after cardiovascular exercise or came back after periods of remission after cardiovascular exercise.

While in India, I experienced a strange cold that never really took hold, but remained at about 20% for 2 weeks. This had never happened to me before. About 4 months later, one night I suddenly started feeling queasy and nauseous and had mild diarrhea once. This was followed by immediate exhaustion. I suddenly, overnight, developed severe ME/CFS. I slowly recovered from the symptoms to about 60% health, and then would get mild diarrhea (once) again and it would come back immediately, in full swing, and I would be in bed, unable to eat anything but liquified white rice soup.

After battling this for about 3 months, I finally wound up with pneumonia in a hospital in Calcutta, India and decided I had to come home. Upon arriving home, I immediately started getting better, and the ups and downs of getting worse-then-better stopped. I slowly recovered to about 80% of my former health. I still could not do cardiovascular exercise, or I would risk the lightheadedness coming back.

I then learned of a doctor from a close friend of mine, who he claimed was a sort of miracle worker and had often cured undiagnosed illnesses in people before. In hopes of regaining my full health, I flew to Guatemala to see this Doctor. After being there for about a month, the same thing that happened in India happened again. After a meal, I got mild diarrhea, my stomach shut down, and I lost all energy. I stayed there in this condition for about a week, barely hanging on, until I decided to return home. Again, upon returning home, I immediately got better. However, I did not return to the same state I was in after returning from India. I was a bit worse. I was weaker and my stomach was less functional.

In 2009, while working for Environment California, fundraising on the streets, there was an extremely cold spell (for the Bay Area). Every other worker wound up getting a cold. I didn't. However, sometime after that, the same symptoms of the strange 20% cold I got in India came back and my health quickly started slipping back into the state I was in in India and Guatemala, though I did not get as severely ill. My health stopped deteriorating when it was about half as severe as the worst I was in India and Guatemala. At this point, I had moderate ME/CFS symptoms. I was mostly housebound but could walk a short distance to get food and do grocery shopping, cook for myself and take care of myself, although I had to spend large portions of the day resting in my room. This state continued for about 2.5 years, slowly getting worse until it was so difficult to take care of myself that I decided to move in with my parents, hoping that if I lightened my required work load and stayed under my energy limits, my condition might improve. It actually did, a little bit (see Appendix A.2).

Then, I took Rituxan (Rituximab) and this wound up permanently changing the way the illness worked in my body [3,4]. ME/CFS patients who survive have to develop acute awareness of their own bodies to monitor their energy limits and how food and various stimul affect them. Before Rituxan, if I went over my energy limits, I experienced a crash that made me exhausted for the rest of the day or multiple days, but I slowly recovered from it, close to, but not quite equal to, where I was before (as I remained slightly worse).

After Rituxan, when I went over my energy limits, I experienced a much more extreme crash. Instead of a steep curve down and slowly back up, it was like going off a cliff and I did not recover: my symptoms permanently worsened. The crash was a downward line that then just leveled off and did not curve back up. It is very difficult to never go over your energy limits, especially when every time you go over them, you get permanently worse and have to relearn your new limit, which often requires going over it once to find where it is. For me, at that point, it meant getting permanently worse, so I very quickly got much, much worse, I developed pain in the muscles in the back of my legs when standing for short periods of time or walking short distances, then I lost my ability to speak, then I could only text a few words and had to use an app with pre-programmed text messages to ask for food so I only had to touch the phone a few times to send a text. I later taught my parents a routine for my food that I stuck to so that I did not have to ask for a specific thing. My diet was an ongoing, constant rotation through the same meals at the same times. This saved me from having to text.

I continued to get worse. For many months, I walked out into the yard, laid down on a lawn chair and listened to music with headphones for several hours before walking back into my room. This, and 2–3 twenty-foot trips to the kitchen, were the most I could walk per day. I later got a wheelchair, so I didn't have to walk to get to the kitchen (see Appendix A.8).

I continued to get worse.

One night, something traumatic happened that led to me texting more than I was capable of, due to the emotions evoked by the event. This was the end of being able to walk outside or use the wheelchair to get into the kitchen. After this one event that put me over my energy limit, I was bedridden and have been ever since (see Appendix A.5).

4. From Severe to Extremely Severe

In my Severe state, I was bedridden and became sensitive to human contact. I could not tolerate people being in my room for more than short periods of time [5]. This got worse, and soon trips into my room to bring me food and basic necessities became too much. Before I could get my caretakers to successfully limit their trips in and out of my room, they came in and out too many times *once* and I crashed and got worse [6]. This was on Christmas Eve. I remember lying in bed on Christmas Day, not knowing how I was going to get help or food because I couldn't tolerate people coming in my room at all anymore (the crash made me permanently worse). I just laid in bed, kind of panicked, trying to think of a solution. I eventually came up with the idea of wearing headphones playing music while they came in. This worked! It eventually morphed into earmuffs with earphones (small earbuds) inside playing white noise very softly, the combination of which did a great job blocking outside noise. From that moment on, I didn't have anyone in my room without earphones and earmuffs. Only when I discovered Ativan and Abilify [7] in the last few months of 2020 and improved was I able to be around people without earmuffs and actually listen to them speak, but only with Ativan or when my body naturally releases adrenaline to enable me to get up to have a bowel movement in the bathroom (a 6-foot walk). I will explain this natural adrenaline release later in this article.

I continued to slowly get worse, mostly because of the fact that it was impossible to never exceed my very low energy limits. The world is not completely predictable. Sometimes, I would think for days about how to communicate something I needed or something my caretakers were (unintentionally) doing that was hurting me in a way that they would understand. I would try to think of every possible way they could interpret the signals I planned to lay out for them, and every possible reaction they could have. Then, I would try it and they would often react in the one way I hadn't thought of, and I would have to have them come in over and over, trying to communicate what I needed in different ways. Each trip into my room hurt me and made me worse. I often used paper towels folded into arrow shapes pointing at things, but there was a lot of room for interpretation. It was extremely stressful and devastating to try so hard to stay below my limits and then have these unexpected, uncontrollable things happen that forced me beyond my limits, when I knew I would be getting worse.

My Stomach Functionality Declining

In 2011, my stomach was in a steady condition; in fact, it was slowly getting better after my trip to Guatemala and continued to get a bit better when I moved in with my parents, although it was still at maybe 60% of my healthy stomach functionality. Then, it suddenly collapsed and got much worse. I think this happened because of a combination of the illness getting worse and because of one single dietary change I made. In order to sleep, I had to eat right before bed, and it had to contain protein. For years, I ate yogurt right before going to sleep and it worked great. However, many ME/CFS patients talked about dairy intolerance and that was the one thing I had never cut out of my diet to see if it made me feel better. Out of desperation, I tried eating turkey patties before bed instead of yogurt. About 3 weeks later, I woke up one morning and my stomach still felt full from the turkey patties I ate the night before. This was the beginning of my stomach no longer functioning [8–11].

After this, I slowly could eat less and less, despite being very, very careful about never eating too much and eating the things that were easiest to digest. Interestingly, my stomach wound up with a very similar pattern to my energy limits. It's as if my stomach had PEM. My stomach had reduced capacity and reduced digesting ability. If I ever ate too much, it would be incredibly uncomfortable, and afterward my stomach's functional limit would go down permanently, so I had to be incredibly careful. I would sometimes take 1/4 sized bites when I got towards feeling full, because one bite too much could be devastating.

It was a horrible, horrible experience, slowly being able to eat less and less. I was slowly starving. It got to the point where I could only eat yogurt and apple juice and I drenched the yogurt in maple syrup for extra calories. I discovered my stomach worked better while asleep, so I extended my sleeping time and would wake up every hour or two and eat another yogurt cup with maple syrup. Then, I'd go back to sleep. At best, doing this, I could eat about 3 cups per day. Still not enough calories. Or nutrients. But it got worse, down to one cup split between sleep. Then, no yogurt at all. I could only handle amino acid pills which I opened and dumped in my mouth. Just a few capsules filled me up. A healthy person would feel nothing from such a tiny bit of amino acid. I also took tiny sips of straight maple syrup for sugar, which helped my brain continue to function. Then, it continued to get worse, and I could only take a few tiny sips of maple syrup spaced throughout the day, just to give my brain a little fuel.

Then, nothing at all. I was extremely weak and lost a huge amount of weight. I weighed 115 lbs (the same as after India) (see Figure 3). I remember being desperate but not being able to communicate. At this point, one way I was able to communicate was by using small index cards with pre-written phrases on them—generic phrases that I could use for anything, like "more/less" "please put it here", and some specific ones as well. I was dizzy and extremely weak from starvation, and all I managed to do was put out a pre-written card on my pillow that said "Nd Hlp". My mother, Janet Dafoe, found an in home PICC line service and they came and installed a PICC line with IV nutrition just before things started to really fall apart [12] (see Figure 4). From that point on, I have not been able to eat even a tiny crumb of food or drink a drop of water.

Figure 3. Extreme Distress and Weight Loss.

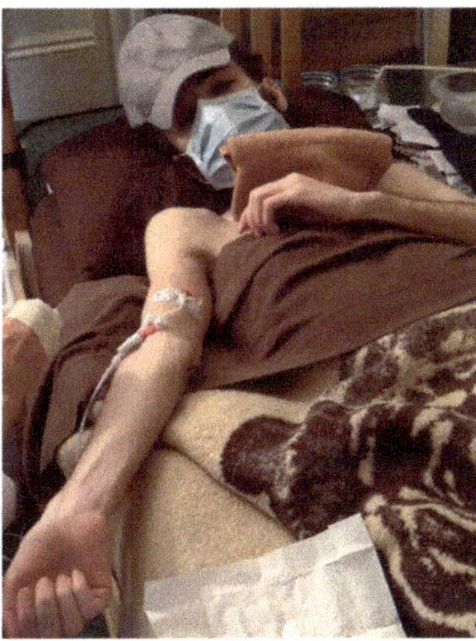

Figure 4. PICC Line.

5. Having Extremely Severe ME/CFS

I fell from Severe to Extremely Severe because I passed my energy limits one day too many in a row. I had gone just over my limit multiple days in a row trying to figure out new tools and new routines to help my stomach, which was still getting worse and more uncomfortable. I felt strongly that I needed to have at least a week of calm days, but the next day, the film crew for Jen Brea's film "Unrest" was scheduled to come. I pushed myself to let them film me despite what my body was telling me I needed. This day was one too many, and afterwards I started going downhill fast, with no bottom in sight.

When I finally leveled out, I could no longer write cards to communicate, or put out pre-written cards. I couldn't communicate in any way. The only thing I figured out I was able to do was fold a paper towel into an arrow shape to point to something, and this only worked because a paper towel was not a tool for communication; I was re-purposing it. All communication tools were too much for me to handle.

These years are very difficult to describe in detail for multiple reasons. When I became housebound, I, at one point, realized that my thoughts were rather negative, and I realized that if I could put a negative tint on everything, I could put a positive tint on everything too. I began practicing and training my mind to think more positively. It was not easy and took practice, but this eventually became integrated into how I saw and thought about things. It was crucial for what I wound up going through (see Appendix A.4).

When I was at my worst in this two-year period, from the filming of "Unrest" until I discovered Ativan in January of 2016, I tried not to think about how bad things were; I really kept my mind focused. Thinking about my reality was extremely distressing and didn't help anything, because it was out of my control.

This was also my least conscious period of time in the illness. I think, in time, science will show that ME/CFS patients are in a kind of hibernation state and are literally less alive. I'm confident there was less activity in my brain during this period and still is to this day compared to my healthy brain [13].

The brain is complex. When you lose a part of it, do you know it's missing? In a way you do, but, similar to state-dependent memory, when my brain was so dysfunctional I don't think I fully realized it.

I also don't remember that time very well because it was so traumatizing that I blocked a lot of it out and just pushed forward, so there are multiple factors at play making that part of my illness a bit hazy.

Before describing this period of my illness, I should explain that, at this point—post Rituxan—I had developed a new kind of crash. A mental crash. When most patients refer to crashing, they are talking about what I call a body crash. A *body crash* is mental and physical exhaustion and worsening of all or most symptoms after going over one's energy limits, followed by a gentle slope back up, but usually not back all the way to where the patient was before: the crash makes the patient permanently worse. A *mental crash* is very different. It can happen from thinking too much, from too much stimulation like noise or light. What was happening is that I got so severe that my energy limit extended into my brain. Anything that caused me to think more than my mental limit permitted caused a mental crash. It got so severe that certain subjects were too much for me to think about, and I had to try to control what my mind thought about. You know the saying that goes "Don't think of a pink elephant". It is very difficult not to think about something, but I had to learn to. I was in a nightmarish situation where my mind started playing tricks on me, flashing subjects I could not tolerate thinking about into my mind at the worst times and causing mental crashes. I was completely lost in a corner of my mind trying to keep my brain activity to a minimum. It was horrific.

The symptoms of this type of mental crash were usually a hot flush starting in the back of my head and moving down through my whole body, followed by an adrenaline release that temporarily made me a little better, but was later followed by my mind getting much worse. After a mental crash, I could not think at all. I was stuck in a thoughtless, feelingless void that you couldn't imagine without experiencing it. It's like being alive but dead at the same time. Alive only to bear witness to the absence of life in your mind and body. This would last for the entire rest of the day. One crash and I lost the only thing I had left—daydreaming of other things, other places, and creative ideas.

Because of the effect crashing had on my life, I had to put a tremendous effort into keeping to my routine as best as possible so I wouldn't overdo it and crash and get worse. As I said, during this time, my brain was extremely sensitive to crashing from the tiniest extra interaction with caregivers or even thinking about the wrong thing, or from thinking about something for too long. I put all my focus on being perfect and then, if nothing went wrong at night when my caregivers were gone for a long period of time, all night, I could think a little bit. I remember after they left for the night, I had a little adrenaline to get fixed on my pillow and get my blankets comfortable and then it would very quickly wear off and I had to hold still. It was often a battle just to get into a position I could stay in comfortably before the adrenaline wore off, and sometimes I crashed just adjusting my pillow too much. Sometimes, I would force myself to stop before getting into a comfortable position, and then I would wind up in sometimes significant pain from this, but would try to ignore it because if I moved even one muscle, I would crash and wouldn't be able to think. If I pulled it off, and didn't crash, it was the best part of the day. I let my mind wander. I usually thought about making things. I have a whole business plan for multiple restaurants, buying and fixing this local natural food store, and lots more. I also thought about art projects in depth, of course. I lived for that time of daydreaming at night and somehow made a sort of life out of it.

It's also important to note that I hadn't been sick for nearly as long then as I have now. I had lots of hope for a recovery in the near future. I thought my father, Ronald Davis, PhD, Professor of Biochemistry and of Genetics and Director of the Stanford Genome Technology Center and now the ME/CFS Collaborative Research Center at Stanford University, would figure it out quickly. It turns out this illness is more complicated than I imagined at the time, but that hope helped carry me onward (see Appendix A.7).

During the day, it was also very difficult for me to move other than unconscious movements like adjusting in bed or scratching an itch. If I thought about any movement too much, it became extremely difficult to do because anything intentional was difficult or impossible. I had to come up with ways of "tricking" my mind into releasing adrenaline to allow me to do things like pick up the electric shiatsu massager I used on my stomach to help with the symptoms of my severe gastroparesis. I broke the movement into steps. I used various methods over time. One was to visualize the movement I was going to make over and over until suddenly my mind released the necessary adrenaline and I could tell that I could do it safely, and then I could pick it up with no problem but had to follow my pre-visualized movement. Then, I did the same for putting it on my stomach and the same for pushing the on button, then moving it to a different spot on my stomach. There were actually more steps than this in order for me to move enough to massage my stomach. It took painstakingly long hours to accomplish simple tasks.

I also became extremely sensitive to, mostly visual but also some audio, stimuli [14]. I couldn't tolerate bright colors and had to remove everything with bright colors from my room. Everything needed to be neutral colors like white, black, brown or shades of gray. My caregivers had to wear all plain black clothing because I couldn't tolerate any colors or patterns on them. I also became sensitive to text like logos or labels on things because it is impossible not to read text that you see; it is something we do instinctually at this age. Reading required more mental energy than I had and caused a mental crash. Due to crashing from the text I could read in my room, I wound up becoming sensitive to text I couldn't read as well. Just knowing it was there was extremely stressful. My caregivers had to slowly and very painstakingly (often with direction from me trying to tell them where the text was, which always hurt me terribly) cover all text with black electrical tape. It remains to this day. I was also sensitive to certain sounds, especially the human voice, and during one period, any noise at all.

When I say that I became extremely sensitive to stimulation, or when you read this about severe ME/CFS, it's not always sensitivity to the stimuli itself. The stimuli, whether it is a sound, a sight, smell, or touch, could connect my mind to something and it was this connection that often pushed my mind over its limit. The sound of people talking, for example, was too much human connection for me to tolerate. Interestingly, it was much easier to tolerate hearing people I didn't know, like neighbors, talking. This is because it caused much less thought, because I didn't know the people. When someone I knew spoke, their whole personality and my memories of them, etc., were forced on my mind and this was much more thought-provoking then an unknown voice.

Sounds or other stimuli that had no mental link to anything could also be too much, simply because they are something for the brain to process. This is why I wear earmuffs and earphones playing white noise, along with a folded towel over my eyes, when someone comes into my room (see Figure 5). I need to isolate myself from the human presence and, in general, I need to isolate myself from the world. This is also why you see severe ME/CFS patients wear eye masks, baseball hats and other apparel or devices to help isolate themselves.

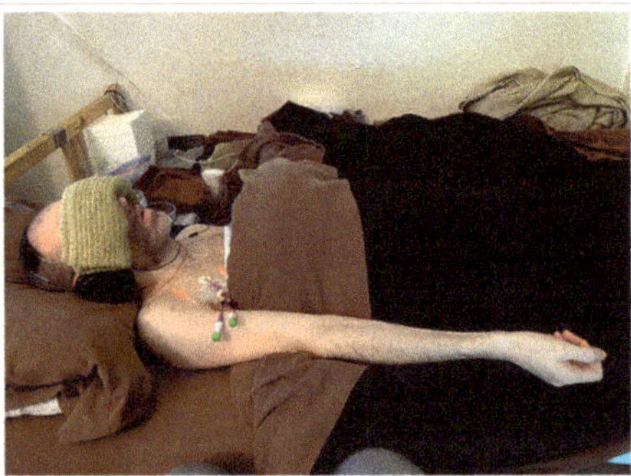

Figure 5. Isolating Myself from Caretaker Presence.

I also suffered from something I call "crash memory". If I crash or get hurt from something, my mind gets what I think is a form of physically induced PTSD caused by my stress or fight/flight response being turned up as high as they could go. When I crashed from something, I developed a stress response to it and became sensitized to it, so I had to be very careful not to crash from the few things that I was able to do or think about. These "crash memories" slowly built up over time. One was getting really sensitive to noise and doing anything at the same time as hearing noise. I couldn't turn on my stomach massager while various noises were happening. The heater air noise, a train or car going by, the click of my in-room heater turning on. I had to wait a certain amount of time after any noise before I could turn on the massager and, if I ignored it, I would crash. I slowly built up more and more sensitivity to noises and it took me forever to massage my stomach or anything else because I had to do so much waiting for gaps in the noises. If I just "did it anyways", it would really hurt my brain and I would be in the worst brain fog of all, which created stress and compounded the whole thing. The Klonapin and Ativan I later took helped me reset these Crash Memories, so they didn't build up. I'm now able to crash from something and let it go and do it again (with the same energy limitations as before, but no added stress or limitations).

When an ME/CFS patient becomes so severe that they are no longer able to communicate, they often start displaying what appears to be emotions like anger or rage [15]. This is a very unfortunate misunderstanding that needs to be clarified for doctors. When I lost the ability to communicate in any way, my caregivers didn't somehow develop telepathic powers. They became out of sync and out of touch with my condition and what was happening in my life. They didn't know when they did something that hurt me, and I had no way to tell them so they would stop doing it or do it in a different way. I was forced to resort to doing things that would connect what they did to a bad experience for them so they would stop doing it, not because they thought it was bad for me, but because they knew what would happen if they did it again. This was unfortunate, but it was the only way to survive. Babies are in a similar situation, where they have needs that don't get met because they can't communicate. They do similar things to what I wound up having to do. I often had to display anger and throw things that would break or otherwise make a mess that was tiring for my caregivers to clean up. I most often would dump a jar of water that was kept by my bed onto the floor, which they then had to soak up so mold wouldn't grow. They sadly thought, at the time, that the illness was making me emotionally unstable and angry. However, I was never actually angry and always felt terrible about forcing them to clean things up, but it was the only thing I could do to change their behavior so that they stopped hurting me. This is important for doctors and caregivers to know, for two reasons—so that caregivers do not take this behavior personally and so that patients are not improperly diagnosed with mental illness by doctors. It is no more a mental illness than a baby's cry for help (see Appendix A.6).

One thing I've thought about is that, despite my caregivers' entirely good intentions and tremendous effort, my actual experience during this period was one of rather extreme abuse. It's still true, though to a lesser extent. I got worse almost entirely because of interactions with my caregivers. This isn't because of anything intentional on their part, but due to my sensitivity to human interaction. If I could have somehow gotten what I needed without people ever coming in my room, I would never have become so severe. I must emphasize that this was despite their good intentions, effort and sacrifices, which I have always acknowledged and been grateful for.

I don't think I ever worried that my brain was permanently damaged. I'm not sure why. I've always been very in touch with my body and most of my conceptual intuitions about the illness have been proven correct by Ronald Davis (molecularly). I do still worry that brain crashes (as opposed to body crashes from overexertion, which last longer and are more of a gentle curve, not a cliff) cause brain damage, but I think the brain is resilient and can rewire itself. I try not to think about it.

My personal theory of a mental crash is as follows. When an ME/CFS patient gets severe enough, the energy limit invades the brain because use of the brain starts exceeding the energy limit. When the brain exceeds this limit, it runs out of oxygen or some other vital element, and the body responds by inducing an emergency release of adrenaline (this is the hot flush I experience) and this adrenaline increases my heart rate, which pumps more blood to my brain to avoid sudden brain death. I don't know what this essential element is, but I feel fairly confident this is an accurate laymen's description of what is happening. It's an automatic emergency brain-saving reaction.

Severe ME/CFS List in Brief—Summarizing My Quality of Life
- I haven't left my room for 7 years, except when I have to go to the hospital to change my J-tube feeding tube out of medical necessity. I am only able to do this without dying by being sedated with Ativan the entire time, as well as Fentanyl and Versed during the procedure;
- I haven't been touched by another human being without it hurting me in 7 years;
- I haven't been able to speak for 7 years. I haven't had a conversation with another human being in 8 years;
- I haven't eaten a crumb of food or felt a drop of water in my mouth in 6 years. I'm alive because of nutrients being pumped into my body with machines and tubes;
- I haven't taken a shower in 7 years. I clean the most necessary parts of myself with baby wipes every day and it absolutely exhausts me. I can't handle having someone else clean me;
- I haven't cut my own toenails in 7 years;

- I haven't been able to hold or even touch my camera in 7 years (photography is my passion and my life);
- I haven't peed standing up in 9 years. I haven't walked to the bathroom to pee in 7 years. I pee in a urinal in bed;
- I haven't made love to a woman in 9 years. I haven't been sexual in any way in 5 years;
- I haven't brushed my teeth in 6 years. It hurts my stomach, making it worse and putting my ability to tolerate the feeding tube at risk, which puts my life at risk;
- I haven't seen a dentist in 9 years;
- I haven't been able to tolerate the sound of another person's voice without being sedated in 7 years. I wear heavy-duty earmuffs whenever my caregivers are in my room for the bare minimum of time. They can't talk and have to be as quiet and gentle as possible;
- I haven't felt like a human being in 7 years. All humanity has been taken from me by ME/CFS. I live only to continue living. There is no love, joy, passion or creation, only endless, numbered days; (See Appendix A.1)
- I fight to survive for all those living and dying in silence and darkness (see Appendix A.3).

6. Slight Improvement—Ativan and Abilify

Discovering Ativan saved me from living on the brink. After taking it for the first time, it had some sort of reset effect on my system, and all my symptoms improved permanently in addition to the temporary benefit of the drug. I had been on Total Parenteral Nutrition (TPN) through a PICC line for 1.5 years, and this is the maximum time a person can be on this type of nutrition. Some things just can't be given through your veins. It bypasses the entire GI system, and risks liver damage. I should have been put on Total Enteral Tube Nutrition with a jejunostomy tube [16], instead of TPN and a PICC line, from the start, but my family, my doctors and I were all scared of what a trip to the hospital would do to me and we didn't realize how much Ativan would help me. I would probably be much healthier today if I had a J tube installed then, because I would have discovered Ativan sooner, and TPN through a PICC line in your veins causes the GI tract to deteriorate and healthy bacteria to die off.

I took Ativan for the first time to try to make my trip to the hospital to have the J tube inserted tolerable, or at least less harmful (see Figure 6). It wound up being a game-changer for me. In addition to somehow resetting my system and permanently improving all of my symptoms, I now had a way to periodically communicate (Ativan can't be taken all the time or you habituate to it, so I took it once every week or two). This meant that I no longer had to figure out how to communicate problems or new needs that arose; I just had to hold out and tolerate them until I could take Ativan.

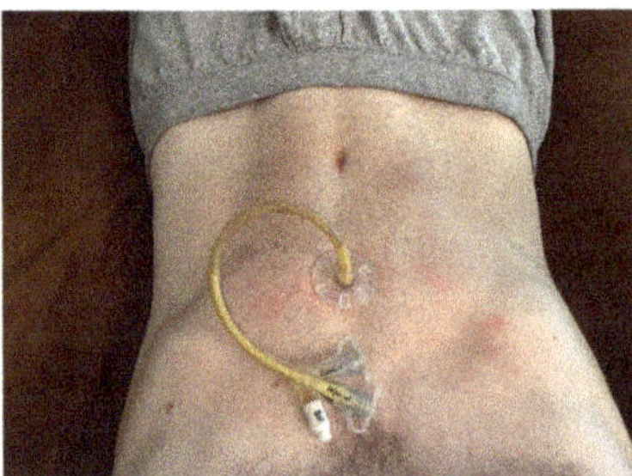

Figure 6. J Tube.

Ativan mainly reduced my sensitivity so that I was able to tolerate being in the presence of people. I still could not speak and certainly could not get out of bed or do anything extra physically, but I figured out that I could gesture to communicate. This was painfully slow and took an enormous amount of energy. In time, I learned that I could also do a limited amount of writing out words in the air with my finger or onto my blanket (I still could not write on paper). I used the combination of gesturing for most things and filling in gaps that I couldn't successfully gesture by writing them in the air. This was still hard on me, though, so I would often reduce the number of letters I had to draw out by playing a sort of "hang man". I would write a few essential letters of a word and draw blank lines with my finger in between and try to use gesturing to help people figure out the word by guessing the other letters. Or I would do the same with a sentence, with the blank being a word I wasn't able to get across. It was a relief to be able to finally communicate directly to people, but also traumatic in how difficult and often imprecise it was. It frequently made me feel pretty desperate.

Going to the hospital, especially for the first time, was incredible. I had no idea Ativan was going to have such a profound effect on me. I was preparing to get way worse and have a terrible time and crash horribly. Instead, I improved and was calm and got to enjoy things like seeing the sky for the first time in 6 years: all the sights of the real world out the window of the ambulance, all the healthy people working at the hospital leading healthy (or at least much healthier) lives with careers and loved ones and goals and things they were looking forward to, etc., and, a few times, seeing women my age and feeling attracted to them, and more. It was all amazing and continues to be, though it's also exhausting and a big disruption to my routine, so it's a mixed bag, especially coming back home and seeing the door to the outside world shut behind me. This is very difficult emotionally. I also let a lot more of myself out while on Ativan because I'm able to, and when it wears off, I have to pull it back in again and suppress myself again. It usually takes a couple days of emotional turmoil to adjust.

In the fall of 2019, I started taking Abilify at a low dose. It did nothing for the first few months because I was adding multiple medications and supplements, so I tapered up very slowly, much slower than most people when they take it now. I think I spent 6 months going from 0.25 mg to 2 mg (February 2020). After being at 2 mg for about a month, I started noticing an effect. It wasn't the same as Ativan. I didn't suddenly feel it like Ativan, which had an instantaneous, noticeable, drug-like effect. Abilify seemed to be changing something at a deeper level. I had more energy and could slowly tolerate more things that used to cause me stress. For the last 6–7 months, I have continued to improve, tolerating more and more things that used to make me crash from stress and over-stimulation. I can't get out of the bed, but I can move around in bed much more than before. I can even work on some hobbies in bed on most days for some time. When I take Ativan now, I can actually listen to people talk to me, so instead of pantomiming both directions, I can listen and then pantomime back. They can say what they think I mean, and I can nod if they've got it. This makes it much faster but still painfully imprecise and slow for me to communicate anything to them.

Soon after my stomach completely shut down and was unable to tolerate even a drop of water, I discovered that ice helped it function better. When I started getting Total Enteral Formula with the J tube, I kept ice on my stomach for basically all of my waking hours. After being on Abilify for 6 months, I discovered through an act of brave experimentation that I could tolerate the food pump with no ice. This was a huge breakthrough, because it allowed me to move much more in bed and avoid the constant replacement of two-gallon-sized ziplock bags of ice on my entire stomach from ribs to waistline. I'm now able to get up on my knees in bed for a moment to move or reach things. I haven't tried standing up in bed.

For my entire time with severe ME/CFS, I've gotten a natural adrenaline boost when I have a bowel movement that improves my condition, and, with the exception of a 6-month period when I used a commode, gives me enough energy to get out of bed and walk the 6 feet to the bathroom to go in there on the real toilet [17]. Which is, of course, a good thing for my sense of humanity and autonomy, and it's just easier for everyone. Since Abilify has started to improve my heath, I've been able to harness this adrenaline to communicate after having a BM in the bathroom. I can't interact with people for a long time, but long enough to communicate some basic things. It's been enough that I haven't needed to take Ativan anymore because I just wait until my next BM to communicate new needs or problems. I've recently added washing my feet, privates, head and face in the shower to having a BM. After the BM, I stand in the shower for a very short period and wash my feet and privates. Then, I lie down on the ground with my head sticking into the shower and my caregiver washes my head and I quickly wash my face. I don't totally understand how my body produces adrenaline to go to the bathroom or why it isn't able to for other things like communication or cleaning myself, but it has something to do with hardwired necessity. There is something hardwired to having a bowel movement that must push the body to get up and move somewhere else. I believe it is something we evolved to have, to help early humans move away from their sleeping place to go to the bathroom because this improved sanitation and reduced illness [18].

7. Important Notes

A doctor recently asked me to describe why I am unable to talk and the process behind that. The answer to this question ties into a core process of ME/CFS that is important for the world to understand because of its significance and because of how much suffering it causes patients. I haven't spoken in 8 years, but I could talk right now if I chose to. The keyword here is *choose*. ME/CFS is not generally defined by inability, but by consequences. Everything is about Post-Exertional Malaise, mild or severe [19]. I could get out of bed and walk out the door and run right now if I chose to. I'm capable of it. The question is not what I am capable of, the question is what will happen to me afterwards (or in severe cases, might leak into the very act because it would take very little running before the reaction of severe PEM started and I might still be running when it hit, causing me to collapse or possibly die, but not from the immediate consequences of the action, but PEM). ME/CFS patients very quickly learn that their actions have consequences that occur after the fact. Patients have to learn to read and listen to their bodies.

I have learned to pre-visualize an action before doing it. When I pre-visualize performing the action, I can feel what the consequences of performing that action will be, and whether it will hurt me or not. I have incorporated this pre-visualization into every single action, big and small, and it is now how I function without having to consciously think about it. I don't speak because, when I do this split-second pre-visualization I feel that it will hurt me.

There are some things I don't use entirely this technique for, though. Some things feel like they might be OK, but I rely on my prior memories of doing them and what the consequences were to steer me in the right direction. One time, in 2012, when I first became bedridden, I got up out of bed to move something that had fallen over. It wound up causing severe PEM that left me exhausted and brain-fogged for days. Right now, when I pre-visualize getting out of bed and walking a few steps to get something, it feels like I maybe could, but I am scared to do it because of what happened in 2012. I can hear psychologists everywhere screaming "deconditioning!" That's not what is happening here. It is simply intelligent learning. When I am able to get out of bed and walk a few steps to pick up something, it will be obvious to me. Getting better is a slow process with ME/CFS because of how careful patients need to be about overdoing it. When getting better, the energy limit is suddenly in unknown territory. Patients must very slowly do new things only when they feel very safe doing so. It is wiser to get better staying a bit under your absolute limits than to try to do as much as possible, and wind up making a mistake, going over the limit and then getting worse and ruining a possible upward spiral toward better health.

7.1. Routines

This brings me to another important part of living with ME/CFS—developing a routine. It is difficult, especially for people new to ME/CFS, to pre-visualize every little thing. It's likely that I am better at this than other people because I am good at spatial thinking—I am a visual artist. It is difficult to always know when you will go over your limits by performing a certain action. Patients soon learn what hurts them and what is OK, and to make life easier, instead of trying to figure these limits out every hour of every day, which leads to making mistakes, patients develop routines. These routines are sets of actions a patient learns that, if performed in a certain way or order, can be done without going over their energy limits. Most people who survive ME/CFS develop routines and stick to them so that they are much less likely to go over their limits. It is a way of living that leads to much better health than constantly guessing or taking chances. It, of course, causes suffering as well because it takes spontaneity out of life. The worse a patient gets, the more every day becomes a chore of endlessly going through the same actions in the same order and in the same way, but if patients don't develop a routine, they get worse, and lose the ability to do things they were once able to do. Do not make the mistake of diagnosing this as OCD behavior. It is a choice that ME/CFS patients make. The choice is obvious to ME/CFS patients: it is preferable to sacrifice spontaneity in order to be healthier and more active, and think more clearly.

7.2. The Great Beyond

Having Severe ME/CFS is so close to being dead. There's really no other way to describe the experience I have had. I don't think it's something that people who haven't had severe ME/CFS can likely understand. Looking back at who I was when I had mild and moderate ME/CFS, I'm not sure it's something that even patients who haven't been in the extremely severe state can fully understand. I was literally barely alive, and I am confident that, in a short time, science will prove that severe ME/CFS patients are barely alive and that ME/CFS patients, in general, are less alive mentally and physically than healthy people.

I think the only time a healthy person maybe experiences anything like this is shortly before actually dying. In that case, the person is generally in this state for a much shorter period of time and so remains much more connected to who they were, and their former lives. This is the state in which healthy people let go of their former lives and accept death, which is probably one of the reasons that suicide is so common for ME/CFS patients.

When I was severely ill, I lost so much of myself. I was holding on to fragmented memories left imprinted in my mind of who I was, but that person, in reality, didn't exist anymore. The thought patterns and emotions and worldviews that created the person I was no longer existed. However, I was still technically alive, just enough to be conscious and bear witness to this state of non-existence.

The suffering this causes is so profound. I can only liken it to one of the hell realms described in Tibetan Buddhism. A world full of nothing but pain, loss, agony and constant never-ending challenges in holding on to what little I had left. Every mistake took me deeper into the void of nothingness.

As you know, I have recently gained back some of my mind and body. It feels like coming back from the dead. I'm in a strange state now, where bits and pieces of Whitney have come back to life but most of me has not. I'm not able to get out of bed, eat or drink water or go out and feel the world again—feel that feeling that is being alive.

I have, so far, just been riding this wave of improvement and the new-found abilities I have, like being able to write and have some semblance of connection with the world again.

However, the honeymoon phase for these improvements wore off, I started realizing how far I actually am from being Whitney again. I've realized that I don't really know who I am anymore. I know who I used to be, but is that who I am? I guess I've realized that it is not.

The experience of being on death's door for never-ending years has changed me permanently. I'm still not well enough to come anywhere close to fully inhabiting my own mind and body again. I don't really know who I am. I'm in a sort of limbo right now, stripped of the person I once was and would have become, but not able to take the experiences I've had and create a new person out of them. I'm still a ghost, suddenly no longer fully transparent, yet, at the same time, unable to actually exist in physical form.

It's so confusing.

While my new capabilities have improved my quality of life a small amount, I realize how much I'm still suffering and how much is still missing from my being a human being again. I've been so focused on my small improvements that I've somewhat lost touch with how far away the world still is. When I think about it now, it's hard for me to even imagine what it would be like to be fully healthy again, out in the world again, alive again.

I don't know who I am going to become. One thing I do know is how much the experience of losing everything has taught me. I think ME/CFS is the greatest teacher I've ever had. I have hope that when better treatments, and then a cure, are found, I will be a much more conscious, wiser, more realized being. That person waiting to be reborn is an incredible person, and I can't wait to see that person and be that person and contribute to the world with my whole being (see Appendix A.9).

I think this is one of the most tragic things about the high rate of suicide among ME/CFS patients. These are people who have been through something completely unique to the rest of society and have a truly unique and profound perspective to offer the human race. When an ME/CFS patient kills themselves, so much is lost from the world.

We have seen the other side. We need to stay alive so that we can join the world again and share what is really out there in the great beyond with the rest of humanity. We have an incredible understanding of what life is. How precious and fleeting it is, how little time we have, and more. These are lessons that most people never learn, and we need to teach the rest of humanity how sacred the life they have truly is.

Funding: This research received no external funding

Informed Consent Statement: Informed consent was obtained from all subjects involved in the study.

Conflicts of Interest: The authors declare no conflict of interest.

Appendix A. Articles I Have Written about ME/CFS

Appendix A.1. What Patients Often Go Through

- Loss of some or all family;
- Loss of all friends;
- Loss of job;
- Loss of hobbies;
- Loss of loved ones/relationships;
- Loss of things that used to define who you were;
- Loss of connection to the world;
- Loss of sense of dignity;
- Loss of ability to do anything physical (this includes chores, sports, outdoor activities, using your legs as transportation, and in more severe patients—self care like showers, keeping clean in general, brushing teeth, changing clothes, changing socks, etc.);
- Loss of ability to think and remember the way you used to, your mind lost from you in what is often called a "fog";
- Accordingly, loss of your personality;
- Loss of your sense of self and sense of humanity;
- Prejudice from everyone in a patient's life, accusations from everyone in a patient's life of the illness being "in their head", even after decades of illness;
- A complete lack of support from society. There is no safety net for ME/CFS patients because most patients aren't diagnosed and even when they are, it is not considered a valid diagnosis [20]. If patients aren't lucky enough to have friends/family to take care of them, they are left on their own. Even patients who do have people in their lives who are willing to make the incredible sacrifices required to take care of them, very few are prepared or trained, or the right kind of person for that job, which is incredibly

difficult. Most Severe ME/CFS patients probably die or commit suicide when there literally is no hope [21]. A great number of them lose hope before this point;

- Lack of funding for research that would give patients something to hope for [22]. All of the above plus no research puts a huge burden of literal hopelessness on patients. Their condition is likely never to get better [23]; their only hope is a cure or treatments, but there's no funding for scientists to do research to find treatments or a cure. The Open Medicine Foundation-funded research, spearheaded by Ronald Davis out of Stanford University, is the first extensive, collaborative research effort into ME/CFS [24], but it is pretty new. For the last 40 years (the illness has likely been around for much, much longer than that, but was even more covered up, prejudiced against and misunderstood, to the point that there was not even recognition of its existence), there has been nothing but small efforts at research, even if a few have been well-meaning and well-conceived. NIH allocates only 15 million dollars per year for ME/CFS research but, just a few years ago, it was only 6 million dollars per year [25]. Multiple Sclerosis is thought to be, on average, much less severe in its impact on patients' quality of life, and affects half the number of people (at least, the number of affected MS patients is likely accurate, the number of estimated ME/CFS patients is likely very inaccurate). However, MS receives 100 million dollars per year from the government for research. HIV receives 28 billion dollars per year [25].
- Some patients are committed to psych wards [26]. This probably happened a lot more in the past. The number is thankfully declining, but it still happens. There is one woman who was forced into a psych ward and, while there, the clinicians, at one point, threw her into a swimming pool to try to force her to "take initiative", or something. She almost drowned. She got much, much worse while kept at the psych ward but did finally get out after relentless help from the ME/CFS community or family/friends (I'm not sure which) [26]. I'm sure there are many diagnosed and even more undiagnosed ME/CFS patients around the world being forced into treatments and forced to take medications that harm them, getting worse and worse and suffering profoundly as a direct result of being locked in psych wards. I recently wrote a letter to a hospital that is currently threatening to lock up an ME/CFS patient in Sweden against his will. It is included in the appendix below (see Appendix A.6).

Appendix A.2. Staying below Energy Limits

My number one piece of health advice for Chronic Fatigue Syndrome/Myalgic Encephalomyelitis patients, more important than any current medication or treatment, is to never exceed your energy limit. Let me explain to be clear. The most unique, best identifying symptom of Chronic Fatigue Syndrome/Myalgic Encephalomyelitis (ME/CFS) is Post-Exertional Malaise. Healthy people can exercise way past the point of exhaustion. They can continue when their bodies scream at them to stop and later that day, they recover and feel fine, if not euphoric. I know this feeling well; I ran cross-country in high school. ME/CFS patients have an energy limit and if we exceed that limit, we get Post-Exertional Malaise, which means we get physically sick afterward, and any ME/CFS symptoms we have get worse. This can last for days, weeks, months or be a permanent worsening of the illness. The most important part is that, when we exceed our energy limit, the limit goes down, so next time we have to stop sooner or the whole process repeats itself as a vicious cycle.

Patients with mild ME/CFS usually only reach their limit with anaerobic exercise, but a more severe patient's limit can be brushing their teeth for too long a time and, for even worse patients like me, the limit is, for example, being touched by another person, being in the same room with someone else, looking at something for too long or even thinking about something for too long, or thinking of something that requires too much mental energy. Having someone in the room, especially, puts me over my limit. The combination of thinking at the same time is extremely overwhelming. I have to meditate on a couple simple ideas or memories, and if my mind strays, even for a moment, it can be devastating.

Most people are completely out of touch with their bodies. ME/CFS patients have to learn to be keenly aware of our bodies and exactly where our limit is. We have to make a choice to stop when we feel ourselves reaching our limit before we go above it. This choice is part of the reason we are judged so harshly by friends, family and even loved ones: we have to choose to stop activities or refrain from activities before they make us symptomatic. People don't see us getting Post-Exertional Malaise—we go home and collapse, suffer, sleep or rest in private. Due to this lack of awareness, they never understand the connection.

It's also extremely difficult for us because it's pleasure first, negative consequences later, and our minds are famously bad at negotiating this. Think of how hard it is to stop using drugs. It's a similar pleasure first, negative consequences later situation. With ME/CFS, it is life itself we are having to refrain from, not a high from life.

A good way to control the urge to indulge the pleasure center of the mind is to think about how you will feel afterward. If you exceed your energy limit pushing yourself to continue engaging in an activity—mental or physical—your limit will go down, you may never be able to do that activity again, and, in the future, you'll have to do even less just to stay below your new limit. It helps to bring the negative consequences into the present and hopefully make it easier for you to stop within your limits.

This happens to be Ronald Davis number one piece of health advice as well, arrived at independently of me. Genius minds think alike.

Please understand, getting worse DOES NOT mean this is your fault. For one thing, ME/CFS is an extremely complicated illness, and its mechanisms are mostly unexplained to date. Some people (like me) get worse without exceeding their energy limits. I went from mild to severe ME/CFS overnight while traveling in India. Then, I slowly recovered and went back to ME/CFS severe over and over until I came home and got better, but remaining worse than before India. No one knows what caused this. The world is also full of chaos, and we can't always accurately predict how much energy something will take. We often get stuck in situations we can't just stop, and we have to push ourselves past our limits. This is why a predictable routine becomes important—the less unexpected energy expenditure, the less likely we are to wind up in one of these situations and overdo it. It's also really hard to know your limit and body well enough to feel it coming on and stop. It takes years of experience.

So don't blame yourself. Just do your best and let go of the rest, and prepare for the very real likelihood that you could get worse.

Appendix A.3. The True Horror Of ME/CFS

From the CDC, "According to an Institute of Medicine (IOM) report published in 2015, an estimated 836,000 to 2.5 million Americans suffer from ME/CFS, but most of them have not been diagnosed." [27].

This number is generally stated as being about 2 million, the higher end of the CDC's estimate [28]. However, I'm not sure how the CDC or anyone else thinks they can accurately guess at the number of people afflicted by an illness for which it is so difficult to simply obtain a diagnosis. What logic or thought process lead them to the "2 million" number?

In honor of all those who lie in silence and darkness, to those whose terrible deaths which were marked as cause "unknown" or "heart attack", etc., and to those who have taken their lives due to unbearable suffering, I must relay to you the unfortunate true horror of ME/CFS. It is much worse than estimates like this and has been for at least forty years. I know this by looking at what we know and using simple logic to extrapolate from there. These are the logical steps:

(1) We only hear from people who are diagnosed. How hard was it for you to get diagnosed? It took me about 7 years of constantly seeing doctors. I was told my symptoms would resolve themselves, or that nothing was actually wrong with me, or there was just no answer and nothing else to test for, or the famous "it was in my head and not a physical illness". How many people have the fortitude to keep going in the face of this, and for how long? It takes a very specific kind of person to be utterly relentless enough to continue

pushing doctors to dig deeper and continue seeing new doctors for second, third, fourth, fifth—and even more—opinions, all the while ignoring the blatant prejudice and disrespect constantly shown to them. Everything pushes ME/CFS patients to give up and try to either live with their "health problems" (what I called it for years) or kill themselves. Everything. What percentage of regular people out there who have ME/CFS have the chutzpah to keep fighting for answers in spite of all of that? 10%? 5%? Less? [29].

(2) Of that small percentage of ME/CFS patients who are diagnosed, how many are lucky enough to remain mildly sick and not get worse, despite holding a job, taking care of kids, feeding themselves, feeding a family and doing all the myriad other things it takes to be an independent adult?

(3) Of those who get diagnosed but become too sick to care for themselves, how many are lucky enough to have family or loved ones who will support them and take care of them? It's worth noting that it's much less likely that undiagnosed ME/CFS patients will have family/friends/loved ones who will be understanding enough to help them or take care of them. Even for diagnosed patients, there are few people in the world who have access to the care that is necessary to keep someone who is severely ill alive. I know how lucky I am. It takes loved ones who are willing to give up their personal and professional lives, with enough money to pay for huge medical bills that aren't covered by insurance, because we don't have a "legitimate illness".

(4) How many are willing or able to carry on emotionally, continuing to survive the horrifying "living death" that defines severe ME/CFS?

(5) Of those who manage to get (1) a diagnosis and (2) are lucky enough to either not get worse (rare, especially if they are trying to maintain a healthy person's workload, which is very common) or (3) have people in their lives willing and able to take care of them and (4) are emotionally able to carry on despite the incredible suffering ME/CFS inflicts, how many of this dwindling percentage are interested in social media/find the forums and social media pages, and then how many are even capable of using computers (many ME/CFS patients are not able to)?

This is what it takes to be "seen" as an ME/CFS patient. The people on Facebook, Twitter, Phoenix Rising and other ME/CFS forums are the very tip of the iceberg. We only hear from people who make it through these five tiers. It's likely that this is a very, very small percentage of ME/CFS patients [29].

What happens to the rest of the diagnosed/undiagnosed ME/CFS patients in the country/world? This is something no one talks about. If you just look at the facts, human nature and how our society functions, it suddenly becomes horrifying. They must wind up on the streets, getting worse and worse until they die a terrible death alone in a gutter somewhere [30]. I believe this happens to a huge number of ME/CFS patients.

How many people are as sick as I am? Only a few of us are publicly known, but surely there are staggering numbers of people as sick as I am. (Even if the estimates are correct and one quarter of the "2 million" is severely ill, that's at least 500,000, just in the U.S.)

I'll ask again: what happens to all the rest of us? My fellow severe ME/CFS patients have either killed themselves or will die alone in a ditch; likely hundreds of thousands of us or more. In a ditch. Alone. We need the support of the federal health agencies to fund research and care programs (ME/CFS wards that house ME/CFS patients who don't have anyone to take care of them, and can cater to their sensitivities) for this disease, and, yet it hasn't been offered. So many of us have just been left to die alone.

Appendix A.4. Adjusting Expectations

I think one important aspect of coping with ME/CFS is lowering our expectations, as sad as that is to do. A discrepancy between expectations and reality is one of the biggest causes of unhappiness, even among healthy people. If your happiness depends on something you don't have, you will be unhappy. Living with ME/CFS is a process of lowering the bar of expectations you once had for your life. You have to do it, or you'll go crazy. Since ME/CFS is often degenerative, as it was for me, it becomes a process of continuing to let go of expectations and continuing to lower the bar until, as in my case, it's practically on the ground. There was a time when I said I would kill myself if I ever had to move in with my parents. Yet here I am alive, after living with them for about 10 years now, since 2011.

One way to do this is to try to be open minded to things you once thought were beneath you, or simply not befitting your personality or the way you wanted to live. I did a lot of this.

When I became housebound while living in Berkeley, California, I realized how difficult it was going to be to meet people while stuck in my house, so I signed up for a dating website, which is something I never would have done before I was sick. I decided I needed to be open minded to the options that were actually on the table to maximize my quality of life. Nothing much came of the dating website, except one really awesome girl, who made it all worthwhile. We had a good, short relationship—a few months—before my illness got in the way and the relationship ran its course, and we ended things on good terms.

I also really worked on overcoming shyness and asking girls out for coffee/tea who I didn't know but met randomly in public. I met a girl this way too—an employee at the Whole Foods I went to for groceries. (One of my few outings—which exhausted me.)

Being creative is hardwired in my existence and the worst parts of this illness have been when I'm too sick to be creative with anything in my life. As long as my health allowed, I've always tried to find creative projects I could handle working on within my energy limits. One thing I did in Berkeley, and, while my health permitted, here at my parents' house, was collect headphones with good drivers that had poor acoustic implementation, resulting in much inferior sound than they were capable of. I learned how to add acoustic implementation that allowed the drivers to operate better so the headphones would sound as good as they could, resulting in some incredible-sounding headphones. I had to give up the kind of photographs I used to make, but things like this somewhat filled the creative void that was left in my life.

For the last 7 years, I haven't had the energy or freedom in my daily routine to be creative at all, and it's been crushing. I've felt adrift and empty. However, I hung on for the ride and now I've had a completely unexpected upswing, and I get to try to be creative with these social media pages and writing. Again, it's not what my healthy self would be doing but I have to change my expectations and adjust and be open-minded, and then I can find happiness in things I wouldn't have before.

I even love my iPhone now, which is something that would have been an anathema to me before. It's now my only connection to the world and only way of engaging with the world. When I get better, hopefully it won't remain attached to my hip, but if I'm better it won't matter—I'll be better!

Appendix A.5. My Whole World Exists in Bed

Something I don't think people who haven't experienced being bedridden understand is that, when you're bedridden, your whole world exists in bed. You don't climb into bed to sleep or nap or get cozy and then get out of bed and live in the rest of the room/house/world.

You are always in bed. It's your whole world. I think this contributes to some of the sensitivity that severe ME/CFS patients experience. Anyone would be particular about their bed if they were bedridden, but it is also exacerbated by the sensitivity that the illness causes.

Half of my bed is dedicated to me and the other half to storing things I need access to because I can't get up to get things. On my bed are: my stomach massagers (for my severe gastroparesis), ice for my stomach (ice helps my stomach feel and function better), a stack of paper towels, remote door bells which I use as call buttons when I need something, a container of water for cleaning or rinsing off baby wipe soap, a basket with odds and ends like the remote control to my A/C, masks for the smoke from the forest fires before I got an air purifier, the towel I use to cover my eyes when people come in, my white noise earphones, my earmuffs, a stack of adult diapers because I got a urine infection once and had trouble holding it in time to get a urinal and now I keep them accessible just in case, a little jar for trash, boogers, etc.—you get the idea. I've packed as much as possible within reach under my bed, like a vibrating massager for my feet and legs, which get restless leg syndrome (tingly feeling in the legs and feet that can be unbearable) from being so still, or nervous system weirdness—I don't know which. I also store extra backup stomach massagers in case one fails, my heating pad, which I use on my feet to keep warm since I have ice on my stomach all the time, and lots more.

All these items are, to me, like all the stuff in your house. You are just as particular about how your house is arranged as I am about how my bed is arranged, and you get to leave the house and get away from all that stuff and move freely with a few possessions. I don't. This bed is where I reside 24/7 and I need access to this stuff 24/7.

Some symptoms of severe ME/CFS are partly just normal reactions to horrid nightmare living conditions. They are, of course, compounded by the sensitivity that severe ME/CFS causes. However, I think it's important for caregivers, doctors and healthcare professionals to understand how challenging living conditions are with severe ME/CFS, and the fact that any healthy person would also react adversely and wind up acting "abnormally" in response to these conditions. It is, in fact, not an abnormal reaction, nor is it a sign of mental or psychological illness. It is a healthy, pro-active response to the limitations imposed by the illness, making it easier to access things with minimal energy expenditure. For example, I have a piece of tape on the floor marking where my bed urinals should be precisely lined up. If they are always in the same place, I can develop muscle memory for the action of reaching down and picking them up and can do it with very little thought or energy. I can even reach them with my eyes closed. I try to have everything in my bed like this. Again, it is not OCD.

Appendix A.6. A Hospital in Sweden Is Threatening to Commit a Severe ME/CFS Patient to a Psych Ward

Holger Klintenberg is a severe ME/CFS patient in Sweden who is being threatened by a local hospital with committing him by force to their psych ward. He is extremely severe and this will kill him.

This is a letter I wrote to the Hospital that is threatening to commit Holger.

Dear *LänssjukhusetRyhov* hospital threatening to commit Holger Klintenberg against his will,

My name is Whitney Dafoe and like Holger I also have severe ME/CFS. I have recently seen some minor improvement from an experimental drug that reduces brain inflammation which is the only reason I'm able to write this now. I spent four long years in a state very similar to the condition Holger is in.

I'm writing you to tell you that if I was committed to a psych ward even now it would without question kill me. It would have killed me faster if I was committed when I was in Holger's condition. Holger will die if you commit him to a psych ward. Period. If you doubt this you should watch this news clip about an ME/CFS patient who died as a direct result of being committed. And ask yourself: do you want that on your conscience? Do you want that kind of publicity? Killing someone? Because you will get it.

https://youtu.be/yrBAlKtroBw (accessed on 22 April 2021)

ME/CFS is not a psychological illness. It never has been and there has never been an acceptable reason to treat ME/CFS patients the way you are threatening to treat Holger. For decades there has been so little research into ME/CFS that there hasn't been a lot of proof of physical illness. However in recent years there has been a surge of research into ME/CFS due to a team of world renowned scientists (all award winning and 3 Nobel prize award winners) at Stanford University taking on the illness full steam, almost entirely privately funded. They have made a number of profound discoveries in only a few years that prove this is a real physical illness and they are only going to find more proof as they move closer to finding a diagnostic test and a cure. The lead researcher is Ronald Davis and he would be happy to speak with you about these discoveries and give you his informed opinion on the consequences of committing Holger. How will you feel after killing someone who could have lived to see a cure discovered and experience a full recovery?

The main, most distinguishing symptom of ME/CFS is something called Post Exertional Malaise (PEM) which refers to symptoms worsening with physical exertion, or with severe patients like Holger and myself, mental exertion as well. Patients with ME/CFS have what is called an energy envelope—in other words—an energy limit. When patients exceed this limit, two things happen. Their symptoms get worse which can last for hours, days, months, or years. And most importantly, the energy limit lowers. I got worse for 4 years due to going over my energy limit for one too many days.

Holger is in an extremely fragile state. Because he is so severe, his energy limit is so low that even the most mild stimulation such as light or noise forces his mind to use more energy than his limit permits and he gets worse. And his energy limit goes down even further. This is an extremely dangerous vicious cycle where every time the energy limit is exceeded, it gets lower and the patient has to figure out how to live, or in Holger's case, survive, while staying under this limit. At some point, if he is not in an environment that allows him to do this, the limit will get so low that he will not be able to stay under it and he will quickly get worse at an exponential rate until he dies.

This is not an idea or theory. This is the reality that millions of people suffering from ME/CFS face every day among other devastating symptoms.

If you commit Holger to a psych ward, it will kill him without question. If you do not commit him and allow him to live in the space he has been living in, he will very likely survive long enough for the research team at Stanford to find treatments that will make him much better or a cure that will return him to a fully healthy, productive member of society.

You have a choice to make and you now know what the consequences of that choice will be. If you have any semblance of humanity or decency you will give Holger a chance to live. That's the least that any human being deserves.

Thank you for your time and consideration.

Sincerely,

Whitney Dafoe, severe ME/CFS patient

www.whitneydafoe.com/mecfs (accessed on 22 April 2021)

www.facebook.com/whitneydafoe (accessed on 22 April 2021)

www.twitter.com/dafoewhitney (accessed on 22 April 2021)

www.instagram.com/whitneydafoe (accessed on 22 April 2021)

Appendix A.7. Good Science Grants Being Turned Down by NIH

Ronald Davis and other good ME/CFS scientists' brilliant grants are currently being turned down by NIH. Part of the reason for this is that the system for grant review is a mess [31]. Grants are reviewed by study sections, whose reviewers give them a score. They then go to Council, which funds the ones with the best scores. The main problems are: (1) There is only money for about 10% to get funded. (2) Reviewers nit-pick the grants so only a few get good scores, when, in fact, a much larger percentage of the grants are good. (3) Reviewers are often underqualified and uninformed about the subject of the grant, so their criticisms are incorrect—at times, ridiculous. (4) Council takes these inaccurate reviews as gospel and just funds the top few, without any evaluation of the competence of the reviews or consideration about the importance/urgency of the science. (5) *The leadership at NIH is obviously not committed to addressing the urgent problem of millions of people suffering from the horrific disease of ME/CFS.*

What is required is that scientists focus on researching things that can actually make a difference and lead to treatments or a cure for this disease. Grant reviewers are looking for a hypothesis that can be researched and lead to an answer that can be published. Ronald Davis isn't thinking about getting published; he's trying to find answers to what's happening inside the bodies of ME/CFS patients and discover an intervention that might help. What he wants to do isn't always so simple as "hypothesis, research, publication" and the grant reviewers only like grants that use existing, well-established methods that will lead to a publication. They can't imagine that anything they have never seen before could actually work. All of Dr. Davis' grants involve things they have never seen before. They don't understand it, which, when combined with what a mess ME/CFS is, makes it even harder for them. A lot of them also probably know nothing about ME/CFS or are prejudiced against it to begin with.

Another thing that gets in our way is, actually, probably a good thing most of the time. NIH has a rule that they are not allowed to communicate with the grant reviewers. I believe it's to try to keep things impartial. However, this rule hinders NIH from intervening and urging acceptance of Dr. Davis' grants and other good ME/CFS grants to try to make it impartial as it should be. I don't believe NIH is allowed to pick who reviews which grants, either. However, the Council's JOB is to make certain the reviews are competent and unbiased, and that the research addresses urgent and nationally important topics. In the case of ME/CFS, their job should be to make sure ME/CFS has adequate funding and to make certain that the research is likely to make progress towards understanding the disease in a way that might lead to treatment and cure. Not just a bunch of random data to publish.

This is all true, but to offer this as the cause of the problem presumes that the various heads of NIH actually want these grants approved in the first place.

What is also true is that NIH is engaged in a duplicitous publicity stunt, trying to continue their 40-year campaign of intentionally ignoring ME/CFS and systematically denying grants simply because they relate to ME/CFS, while, at the same time, trying to cultivate a public image of supporting ME/CFS. NIH has recently been saying things like "we want to and are ready to fund ME/CFS grants 'based on good science' so turn in grants and we'll fund them". Sounds good, right?

However, when good science grants about ME/CFS are submitted to NIH, these scientists review them and find absurd reasons to give them bad scores so they then get dismissed as "bad science".

I've got news for you, Francis Collins (the Director of NIH). Ronald Davis doesn't write, speak or think "bad science". We see the game you're playing, and we think you are an even more depraved human being for playing it. Either do the right thing and fund worthwhile ME/CFS grants, or publicly face the consequences of the blatant prejudice you are enacting.

We know your system is difficult, but we also know that you are the Director and you are capable of intervening when there is a severe health crisis, so that it gets addressed. It's been done before. You just have to believe that we have a real disease, that we are suffering, that more people will suffer, and that science needs significant funding to end the disease and end the suffering. You have to care. You told us "We are the National Institutes of Hope", "We are a family, in this together" and "We are ready to fund good grants". You need to put your money where your mouth is. You know what a good grant looks like.

This isn't something you're going to get away with. We all see what you are doing, will remember it, and history will record it.

Appendix A.8. My Experience in a Wheelchair

Have you ever had to use a wheelchair because of ME/CFS? I have and found it to be an unexpected experience. My legs slowly got worse because of circulation problems (I think) to the point where I could only walk into the kitchen once per day (15 feet or so). Then it got worse, and I had to crawl. I couldn't get a wheelchair from my insurance company because, even though I couldn't walk, I had no diagnosis they considered valid. When I asked my primary doctor (who I'd been seeing for years, trying to get a diagnosis, before I got diagnosed) for help, he said he thought it would be bad for me and he couldn't in good conscience help me get a wheelchair (because he thought getting me a wheelchair would reinforce my "non-existent illness that was in my head" and that I "needed to get over"). I finally got one on Craigslist for cheap. At first, I just used it to get around the house, which was a huge help. No more crawling to the kitchen for ice cream. It also gave me more independence to microwave my own food from leftovers. The real surprise came when I went out in public with it—mostly to Drs. appointments.

The way people treated me was a revelation. They instantly knew there was something dysfunctional in my body and treated me with respect, let me go first, and kind of bowed in respect to the hardship I was facing. It wasn't that I was craving attention, but after the way my friends and doctors treated me and the lack of funding and support from society, it was a shocking polar opposite that honestly felt really, really good. It was amazing to feel instant recognition from people of a real illness.

Most people don't feel good about being seen in public in a wheelchair, so this illuminates just how badly I was treated, and how many people in my life were constantly questioning the validity of my illness. When seen in a wheelchair, there was no question, just instant recognition and understanding.

Appendix A.9. When Life Gives You Rotten Lemons

They say, "When life gives you lemons, you make lemonade". What do you do when life gives you rotten lemons?

First, you are overwhelmed with anger that you didn't even get fresh lemons. In time, the anger turns to sadness, and slowly you start longing for fresh lemons.

You spend all your available energy thinking about making lemonade. How you would squeeze them, all the ingredients you could use. You become the most incredible lemonade maker in the world, only you're stuck at home, or in bed.

I can only imagine the torrent of knowledge and wisdom that will be unleashed upon the world when we are all cured.

References

1. Van Campen, C.; Rowe, P.; Visser, F. Validation of the Severity of Myalgic Encephalomyelitis/Chronic Fatigue Syndrome by Other Measures than History: Activity Bracelet, Cardiopulmonary Exercise Testing and a Validated Activity Questionnaire: SF-36. *Healthcare* **2020**, *8*, 273. [CrossRef] [PubMed]
2. Available online: https://www.cdc.gov/me-cfs/healthcare-providers/presentation-clinical-course/index.html (accessed on 1 April 2021).
3. Eaton, N.; Cabanas, H.; Balinas, C.; Klein, A.; Staines, D.; Marshall-Gradisnik, S. Rituximab impedes natural killer cell function in Chronic Fatigue Syndrome/Myalgic Encephalomyelitis patients: A pilot in vitro investigation. *BMC Pharmacol. Toxicol.* **2018**, *19*, 12. [CrossRef] [PubMed]
4. Strayer, D.R.; Young, D.; Mitchell, W.M. Effect of disease duration in a randomized Phase III trial of rintatolimod, an immune modulator for Myalgic Encephalomyelitis/Chronic Fatigue Syndrome. *PLoS ONE* **2020**, *15*, e0240403. [CrossRef] [PubMed]
5. Gabbey, E. What is Aggressive Behavior? Healthline. 18 September 2019. Available online: https://www.healthline.com/health/aggressive-behavior (accessed on 5 April 2021).
6. Surian, A.A.; Baraniuk, J.N. Systemic Hyperalgesia in Females with Gulf War Illness, Chronic Fatigue Syndrome and Fibromyalgia. *Sci. Rep.* **2020**, *10*, 1–12. [CrossRef] [PubMed]
7. Crosby, L.D.; Kalanidhi, S.; Bonilla, A.; Subramanian, A.; Ballon, J.S. Off label use of Aripiprazole shows promise as a treatment for Myalgic Encephalomyelitis/Chronic Fatigue Syndrome (ME/CFS): A retrospective study of 101 patients treated with a low dose of Aripiprazole. *J. Transl. Med.* **2021**, *19*, 1–4. [CrossRef] [PubMed]
8. Mandarano, A.H.; Giloteaux, L.; Keller, B.A.; Levine, S.M.; Hanson, M.R. Eukaryotes in the gut microbiota in myalgic encephalomyelitis/chronic fatigue syndrome. *PeerJ* **2018**, *6*, e4282. [CrossRef] [PubMed]
9. Giloteaux, L.; Goodrich, J.K.; Walters, W.A.; Levine, S.M.; Ley, R.E.; Hanson, M.R. Reduced diversity and altered composition of the gut microbiome in individuals with myalgic encephalomyelitis/chronic fatigue syndrome. *Microbiome* **2016**, *4*, 1–12. [CrossRef] [PubMed]
10. Navaneetharaja, N.; Griffiths, V.; Wileman, T.; Carding, S.R. A Role for the Intestinal Microbiota and Virome in Myalgic Encephalomyelitis/Chronic Fatigue Syndrome (ME/CFS)? *J. Clin. Med.* **2016**, *5*, 55. [CrossRef] [PubMed]
11. Proal, A.; Marshall, T. Myalgic Encephalomyelitis/Chronic Fatigue Syndrome in the Era of the Human Microbiome: Persistent Pathogens Drive Chronic Symptoms by Interfering with Host Metabolism, Gene Expression, and Immunity. *Front. Pediatr.* **2018**, *6*. [CrossRef] [PubMed]
12. Cunliffe, R.N.; Bowling, T.E. Artificial Nutrition Support in Intestinal Failure: Principles and Practice of Parenteral Feeding. *Clin. Colon Rectal Surg.* **2004**, *17*, 99–105. [CrossRef] [PubMed]
13. Stevens, R.D.; Nyquist, P.A. Types of Brain Dysfunction in Critical Illness. *Neurol. Clin.* **2008**, *26*, 469–486. [CrossRef] [PubMed]

14. Bileviciute-Ljungar, I.; Schult, M.; Borg, K.; Ekholm, J. Preliminary ICF core set for patients with myalgic encephalomyelitis/chronic fatigue syndrome in rehabilitation medicine. *J. Rehabil. Med.* **2020**, *52*, jrm00074. [CrossRef] [PubMed]

15. O'Brien, C.P. Benzodiazepine use, abuse, and dependence. *J. Clin. Psychiatry* **2005**, *2* (Suppl. 2), 28–33.

16. Thomas, D.R. Enteral Tube Nutrition. Merck Manual Professional Edition. Review/Revision July. 2020. Available online: https://www.merckmanuals.com/professional/nutritional-disorders/nutritional-support/enteral-tube-nutrition (accessed on 5 April 2020).

17. Furness, J.B. Integrated Neural and Endocrine Control of Gastrointestinal Function. *Adv. Exp. Med. Biol.* **2016**, *891*, 159–173. [CrossRef] [PubMed]

18. Furness, J.B.; Stebbing, M.J. The first brain: Species comparisons and evolutionary implications for the enteric and central nervous systems. *Neurogastroenterol. Motil.* **2018**, *30*, e13234. [CrossRef] [PubMed]

19. Stussman, B.; Williams, A.; Snow, J.; Gavin, A.; Scott, R.; Nath, A.; Walitt, B. Characterization of Post–exertional Malaise in Patients with Myalgic Encephalomyelitis/Chronic Fatigue Syndrome. *Front. Neurol.* **2020**, *11*, 1025. [CrossRef] [PubMed]

20. Green, J.; Romei, J.; Natelson, B.H. Stigma and Chronic Fatigue Syndrome. *J. Chronic Fatigue Syndr.* **1999**, *5*, 63–75. [CrossRef] [PubMed]

21. Johnson, M.L.; Cotler, J.; Terman, J.M.; Jason, L.A. Risk factors for suicide in chronic fatigue syndrome. *Death Stud.* **2020**, 1–7. [CrossRef] [PubMed]

22. The Chronic Lack of Funds for ME/CFS Research. The Hidden. April Qw. 2015. Available online: https://debortgjemteinternational.wordpress.com/2015/04/12/the-chronic-lack-of-funds-for-mecfs-research/ (accessed on 6 April 2021).

23. Devendorf, A.R.; Jackson, C.T.; Sunnquist, M.; Jason, L.A. Defining and measuring recovery from myalgic encephalomyelitis and chronic fatigue syndrome: The physician perspective. *Disabil. Rehabil.* **2019**, *41*, 158–165. [CrossRef] [PubMed]

24. Kashi, A.A.; Davis, R.W.; Phair, R.D. The IDO Metabolic Trap Hypothesis for the Etiology of ME/CFS. *Diagnostics* **2019**, *9*, 82. [CrossRef] [PubMed]

25. NIH RePORT. Report Portfolio Online Reporting Tools. Available online: https://report.nih.gov/funding/categorical-spending#/ (accessed on 7 April 2021).

26. Boulton, N. Reflections on "Voices in the Shadows". August 2013. Available online: https://voicesfromtheshadowsfilm.co.uk/welcome/reflections/ (accessed on 7 April 2017).

27. Institute of Medicine. *Beyond Myalgic Encepahlomyelitis/Chronic Fatigue Syndrome: Redefining an Illness*; The National Academies Press: Washington, DC, USA, 2015. [CrossRef]

28. Centers for Disease Control and Prevention. *Myalgic Encephalomyelitis/Chronic Fatigue Syndrome. What is ME/CFS?* CDC: Atlanta, GA, USA. Available online: https://www.cdc.gov/me-cfs/about/index.html (accessed on 7 April 2021).

29. The Mild, Moderate, Severe and Very Severe Levels of ME/CFS. 12 July 2018. Phoenix Rising. Available online: https://forums.phoenixrising.me/threads/the-mild-moderate-severe-and-very-severe-levels-of-me-cfs.60746/ (accessed on 7 April 2021).

30. Institute of Medicine (US); Committee on Health Care for Homeless People. Health Problems of Homeless People. In *Homelessness, Health, and Human Needs*; National Academies Press: Washington, DC, USA, 1988; p. 3. Available online: https://www.ncbi.nlm.nih.gov/books/NBK218236/ (accessed on 22 April 2021).

31. Nakamura, R.; Rockey, S. New Efforts to Maximize Fairness in NIH Peer Review. 29 May 2014. National Institutes of Health. Extramural Nexus. Available online: https://nexus.od.nih.gov/all/2014/05/29/new-efforts-to-maximize-fairness-in-nih-peer-review/ (accessed on 7 April 2021).

Article

Experiences of Living with Severe Chronic Fatigue Syndrome/Myalgic Encephalomyelitis

Victoria Strassheim [1,2,*], Julia L. Newton [1,2,3] and Tracy Collins [4]

[1] CRESTA Fatigue Clinic, Newcastle upon Tyne Hospitals NHS Foundation Trust, Newcastle upon Tyne NE4 6BE, UK; Julia.Newton@newcastle.ac.uk

[2] Academic Health Science Network–North East and North Cumbria, Newcastle upon Tyne NE4 5PL, UK

[3] Population Health Science Institute, Newcastle University, Newcastle upon Tyne NE1 7RU, UK

[4] Department of Social Work, Education and Community Wellbeing, Coach Lane Campus, Northumbria University, Newcastle upon Tyne NE7 7AX, UK; tracy.l.collins@northumbria.ac.uk

* Correspondence: victoria.strassheim@newcastle.ac.uk

Abstract: Chronic Fatigue Syndrome/Myalgic Encephalomyelitis (CFS/ME) is a rare disease with no known etiology. It affects 0.4% of the population, 25% of which experience the severe and very severe categories; these are defined as being wheelchair-, house-, and bed-bound. Currently, the absence of biomarkers necessitates a diagnosis by exclusion, which can create stigma around the illness. Very little research has been conducted with the partly defined severe and very severe categories of CFS/ME. This is in part because the significant health burdens experienced by these people create difficulties engaging in research and healthcare provision as it is currently delivered. This qualitative study explores the experiences of five individuals living with CFS/ME in its most severe form through semi-structured interviews. A six-phase themed analysis was performed using interview transcripts, which included identifying, analysing, and reporting patterns amongst the interviews. Inductive analysis was performed, coding the data without trying to fit it into a pre-existing framework or pre-conception, allowing the personal experiences of the five individuals to be expressed freely. Overarching themes of 'Lived Experience', 'Challenges to daily life', and 'Management of the condition' were identified. These themes highlight factors that place people at greater risk of experiencing the more severe presentation of CFS/ME. It is hoped that these insights will allow research and clinical communities to engage more effectively with the severely affected CFS/ME population.

Keywords: ME/CFS; severe; very severe; housebound; qualitative; interview; experience

Citation: Strassheim, V.; Newton, J.L.; Collins, T. Experiences of Living with Severe Chronic Fatigue Syndrome/ Myalgic Encephalomyelitis. *Healthcare* **2021**, *9*, 168. https://doi.org/10.3390/healthcare9020168

Academic Editors: Christopher R. Cogle, Kenneth Friedman, Kenny Leo De Meirleir and Lucinda Bateman

Received: 17 December 2020

Accepted: 28 January 2021

Published: 5 February 2021

1. Introduction

Chronic Fatigue Syndrome/Myalgia Encephalomyelitis (CFS/ME) is a rare disease with no known etiology [1,2]. Its cause is unknown [3], and studies suggest it affects 0.4% of the population [4]. Criteria have been produced to identify clinical characteristics [5]; however, diagnosis remains by exclusion due to the absence of biomarkers [6], which has led to significant stigma [7].

Diagnostic criteria have evolved with better understanding of the condition. The International Consensus in 2011 [8] developed from a growing understanding of the condition and has led to the identification of heterogeneous groups within the CFS/ME population.

Subgroups had previously been defined by Cox et al. [9,10]. The categories are mild, moderate, severe, and very severe, they and were implemented in the National Institute for Health and Care Excellence guidelines. Severe and very severe CFS/ME individuals are wheelchair, house, or bedbound, due to the severity of their symptom burden.

The severe and very severe CFS/ME population find it difficult to access their wider environments. This creates difficulty for them to engage in research and healthcare provision as it is currently delivered. Therefore, this group are classed as hard to reach [11].

It is suspected that 10–25% of the 0.4% CFS/ME population are in the severe to very severe category. This figure has been supported by the CFS/ME charities and evidenced in the research literature [12,13]. It has been estimated that there are up to 40 people with CFS/ME in each GP practice across England, 25% of which are severe and very severe [14]. Therefore, across England alone, there are approximately 82,000 very vulnerable, severe, and very severely affected CFS/ME individuals housebound and bedbound that may not be currently not receiving appropriate treatment.

Quantitative and qualitative data are limited due to the hard to reach nature of the severely affected CFS/ME population. However, this is slowly changing, and texture is being added to the objective evidence published [15–18]. In recent years, more qualitative research has been performed with the CFS/ME population. However, the gap in knowledge regarding severely and very severely CFS/ME remains. Reviewing the research and evidence as it is published and reflecting on previous trials leads to a better understanding of how variables impact each other. The PACE trial (A randomized controlled trial of adaptive pacing, Cognitive Behavioural Therapy and graded exercise) was one such trial [19,20]. It used operational diagnostic criteria to identify 600 patients to be randomised to one of four treatment pathways: specialist care, plus or minus adaptive pacing therapy, Cognitive Behavioural Therapy (CBT), and Graded Exercise Therapy (GET). The aim was to gather scientific evidence to demonstrate the outcomes of each treatment. The research was planned and executed prior to the 2011 International Consensus Criteria [8], which identified the heterogeneity of the condition, categorising mild, moderate, severe, and very severe [9,10]. Within the PACE trial, the eligibility assessment and consent for treatment was the outcome 6-min walk test, which would have precluded severe and very severe patients from taking part in the study [21].

In addition to exclusion from major research, it is emerging that people with CFS/ME experience discrimination from healthcare professionals and wider society [14,22]. This can be partially attributed to a lack of understanding around the condition. CFS/ME is an illness, not a disease, and as such, it currently has no identifiable pathology [23]. This does not sit well within the current biomedical illness model that has been historically taught within medical schools [23]. Large institutions such as universities, hospitals, health, and social care agencies function within clear objective, evaluative models in which theories and ideas are critiqued and scrutinised. Evaluative models shift the physiological state to defence, which is incompatible with creativity and expansive theories [24].

People with CFS/ME that take a biomedical approach [23] may feel vulnerable without biomarkers to authenticate their illness or referral to specialist services to endorse complex overlapping symptoms [25]. The lack of identifiable pathology can result in psychological labelling or somatisation [23]. This lack of understanding can contribute to a sense of blame shifting, where patients may feel that they are held accountable for their poor health [23].

For therapy to work, an individual must first accept their situation [26] and feel believed [18]. This can be impeded when healthcare professionals, the structure they work within, and wider society do not acknowledge the limits of medical science and so continue to require hard evidence to support a diagnosis, rather than treat and manage symptoms [23,25,27,28]. The therapeutic relationship may be further jeopardised if an individual with CFS/ME does not have the energy required to express themselves effectively, particularly in a stressful situation where their illness may not be believed [18].

This article expands on a 2016 two-phase pilot study. The project aimed to understand the feasibility of severe and very severe CFS/ME individuals engaging with research, whilst scoping and defining the prevalence of CFS/ME in the region. In that study, 2.5% of the 2500 severe CFS/ME population in the northeast of England were identified and characterised. The study also explored the quality of life, symptom burden, and impact of severe CFS/ME, collating a database from a postal survey in phase 1 [29]. Using the database, five individuals with severe and very severe CFS/ME were identified for phase 2 of the study. Attending the participant in their own home to limit the research burden on the CFS/ME individual and aid pacing strategies is novel to this research area. The patient and public involvement and methodology for the entire study is presented elsewhere [29]. This paper presents findings from the five qualitative interviews.

Aims

To explore the personal experience and understanding of individuals with CFS/ME.

To identify overarching themes that may highlight factors putting people at greater risk of experiencing the more severe presentation of CFS/ME.

To provide a better understanding of this population to allow healthcare and research communities to engage with individuals more effectively.

2. Materials and Methods

2.1. Methodology

A qualitative methodology involving in-depth interviews and drawing on phenomenology was selected. The interviews followed a semi-structured format [30], with interview schedules designed and worded to establish rapport and explore the area of concern in an open and flexible manner [31].

The aim of the interviews was to understand the perspective of the individual experience [32], to uncover personal meaning [31,33].

This group have limited presence in the research literature, and the methodology was explorative. By understanding the personal experience of individuals within the severe and very severe CFS/ME community, future qualitative research may be better focused with an improved understanding of practice, service, and research delivery to this population.

2.1.1. Data Sampling

The sample was purposive [34], being defined as having severe CFS/ME as identified from the database of self-reported CFS/ME individuals. How the participants were recruited is described in some detail elsewhere [29]. However, for completeness, 483 questionnaire packs with an expression of interest (EOI) to be involved in future studies were posted out. There were 425 packs sent via the charity ME North East. Of those 483 packs, 63 were returned in various stages of completion. Within the returns were over 40 completed expression of interest forms to be involved in future studies.

Resource was a consideration when recruiting the five participants to understand the feasibility of engaging severely affected CFS/ME individuals with research. The returned EOI came from a large geographical area, which was to be covered by one researcher. The fluctuating and unpredictable health of the cohort was also considered. Therefore, participants were approached who lived near the research base to limit disruption if appointments had to be rescheduled at short notice. Participants within the designated area were contacted until five had been recruited. These participants each agreed to four home visits over a three-month period. In each case, the first visit was to obtain consent and perform autonomic tests; the second visit was cognitive testing; the third was an interview; and finally, there was a physical assessment.

The five participants completed a second consent form and patient information sheet, which was specific to phase 2 and detailed the agreement to have results published in a peer-reviewed journal.

The ethical principle of non-maleficence to do no harm [35] was a primary aim in planning the research.

2.1.2. Data Collection

A pre-organised date and time was agreed, with the understanding that it could be rescheduled if necessary. Some participants gave specific instructions as to how to arrive at the home in order to reduce noise and stimulation.

Each interview lasted approximately one to two hours. Two patients had planned in advance and had compiled additional information to support the interview process and to reduce cognitive strain. A third patient found the study following a conversation too difficult and supplied a report that they had compiled over the course of their illness, which VS (author/researcher) read aloud as part of the interview, and the individual corrected or expanded on as they felt necessary. The youngest participant, who had been ill since her teens, requested that her mother be present to help answer any questions.

The interviews were conducted and recorded in conjunction with the collection of field notes [36–38]. This was to increase the trustworthiness and rigour of the data as well as the transferability of the findings [38].

2.1.3. Data Analysis

Thematic analysis drew on a grounded theory approach, with a method of inductive analysis [39]. Data was coded without trying to fit it into a pre-existing framework. The researchers were implicit in the process, taking a constructivist style to develop theories based on the data, not a pre-defined question to answer. The aim was to explore the individuals' experiences with severe CFS/ME.

Braun and Clark describe a six-phased approach that was followed [40]: initial familiarisation, generating codes, searching for themes, reviewing the themes and codes, defining and naming the codes, and finally writing the report. The software package NVivo version 12 was used to organise and analyse the data.

The transcripts were reviewed, and codes were created and grouped into similar themes. Then, themes were grouped for similarities and reduced to a manageable number.

Reflexivity was employed through conducting the analysis and interpretation of the findings with a second (qualitative) researcher (TC). The process was inductive, and the data produced were broad and rich.

2.2. Rigour

A research team with diverse background, knowledge, and skills was created to collect and analyse the data. The team had the collective ability to (1) access this hard-to-reach community, (2) collect the data whilst monitoring and limiting the impact the research might have on the participants, and (3) analyse the output. It was through this pooling of skills that this research was made possible. The potential for research bias is recognised and was limited through co-author collaboration of the research team to increase the trustworthiness of the analysis [38].

A clinically reasoned decision was taken not to have participants validate the researcher's transcripts. It is understood that "member checking" increases the internal validity and credibility of research [38]. However, this was outweighed by the need to adhere to the ethical principle of non-maleficence, to do no harm [35]. Participants were very fragile, and so it was felt that additional home visits would impact health and function too greatly. However, the Chief Executive Officer of the charity ME North East, who was instrumental in accessing the interviewed individuals, was sent a summary of the findings on behalf of the participants.

2.3. Ethical Considerations

Full ethics approval was granted by North East-Newcastle and North Tyneside 2 Research Ethics Committee. The participants had provided separate informed consent for both phase one and two of the study.

Pseudonyms have been given to each of the participants to protect their identities and maintain confidentiality

3. Results

3.1. Findings

The characteristics of the five participants are presented in Table 1. Individual circumstances created very diverse presentations, despite each participant being within the severe or very severe CFS/ME category.

Table 1. Participant characteristics.

Pseudonym	Age Range	Length of Illness	Gender	Living Arrangements	Support	Dependents	Background
JANE:	36–44	7 years	Female	Terraced house, predominantly bedbound, upstairs toilet.	Husband and mother	2 primary aged children	Very physically active prior to illness, professional, mother of 2 married.
DAVID:	36–45	18 years	Male	Semi-detached house. Predominantly housebound, living mostly in his bedroom.	Ageing parents	0	Single. Completed A-levels and managed to do office work between episodes of ill health, until early 20s. Had been very active prior to ill health.
ABI:	36–45	20 years	Female	Lived in a bungalow. Predominantly house bound.	Husband and ageing parents nearby	0	Married. Had been very active prior to ill health.
LORRAINE:	56–65	37 years	Female	Bungalow. Completely bedbound in a bedroom.	24-h social care	0	Single, professional, unable to work. Had been very active academically prior to ill health.
HELEN:	16–25	6 years	Female	Lived in a house with parents and older sister. Housebound, except for occasional outing assisted by family.	Working age parents and an older sister	0	Single, education abandoned due to ill health. Not able to work or continue education due to symptoms burden.

Three overarching themes were identified from the initial codes. Within two of these themes, subthemes were identified (see Table 2).

Each of these themes is now discussed in turn.

Table 2. Themes and subthemes.

Theme	Subtheme	Concepts within Subtheme
Lived experience	History and initial presentation Impact of illness	
Challenges to Everyday Life	Intrinsic	Physical Processing/Psychological Cognitive
	Extrinsic	
Management		

3.1.1. Theme 1: Lived Experience

The illness experiences of each of the five participants were very different, with one shared feature: the impact the illness had had on their lives. This overarching 'Lived Experience' theme explores the experience of living with severe CFS/ME and incorporates two subthemes: history and initial presentation and the impact the illness has had on each participant.

3.1.2. History and Initial Presentation

This subtheme considers family history, comorbidities, life events, age at onset of illness, initial presentation, and advice. The participants became ill at different points in their life, with different resources and burdens to manage their illness with. This led to different expressions of the illness, reflecting diverse lives and values and unique personal biopsychosocial frameworks.

For all the individuals, there was a recognised pre-existing vulnerability to becoming ill. Then, a trigger led to the development of multiple symptoms. This is illustrated in Table 3.

There was also evidence that for some participants, precipitating behaviours and circumstances made managing this complex cluster of symptoms that define the illness difficult. For several, it was an active life. For example, Lorraine became ill after completing her honours degree. She continued to struggle, living alone whilst trying to establish a new career and complete a postgraduate degree. Similarly, Jane was a busy full-time working mum of a sick baby who required regular hospital admissions.

For the individuals who became ill as young adults, multiple burdens were not a factor; however, they had not created robust coping strategies to manage a debilitating long-term condition before the illness was triggered. As can be seen from Table 1, David and Helen did not have the opportunities to establish themselves in a workplace or higher education to gain life experience before circumstances called on them to manage their illness.

Two of the five participants expressed how the initial presentation of this illness was difficult to describe to healthcare professionals, as there were often multiple competing and overlapping symptoms. For example, Abi described feeling *"tired exhausted, muscles were hurting. Felt poisoned, more than ill, horrible feeling all over, like that, but I was just feeling weird and wrong."*

When the illness was triggered in a transitionary phase of life e.g., new jobs/careers, new parents, it was difficult to understand the cause and effect. Jane stated: *"In hindsight, it wasn't a normal type of tiredness. I just didn't have the words or assertiveness to convince anyone."*

The initial presentation was often vague, and the initial advice received from healthcare professionals was often to keep going, stay active and get fitter.

Table 3. Characteristics of early illness.

Pseudonym	Pre-Existing Vulnerability	Trigger	Initial Symptoms	Time to Diagnose	Transitions of Coping Strategy
Jane	Young mum, full-time worker, transitioning point in life.	Complex. Viral infections, birth of son.	Myalgia, pain, viral infections.	3 years	Persisting to Acceptance
David	Never returned to pre-glandular fever energy levels. Began working full time. Contracted influenza B. Transitioning point in his life.	Influenza B.	Post viral fatigue, aches and pains.	3 months to diagnose post viral fatigue.	Avoidance 'Head in the sand like an ostrich'.
Abi	Not listening to body.	Tick bite. Working as a gym and aerobics instructor. Just kept going.	Overly hot and sweating during routine exercise. Exhausted, muscles hurting, fatigue.	Unknown	Persisting to acceptance of situation.
Lorraine	Limited support network. Establishing herself in her new career. Transitioning point in her life	Viral.	Orthostatic intolerance, fatigue, weakness.	Unknown	Persisting and boom and busting towards acceptance.
Helen	Exam time, transitioning point in her life.	Migraines and hypersomnia	Headaches, migraines, and hypersomnia	Unknown	Degree of persisting, boom and busting due to limited experience.

Most participants described a "persisting" or "boom or bust" behaviour pattern during a sustained period in which they attempted, unsuccessfully to regain their previous "normal". For example, Jane changed jobs, trying to get a healthier balance: *"I took a month off, when I was to get myself well, get myself fit. Doctors' advice was to exercise in the hope that, you know, start of the new job physically better, physically fitter."*

Helen was given similar advice, and being younger, it was her mum who directed the activity: *"I did try taking her to the park, running around, desperately thinking she needs to exercise, she needs exercise. And then she was wiped out for days and days and days."* Exercise intolerance appears to be present for multiple individuals; however, without an objective physical cause, it is difficult to identify. Each of the participants deteriorated whilst trying to get fitter or be "normal".

There appears to be a consistent pattern of boom/bust and persistence leading to deterioration, with consequent periods of bedrest to alleviate symptoms, resulting in deconditioning and limited function, which can be profound.

At best, energy fluctuations allow individuals to experience only fleeting moments of "normal". For example, Jane stated she was active for 2% of the day when she washed and toileted, whereas Lorraine was completely bedbound and dependent on assistance just to sit up in bed.

For all or many of the participants, tasks such as personal hygiene and food preparation are limited to basic needs. Overlaying complexities of orthostatic symptoms, allergies, and sensitivities all impact activities of daily living. This was demonstrated by David being limited to a bowl of cornflakes on the days his ageing mother could not assist him to make a meal, which was becoming more frequent. Jane reluctantly admitted she could possibly make a meal for herself twice a month and at times had gone days without food when her carers had been away; she was simply too ill to do any more than the essentials. Nutrition had been prioritised out of that energy calculation.

Sleep was also affected. Two individuals described parasomnias, sleeping 22 h per day for long periods of time or being awake but unable to move. Lorraine related her poor sleep to autonomic disturbance: *"hyperadrenergic-over heating and waking hourly. Bad dreams, hallucinations"*.

Jane expressed that sleep was beyond her control due to family circumstances, with *"two small children who can be ill and climb into bed"*.

3.2. Impact on Life

This subtheme illustrates the impact on life including function, nourishment, sleep, and social isolation. Individuals acknowledge the confining effects of their illness. Some found it difficult to live with meaning, as this required energy and effort that they did not have. Therefore, life was a passage of time, without having the resources to take action. For David, *"The biggest problem for people with ME, you are in limbo, nobody knows what to do with you. What life, because it's more about filling time than actually living, because living requires doing things with much effort."*

However, Jane had adapted and altered her thoughts to accept her situation, prioritising her energy for the activities that gave her life meaning and value, demonstrating resilience to her situation:

> *"I have all the people and things that are important in life and you know, like, relationships and the love you have got in your family, I still have that so my life has been boiled down to the most important bits that are still here. If I was going to lose things from my life, it would be my work, and yeah, it would be reluctantly to be able to go outside and have a life, but it's these special relationships that I cherish that make me feel happy and content."*

3.2.1. Theme 2: Challenges to Everyday life

This overarching theme comprises two subthemes: Intrinsic—those elements that were fundamental to the individual's physiology or psychology and extrinsic—those elements that operated from outside the individual.

3.2.2. Intrinsic Concepts

Within this subtheme were three distinct concepts. The physical concept included the effect of activity, adrenaline response/orthostatic intolerance, allergies, and sensitivities and baseline level of activity/unpredictable fluctuations. Processing/psychological concept encompassed belief about cure, personal views, views about management, communication, and energy balance. Finally, there were concepts comprising cognitive impairment, dissociation/detachment, and mental health.

Each individual's ability to function was limited to their personal capacity, which fluctuated extremely and created difficulties to plan and manage basic day-to-day activities such as washing, toileting, and in some instances feeding, particularly with the added symptom of allergies and sensitivities. For example, Lorraine said, *"Digesting food is an energy challenge,"* and Abi found, *"In the food department (allergies), the big one is nickel, which limits food diversity."*

It appears that when baseline energy levels are so low and capacity is so limited, there is a frequent tendency to experience the challenge of everyday activities as a physiological threat. Participants often access the sympathetic nervous system just to carry out basic functions such as personal hygiene and eating. It seems this often leads to dysautonomia, which is a disruption of the autonomic nervous system that regulates the heart, blood vessels, digestion, and breathing. This disruption could be due to activity and/or orthostatic challenge. For example, Jane said, *"I am in bed 100% of the time. When active, pain feels worse, feel weak, feeling hot, cold, fevery, kind of heart palpitation, breathlessness, shaky."*

The management of physiological and physical limitations was often bound by thoughts and perceptions of the participant's experience, which was in turn often limited by their condition. The ability to process and communicate beliefs about their situation were limited by their energy levels. Lorraine said:

> *"You get to a point in a relationship where you actually need to say, 'This is what is going on with my illness...' And then you have to eat and we never get a chance, there's no time for conversation. All my emotions are around, are set to one side, all the loss, all the bereavement, loss of self, loss of life, loss of opportunity, loss of the living, so family relationships are just set to one side, there isn't time to process those emotions to ever have them."*

The processing ability appeared to be further impacted by cognitive impairments. Individuals recognised this; however, the strategies they put in place created increased energy expenditure.

Compounding these challenges was the concerning presence of dissociation and detachment, which was expressed by four of the five participants, along with, in some cases, loss of identity. For example: *"I feel detached and confused. I suffer from disconnectedness, so I don't feel physically present—don't concentrate on that—having a blister and gritting your teeth. I don't know, I can't remember"* (Lorraine). *"There is numbness most of the time, a contentedness. Then the rest of the time—frustration, disappointment, fear. My body being unresponsive. Just staring into space. I don't feel myself anymore, sense of self or identity... I have accepted, peaceful, but I struggle with I don't feel like myself anymore"* (Jane).

3.2.3. Extrinsic Concepts

This subtheme relates to the influences outside of the individual's control that harm their ability to function in the world given their illness. Such influences include the benefits system, lack of professional understanding, and prejudicial views.

Most social challenges came from a lack of understanding within society as to the nature of CFS/ME or the extent to which it can impact. This lack of understanding can lead to prejudicial views and preconceptions that impede an individual's access to healthcare and social services. For example, Abi conveyed an experience she had had in an accident and emergency department, having been taken from her health centre to hospital in an ambulance. Once she arrived, the attitude of the staff changed when they realised she was a frequent visitor: *"I passed out. I woke up and was being dragged along the floor by two orderlies and this nurse screaming at me."*

Abi received the care she required once her tachycardia was identified. Lorraine, too, recognised this engrained culture. *"It takes years to erode institutionalised discrimination and prejudice which are unhealthy and negative for both victim and perpetrator."*

Health and social care require a diagnosis to be current in order for an individual to be eligible for social support. This support is binary: there is no grading within the system. This creates a difficulty when there is not a biomedical marker to identify the illness. The repeated cycle of having to demonstrate ill health impacts an individual's ability to manage and improve from that health issue. It also undermines continuity of recognition of that ongoing issue. For example, Lorraine said, *"I lack a current diagnosis so I can't get my benefits. They keep saying no current diagnosis."* Similarly, for David, *"After probably a year to a year and a half of having ME, I was improving, doing much better, but then the benefits agency reviewed and decided I was fit for work. I lasted 6 months working as admin in a restaurant before I crumbled … [following repeated appeals the benefits withdrawal decision was repealed]. I should never have been taken off (benefits) in the first place."*

3.2.4. Theme 3: Management of the Condition

This theme incorporates managing and coping strategies, relief from symptoms, understanding acceptance, acceptance of professional lack of understanding, social media, GP attitudes, and healthy carer beliefs.

The five individuals interviewed managed their condition within the confines of the means at their disposal, both intrinsic and extrinsic. Their strategies were founded on a subjective understanding of their capabilities, their illness, and their resources. Participants relied on support from family members, social media, and the internet to gain information they required. Social media was used to maintain contact with friends. For example, Abi used *"Twitter with friends"* and Helen used the Internet as she transitions from sleep to wakefulness, *"In bed 30 min am on phone waking up, googling".*

Whilst the internet alleviates the social isolation, it can distil and reinforce beliefs. For example, Abi reported, "I have never been to a CFS clinic and I am glad, because I won't want to do GET (Graded Exercise Therapy). I know from experience, physically pushing past what you feel, it made me worse, so I wouldn't want to entertain that."

In terms of social support, Helen described how she was reliant on the availability of her working parents to take her out to socialise. However, their availability also had to coincide with when she was well enough. Helen's parents appeared to have had sacrificed their social life to prevent her from being left alone and isolated. Her family appeared to plan their lives around her illness, as far as they could whilst also maintaining their income. The focussed pressure on caregivers was substantiated by Jane, who explained it was her husband, the income provider, who worked full time, was the main caregiver to two primary aged children, and was also her care giver. Therefore, it was not only the CFS/ME participants who had reduced their lives to necessary priorities but also their caregivers and families.

In addition to social support, healthcare professionals were acknowledged as people who could help.

Abi reported the occasional doctor who understands, whilst Helen and her mother were very keen to praise their current GP as "fabulous" because they understood Helen's situation. This relieved a lot of stress. Lorraine found that individuals who were not fixed in their beliefs of the condition were most beneficial: *"People who are genuine, non-judgemental and open minded"* (Lorraine).

This often left a very narrow path to navigate. In turn, this may limit the potential for an individual with CFS/ME to improve. For example, Abi stated "I have learnt it is pointless, some people won't listen and there is no point. I have tried to fight back and been called neurotic. I was really frustrated, not neurotic." Similarly, for David reported:

"Resignation—it's not a great surprise after all these years. You hear about how much progress they make with this and that and the other and you think yeah, but there is obviously an awful lot of conditions where nothing changes for decades."

4. Discussion

Our first aim was to explore the personal experience and understanding of individuals with CFS/ME. This was achieved through open questioning and exploration of the participant's views of their reality.

The second aim was to identify overarching themes that may help identify risk factors that place people at greater threat of experiencing the more severe presentation of CFS/ME. Many of the participants demonstrated previously identified risk factors for expressing the severe form of CFS/ME: a delayed diagnosis [22,28]; problems accessing social security [22] and poor relationships with doctors or health professionals [14,22,23].

This small qualitative study has identified other common factors, which need further research to clarify and confirm. For example, demonstration of deterioration as the individual initially attempted to get fitter or remain "normal". This may be an indication of an unidentified exercise intolerance. It appears that the point in a person's life when the illness presents is of importance. Several participants were moving from one phase of life to another. For example, school exam time, moving from school or higher education to work life, or following the birth of a child. It appears that severe presentation may manifest when the illness coincides with a transitioning time in a person's life. Another common factor is the relationship between burden and resource. Those with dependents or many responsibilities and a limited support network appear to be more vulnerable to the severe expression of the illness. In addition, those individuals who had a support network but remained to some degree dependent on carers were not able to establish independence due to the illness.

The final aim of this study was to provide a better understanding of this population to allow a research community to engage with them more effectively. This has been addressed to an extent by Kingdon [14]. We have taken a phenomenological approach to report the lived experience of five individuals with CFS/ME. These findings cannot be generalised; however, it is possible that they are transferable to other individuals in a similar position. Here, we will expand on how these findings may be applied to the evolving understanding.

All five of the participants had vague initial presentations that they found difficult to explain, illustrating the experience of living with poorly understood illness. Despite a fatigue presentation, they were actively encouraged to keep going and push through or had themselves tried to regain their former life. Maladaptive sickness patterns have been recognised in chronic illness [41], and the recommendation of exercise in the presence of fatigue is increasingly acknowledged as detrimental. Inappropriate advice may promote unhealthy pacing behaviours of "boom and bust" and persistence [42]. It is suggested that this ultimately leads to deconditioning through the over training exercise curve, which is recognised in athletes but remains an under researched area [43]. Fatigue self-efficacy improves outcomes [41]; however, the confidence to self-manage fatigue must be fostered gradually and immediately if unhealthy adaptive behaviours are to be avoided.

The timing of the illness appears to have importance. When illness occurs at a young age, school attendance is reduced, seriously affecting intellectual and social development [16,44]. This is illustrated by Helen and to some extent David, who continued to be dependent on their parents, who were their carers into adulthood. This combination can further impact managing this complex illness [15]. Another critical factor is if the illness occurs during a transition, e.g., from professional to working mother, when tiredness is expected. This may make diagnosis more difficult: transitioning life stages produce confounding factors that confuse a biomedical assessment.

All the participants followed a deteriorating pattern. It appears there comes a point when burdens exceed resources and the opportunity to improve is extinguished. Then, people experience physiological threats, resulting in fight or flight reactions or dissociative responses. Dissociation is described within polyvagal theory as losing a sense of presence resulting in experiencing a disconnection and a lack of continuity between thoughts, memories, surroundings, and actions [45].

It is concerning that at least two of the participants reported symptoms of dissociation. It is suspected in the three others. Benign aspects of existence were experienced as threats so extreme that the ability to be present was lost. Acceptance has been identified as a precursor for any therapeutic intervention to succeed. However, it is proposed that for acceptance to occur, a person must feel safe and present within their physical environment. This has significant implications for management and rehabilitation.

Intrinsic challenges to everyday life are further compounded by the extrinsic burdens. All five of the participants reported poor or limited interactions with healthcare professionals during their illness. Negative attitudes towards CFS/ME by medical professionals are repeatedly reported [23,46]. After many attempts at trying and failing to navigate the health and social care systems, with an imbalance of energy, resources, and burdens, some individuals experiencing CFS/ME eventually appear to accept their limitations and those of their health professionals.

All of the participants were forced to give up education or employment. In work and educational institutions, the lack of understanding and provision for people with CFS/ME creates obstacles for people with the illness to remain in those environments. CBT and GET do not restore the ability of a person with CFS/ME to work [47]. People with CFS/ME who cannot remain in employment need to access the benefits system. As CFS/ME does not sit within the current biomedical model of health and social care, this creates issues navigating the benefits system. This was reported by two of the five participants. It has been recorded that the benefits system in the UK does not meet the needs of people with CFS/ME, leaving them socially isolated and/or increasingly dependent on friends and family. The distress of navigating the system often exacerbates health conditions [40].

The five participants managed their condition as well as resources allowed, both intrinsic and extrinsic. At times, this unfortunately meant accepting the limits of the system in which they found themselves.

Four of the five study participants had received specialist support during their illness. However, this support was not always valued. One participant was receiving support at the time of the study. All the participants presented with complex multi-faceted issues that impacted every component of the biopsychosocial model. Their ability had declined to the extent where it impacted every aspect of their life: physical function, diet, sleep, and social interaction.

The participants were heavily reliant on the internet to source management strategies. This often distilled illness beliefs. Health literacy has been shown to be a challenge in vulnerable groups [48]. However, we do not understand how severely affected CFS/ME individuals use health literature, because they are so under researched.

It appears that severely affected CFS/ME individuals must lead a very disciplined and limited existence in order to manage symptom burden within their intrinsic and extrinsic limitations. It is an open question as to whether such limits impact their ability to be psychologically flexible and resilient in their outlook.

5. Conclusions

This study is novel, as it has accessed this hard to reach population group and recorded their experience. Most of the participants had received some form of specialist CFS/ME support or had access to the healthcare services. However, their experiences ranged from accepting the limitations of the service to having a very negative view.

CFS/ME is a medically unexplained illness lying at the boundaries of understanding within the legacy biomedical model. An illness where there is no single, simple cause or theoretical model, no clear mind/body division, and no definitive classification [1] does not sit easily in the current healthcare system. The CFS/ME presentation conflicts with the current health and social care model [1,2]. The severe CFS/ME presentation sits outside the model and therefore is not acknowledged.

This illness ranks low within primary care, as it is not life threatening [23]. However, it is potentially life shortening [14]. There are certainly physical and mental health symptoms that are often disregarded or missed within the complex presentation [14], and reports suggest that 88 suicides have been partly attributed to CFS/ME between 2001 and 2016. However, it has been noted that it is not necessarily intrinsic factors that lead to suicide, but a combination of extrinsic factors, which include a lack of medical care and social support, failure to control key symptoms, and inadequate financial help. Depression is not always a feature in CFS/ME-related suicide [49].

Pathway-focused institutional cultures are not predisposed to embrace the ambiguities inherent in adopting the more holistic biopsychosocial model, where outcomes are more difficult to define and evaluate. The resulting continued narrow biomedical focus of the current social care system results in neither the healthcare professional nor the CFS/ME patient feeling safe with each coming from a position of defence when they communicate [23,27,45]. People with the severe expression of CFS/ME appear to avoid the harm of the current health and social care system by purposely withdrawing from it. This reduces opportunities for rehabilitation and is an area for further study.

Individuals with severe CFS/ME live on the peripheries of society, at the edges of the research bell curve [50]. They do not belong within "normal" expectations and they do not have the energy to try to fit [51]; therefore, they remain socially, medically, and financially isolated. The role of environment has been discussed within the international classification of function. Disability has been acknowledged as a socially created problem that can limit freedom by failure to provide the resources and opportunities needed to make participation feasible [52]. This paradigm must be explored further if we are to better understand and provide adequate health and social care for the severe CFS/ME population or other people experiencing "illness" that does not fall into the biomedical model.

The findings of this study aim to assist understanding of the needs of the severe CFS/ME population. Currently, the healthcare system and research community are failing to provide resources and opportunities for this group to engage, and so enable the positive outcome of increased independence. Longer periods of intervention, home visits and telephone consultations and in extreme cases inpatient rehabilitation in specialist services are effective evidenced interventions in the research literature [10,16,44,53]. Such services would meet the needs of CFS/ME individuals much better than the status quo which often forces patients to meet the needs of the system in order to secure the care that they need.

A re-evaluation of the approach taken to CFS/ME and other unexplained illness is ever more urgent given the upcoming surge in numbers of long-haul COVID-19 individuals. A major symptom of such long-haul COVID-19 is fatigue [54,55]. Research and healthcare communities have much experience to share and further research to perform, particularly in the area of health, social care, and societal attitudes allowing vulnerable ill people to remain valued members of society.

Limitations

The thematic analysis aspect of this research studies a small number of participants in depth, giving a rich presentation. The participants were from a small geographical area and may not be representative of the wider CFS/ME community.

It is recommended that further research is conducted with a larger sample of participants across a wider geographical area of the United Kingdom. Adequate financial and time provision must be allocated to allow severe and very severe CFS/ME individuals to engage in future projects. Part of future research regarding CFS/ME must explore the wider biopsychosocial factors that lead to the severe expressions of fatigue. The goal is to identify risk factors that affect the deterioration of the condition within different life phases and aid earlier detection of those at risk of the severe and very severe expression of CFS/ME and adequate provision of healthcare.

Author Contributions: Conceptualisation, J.L.N.; V.S. and T.C.; methodology, T.C. and V.S. validation, V.S. and T.C.; formal analysis, V.S. and T.C.; investigation, J.L.N.; V.S. and T.C.; data curation, V.S. and T.C.; writing—original draft preparation, V.S.; writing—review and editing, V.S., T.C. and J.L.N.; supervision, T.C. and J.L.N.; project administration, V.S.; funding acquisition, V.S. All authors have read and agreed to the published version of the manuscript.

Funding: This study was supported by ME Research UK (SCIO charity number SC036942).

Institutional Review Board Statement: The study was conducted according to the guidelines of the Declaration of Helsinki, and approved by the NRES Committee North East: Newcastle and North Tyneside 2 REC Ref 15/NE/0103, IRAS 154422. Date of Approval 13.01.2016.

Informed Consent Statement: Informed consent was obtained from all subjects involved in this study.

Data Availability Statement: The data presented in this study are available on request from the corresponding author. The data are not publicly available due to privacy and data protection protocols.

Acknowledgments: We thank ME Research UK for funding the study that allowed this pilot project. Researcher and Health Psychologist Practitioner, Vincent Deary for his guidance and insights into the presentations of some of the participants. We would also like to thank ME North East, all the local support groups and in particular, the participants who contributed to this study.

Conflicts of Interest: The authors declare no conflict of interest.

References

1. Eriksen, T.E.; Kerry, R.; Mumford, S.; Lie, S.A.N.; Anjum, R.L. At the borders of medical reasoning: Aetiological and ontological challenges of medically unexplained symptoms. *Philos. Ethicsand Humanit. Med.* **2013**, *8*, 1. [CrossRef] [PubMed]
2. Deary, V. Explaining the unexplained? Overcoming the distortions of a dualist understanding of medically unexplained illness. *J. Ment. Health* **2005**, *14*, 213–221. [CrossRef]
3. Baker, R.; Shaw, E.J. Guidelines: Diagnosis and management of chronic fatigue syndrome or myalgic encephalomyelitis (or encephalopathy): Summary of NICE guidance. *BMJ Br. Med. J.* **2007**, *335*, 446–448. [CrossRef]
4. NICE. Chronic Fatigue Syndrome/Myalgic Encephalomyelitis Diagnosis and Management in Adults and Children. Available online: http://www.nice.org.uk/guidance/cg53 (accessed on 15 August 2020).
5. Fukuda, K.; Straus, S.E.; Hickie, I.; Sharpe, M.C.; Dobbins, J.G.; Komaroff, A. The chronic fatigue syndrome: A comprehensive approach to its definition and study. *Ann. Intern. Med.* **1994**, *121*, 953–959. [CrossRef]
6. Brenu, E.W.; Van Driel, M.L.; Staines, D.R.; Ashton, K.J.; Ramos, S.B.; Keane, J.; Klimas, N.G.; Marshall-Gradisnik, S.M. Immunological abnormalities as potential biomarkers in chronic fatigue syndrome/myalgic encephalomyelitis. *J. Transl. Med.* **2011**, *9*, 1. [CrossRef]
7. McManimen, S.; McClellan, D.; Stoothoff, J.; Gleason, K.; Jason, L.A. Dismissing chronic illness: A qualitative analysis of negative health care experiences. *Health Care Women Int.* **2019**, *40*, 241–258. [CrossRef] [PubMed]
8. Carruthers, B.M.; Van de Sande, M.I.; De Meirleir, K.L.; Klimas, N.G.; Broderick, G.; Mitchell, T.; Staines, D.; Powles, A.C.P.; Speight, N.; Vallings, R. Myalgic encephalomyelitis: International consensus criteria. *J. Intern. Med.* **2011**, *270*, 327–338. [CrossRef] [PubMed]
9. Cox, D.L.; Findley, L.J. The management of chronic fatigue syndrome in an inpatient setting: Presentation of an approach and perceived outcome. *Br. J. Occup. Ther.* **1998**, *61*, 405–409. [CrossRef]
10. Cox, D.L.; Findley, L.J. Severe and very severe patients with chronic fatigue syndrome: Perceived outcome following an inpatient programme. *J. Chronic Fatigue Syndr.* **2000**, *7*, 33–47. [CrossRef]
11. Pathways Through Participation. Briefing Paper No. 3—Who Participates? The Actors of Participation? 2009. Available online: http://pathwaysthroughparticipation.org.uk/wp-content/uploads/2009/09/Briefing-paper-3-Who-participates1.pdf (accessed on 13 September 2020).
12. ME Research UK. Severely Affected ME/CFS Patients a Geographically Defined Study. Available online: http://www.meresearch.org.uk/our-research/ongoing-studies/severely-affected-me-patients/ (accessed on 21 January 2020).
13. Pendergrast, T.; Brown, A.; Sunnquist, M.; Jantke, R.; Newton, J.L.; Strand, E.B.; Jason, L.A. Housebound versus nonhousebound patients with myalgic encephalomyelitis and chronic fatigue syndrome. *Chronic Illn.* **2016**, *12*, 292–307. [CrossRef] [PubMed]
14. Kingdon, C.; Giotas, D.; Nacul, L.; Lacerda, E. Health Care Responsibility and Compassion-Visiting the Housebound Patient Severely Affected by ME/CFS. *Healthcare* **2020**, *8*, 197. [CrossRef]

15. Ali, S.; Adamczyk, L.; Burgess, M.; Chalder, T. Psychological and demographic factors associated with fatigue and social adjustment in young people with severe chronic fatigue syndrome/myalgic encephalomyelitis: A preliminary mixed-methods study. *J. Behav. Med.* **2019**, *42*, 898–910. [CrossRef] [PubMed]
16. Burgess, M.; Lievesley, K.; Ali, S.; Chalder, T. Home-based family focused rehabilitation for adolescents with severe Chronic Fatigue Syndrome. *Clin. Child. Psychol. Psychiatry* **2019**, *24*, 19–28. [CrossRef] [PubMed]
17. Broughton, J.; Harris, S.; Beasant, L.; Crawley, E.; Collin, S.M. Adult patients' experiences of NHS specialist services for chronic fatigue syndrome (CFS/ME): A qualitative study in England. *BMC Health Serv. Res.* **2017**, *17*, 384. [CrossRef]
18. Adamowicz, J.; Caikauskaite, I.; Friedberg, F.; Seva, V. Patient change attributions in self-management of severe chronic fatigue syndrome. *Fatigue Biomed. Health Behav.* **2017**, *5*, 21–32. [CrossRef]
19. White, P.D.; Sharpe, M.C.; Chalder, T.; DeCesare, J.C.; Walwyn, R. Protocol for the PACE trial: A randomised controlled trial of adaptive pacing, cognitive behaviour therapy, and graded exercise as supplements to standardised specialist medical care versus standardised specialist medical care alone for patients with the chronic fatigue syndrome/myalgic encephalomyelitis or encephalopathy. *BMC Neurol.* **2007**, *7*, 6.
20. White, P.D.; Goldsmith, K.A.; Johnson, A.L.; Potts, L.; Walwyn, R.; DeCesare, J.C.; Baber, H.L.; Burgess, M.; Clark, L.V.; Cox, D.L. Comparison of adaptive pacing therapy, cognitive behaviour therapy, graded exercise therapy, and specialist medical care for chronic fatigue syndrome (PACE): A randomised trial. *Lancet* **2011**, *377*, 823–836. [CrossRef]
21. Bavinton, J. GET Manual Version 7. Available online: http://www.pacetrial.org/docs/get-therapist-manual.pdf (accessed on 17 October 2020).
22. Pheby, D.; Saffron, L. Risk factors for severe ME/CFS. *Biol. Med.* **2009**, *1*, 50–74.
23. Bayliss, K.; Goodall, M.; Chisholm, A.; Fordham, B.; Chew-Graham, C.; Riste, L.; Fisher, L.; Lovell, K.; Peters, S.; Wearden, A. Overcoming the barriers to the diagnosis and management of chronic fatigue syndrome/ME in primary care: A meta synthesis of qualitative studies. *BMC Fam. Pract.* **2014**, *15*, 1. [CrossRef]
24. Wernham, W.; Pheby, D.; Saffron, L. Risk Factors for the Development of Severe ME/CFS—A Pilot Study. *J. Chronic Fatigue Syndr.* **2004**, *12*, 47–50. [CrossRef]
25. McDermott, C.; Lynch, J.; Leydon, G.M. Patients' hopes and expectations of a specialist chronic fatigue syndrome/ME service: A qualitative study. *Fam. Pract.* **2011**, *28*, 572–578. [CrossRef]
26. Van Damme, S.; Crombez, G.; Van Houdenhove, B.; Mariman, A.; Michielsen, W. Well-being in patients with chronic fatigue syndrome: The role of acceptance. *J. Psychosom. Res.* **2006**, *61*, 595–599. [CrossRef]
27. Åsbring, P.; Närvänen, A.-L. Ideal versus reality: Physicians perspectives on patients with chronic fatigue syndrome (CFS) and fibromyalgia. *Soc. Sci. Med.* **2003**, *57*, 711–720. [CrossRef]
28. De Lourdes Drachler, M.; De Carvalho Leite, J.C.; Hooper, L.; Hong, C.S.; Pheby, D.; Nacul, L.; Lacerda, E.; Campion, P.; Killett, A.; McArthur, M. The expressed needs of people with chronic fatigue syndrome/myalgic encephalomyelitis: A systematic review. *BMC Public Health* **2009**, *9*, 458.
29. Strassheim, V.J.; Sunnquist, M.; Jason, L.A.; Newton, J.L. Defining the prevalence and symptom burden of those with self-reported severe chronic fatigue syndrome/myalgic encephalomyelitis (CFS/ME): A two-phase community pilot study in the North East of England. *BMJ Open* **2018**, *8*, e020775. [CrossRef] [PubMed]
30. Patton, M.Q. *Qualitative Research & Evaluation Methods: Integrating Theory and Practice*; Sage Publications: San Francisco, CA, USA, 2014.
31. Ten Have, P. David Silverman (2006). Interpreting Qualitative Data: Methods for Analysing Talk, Text and Interaction. *FQS Forum Qual. Soc. Res.* **2008**, *9*. [CrossRef]
32. Green, J.; Thorogood, N. *Qualitative Methods for Health Research*; Sage: San Francisco, CA, USA, 2018.
33. Guba, E.G. *The Paradigm Dialog*; Sage Publications: San Francisco, CA, USA, 1989.
34. Barbour, R. *Doing Focus Groups*; Sage: San Francisco, CA, USA, 2008.
35. Gilhooly, M. Ethical issues in researching later life. In *Researching Ageing and Later Life*; Open University Press: Buckingham, UK, 2002; pp. 211–225.
36. Mays, N.; Pope, C. Assessing quality in qualitative research. *BMJ* **2000**, *320*, 50–52. [CrossRef]
37. Smith, S. Encouraging the use of reflexivity in the writing up of qualitative research. *Int. J. Ther. Rehabil.* **2006**, *13*, 209–215. [CrossRef]
38. Finlay, L. 'Rigour','ethical integrity'or 'artistry'? Reflexively reviewing criteria for evaluating qualitative research. *Br. J. Occup. Ther.* **2006**, *69*, 319–326. [CrossRef]
39. Braun, V.; Clarke, V. Using thematic analysis in psychology. Qualitative research in psychology. *Qual. Res. Psychol.* **2006**, *3*, 77–101. [CrossRef]
40. A Deeply Dehumanising Experience ME/CFS Journeys Through the PIP Claim Process in Scotland. Report for Action for ME. 2016. Available online: https://www.actionforme.org.uk/uploads/pip-report-scotland.pdf (accessed on 20 September 2020).
41. Dantzer, R.; Heijnen, C.J.; Kavelaars, A.; Laye, S.; Capuron, L. The neuroimmune basis of fatigue. *Trends Neurosci.* **2014**, *37*, 39–46. [CrossRef] [PubMed]
42. Antcliff, D.; Keeley, P.; Campbell, M.; Woby, S.; McGowan, L. Exploring patients' opinions of activity pacing and a new activity pacing questionnaire for chronic pain and/or fatigue: A qualitative study. *Physiotherapy* **2015**, *102*, 300–307. [CrossRef]
43. Halson, S.L.; Jeukendrup, A.E. Does overtraining exist? *Sports Med.* **2004**, *34*, 967–981. [CrossRef] [PubMed]

44. Burgess, M.; Chalder, T. Adolescents with severe chronic fatigue syndrome can make a full recovery. *BMJ Case Rep.* **2011**. [CrossRef] [PubMed]
45. Porges, S.W. *The Pocket Guide to the Polyvagal Theory: The Transformative Power of Feeling Safe*; WW Norton & Co: New York, NY, USA, 2017; ISBN 978-0393707878.
46. Venter, M.; Tomas, C.; Pienaar, I.S.; Strassheim, V.; Erasmus, E.; Ng, W.-F.; Howell, N.; Newton, J.L.; Van der Westhuizen, F.H.; Elson, J.L. MtDNA population variation in Myalgic encephalomyelitis/Chronic fatigue syndrome in two populations: A study of mildly deleterious variants. *Sci. Rep.* **2019**, *9*, 2914. [CrossRef]
47. Vink, M.; Vink-Niese, F. Graded exercise therapy doesn't restore the ability to work in ME/CFS. Rethinking of a Cochrane review. *Work* **2020**, *66*, 283–308. [CrossRef]
48. Sørensen, K.; Van den Broucke, S.; Fullam, J.; Doyle, G.; Pelikan, J.; Slonska, Z.; Brand, H. Health literacy and public health: A systematic review and integration of definitions and models. *BMC Public Health* **2012**, *12*, 80. [CrossRef] [PubMed]
49. Shepherd, C.B. Severely affected, severely neglected. *BMJ* **2010**, *340*, c1181. [CrossRef]
50. Petrie, A.; Sabin, C. *Medical Statistics at a Glance*; John Wiley & Sons: Hoboken, NJ, USA, 2019.
51. Brown, B. *Braving the Wilderness: The Quest for True Belonging and the Courage to Stand Alone*; Random House: New York, NY, USA, 2017; ISBN 978-0812995848.
52. Chapireau, F. The environment in the international classification of functioning, disability and health. *J. Appl. Res. Intellect. Disabil.* **2005**, *18*, 305–311. [CrossRef]
53. Burley, L.; Cox, D.L.; Findley, L.J. Severe chronic fatigue syndrome (CFS/ME): Recovery is possible. *Br. J. Occup. Ther.* **2007**, *70*, 339–344. [CrossRef]
54. Islam, M.F.; Cotler, J.; Jason, L.A. Post-viral fatigue and COVID-19: Lessons from past epidemics. *Fatigue Biomed. Health Behav.* **2020**, *8*, 61–69. [CrossRef]
55. Perrin, R.; Riste, L.; Hann, M.; Walther, A.; Mukherjee, A.; Heald, A. Into the looking glass: Post-viral syndrome post COVID-19. *Med. Hypotheses* **2020**, *144*, 110055. [CrossRef] [PubMed]

Case Report

Chronic Fatigue Syndrome: A Case Report Highlighting Diagnosing and Treatment Challenges and the Possibility of Jarisch–Herxheimer Reactions If High Infectious Loads Are Present

Rachel K. Straub * and Christopher M. Powers

Division of Biokinesiology and Physical Therapy, University of Southern California, 1540 E. Alcazar St., CHP-155, Los Angeles, CA 90089-9006, USA; powers@pt.usc.edu
* Correspondence: rachelks628@gmail.com

Abstract: Myalgic encephalomyelitis/chronic fatigue syndrome (ME/CFS) is a complex multi-system disease with no cure and no FDA-approved treatment. Approximately 25% of patients are house or bedbound, and some are so severe in function that they require tube-feeding and are unable to tolerate light, sound, and human touch. The overall goal of this case report was to (1) describe how past events (e.g., chronic sinusitis, amenorrhea, tick bites, congenital neutropenia, psychogenic polydipsia, food intolerances, and hypothyroidism) may have contributed to the development of severe ME/CFS in a single patient, and (2) the extensive medical interventions that the patient has pursued in an attempt to recover, which enabled her to return to graduate school after becoming bedridden with ME/CFS 4.5 years prior. This paper aims to increase awareness of the harsh reality of ME/CFS and the potential complications following initiation of any level of intervention, some of which may be necessary for long-term healing. Treatments may induce severe paradoxical reactions (Jarisch–Herxheimer reaction) if high infectious loads are present. It is our hope that sharing this case will improve research and treatment options for ME/CFS.

Keywords: myalgic encephalomyelitis (ME); chronic fatigue syndrome (CFS); post-exertional malaise; die-off reactions; chronic illness; Lyme disease; Epstein–Barr virus; Mycoplasma pneumonia; candida; orthostatic intolerance; light therapy; eye movement desensitization and reprocessing (EMDR); emotional freedom technique (EFT)

Citation: Straub, R.K.; Powers, C.M. Chronic Fatigue Syndrome: A Case Report Highlighting Diagnosing and Treatment Challenges and the Possibility of Jarisch–Herxheimer Reactions If High Infectious Loads Are Present. *Healthcare* **2021**, *9*, 1537. https://doi.org/10.3390/healthcare9111537

Academic Editor: Roberto Nuño-Solinís

Received: 13 September 2021
Accepted: 9 November 2021
Published: 10 November 2021

Publisher's Note: MDPI stays neutral with regard to jurisdictional claims in published maps and institutional affiliations.

1. Background

Myalgic encephalomyelitis/chronic fatigue syndrome (ME/CFS) is a debilitating and life-altering disease that affects people around the world, including as many as 2.5 million Americans [1]. Clinically, ME/CFS manifests as debilitating fatigue that worsens with physical or mental activity that is not relieved by rest and is not caused by excessive exertion [1]. Because of the severe impact of ME/CFS on general function (both mental and physical) and no accepted treatment, approximately 25% of patients are house or bedbound [1]. Some are so severe in function that they require tube-feeding and are unable to tolerate light, sound, and human touch [2,3].

Diagnosing ME/CFS remains a challenge, and it has been estimated that approximately 85% of patients remain undiagnosed [1]. In addition, information about ME/CFS is not taught in the majority of the nations' medical schools [1], which has contributed to widespread disbelief and uncertainty among health care providers, many who do not accept ME/CFS as a genuine clinical entity [1,4,5]. As such, patients are frequently misdiagnosed with a psychological condition [1,4,5].

For those diagnosed with ME/CFS, long-term prognosis remains poor. Patients with ME/CFS have a lower quality of life compared to patients with other chronic diseases, such as cancer, multiple sclerosis, and stroke [6–8]. In addition, little progress has been made on developing diagnostics and treatments for ME/CFS in recent decades. Currently, there is no cure and no FDA-approved treatment for ME/CFS. ME/CFS is the most underfunded disease relative to disease burden among all the diseases funded by the United States National Institutes of Health (NIH) [8]. Although studies have documented a wide range of abnormalities in patients with ME/CFS (e.g., central and autonomic nervous systems, metabolic dysfunctions, compromised immunity, and chronic infections), most patients with ME/CFS have "normal" standard lab tests, making a definitive diagnosis difficult [9–11].

In this case report, we describe the complex medical history of one severe ME/CFS patient and her efforts to recover over a 4-year period, which resulted in her return to graduate school. The overall goal of this case report was to (1) describe how past events may have contributed to the development of severe ME/CFS in a single patient, and (2) the extensive medical interventions that the patient has pursued in an attempt to recover, which enabled her to return to graduate school after becoming bedridden with ME/CFS 4.5 years prior. This paper aims to increase awareness of the harsh reality of ME/CFS and the potential complications following initiation of any level of intervention, some of which may be necessary for long-term healing. It is our hope that sharing this case will improve research and treatment options for ME/CFS.

2. Case Presentation

Upon presentation to the rheumatologist's office (August 2013), the patient's vitals were normal [BMI: 21.1, BP: 95/47, pulse: 66, RR: 13, Temperature: 97.8°]. She was a graduate student (age 28 and Caucasian) but had been on medical leave since January 2013 (7 months). She had to move home with her parents because she was no longer capable of caring for herself. Previously an avid exerciser (running and weight training), she was no longer capable of any physical activity. Simple activities such as showering would force her back to bed (post-exertional malaise [1,12,13]) (Table 1). She reported that she was no longer capable of eating most foods, as her digestive tract felt like it was shutting down. Her diet consisted of easily digestible foods proposed to be helpful in treating chronic diseases [14]. She also reported taking a wide range of supplements (~15 total), including probiotics, zinc, vitamin D, and fish oil.

From January 2013 to July 2013, the patient had been seen by four physicians with a complaint of profound fatigue and general malaise (urgent care, internal medicine, infectious disease, and family medicine). The most recent family medicine physician diagnosed her with ME/CFS in July 2013 (Figure 1) based on recognized criteria from the Centers for Disease Control and Prevention (CDC), which included profound disabling fatigue for at least 6 months that remained unexplained and was accompanied by frequent sore throats, impaired cognitive function, post-exertional malaise (Table 1) [1,12,13], unrefreshing sleep, headaches, and joint/muscle pain [15]. The patient also met the definition for ME/CFS as described by the ME International Consensus Criteria (ICC) based on energy production/transportation impairments (e.g., thermostatic instability and dizziness), immune/gastrointestinal/genitourinary impairment (e.g., chronic flu-like symptoms that worsened with exertion), neurological impairment (e.g., cognitive dysfunction and unrefreshing sleep), and post-exertional neuroimmune exhaustion (e.g., post-exertional malaise) [16]. The patient was then referred to the present physician (rheumatologist with a PhD in immunology) for further workup and treatment. The family medicine physician was concerned by a positive test for Antinuclear Antibody (ANA) (suggesting a possible autoimmune disorder). In addition, the family medicine physician viewed recent lab tests as inconclusive for Epstein–Barr virus, Mycoplasma pneumonia, and Chlamydia pneumonia.

Table 1. Jarisch–Herxheimer Reaction vs. Post-Exertional Malaise.

	Jarisch–Herxheimer Reaction [17–26]	Post-Exertional Malaise [1,12,13]
Definition	Worsening of existing symptoms (and appearance of new symptoms) following treatment in several infectious diseases (including viral, bacterial, and fungal). It should not be confused with a drug allergy or adverse reaction to treatment.	Worsening of existing symptoms following excessive cognitive, physical, orthostatic, emotional, or sensory challenges that were previously tolerated.
Onset	Typically occurs within 24 h but may be delayed by 7–14 days.	Typically occurs within 48 h of excessive exertion.
Duration	Hours or days	Hours, days, weeks, or months
Specific to ME/CFS	No	Yes
Controlling	Patients with ME/CFS may need to commence treatment with very low dosages and titrate upwards with caution [27].	Pacing (i.e., staying within energy envelope) is necessary to avoid.
Triggers	Should be expected by all patients receiving treatment (or related herbal treatment) for infectious diseases if an adequate infectious load is present.	Can be triggered by the most mundane activities (conversation and showering), depending on ME/CFS severity.
Example of Symptoms Experienced by Patient over Course of 4+ Years *	<ul><li>Severe hypotension (systolic blood pressure would frequently drop below 80 mm Hg)</li><li>Severe musculoskeletal and joint pain (often uncontrollable)</li><li>Migraines</li><li>Sore throat</li><li>Severe bloating and intestinal cramps</li><li>Extreme fatigue (bedridden)</li><li>Sweating and chills</li><li>Nausea</li><li>Brain Fog</li></ul>	<ul><li>Hypotension</li><li>General body aches</li><li>Headaches</li><li>Sore throat</li><li>Extreme fatigue (bedridden)</li><li>Nausea</li><li>Brain fog</li></ul>

* Symptoms of post-exertional malaise vs. Jarisch–Herxheimer reaction commonly overlap. If the patient was in a Jarisch–Herxheimer state and overexerted, then the Jarisch–Herxheimer reaction would simply worsen. The Jarisch–Herxheimer reaction is more commonly known as die-off.

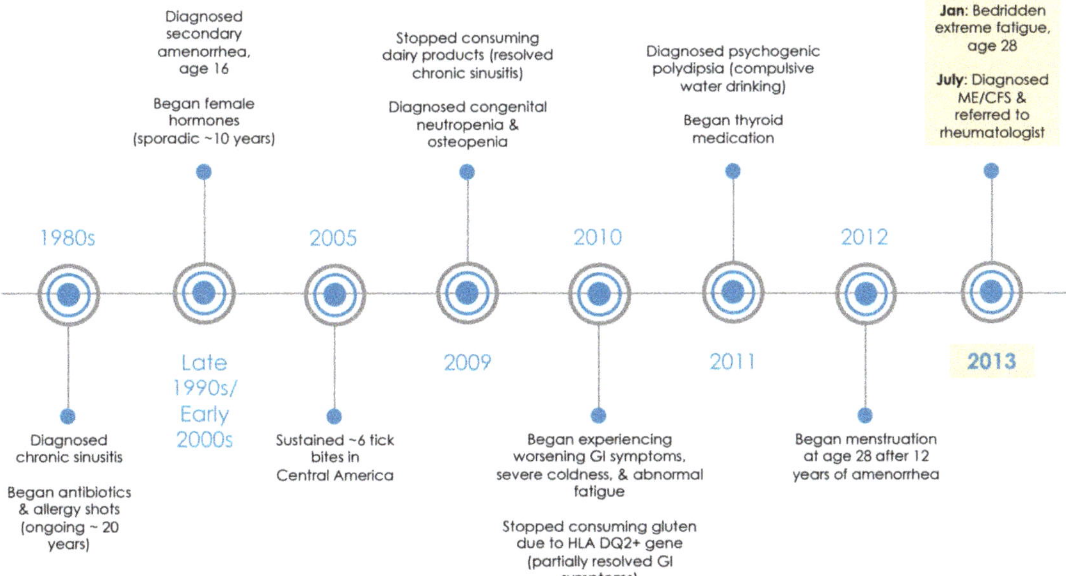

Figure 1. Medical history timeline. Patient saw over 20 medical specialists (i.e., pediatrics, allergy and immunology, gynecology, gastroenterology, endocrinology, hematology, infectious disease, internal medicine, alternative medicine, urgent care, and family medicine) from 1980s through 2013. In 2013, she was diagnosed with ME/CFS at age 28 by a family medicine physician and referred to a rheumatologist (MD, PhD).

3. Past Medical History

The patient had been struggling medically since early childhood and had seen over 20 medical specialists (Figure 1). She had a medical history of chronic sinusitis with no food allergies, amenorrhea, osteopenia, gluten intolerance (HLA DQ2+ gene), congenital neutropenia, polyuria, polydipsia, and hypothyroidism.

For chronic sinusitis, the patient had a 20+ history of long-term antibiotics, steroids, and allergy shots. She reported that despite these treatments, she spent most of her childhood chronically sick and was frequently absent from school. Her chronic sinusitis resolved in her early 20s once she stopped consuming dairy products.

The patient was diagnosed in her 20s with congenital neutropenia by a hematologist and psychogenic polydipsia (compulsive water drinking) by an endocrinologist. In addition, she had a 10+ year history of amenorrhea until recently and had seen over 10 doctors of various specialties for this condition alone (including 5 different gynecologists). She had 1–2 instances of spotting at age 16 and no menses thereafter. All lab tests were reported normal (including an MRI of her pituitary and an ultrasound of her ovaries). She reported that she was on female hormones for at least 10 years, but she was not compliant as she suffered ongoing negative side effects (including migraines). Apparently, all 3 conditions (polydipsia/polyuria, neutropenia, and amenorrhea) were resolved after she was prescribed Amour Thyroid at age 28 for hypothyroidism despite "normal" labs.

The patient reported being bitten by several ticks in Central America 8 years ago. When she returned home, she began experiencing flu-like symptoms. However, she has never tested positive for Lyme disease nor displayed a bullseye rash. The patient also described that she has had declining health over the past 6 years, but she was semi-stable until recently. During the fall of 2012, the patient was continuously fighting ongoing infections and had several bouts of the flu. She was working on a limited basis while going to school, but she had to quit work completely. She no longer had the capacity to exercise, and minimal time away from home would force her to bed for several hours.

4. Differential Diagnosis

A physical exam was generally normal, except for yellow hyperpigmentation of the palms, dry eyes, and dry mouth. A salivary gland ultrasound indicated enlarged intra-parenchymal lymph nodes with increased cortex to hilum ratio (right and left parotid glands) and multiple hypoechoic intraparenchymal areas (right and left submandibular glands). Based on the examination, the patient was diagnosed with Sicca syndrome (dry eyes/mouth). An unspecified disease of the salivary glands (high probability of Sjogren's syndrome, an autoimmune disease) and unspecified inflammatory spondylopathy also were suspected. The patient's medical history suggested that hypothyroidism needed to be addressed, along with possible causes for ME/CFS (including a reactive post-infectious process due to Mycoplasma pneumonia/Chlamydia pneumonia infection). Gluten intolerance (HLA DQ2+) was also noted. Laboratory serology tests ordered are presented in Table 2.

The patient's labs indicated low T4 (55.03 nmol/L, normal: 60–120), low serum iron (36 ug/dL, normal: 37–160), low free lambda chains (5.41 mg/L, normal: 5.71–26.3) with increased kappa to lambda ratio (2.78, normal: 0.26–1.65), a negative extractable nuclear antigen (ENA) panel, elevated ammonia (47 umol/L, normal: 11–35), borderline elevated Mycoplasma pneumonia IgG (193 U/mL, indeterminate: 100–320), borderline elevated Chlamydia trachomatis IgM (0.8, borderline: 0.8–1.0), and elevated Chlamydia pneumonia IgG (1:128, negative: <1:16). The patient was considered to be suffering from a post-infectious process due to Chlamydia pneumonia. IgG titers can be elevated from past exposure, as opposed to a post-infectious process [28,29]. A repeat sample drawn weeks later that demonstrated a significant rise of IgG titers would provide increased evidence of a post-infectious process [28,29]. Although Mycoplasma pneumonia and Chlamydia pneumonia (bacterial infections) are usually self-limiting, clinical manifestations can range from self-limiting to life-threatening, from pulmonary to extrapulmonary [28,30]. In addition, intestinal Candida was suspected based on elevated ammonia levels and the patient's medical history of long-term antibiotics.

Table 2. Serology Laboratory Tests Ordered at Initial Visit to Rheumatologist (August 2013) after ME/CFS Diagnosis.

ACL (Anti-Cardiolipin) Antibodies (IgM, IgG, IgA)
Ammonia
Amylase
ANA (Antinuclear Antibodies) with reflex to 11
Anti-CCP (Cyclic Citrullinated Peptide) lgG Semi-Quantitative
ASO (Antistreptolysin O) Antibodies
B2M (Beta 2 Microglobulin) Tumor Marker
Bilirubin, Direct
Chlamydia pneumonia (IgG/IgM)
Chlamydia trachomatis (IgM)
Complement C3a/C4a
CBC (Complete Blood Count) with Differential/Platelet
CK (Creatine Kinase)
Complete Metabolic Panel
CRP (C-Reactive Protein)
Ferritin
Free Kappa Light Chains
Free Lambda Light Chains
GGT (Gamma-Glutamyl Transferase)
G6PD (Glucose-6-phosphate dehydrogenase) Enzyme
Hemoglobin A1c
HNK-1 (Human Natural Killer-1) CD57
Immunoglobulins (IgA, IgG, IgM)
LDH (Lactic Acid Dehydrogenase)
Lipase
Lyme Western Blot
Magnesium
Mycoplasma pneumonia (IgG/IgM)
Natural Killer Cell Surface Antigen (CD56/16)
Phosphorus
Rheumatoid Factor
Rheumatoid Factors (IgM, IgG, IgA)
Sedimentation rate
Serum Iron
Serum Protein Electrophoresis
Thyroid Panel (TSH, T3, T4, Free T4)
Uric Acid
Urinalysis
Vitamin D (25-Hydroxy)

5. Treatment

For hypothyroidism, the patient was placed on Levothyroxine and was told to continue Armour Thyroid. For possible chronic bacterial infections, she was placed on antibiotics (doxycycline). She was also placed on Nystatin to combat possible Candida (fungal) overgrowth. She was expected to begin feeling better within 2–3 months of treatment, but the treating physician warned her that she may get worst before she gets better due to the Jarisch–Herxheimer reaction (die-off). The Jarisch–Herxheimer reaction is the worsening of existing symptoms (and the appearance of new symptoms) following treatment of several infectious diseases (including viral, bacterial, and fungal). It is an immunologic response that should not be confused with a medication allergy or an adverse reaction to treatment (Table 1) [17–26]. Authors have distinguished a Jarisch–Herxheimer reaction from a drug allergy based on resolution of symptoms despite continuation of therapy [21] and absence of liver test abnormalities [26]. An overview of the symptom changes observed over the course of treatment is shown in Figure 2. During treatment, it was not uncommon for the patient to revert backwards for days, weeks, or months at a time due to Jarisch–Herxheimer reactions. Any perturbation (e.g., new medication, dose increase, and reintroduction of past medication) would induce a Jarisch–Herxheimer reaction. During such periods, the patient's tolerance to any of level of exertion (post-exertional malaise; Table 1) also decreased.

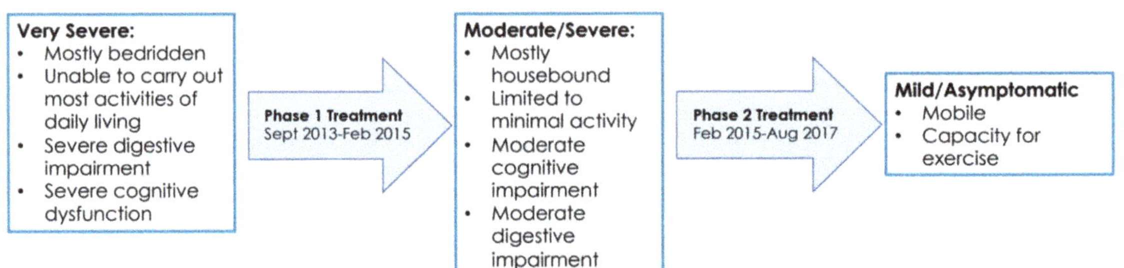

Figure 2. Symptom timeline over the course of treatment for ME/CFS (September 2013–August 2017).

6. Outcome and Follow-Up

6.1. September 2013 to February 2015

The patient was seen every 3–6 weeks over the course of 1.5 years. Prior to her first follow-up visit, she reported symptoms characteristic of Jarisch–Herxheimer reactions (severe joint and musculoskeletal pain, worsening fatigue, worsening cognitive function, migraines, drops in blood pressure, etc.) that she was struggling to control (Table 1). Based on this information, the doxycycline was discontinued. At her first follow-up, she was changed to a different antibiotic and was recommended various herbs for detoxification support (such as succinic acid, N-acetyl cysteine, bromelain, and a liver detox blend). Follow-up labs indicated that with the initiation of treatment, IgG titers rose for Mycoplasma pneumonia but fell for Chlamydia pneumonia (Figure 3). Therefore, Mycoplasma pneumonia was viewed by the treating physician as the more probable factor contributing to ME/CFS. In addition, the Jarisch–Herxheimer reactions she continued to experience were further suggestive of a chronic infection (Table 1) [17–26].

Over the course of 1.5 years, the patient was treated primarily for Mycoplasma pneumonia and was cycled among various antibiotics (including nebulized Gentamycin), in addition to synergists (such as Hydroxychloroquine and Dipyridamole [31]) to enhance antibiotic potency. Immune modulators (such as Colostrum, Astragalus, Andrographis, and Cordyceps) were also recommended to strengthen the patient's immune system. Because of increasing ammonia levels (49 umol/L, normal: 11–35), the patient was changed to Fluconazole (antifungal) in December 2013 to treat suspected Candida. Her labs worsened in certain areas as treatment progressed, which was likely due to high levels of Jarisch–Herxheimer reactions [20,26,32]. C-Reactive Protein (CRP) reached a high of 9.7 mg/L in July 2014 (Figure 4).

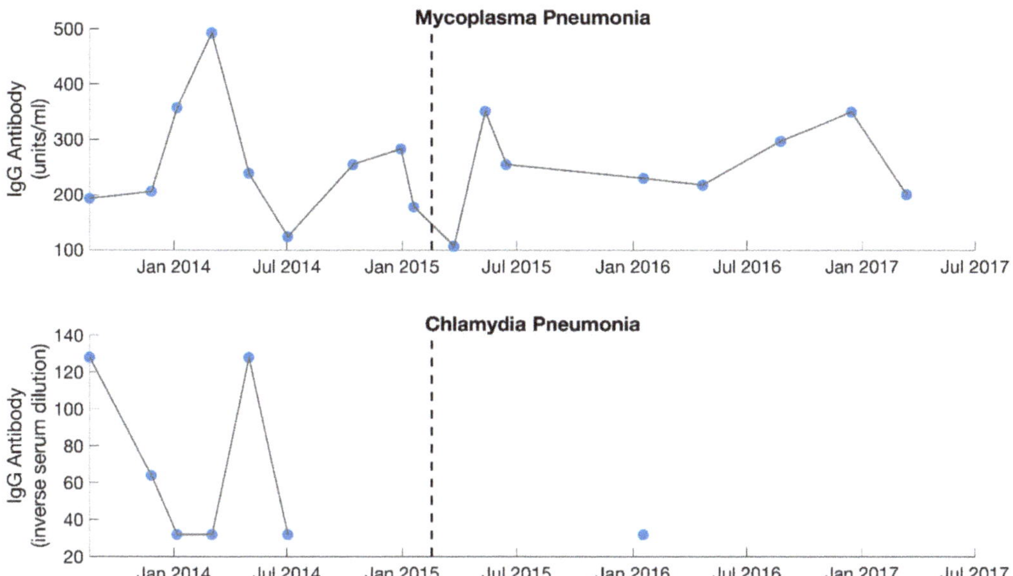

Figure 3. Serology results from IgG specific antibody levels for both Mycoplasma pneumonia and Chlamydia pneumonia. Dotted lines separate the 2 different phases of treatment (September 2013–February 2015, February 2015–August 2017). With the initiation of treatment, IgG levels rose for Mycoplasma pneumonia but fell for Chlamydia pneumonia. Mycoplasma pneumonia was viewed by the treating physician (MD, PhD) as the dominant contributing factor to the patient's illness during the first phase of treatment. Large gaps are present for Chlamydia pneumonia as this was not measured consistently.

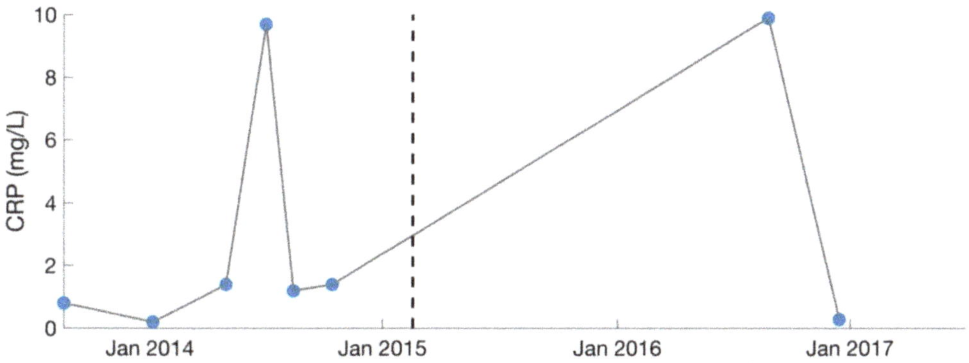

Figure 4. Serology results for CRP (inflammatory marker). Dotted lines separate the two different phases of treatment (September 2013–February 2015, February 2015–August 2017). CRP levels were deemed too high by the treating physician (MD, PhD) in July 2014 and August 2016. Large gaps are present as CRP was not measured consistently. C-Reactive Protein: CRP.

Low iodine (October 2014; 37.1 ug/L, normal: 40–92) and high cortisol levels (January 2014; 27.8 ug/dL, normal: 2.3–19.4) were also identified, so Kelp and adaptogen herbs (such as Ashwaganda and Eleutherococcus) were advised. The patient's thyroid medication (Armour Thyroid and Levothyroxine) was also increased over the course of several months due to low T4 and/or low T3 levels. During the Spring of 2014 (~6 months after onset of treatment), she had improved enough to take a college course 1 day/week. However, she was not well enough to drive. In addition, too much mental exertion would cause her to crash to bed (post-exertional malaise [1,12,13]) (Table 1). She began to struggle with severe orthostatic intolerance and hypotension. Low aldosterone (which helps regulate blood pressure) was contributory, based on non-detectable blood levels. The herb Licorice Root was advised. When this alone was not adequate, Fludrocortisone was prescribed. By February 2015, she was still predominantly housebound, and although she was doing better cognitively, simple activity like walking was still too difficult on most days. Her overseeing physician at this point deemed her a mystery patient. Mycoplasma pneumonia levels appeared to be stabilizing (Figure 3) and frequent labs were not capable of identifying any additional abnormalities or infections. In addition, she remained abnormally sensitive to any level of treatment. An overview of the medications prescribed during phase 1 of treatment (September 2013 to February 2015) is provided in Table 3.

Table 3. Medications Prescribed during Phase 1 of Treatment (September 2013–February 2015).

Name	Reason Prescribed
Prescriptions	
Doxycycline	Antibiotic (Mycoplasma pneumonia)
Clarithromycin	Antibiotic (Mycoplasma pneumonia)
Azithromycin	Antibiotic (Mycoplasma pneumonia)
Dipyridamole [31]	Increase antibiotic potency
Nystatin	Antifungal
Fluconazole	Antifungal
Gentamycin	Nebulized antibiotic (Mycoplasma pneumonia)
Glutathione	Nebulized antioxidant for detoxification
Hydroxychloroquine	Pain and inflammation; increase antibiotic potency
Fludrocortisone	Raise aldosterone to improve hypotension
Armour Thyroid	Hypothyroidism
Levothyroxine	Hypothyroidism
Supplements	
Succinic Acid	Detoxification
N-Acetyl Cysteine	Detoxification; biofilm disruptor
Liver Detox Blend [†]	Detoxification
Modified Citrus Pectin	Detoxification
Bromelain	Pain and inflammation; biofilm disruptor
Boswellia/Curcumin	Pain and inflammation
Colostrum	Immune
Cordyceps	Immune
Kelp (iodine)	Immune and thyroid
Ashwagandha	Adaptogen
Rhodiola Extract	Adaptogen
Eleutherococcus	Adaptogen
Licorice Root	Raise aldosterone to improve hypotension
Phosphatidyl Serine	Brain fog
Artemisian	Antimicrobial
Berberine	Antifungal
Silver Hydrosol	Antimicrobial (Mycoplasma pneumonia)
Olive Leaf Extract	Antimicrobial (Mycoplasma pneumonia)
Anantamul	Antimicrobial (Mycoplasma pneumonia)

Prescribing physician MD, PhD. All oral unless otherwise indicated. Patient was also taking several supplements on her own that the physician approved (e.g., probiotics, zinc, vitamin D, and fish oil). [†] Wide range of herbs that includes milk thistle, schisandra, bupleurum, dandelion, scute, N-acetyl cysteine, methionine, barberry, turmeric and more.

6.2. February 2015 to August 2017

To improve the chances of recovery, the patient made three major shifts in her treatment plan: (1) she began seeing other medical professionals and exploring non-traditional therapies. Most importantly, she began seeing a PhD clinical psychologist with expertise in energy psychology and various non-invasive techniques, such as eye movement desensitization and reprocessing (EMDR) and emotional freedom technique (EFT); (2) she started light therapy; and (3) she shifted her medication protocol from dominantly antibiotics to dominantly herbs.

EFT is a non-invasive method that involves purely tapping on various acupressure points and stating (or thinking) specific statements, while EMDR is a non-invasive method that involves simply moving the eyes in specific patterns while stating (or thinking) specific statements. Systematic reviews and meta-analyses have shown that EFT and EMDR are both effective for the treatment of depression [33,34], post-traumatic stress disorder (PTSD) [34,35], and anxiety [34,36]. In addition, both EFT and EMDR have been shown to improve chronic pain [37,38]. EFT also improves multiple physiological markers of health (such as blood pressure and cortisol) [37].

The patient saw the clinical psychologist approximately 1 day/week and spent 25–100% of her 1-h session on EFT and EMDR, with the focus on various topics related to her specific situation, such as improving health and well-being, clearing toxins and inflammation, and killing specific pathogens. Unfortunately, the patient experienced severe discomfort during her treatment sessions, including joint pain, headaches, excessive yawning, and flu-like symptoms. It would often take her several days to stabilize. Though the psychologist would frequently recommend that his patients perform EFT/EMDR at home (several times a day), the current patient was not capable.

In addition, light therapy was initiated. The patient purchased a LED face light with seven different colors for personal usage and began using it off label, all over her body (most frequently on top of her head). She used all seven colors, but the dominant colors were blue, red, and green. Blue light is effective in treating antibiotic-resistant strains of bacteria [39,40], in addition to acne (where blue light has FDA approval) [41,42]. Furthermore, red light therapy has been reported as a potential neuroprotective treatment for both Alzheimer's and Parkinson's patients [43], which has led to the emergence of red light bucket hats as a potential treatment for those with Parkinson's disease [44]. Moreover, green light therapy has been shown to have anti-inflammatory effects in animal models [45] and decrease pain and improve quality of life in Fibromyalgia patients [46]. When the patient commenced treatment (initially with blue light on the face), she was not able to tolerate it more than 1–2 min without developing severe migraines. In time, light therapy became a daily crucial treatment (up to 6 h/day).

Lastly, the patient shifted her medication protocol from dominantly antibiotics to dominantly herbs and began treating other possible infections (despite the lack of positive test results). She saw the overseeing physician (rheumatologist) every 1–3 months. The focus was no longer on Mycoplasma pneumonia, though this was still monitored (Figure 3). The overseeing physician recommended herbs for other possible conditions and infections for the patient to try on a trial basis, which included herbs for Epstein–Barr virus (Inosine, PABA, DMAE) and Lyme disease (Cat's Claw). Broad-spectrum antimicrobials (such as Silver Hydrosol, Olive Leaf Extract, Anantamul, and Neem) were also recommended. Not surprisingly, the patient experienced severe reactions from all supplements. Therefore, the process of introducing new herbs (or increasing dosages of old ones) was performed with extreme care. Extensive labs were performed regularly, and although no novel infections were identified, inflammatory markers were sometimes elevated. CRP reached a high of 9.9 mg/L in August 2016 (Figure 4), which was during the period the patient was on an incremental light therapy protocol. The abnormal reactions to all forms of treatment (including non-invasive therapies described above) were suggestive of Jarisch–Herxheimer reactions [17–26] (Table 1). An overview of the medications prescribed during phase 2 of treatment (February 2015 to August 2017) is provided in Table 4.

Table 4. Medications Prescribed during Phase 2 of Treatment (February 2015–August 2017).

Name	Reason Prescribed
Prescriptions	
Armour Thyroid	Hypothyroidism
Levothyroxine	Hypothyroidism
Cholestyramine	Detoxification
Supplements	
Cat's Claw (Uncaria Tomentosa)	Antimicrobial (Lyme disease)
Neem	Broad-Spectrum Antimicrobial
Inosine	Antiviral (Epstein–Barr virus)
PABA (Para-aminobenzoic Acid)	Antiviral (Epstein–Barr virus)
DMAE (Dimethylaminoethanol)	Antiviral (Epstein–Barr virus)
L-Lysine	Antiviral (Herpes simplex virus)
Drynaria	Osteopenia
Bamboo Extract	Osteopenia
Andrographis	Immune
Astragalus	Immune
Iporuru	Pain and inflammation
Zeobind	Heavy Metal Chelation
Modified Citrus Pectin	Heavy Metal Chelation

Prescribing physician MD, PhD. All oral unless otherwise indicated. Patient was also taking several supplements on her own that the physician approved (e.g., probiotics, zinc, vitamin D, and fish oil), in addition to several supplements shown in Table 3.

6.3. August 2017

After 2.5 years of a revised treatment protocol, the changes made in the treatment plan were considered successful. In August 2017, the patient was well enough to move out from her parents' home and resume graduate school, after becoming bedridden with ME/CFS 4.5 years prior. Prior to returning to graduate school in August 2017, the patient was stable for at least 6 months. Her diet had also expanded considerably for several months, and other than dairy and gluten, she had no restrictions. She was even back at the gym by April 2017 and was going for walks daily (something she was not capable of doing in over 4 years). She resumed graduate school with a very light schedule and maintenance protocol comprised of prescription medications (Armour Thyroid and Levothyroxine), extensive herbs (e.g., antimicrobials, antivirals, and immune modulators), and non-invasive therapies (EFT, EMDR, and light). A summary of the treatments used by the patient from September 2013 to August 2017 is provided in Table 5.

Table 5. Summary of Treatments Used by Patient Over 4+ Year Period.

	Phase 1	Phase 2
	(September 2013–February 2015)	(February 2015–August 2017)
Long-Term Antibiotics [1]	x	
Long-Term Antifungals [1]	x	
Long-Term Thyroid Medication [1]	x	x
Miscellaneous Prescriptions [1]	x	x
Very Restricted Specialized Diet for Seriously Ill [2]	x	x
Herbs [1]	x	x
Eye Movement Desensitization and Reprocessing (EMDR) [3]		x
Emotional Freedom Techniques (EFT) [3]		x
Light Therapy [3,4]		x

[1] Prescribed by MD, PhD (overseeing physician). Miscellaneous prescriptions included Dipyridamole, Nebulized Glutathione, Hydroxychloroquine, Fludrocortisone, and Cholestyramine. [2] Based on recommendations from the book "Food is Your Best Medicine", by Henry Bieler, MD [14]. Started immediately when health seriously declined, several months before initiation of formal treatments. [3] Core treatment during sessions with PhD clinical psychologist. [4] Light therapy was regularly performed at home (up to 6 h/day). In time, this was incorporated with EFT and EMDR.

7. Discussion

The purpose of this case report was to describe the complex medical history of one severe ME/CFS patient and her efforts to recover over 4 years, which enabled her to return to graduate school. The overall goal of this case report was to (1) describe how past events may have contributed to the development of severe ME/CFS in a single patient, and (2) the extensive medical interventions that the patient has pursued in an attempt to recover. This paper aims to increase awareness of the harsh reality of ME/CFS and the potential complications following initiation of any level of intervention, some of which may be necessary for long-term healing. It is our hope that sharing this case will improve research and treatment options for ME/CFS.

Given the patient's complex case history (e.g., chronic sinusitis, amenorrhea, tick bites, congenital neutropenia, psychogenic polydipsia, and hypothyroidism), it is possible that these events contributed to her onset of severe ME/CFS in her late 20s. The fact that she began menstruation at age 28 after 12 years of amenorrhea (as noted above, the patient had only 1–2 instances of spotting at age 16 and no menses thereafter) with administration of thyroid hormones implies she had serious, long-term hypothyroidism (as absence of menarche by age 15 is statistically uncommon) [47]. Hypothyroidism can result in a wide range of medical problems that were reported by this female patient, including amenorrhea [48], compromised immune function (including neutropenia) [49,50], and severe sensitivity to cold [51]. In addition, thyroid hormones are necessary for normal kidney function [52], which could explain the resolution of polydipsia following administration. It is possible that 10 years of treatment with female hormones was unnecessary, which explains why the patient reported low compliance due to ongoing negative side effects. Given that administration of thyroid hormones also resolved neutropenia, polyuria, and polydipsia suggests that the patient was also misdiagnosed with psychogenic polydipsia and congenital neutropenia. The fact that the severity of her hypothyroidism negatively affected so many different organ systems likely made her more susceptible to ME/CFS.

The patient's 20+ year struggle with chronic sinusitis may have also been implicated in the patient's development of ME/CFS. Although she was on long-term antibiotics, she reported she was still chronically sick. Therefore, it was logical that something else was contributory (dairy intolerance as noted in medical history). Adverse reactions to food can be the result of an immune-mediated reaction (i.e., food allergy) or non-immune reaction (i.e., food intolerance) [53]. The gold standard for the diagnosis of a food intolerance is a food challenge with the suspect food after elimination for several weeks [54,55]. Unfortunately, diagnostic tools available for suspected food allergies cannot accurately predict food intolerances [55], which explains why it took the patient decades to discover a dairy intolerance (as she had no food allergies). True food allergies typically occur within minutes to hours after exposure, while food intolerances typically occur hours to days after exposure [56].

The fact that the patient was treated with long-term antibiotics for decades for chronic sinusitis may have contributed to Candida overgrowth [57–60]. Some authors have called Candida: "a disease of antibiotics" [57]. Although the presence of Candida organisms is generally benign, chronic intestinal Candida (putatively caused by overgrowth of Candida albicans) has been cited as a possible contributor to ME/CFS [61,62] and is associated with several diseases of the gastrointestinal tract [63].

Prior to developing ME/CFS, the patient also reported being bitten by several ticks in Central America in her early 20s. However, she was never diagnosed or treated for Lyme disease. Untreated or inadequately treated Lyme disease can progress to a late disseminated disease after initial infection that can result in substantial disability [64,65]. Lyme disease can result in neurological manifestations [64,66], in addition to chronic fatigue [64,67]. By the time the patient was bitten by ticks, she had been subjected to years of female hormones and antibiotics (which were likely unnecessary), along with daily consumption of dairy (which was likely contributory to chronic sinusitis). Therefore, years of potentially harmful and misdirected treatments may have created an environment more susceptible to disease. Lyme disease tests are falsely negative in 40.5% of cases (accordingly to a recent meta-analysis) [68], which further highlights the challenges in diagnosing Lyme disease.

Research studies have shown that patients with ME/CFS have abnormalities of the central and autonomic nervous systems, metabolic dysfunctions, compromised immunity, and chronic infections [9–11]. However, the overseeing physician was unable to provide an active diagnosis throughout the duration of treatment due to limitations with available clinical testing. Recent research has shown that there are no significant differences between antibody/antigen serology tests against common viral and bacterial pathogens in patients with severe ME/CFS compared to healthy controls [69]. As such, the patient's past medical history and suspected causes of ME/CFS were often used to guide treatment recommendations. For example, although the overseeing physician initially believed Mycoplasma pneumonia (which has been documented in patients with ME/CFS) [70] was the primary contributor to the patient's illness, after 1.5 years of minimal progress on a Mycoplasma pneumonia focused protocol (dominantly antibiotics), the overseeing physician recommended herbs for the patient to try for other possible infections (including Lyme disease and Epstein–Barr virus). This more comprehensive herbal approach for 2.5 years, combined with non-invasive therapies (light, EFT, and EMDR), was successful in getting the patient back to graduate school after becoming bedridden with ME/CFS 4.5 years prior. However, it was not possible to determine whether the improvements observed were mediated by immunologic, antifungal, antiviral, antimicrobial, or neurological effects of various treatments vs. simply a spontaneous remission (or placebo effect) over time. Studies have shown that after a period of 15 months, spontaneous recovery from ME/CFS rarely occurs [71], highlighting that the patient's progress was likely attributed to various aspects of treatment.

It could be argued that a combination of herbs that helped disintegrate drug-resistant biofilms, strengthen immunity, enhance detoxification, and target a wide range of possible infections (including Mycoplasma pneumonia, Candida, Lyme disease, and Epstein–Barr virus) likely played a critical role in improving the patient's symptoms over time. Although herbs were used throughout treatment, a more comprehensive approach was taken in the latter half of treatment after antibiotics were abandoned. In addition, non-invasive therapies (i.e., light, EFT, and EMDR) added during the latter part of treatment may have helped create an environment more conducive for healing. However, given that all changes to the patient's protocol resulted in Jarisch–Herxheimer reactions (with severe hypotension being the most dangerous) and that no meaningful shifts in the patient's laboratory results occurred over a 4+ year period, further complicated determining what was most (or least) effective.

Patients with ME/CFS may have chronic infections that will increase the likelihood of paradoxical reactions to treatment (Jarisch–Herxheimer reaction) (Table 1). Throughout treatment (September 2013 to August 2017), the patient experienced paradoxical responses (Jarisch–Herxheimer reaction) to all forms of treatment, which were suggestive of unresolved chronic infections. Indeed, authors have emphasized that the Jarisch–Herxheimer reaction is a necessary adverse reaction for achieving a cure from various infections [17–26]. Although the Jarisch–Herxheimer reaction is generally a transient reaction, the patient experienced ongoing Jarisch–Herxheimer reactions that lasted for years. Any perturbation in her treatment protocol was often enough to reinstate a Jarisch–Herxheimer reaction. To manage Jarisch–Herxheimer reactions, the patient relied primarily on herbal recommendations from her overseeing physician for detoxification support. In addition, she frequently had to pause (or slow treatment), and daily light therapy was often crucial.

It could be argued that the patient's long-term struggles with Jarisch–Herxheimer reaction have been a necessary process for healing. However, care must be taken to distinguish the Jarisch–Herxheimer reaction (which can be beneficial and is not specific to the ME/CFS patient) from post-exertional malaise (which is the hallmark symptom of ME/CFS and detrimental to the health of the ME/CFS patient) (Table 1). Recent recommendations from the Mayo Clinic acknowledge that graded exercise therapy is contraindicated for patients with ME/CFS [13]. If a patient improves with graded exercise, he/she does not have post-exertional malaise, and thus he/she does not have ME/CFS. Graded exercise protocols have been shown to have detrimental effects on patients with ME/CFS due to mitochondrial dysfunctions, low oxygen update, abnormal autonomic responses, and immunological abnormalities to name a few [1,10,72]. Therefore, it is not surprising that the patient reported a decreased tolerance to exertion when under treatment. This sheds light on the overlap between the two responses (i.e., post-exertional malaise and Jarisch–Herxheimer reaction), as a temporary worsening of symptoms from treatment may require decreased exertion to avoid long-term setbacks. Excessive exertion can result in an irreversible decline in function (which has been reported by patients with severe ME/CFS) [2].

In summary, this case report documents the progression of a patient with ME/CFS over 4 years and her continuous paradoxical reactions to treatment (which we propose have been a necessary aspect for healing). Patients with ME/CFS tend to have severe sensitivities to treatment (as documented elsewhere [27] and in the current report), highlighting that potential therapies need to be performed with extreme care to avoid detrimental results. Given the patient's case history (e.g., chronic sinusitis, amenorrhea, tick bites, congenital neutropenia, psychogenic polydipsia, food intolerances, and hypothyroidism), we hypothesize that these events contributed to her development of severe ME/CFS in her late 20s. Although the patient improved on a protocol combining herbs, traditional pharmaceuticals, and non-invasive therapies (LED colored lights, EFT, and EMDR), these treatments were experimental as the overseeing physician was unable to provide an active diagnosis (which was complicated by limitations with available clinical testing). From a clinical standpoint, this report aims to alert health care providers to the complications in treating patients with ME/CFS and provide possible guidance on future research and treatment options for ME/CFS.

Author Contributions: R.K.S. reviewed the details of the medical records (400+ pages) and drafted the case report. C.M.P. has been involved in the long-term support of the patient and assisted significantly in the editing of this report. All authors have read and agreed to the published version of the manuscript.

Funding: This case report received no external funding.

Institutional Review Board Statement: Not applicable.

Informed Consent Statement: Written informed consent was obtained from the patient to publish this paper.

Data Availability Statement: Not applicable.

Conflicts of Interest: The authors declare no conflict of interest.

References

1. Institute of Medicine (US). *Committee on the Diagnostic Criteria for Myalgic Encephalomyelitis/Chronic Fatigue Syndrome. Beyond Myalgic Encephalomyelitis/Chronic Fatigue Syndrome: Redefining an Illness*; National Academies Press: Washington, DC, USA, 2015.
2. Dafoe, W. Extremely Severe ME/CFS-A Personal Account. *Healthcare* **2021**, *9*, 504. [CrossRef] [PubMed]
3. Baxter, H.; Speight, N.; Weir, W. Life-Threatening Malnutrition in Very Severe ME/CFS. *Healthcare* **2021**, *9*, 459. [CrossRef] [PubMed]
4. Pheby, D.F.H.; Araja, D.; Berkis, U.; Brenna, E.; Cullinan, J.; de Korwin, J.D.; Gitto, L.; Hughes, D.A.; Hunter, R.M.; Trepel, D.; et al. A Literature Review of GP Knowledge and Understanding of ME/CFS: A Report from the Socioeconomic Working Group of the European Network on ME/CFS (EUROMENE). *Medicina* **2020**, *57*, 7. [CrossRef] [PubMed]
5. Bayliss, K.; Goodall, M.; Chisholm, A.; Fordham, B.; Chew-Graham, C.; Riste, L.; Fisher, L.; Lovell, K.; Peters, S.; Wearden, A. Overcoming the barriers to the diagnosis and management of chronic fatigue syndrome/ME in primary care: A meta synthesis of qualitative studies. *BMC Fam. Pract.* **2014**, *15*, 44. [CrossRef]
6. Falk Hvidberg, M.; Brinth, L.S.; Olesen, A.V.; Petersen, K.D.; Ehlers, L. The Health-Related Quality of Life for Patients with Myalgic Encephalomyelitis/Chronic Fatigue Syndrome (ME/CFS). *PLoS ONE* **2015**, *10*, e0132421. [CrossRef]

7.   Nacul, L.C.; Lacerda, E.M.; Campion, P.; Pheby, D.; Drachler Mde, L.; Leite, J.C.; Poland, F.; Howe, A.; Fayyaz, S.; Molokhia, M. The functional status and well being of people with myalgic encephalomyelitis/chronic fatigue syndrome and their carers. *BMC Public Health* **2011**, *11*, 402. [CrossRef]

8.   Mirin, A.A.; Dimmock, M.E.; Jason, L.A. Research update: The relation between ME/CFS disease burden and research funding in the USA. *Work* **2020**, *66*, 277–282. [CrossRef]

9.   Komaroff, A.L. Advances in Understanding the Pathophysiology of Chronic Fatigue Syndrome. *JAMA* **2019**, *322*, 499–500. [CrossRef] [PubMed]

10. Komaroff, A.L.; Cho, T.A. Role of infection and neurologic dysfunction in chronic fatigue syndrome. *Semin. Neurol.* **2011**, *31*, 325–337. [CrossRef]

11. Sweetman, E.; Noble, A.; Edgar, C.; Mackay, A.; Helliwell, A.; Vallings, R.; Ryan, M.; Tate, W. Current Research Provides Insight into the Biological Basis and Diagnostic Potential for Myalgic Encephalomyelitis/Chronic Fatigue Syndrome (ME/CFS). *Diagnostics* **2019**, *9*, 73. [CrossRef]

12. Jason, L.A.; Evans, M.; So, S.; Scott, J.; Brown, A. Problems in defining post-exertional malaise. *J. Prev. Interv. Community* **2015**, *43*, 20–31. [CrossRef] [PubMed]

13. Bateman, L.; Bested, A.C.; Bonilla, H.F.; Chheda, B.V.; Chu, L.; Curtin, J.M.; Dempsey, T.T.; Dimmock, M.E.; Dowell, T.G.; Felsenstein, D.; et al. Myalgic Encephalomyelitis/Chronic Fatigue Syndrome: Essentials of Diagnosis and Management. *Mayo Clin. Proc.* **2021**, *96*, 2861–2878. [CrossRef] [PubMed]

14. Bieler, H.G. *Food Is Your Best Medicine: The Pioneering Nutrition Classic*; Ballantine Books: New York, NY, USA, 2010.

15. Fukuda, K.; Straus, S.E.; Hickie, I.; Sharpe, M.C.; Dobbins, J.G.; Komaroff, A. The chronic fatigue syndrome: A comprehensive approach to its definition and study. International Chronic Fatigue Syndrome Study Group. *Ann. Intern. Med.* **1994**, *121*, 953–959. [CrossRef] [PubMed]

16. Carruthers, B.M.; van de Sande, M.I.; De Meirleir, K.L.; Klimas, N.G.; Broderick, G.; Mitchell, T.; Staines, D.; Powles, A.C.; Speight, N.; Vallings, R.; et al. Myalgic encephalomyelitis: International Consensus Criteria. *J. Intern. Med.* **2011**, *270*, 327–338. [CrossRef]

17. Belum, G.R.; Belum, V.R.; Chaitanya Arudra, S.K.; Reddy, B.S. The Jarisch-Herxheimer reaction: Revisited. *Travel Med. Infect. Dis.* **2013**, *11*, 231–237. [CrossRef]

18. Bryceson, A.D. Clinical pathology of the Jarisch-Herxheimer reaction. *J. Infect. Dis.* **1976**, *133*, 696–704. [CrossRef] [PubMed]

19. Dhakal, A.; Sbar, E. *Jarisch Herxheimer Reaction*; StatPearls: Treasure Island, FL, USA, 2021.

20. Nykytyuk, S.; Boyarchuk, O.; Klymnyuk, S.; Levenets, S. The Jarisch-Herxheimer reaction associated with doxycycline in a patient with Lyme arthritis. *Reumatologia* **2020**, *58*, 335–338. [CrossRef]

21. Muscianese, M.; Magri, F.; Pranteda, G.; Pranteda, G. A case of Jarisch-Herxheimer reaction in candidiasis treated with systemic fluconazole. *Dermatol. Ther.* **2020**, *33*, e13244. [CrossRef]

22. Butler, T. The Jarisch-Herxheimer Reaction after Antibiotic Treatment of Spirochetal Infections: A Review of Recent Cases and Our Understanding of Pathogenesis. *Am. J. Trop. Med. Hyg.* **2017**, *96*, 46–52. [CrossRef]

23. Lacout, A.; Marcy, P.Y.; El Hajjam, M.; Thariat, J.; Perronne, C. Dealing with Lyme Disease Treatment. *Am. J. Med.* **2017**, *130*, e221. [CrossRef]

24. Kadam, P.; Gregory, N.A.; Zelger, B.; Carlson, J.A. Delayed onset of the Jarisch-Herxheimer reaction in doxycycline-treated disease: A case report and review of its histopathology and implications for pathogenesis. *Am. J. Dermatopathol.* **2015**, *37*, e68–e74. [CrossRef] [PubMed]

25. Montoya, J.G.; Kogelnik, A.M.; Bhangoo, M.; Lunn, M.R.; Flamand, L.; Merrihew, L.E.; Watt, T.; Kubo, J.T.; Paik, J.; Desai, M. Randomized clinical trial to evaluate the efficacy and safety of valganciclovir in a subset of patients with chronic fatigue syndrome. *J. Med. Virol.* **2013**, *85*, 2101–2109. [CrossRef]

26. Hryncewicz-Gwozdz, A.; Wojciechowska-Zdrojowy, M.; Maj, J.; Baran, W.; Jagielski, T. Paradoxical Reaction during a Course of Terbinafine Treatment of Trichophyton interdigitale Infection in a Child. *JAMA Dermatol.* **2016**, *152*, 342–343. [CrossRef] [PubMed]

27. Speight, N. Severe ME in Children. *Healthcare* **2020**, *8*, 211. [CrossRef] [PubMed]

28. Daxboeck, F.; Krause, R.; Wenisch, C. Laboratory diagnosis of Mycoplasma pneumoniae infection. *Clin. Microbiol. Infect.* **2003**, *9*, 263–273. [CrossRef] [PubMed]

29. Dowell, S.F.; Peeling, R.W.; Boman, J.; Carlone, G.M.; Fields, B.S.; Guarner, J.; Hammerschlag, M.R.; Jackson, L.A.; Kuo, C.C.; Maass, M.; et al. Standardizing Chlamydia pneumoniae assays: Recommendations from the Centers for Disease Control and Prevention (USA) and the Laboratory Centre for Disease Control (Canada). *Clin. Infect. Dis.* **2001**, *33*, 492–503. [CrossRef]

30. Johnson, D.H.; Cunha, B.A. Atypical pneumonias: Clinical and extrapulmonary features of Chlamydia, Mycoplasma, and Legionella infections. *Postgrad. Med.* **1993**, *93*, 69–82. [CrossRef]

31. Sun, R.; Wang, L. Inhibition of Mycoplasma pneumoniae growth by FDA-approved anticancer and antiviral nucleoside and nucleobase analogs. *BMC Microbiol.* **2013**, *13*, 184. [CrossRef]

32. Kindmark, C.O.; Moller, H.; Persson, K. C-reactive protein, C3, C4 and properdin during the Jarisch-Herxheimer reaction in early syphilis. *Acta Med. Scand.* **1978**, *204*, 287–290. [CrossRef]

33. Nelms, J.A.; Castel, L. A Systematic Review and Meta-Analysis of Randomized and Nonrandomized Trials of Clinical Emotional Freedom Techniques (EFT) for the Treatment of Depression. *Explore* **2016**, *12*, 416–426. [CrossRef]

34. Chen, Y.R.; Hung, K.W.; Tsai, J.C.; Chu, H.; Chung, M.H.; Chen, S.R.; Liao, Y.M.; Ou, K.L.; Chang, Y.C.; Chou, K.R. Efficacy of eye-movement desensitization and reprocessing for patients with posttraumatic-stress disorder: A meta-analysis of randomized controlled trials. *PLoS ONE* **2014**, *9*, e103676. [CrossRef] [PubMed]
35. Sebastian, B.; Nelms, J. The Effectiveness of Emotional Freedom Techniques in the Treatment of Posttraumatic Stress Disorder: A Meta-Analysis. *Explore* **2017**, *13*, 16–25. [CrossRef] [PubMed]
36. Clond, M. Emotional Freedom Techniques for Anxiety: A Systematic Review with Meta-analysis. *J. Nerv. Ment. Dis.* **2016**, *204*, 388–395. [CrossRef] [PubMed]
37. Bach, D.; Groesbeck, G.; Stapleton, P.; Sims, R.; Blickheuser, K.; Church, D. Clinical EFT (Emotional Freedom Techniques) Improves Multiple Physiological Markers of Health. *J. Evid. Based Integr. Med.* **2019**, *24*, 2515690X18823691. [CrossRef]
38. Gerhardt, A.; Leisner, S.; Hartmann, M.; Janke, S.; Seidler, G.H.; Eich, W.; Tesarz, J. Eye Movement Desensitization and Reprocessing vs. Treatment-as-Usual for Non-Specific Chronic Back Pain Patients with Psychological Trauma: A Randomized Controlled Pilot Study. *Front. Psychiatry* **2016**, *7*, 201. [CrossRef]
39. Wang, Y.; Wu, X.; Chen, J.; Amin, R.; Lu, M.; Bhayana, B.; Zhao, J.; Murray, C.K.; Hamblin, M.R.; Hooper, D.C.; et al. Antimicrobial Blue Light Inactivation of Gram-Negative Pathogens in Biofilms: In Vitro and In Vivo Studies. *J. Infect. Dis.* **2016**, *213*, 1380–1387. [CrossRef]
40. Nour El Din, S.; El-Tayeb, T.A.; Abou-Aisha, K.; El-Azizi, M. In vitro and in vivo antimicrobial activity of combined therapy of silver nanoparticles and visible blue light against Pseudomonas aeruginosa. *Int. J. Nanomed.* **2016**, *11*, 1749–1758. [CrossRef]
41. Antoniou, C.; Dessinioti, C.; Sotiriadis, D.; Kalokasidis, K.; Kontochristopoulos, G.; Petridis, A.; Rigopoulos, D.; Vezina, D.; Nikolis, A. A multicenter, randomized, split-face clinical trial evaluating the efficacy and safety of chromophore gel-assisted blue light phototherapy for the treatment of acne. *Int. J. Dermatol.* **2016**, *55*, 1321–1328. [CrossRef]
42. Gold, M.H.; Sensing, W.; Biron, J.A. Clinical efficacy of home-use blue-light therapy for mild-to moderate acne. *J. Cosmet. Laser Ther.* **2011**, *13*, 308–314. [CrossRef]
43. Johnstone, D.M.; Moro, C.; Stone, J.; Benabid, A.L.; Mitrofanis, J. Turning on Lights to Stop Neurodegeneration: The Potential of Near Infrared Light Therapy in Alzheimer's and Parkinson's Disease. *Front. Neurosci.* **2015**, *9*, 500. [CrossRef]
44. Hamilton, C.L.; El Khoury, H.; Hamilton, D.; Nicklason, F.; Mitrofanis, J. "Buckets": Early Observations on the Use of Red and Infrared Light Helmets in Parkinson's Disease Patients. *Photobiomodul. Photomed. Laser Surg.* **2019**, *37*, 615–622. [CrossRef]
45. Catao, M.H.; Costa, R.O.; Nonaka, C.F.; Junior, R.L.; Costa, I.R. Green LED light has anti-inflammatory effects on burns in rats. *Burns* **2016**, *42*, 392–396. [CrossRef]
46. Martin, L.; Porreca, F.; Mata, E.I.; Salloum, M.; Goel, V.; Gunnala, P.; Killgore, W.D.S.; Jain, S.; Jones-MacFarland, F.N.; Khanna, R.; et al. Green Light Exposure Improves Pain and Quality of Life in Fibromyalgia Patients: A Preliminary One-Way Crossover Clinical Trial. *Pain Med.* **2021**, *22*, 118–130. [CrossRef]
47. Adams Hillard, P.J. Menstruation in adolescents: What's normal, what's not. *Ann. N. Y. Acad. Sci.* **2008**, *1135*, 29–35. [CrossRef] [PubMed]
48. Urmi, S.J.; Begum, S.R.; Fariduddin, M.; Begum, S.A.; Mahmud, T.; Banu, J.; Chowdhury, S.; Khanam, A. Hypothyroidism and its Effect on Menstrual Pattern and Fertility. *Mymensingh Med. J.* **2015**, *24*, 765–769.
49. De Vito, P.; Incerpi, S.; Pedersen, J.Z.; Luly, P.; Davis, F.B.; Davis, P.J. Thyroid hormones as modulators of immune activities at the cellular level. *Thyroid* **2011**, *21*, 879–890. [CrossRef] [PubMed]
50. Kyritsi, E.M.A.; Yiakoumis, X.; Pangalis, G.A.; Pontikoglou, C.; Pyrovolaki, K.; Kalpadakis, C.; Mavroudi, I.; Koutala, H.; Mastrodemou, S.; Vassilakopoulos, T.P.; et al. High Frequency of Thyroid Disorders in Patients Presenting with Neutropenia to an Outpatient Hematology Clinic STROBE-Compliant Article. *Medicine* **2015**, *94*, e886. [CrossRef]
51. Surks, M.I.; Ross, D.; Mulder, J.; Clinical Manifestations of Hypothyroidism. Uptodate. *Waltham, MA, USA.* 2012. Available online: http://www.uptodate.com (accessed on 5 June 2021).
52. Iglesias, P.; Bajo, M.A.; Selgas, R.; Diez, J.J. Thyroid dysfunction and kidney disease: An update. *Rev. Endocr. Metab. Disord.* **2017**, *18*, 131–144. [CrossRef] [PubMed]
53. Guandalini, S.; Newland, C. Differentiating food allergies from food intolerances. *Curr. Gastroenterol. Rep.* **2011**, *13*, 426–434. [CrossRef] [PubMed]
54. Patriarca, G.; Schiavino, D.; Pecora, V.; Lombardo, C.; Pollastrini, E.; Aruanno, A.; Sabato, V.; Colagiovanni, A.; Rizzi, A.; De Pasquale, T.; et al. Food allergy and food intolerance: Diagnosis and treatment. *Intern. Emerg. Med.* **2009**, *4*, 11–24. [CrossRef]
55. Niggemann, B.; Beyer, K. Diagnosis of food allergy in children: Toward a standardization of food challenge. *J. Pediatr. Gastroenterol. Nutr.* **2007**, *45*, 399–404. [CrossRef] [PubMed]
56. Mullin, G.E.; Swift, K.M.; Lipski, L.; Turnbull, L.K.; Rampertab, S.D. Testing for food reactions: The good, the bad, and the ugly. *Nutr. Clin. Pract.* **2010**, *25*, 192–198. [CrossRef] [PubMed]
57. Homei, A.; Worboys, M. *Fungal Disease in Britain and the United States 1850–2000: Mycoses and Modernity*; Wellcome Trust-Funded Monographs and Book Chapters; Springer Nature: Basingstoke, UK, 2013.
58. Yan, L.; Yang, C.; Tang, J. Disruption of the intestinal mucosal barrier in Candida albicans infections. *Microbiol. Res.* **2013**, *168*, 389–395. [CrossRef] [PubMed]
59. Kullberg, B.J.; Arendrup, M.C. Invasive Candidiasis. *N. Engl. J. Med.* **2015**, *373*, 1445–1456. [CrossRef] [PubMed]
60. Martins, N.; Ferreira, I.C.; Barros, L.; Silva, S.; Henriques, M. Candidiasis: Predisposing factors, prevention, diagnosis and alternative treatment. *Mycopathologia* **2014**, *177*, 223–240. [CrossRef] [PubMed]

61. Cater, R.E., 2nd. Chronic intestinal candidiasis as a possible etiological factor in the chronic fatigue syndrome. *Med. Hypotheses* **1995**, *44*, 507–515. [CrossRef]
62. Evengard, B.; Grans, H.; Wahlund, E.; Nord, C.E. Increased number of Candida albicans in the faecal microflora of chronic fatigue syndrome patients during the acute phase of illness. *Scand. J. Gastroenterol.* **2007**, *42*, 1514–1515. [CrossRef]
63. Kumamoto, C.A. Inflammation and gastrointestinal Candida colonization. *Curr. Opin. Microbiol.* **2011**, *14*, 386–391. [CrossRef] [PubMed]
64. Lantos, P.M. Chronic Lyme disease. *Infect. Dis. Clin. N. Am.* **2015**, *29*, 325–340. [CrossRef]
65. Donta, S.T. Issues in the diagnosis and treatment of lyme disease. *Open Neurol. J.* **2012**, *6*, 140–145. [CrossRef] [PubMed]
66. Halperin, J.J. Neurologic manifestations of lyme disease. *Curr. Infect. Dis. Rep.* **2011**, *13*, 360–366. [CrossRef]
67. Treib, J.; Grauer, M.T.; Haass, A.; Langenbach, J.; Holzer, G.; Woessner, R. Chronic fatigue syndrome in patients with Lyme borreliosis. *Eur. Neurol.* **2000**, *43*, 107–109. [CrossRef]
68. Cook, M.J.; Puri, B.K. Commercial test kits for detection of Lyme borreliosis: A meta-analysis of test accuracy. *Int. J. Gen. Med.* **2016**, *9*, 427–440. [CrossRef]
69. Chang, C.J.; Hung, L.Y.; Kogelnik, A.M.; Kaufman, D.; Aiyar, R.S.; Chu, A.M.; Wilhelmy, J.; Li, P.; Tannenbaum, L.; Xiao, W.; et al. A Comprehensive Examination of Severely Ill ME/CFS Patients. *Healthcare* **2021**, *9*, 1290. [CrossRef] [PubMed]
70. Nicolson, G.L.; Gan, R.; Haier, J. Multiple co-infections (Mycoplasma, Chlamydia, human herpes virus-6) in blood of chronic fatigue syndrome patients: Association with signs and symptoms. *APMIS* **2003**, *111*, 557–566. [CrossRef] [PubMed]
71. van der Werf, S.P.; de Vree, B.; Alberts, M.; van der Meer, J.W.; Bleijenberg, G.; Netherlands Fatigue Research Group Nijmegen. Natural course and predicting self-reported improvement in patients with chronic fatigue syndrome with a relatively short illness duration. *J. Psychosom. Res.* **2002**, *53*, 749–753. [CrossRef]
72. Twisk, F.N. Dangerous exercise. The detrimental effects of exertion and orthostatic stress in myalgic encephalomyelitis and chronic fatigue syndrome. *Phys. Med. Rehabil. Res.* **2017**, *2*, 2–3. [CrossRef]

healthcare

Case Report

Three Cases of Severe ME/CFS in Adults

Leah R. Williams * and Carol Isaacson-Barash

Massachusetts ME/CFS & FM Association, Quincy, MA 02269, USA; cibarash@helixhealthadvisors.com
* Correspondence: l.r.williams@comcast.net

Abstract: Myalgic encephalomyelitis/chronic fatigue syndrome (ME/CFS) is a complex, only partially understood multi-system disease whose onset and severity vary widely. Symptoms include overwhelming fatigue, post-exertional malaise, sleep disruptions, gastrointestinal issues, headaches, orthostatic intolerance, cognitive impairment, etc. ME/CFS is a physiological disease with an onset often triggered by a viral or bacterial infection, and sometimes by toxins. Some patients have a mild case and are able to function nearly on a par with healthy individuals, while others are moderately ill and still others are severely, or even, very severely ill. The cohort of moderately to very severely ill is often housebound or bedbound, has lost employment or career, and has engaged in a long, and often futile, search for treatment and relief. Here, we present three case studies, one each of a moderately ill, a severely ill, and a very severely ill person, to demonstrate the complexity of the disease, the suffering of these patients, and what health care providers can do to help.

Keywords: myalgic encephalomyelitis (ME); chronic fatigue syndrome (CFS); severe ME/CFS; very severe ME/CFS; post-exertional malaise (PEM)

Citation: Williams, L.R.; Isaacson-Barash, C. Three Cases of Severe ME/CFS in Adults. *Healthcare* 2021, 9, 215. https://doi.org/10.3390/healthcare9020215

Academic Editor: Kenneth J. Friedman

Received: 16 January 2021
Accepted: 10 February 2021
Published: 16 February 2021

Publisher's Note: MDPI stays neutral with regard to jurisdictional claims in published maps and institutional affiliations.

1. Introduction

Myalgic encephalomyelitis/chronic fatigue syndrome (ME/CFS) is a complex multi-system disease that impacts the immune, endocrine, neurological, and energy production pathways in the body. Current diagnostic criteria include the hallmark symptom of post-exertional malaise (PEM), meaning a prolonged exacerbation of symptoms following mental or physical exertion, and fatigue, unrefreshing sleep, cognitive impairment, and/or orthostatic intolerance that has persisted for more than six months and that reduce or impair the ability to engage in pre-illness activities [1,2]. Defined as a neurological disease by the World Health Organization since 1969 [3], ME/CFS has been mistakenly characterized as a mental disorder by some groups, leading to stigmatization and a lack of appropriate care [4]. There are no biomarkers or validated diagnostic tests and no US Food and Drug Administration (FDA)-approved treatments.

ME/CFS is not a rare disease. There are at least an estimated 1.5 million people with ME/CFS (pwME) in the US [1,5,6]. Roughly three times as many women as men are affected [1]. Although the etiology is unclear, many cases follow a viral infection [7]. In a prospective study in Australia, 11% of patients with acute infections of Epstein–Barr virus (EBV), Q-fever (*Coxiella burnetii*), or Ross River virus (an RNA alphavirus) met the criteria for ME/CFS at six months [8]. Studies of the long-term sequelae of severe acute respiratory syndrome (SARS) found that 27% met the diagnostic criteria for ME/CFS after four years [9]. The current SARS-CoV-2 pandemic is expected to lead to a large increase in the number of pwME [10–14].

The frequency and severity of ME/CFS symptoms can vary from day-to-day and week-to-week, and symptoms can range from mild to severe. Carruthers et al. 2011 [15] defined a severity scale for ME/CFS of mild (at least a 50% reduction in pre-illness activity level), moderate (mostly housebound), severe (mostly bedridden), and very severe (completely bedridden and requiring assistance with basic functions). Studies of disability in pwME have estimated that 75% are housebound most of the time, 50% are unemployed, and 25%

are bedbound most of the time [7,16,17]. Full recovery is rare (estimated at less than 5%), although some pwME experience remission of symptoms for extended periods, followed by relapse [7,18].

In this paper, we present three case studies of pwME, one moderately ill, one severely ill, and one very severely ill. Although children and adolescents get ME/CFS, in this paper, we focus on adults. Our goal is to describe the course of the disease and the extensive search for treatments undertaken by pwME and their health care providers, highlighting the complexity of ME/CFS and the suffering of patients. Several common themes emerge, including the difficulty of obtaining a diagnosis, the difficulty of finding supportive doctors, and the lack of treatments.

2. Case A

This 60-year-old white male was diagnosed with chronic EBV syndrome in 1986 by an internist. Symptoms leading to this diagnosis appeared following a 1983 martial arts injury and exacerbating spondylolysis at L5/S1. The injury was treated with a Boston Bucket back brace, which was worn for 11 months. During this time, the patient was cared for by a team of physiatrists, neurologists, and orthopedists. Six months into the treatment, the patient developed recurring bouts of debilitating fatigue, muscle pain, and painful bronchial and sinus discharge lasting for two to five days at roughly six-week intervals. The patient was seen by pulmonary, allergy, and infectious disease doctors who could find no explanation for the symptoms. The consensus among the care team was that the patient's "... chronic complaint of fatigue and muscle ache may be secondary to lack of exercise, secondary to his low back injury." The same team later diagnosed "chronic bronchitis with upper respiratory tract infection of uncertain etiology." A year-long series of endocrine and immunologic testing revealed no abnormalities. The patient was referred for a psychiatric consult. The psychiatrist's notes stated the following: "I believe this young man is malingering or has Munchausen syndrome." At this point, the patient sought a second psychiatric opinion. After three visits, the patient was told "you need a good internist who can help figure this out." This led to consultation with the internist and the previously noted diagnosis of chronic Epstein–Barr virus syndrome.

The patient was treated with low dose tricyclics and antibiotics. He also began acupuncture and chiropractic care and psychiatric counseling. Over several years, the medical diagnosis changed to chronic fatigue immune dysfunction syndrome (CFIDS) and then chronic fatigue syndrome (CFS). The patient became active in a local support group and participated in several research studies. With these supports, he continued to work full-time, including extensive travel, and he started a family, although episodic flare-ups of exhaustion and brain fog (or "crashes") and recurring bouts of bronchitis/sinusitis resulted in frequent sick days and the need to work from home, and a three-month leave of absence.

Over the next 14 years, the frequency and severity of crashes diminished. By 1999, the limitation of the illness became a secondary factor in the patient's daily life and he began to consider himself recovered. In 2012, he contracted pneumonia, and then, six months later, following removal of a tick, he was diagnosed with Lyme disease and anaplasmosis. He developed constant, disabling fatigue, cognitive impairment, severe post-exertional malaise, light-headedness, muscle pain, and gastrointestinal problems. The patient was bedridden for several weeks and then began an aggressive search for symptom relief, coordinated jointly by his internist and an ME/CFS specialist. He was misdiagnosed with sleep apnea in 2013, leading to the identification of deviated septum, concha bellosa, and turbinate hypertrophy, which were corrected by endoscopic sinus surgery in 2014. Surgery did not ameliorate symptoms. In a search for the cause of the sinus problems, he was evaluated for immune deficiencies by an ME/CFS specialist in 2014, revealing a mild deficiency in immunoglobulin G (IgG1). A further study that same year by an immunologist found an absence of 10 pneumococcal serotypes. He received a Pneumovax vaccination, which produced a significant decrease in the occurrence of both sinus and bronchial infections. Later in 2014, in pursuit of an explanation for the debilitating fatigue, he was evaluated by a geneticist who identified a heterozygous mutation in PSTPIP1 and recommended mitochondrial cofactor support as a way to decrease oxidative stress. A mitochondrial cocktail was prescribed. In consultation with the patient's psychiatrist, Ritalin was added to the cocktail. After six months, this regimen resulted in improved energy levels with less dramatic highs and lows. In search of better indicators of mitochondrial dysfunction, the patient underwent a mitochondrial function test [19], which produced a mitochondrial score of 0.24 (24% energy availability at the cellular level) and revealed significant blocking of mitochondrial active sites and translocation proteins. Adjustments to the patient's mitochondrial cocktail were recommended. However, these adjustments had no bearing on symptomology.

In 2014, the patient was also referred to a pulmonologist who suspected dysautonomia and administered an invasive cardiopulmonary exercise test (iCPET) [20]. This testing revealed pronounced deficiencies in the anaerobic threshold, ventricular refill, and oxygen uptake by muscles, resulting in the diagnoses of preload failure, mitochondrial myopathy, and disordered ventilatory control. Treatment with Mestinon (pyridostigmine bromide) and mild exercise with a recumbent bicycle helped the patient recover enough stamina to resume a modest work schedule. Endocrine, neurologic, and hematology assessments between 2017 and the present uncovered De Quervain's thyroiditis, treated with Synthroid and resolved within 24 weeks; generalized autonomic failure and orthostatic hypertension, for which no effective treatments were identified; vestibular migraine, treated with some benefit by vitamin B2; and monoclonal gammopathy (MGUS), which is being monitored.

Presently, the patient is able to work approximately 16 hours a week and, while frequently resting at home, is no longer housebound. His treatment regimen includes monthly talk therapy, weekly shiatsu and chiropractic treatments, daily Qi Gong, supplemental oxygen as needed, and the following medicines and supplements: Ritalin, Mestinon, Glutathione, Co-Q10, Magnesium, vitamins B2, B12, C, and D, HSO probiotics and Enhance(r), a Chinese herbal antiviral.

This case demonstrates that aggressive pursuit of treatable symptoms can yield benefits for the patient. While there are no FDA-approved treatments for ME/CFS itself, several comorbid conditions were identified and were successfully treated. In addition, some symptoms, such as pain, sleep difficulties, and autonomic problems, can be addressed with medication. Alternative therapies and major lifestyle changes also contributed to improved quality of life. Finally, this case shows the heterogeneous nature of ME/CFS, with two distinct periods of disease separated by over a decade of good health, each period with different etiology and symptoms.

3. Case B

This case involves a 44-year-old white female who has had ME/CFS symptoms for almost 20 years, although she did not receive a diagnosis until 2014. Her symptoms started gradually in 2001 after blunt head trauma and a tick bite, with exhaustion being the main complaint. Consultation with a doctor in 2002 was unhelpful because routine bloodwork revealed nothing outside normal limits. This doctor told her she was the "epitome of good health." Her symptoms of fatigue and post-exertional malaise increased in severity over the next few years. She consulted with internal medicine doctors, gynecologists, psychologists, and psychiatrists. In 2004, she was treated by a gynecologist who observed normal results on a blood workup and misdiagnosed her with depression. She was prescribed an antidepressant, which did not improve symptoms. Several chiropractors and acupuncturists tried adjustments, acupuncture, dietary changes, and traditional Chinese medicine (TCM) to improve her energy, none of which helped. Determined to heal herself, she became a certified yoga instructor in 2010, after which she participated in several advanced training workshops, which focused on improving energy in the body. She reports that the training and practice of the methods taught during training only led to decreases in her energy level.

In 2011, an endocrinologist determined that she was post-menopausal (at age 35) and prescribed hormone replacement therapy (HRT). Although HRT relieved night sweats and hot flashes, it did nothing to improve energy levels or reduce fatigue and PEM. Medical doctors continued to dismiss or misdiagnose her symptoms of overwhelming fatigue, PEM, cognitive dysfunction, unrefreshing sleep, light sensitivity, headaches, sore throat, and irritable bowel syndrome, all of which continued to worsen. From her own research, she realized that these symptoms aligned with the symptoms of ME/CFS. In 2014, a doctor of osteopathic medicine diagnosed her with ME/CFS after learning of her self-diagnosis and researching the disease himself. He recommended supplements, which did not improve her condition. Unable to find a local doctor with expertise in ME/CFS, she found a doctor who could treat her remotely. This doctor also recommended supplements, which did not help, and a Paleo Diet, which helped alleviate the irritable bowel syndrome but did not give relief from the fatigue or PEM.

By 2014, she was unable to continue performing the physical aspects of her job as a wellness coach and yoga instructor and instead enrolled in a master's degree program for counseling psychology, taking one class per semester. At this point, she was mostly housebound. Two years later, she lost the ability to read and had depleted her savings paying rent and other bills. She was forced to drop out of graduate school and move back into her parents' home. By then, she was so weak she spent more than 23 hours a day in bed. Her cognitive function was so poor that she was unable to even read a magazine article. She could not practice guitar, sing, or do art. She experienced such severe nausea that she sometimes could not eat. She was so sensitive to light and sound that she rarely left her darkened bedroom. She could not have visitors because talking was too taxing.

In 2017, she applied for Social Security Disability Insurance (SSDI). Her application was denied due to the inaccurate assessment in the denial letter that she was able to "be on (her) feet most of the day" and "lift up to 10 pounds frequently." She appealed the denial and was again turned down, even with extensive documentation of her level of disability by her primary care physician. She requested a hearing by an administrative law judge. In the meantime, she underwent an invasive cardiopulmonary exercise test (iCPET) and was diagnosed by a pulmonologist with severe autonomic dysfunction in the form of preload failure [21]. She was also tested for and diagnosed with small fiber polyneuropathy by a neurologist [22]. These two diagnoses helped convince the administrative law judge to approve her SSDI application.

Treatment with Mestinon (pyridostigmine bromide) for preload failure and with midodrine for low blood pressure helped to slightly improve her functioning. In 2018, she started seeing an integrative medicine doctor who specializes in treating ME/CFS patients. He treated her with an antiviral (for trace findings of Epstein–Barr virus), an antibiotic (for trace findings of tick-borne illnesses), and vitamin B-12 injections. The first two treatments had no effect, while the B-12 seemed to help with getting out of bed and walking around. She currently spends 21 to 23 hours per day in bed and is able to listen to an audiobook or watch TV for up to two hours per day.

The increased availability of telemedicine in the US during the COVID-19 pandemic has dramatically improved the ability of this patient to access medical care. Previously, travel to and from a medical appointment, combined with waiting in a doctor's office and the appointment itself, caused severe PEM that lasted for weeks. While virtual appointments still cause PEM, it only lasts for a few days. In addition, because the virtual appointments are more manageable, she has not had to reschedule as often.

This case demonstrates the harm that disbelieving health care providers inflict on patients with ME/CFS. Had this patient received an early diagnosis and education on managing her illness with pacing and self-care [2,23], she might not be as severely disabled today. In addition, this case shows the enormous losses suffered by the severely ill, including losing a career, not being able to live independently, and giving up the dream of having children. These losses are compounded by fear about a future when her parents are no longer available to be her caregivers.

4. Case C

This white female was 38 years old when she died in 2019 after years of suffering from very severe ME/CFS. She first developed symptoms of ME/CFS as a teenager, although she was not diagnosed for nearly 12 years. In 1999, at age 18, she had a serious case of EBV (mononucleosis) that kept her home from school for three months and required treatment with steroids for lymphadenopathy. Subsequently, she experienced variable episodes of fatigue and daytime sleepiness that gradually increased in severity. The corresponding increase in brain fog and decrease in energy made university level study difficult. Having been an athlete in high school, she tried to maintain an exercise regime for many years, despite worsening symptoms and PEM. Routine blood workups at this time, and at many times in the next 20 years, revealed nothing outside of normal limits.

As a junior in college, she was evaluated by a psychiatrist, who ruled out anxiety, depression, and somatization as sources for the fatigue. She also underwent a neuropsychological evaluation that ruled out any innate learning disability or attention deficit disorder. She was rated in the 99th percentile for intellectual ability, but only in the 39th percentile for reading rate, consistent with self-reported difficulty concentrating and processing information. Despite these challenges, she earned a BA in Biology and started a master's degree program in Environmental Science.

In 2004, at age 23, she developed pain and weakness in her hands, arms, neck, and shoulders to the point where she could no longer type and was forced to take a one-and-a-half year leave of absence from graduate school. She was diagnosed with fibromyalgia and myofascial pain syndrome by her primary care physician. Treatments included physical therapy, sports massage, myofascial release therapy, and acupuncture; these provided some relief from the pain. In 2005 and 2006, she had two bouts of diverticulitis with micro-perforations, requiring hospitalization and IV antibiotics but not requiring surgery. Having noticed that her fatigue seemed worse during the second half of her menstrual cycle, she consulted an endocrinologist, who treated her with oral contraceptives and synthetic progesterone, neither of which helped with the fatigue or PEM.

By 2008, she had finished her master's program and started working part-time as an environmental consultant. However, her condition continued to worsen, and she developed new symptoms of dizziness, chronic light-headedness, unstable blood pressure and heart rate, sound and vibration sensitivity, nausea, and extreme thirst, in addition to the ongoing fatigue, PEM, and pain. She consulted an integrative medicine doctor, who suggested, based on her symptoms, that she might have persistent Lyme disease, even though tests for tick-borne infections were inconclusive or negative. He treated her with a four-month trial of doxycycline, which did not ameliorate symptoms. She subsequently developed a severe sinus infection in 2010 that caused a three-month-long crash and left her bedbound. She had to abandon her career and was unable to drive after this episode.

She had sleep studies in 2005, and again in 2011, that ruled out narcolepsy, sleep apnea, and periodic leg movements. No abnormalities were observed on the electrocardiogram (EKG) or electroencephalogram (EEG). The first sleep study diagnosed atypical sleep disturbances and the second a shifted sleep cycle and poor sleep efficiency. For many years she took a small dose of Ritalin (5–10 mg) during the day to obtain a few hours of ability to focus, and occasionally Ambien at night to help her stay asleep. She also tried Provigil, without benefit.

Still in search of a treatment that would help, she started working with immunologists in other parts of the country. She was officially diagnosed with ME/CFS in 2011 by an immunologist and ME/CFS expert. Immune function testing revealed low levels of some natural killer cells and an abnormal cytokine profile, along with traces of an immune response to EBV and human herpesvirus 6 (HHV-6). Tests for other common viruses were negative. She was treated with the antiviral famciclovir (Famvir) for nine months but had no improvement in her physical condition. She was treated with immunomodulators, including Nexavir and low-dose naltrexone, but they provided no benefit, and the side effects of increased dizziness and neurological symptoms were not tolerable.

She consulted a cardiologist in 2011 and was diagnosed with neurally mediated hypotension (NMH) and postural tachycardia syndrome (POTS), common autonomic problems in pwME. Treatment with fludrocortisone (Florinef) did not help. A second cardiologist confirmed the POTS diagnosis in 2015. Treatment with extra salt and midodrine helped a little. However, her overall condition continued to deteriorate, and by 2013, she was no longer able to travel, thus ending her ability to seek treatment with doctors in other parts of the country. Even for local appointments, some requirements such as early morning appointments or long waits in noisy, brightly lit rooms, made visits impossible.

In 2015, she applied for SSDI. Her application was denied at the initial and the reconsideration levels because the vocational experts provided by the Social Security Administration claimed that she was able to work part-time, ignoring the fact that she was completely bedbound. She was unable to find a lawyer to take her case but did find a vocational expert who would confirm her level of disability. Disability payments were finally approved by an administrative law judge two weeks before she died.

She spent the last four years of her life completely bedbound. She suffered immensely from severe exhaustion, body-wide muscle and joint pain, a stiff neck, unstable blood pressure and heart rate, muscle twitches and spasms, chronic lightheadedness, sound/vibration sensitivity, nausea, food intolerances, and extreme thirst, among other symptoms. She developed mold and chemical sensitivities that made it very difficult to find housing that did not exacerbate her symptoms. She was often unable to speak and only had a brief period on rare good days when she had the cognitive energy to focus. She used her limited amount of functional time to maintain her online connections to other people and to advocate for fellow ME/CFS sufferers and environmental causes.

Since she was unable to leave her bed, she tried to find a doctor who could attend to her in her home, but this proved impossible. She had a day-time caregiver for 15 hours per week, provided by a state agency, but turnover was high. The final unbearable symptom was repeatedly waking up at night and feeling like her heart had stopped and she could not breathe. After some time, her heart would start beating erratically and she would catch a breath, but the experience was terrifying. Worn out by her prolonged struggle with pain, isolation, abysmally low quality of life, and her futile search for some possibility that her condition would improve, and although she felt loved and loving, she ended her life.

This case demonstrates the high risk of suicide for patients with a misunderstood and difficult-to-treat disease such as ME/CFS. Similar to other chronic illnesses, suicide is many times more common in pwME than in the general population [24] and occurs at a younger average age [25]. In addition to unsupportive peer and medical interactions, risk factors for suicide include the stigma associated with the disease name "chronic fatigue syndrome," the continual presence of pain, and often significantly decreased functionality [26].

5. Discussion

One of the common themes in these three case studies is the difficulty pwME encounter in finding supportive doctors and in receiving an accurate diagnosis. All three consulted tens of doctors across many specialties, and all three reported at least some interactions with dismissive or hostile health care providers. All three had the experience of being told by a doctor that they were perfectly healthy, often based on normal laboratory results, despite debilitating symptoms. The length of time to receive a diagnosis of ME/CFS was 3 years, 13 years, and 12 years, respectively. This theme highlights the need for better education of health care providers about ME/CFS. Early diagnosis and support can help pwME understand how to manage their disease through pacing and symptom management and to avoid activities, such as strenuous exercise, which exacerbate symptoms. In addition, treatment of symptoms and comorbidities can greatly improve quality of life. Recent developments, such as guidelines for treatment from the US Clinician's Coalition [2], updated guidelines from the UK National Institute of Health and Care Excellence (NICE) [27], and updated information on the US Center for Disease Control and Prevention (CDC) website [28] are encouraging, but much more is needed.

A second theme is the difficulty imposed on pwME by the lack of a diagnostic test for ME/CFS and the lack of treatments. These deficiencies are directly related to the extremely low level of research funding for ME/CFS in the US [29] and worldwide [30]. In the US, ME/CFS research is underfunded by about a factor of 14 relative to its burden of disease [31,32]. One reason for this may be that pwME are predominantly female, and, as recently demonstrated in Mirin (2020) [33], female-dominated diseases tend to be underfunded relative to male-dominated diseases. A second reason may be that the longstanding pattern of stigmatization and psychologization of ME/CFS has discouraged funding agencies from allocating resources or positively reviewing research proposals. High profile reports, e.g., from the National Academy of Medicine in 2015, have emphasized that ME/CFS is biological in origin and that research is desperately needed [1]. The National Institutes of Health (NIH) has doubled annual funding, created three collaborative research centers, and started an intramural research project [34], but much more is needed.

A third theme is the general lack of support for pwME who are homebound or bedbound due to the fragmented and ineffective social support networks in the US. Despite the fact that most pwME cannot work full-time, or even part-time, disability support is very difficult to obtain. Applications to the US Social Security Disability Insurance program are routinely denied at the initial level and at the reconsideration level, as experienced by both Case B and Case C. In 2017, only 13,000 people received disability payments from Social Security for ME/CFS, out of potentially hundreds of thousands needing assistance [35], leading to severe financial stress and often homelessness. Medical care at home is very limited in the US, as experienced by Case C, meaning that many pwME receive no medical care at all. Better models for caring for this community exist (e.g., Kingdon et al. (2020) [36]) but would require a serious commitment from the US government in funding and resources. However, the dramatic increase in the availability of telemedicine during the COVID-19 pandemic has improved access and will hopefully remain in place.

6. Conclusions

We have presented three cases of people with moderate to very severe ME/CFS, demonstrating the debilitating nature of the disease and the difficulty in obtaining diagnosis and treatment. ME/CFS is not rare and the number of cases is expected to increase dramatically over the next few years due to the long-term consequences of the COVID-19 pandemic. Colloquially termed the COVID-19 long haulers, a large fraction of survivors exhibit ongoing symptoms that closely resemble ME/CFS, including fatigue, PEM, orthostatic intolerance, and cognitive difficulties [14,37]. We owe it to pwME and the COVID-19 long haulers to invest in the research to discover causes, treatments, and hopefully a cure, and to develop the medical and social support networks to improve their lives.

Author Contributions: Conceptualization, C.I.-B.; writing—original draft preparation, L.R.W.; writing—review and editing, L.R.W. and C.I.-B. All authors have read and agreed to the published version of the manuscript.

Funding: This research received no external funding.

Institutional Review Board Statement: Ethical review and approval were waived for this study because it is not a research study.

Informed Consent Statement: Informed consent was obtained from all subjects involved in the study.

Data Availability Statement: No research data was collected.

Acknowledgments: The authors are grateful to Cases A and B, and to the mother of Case C, for sharing their medical histories.

Conflicts of Interest: The authors declare no conflict of interest.

References

1. Institute of Medicine. *Beyond Myalgic Encephalomyelitis/Chronic Fatigue Syndrome: Redefining an Illness*; The National Academies Press: Washington, DC, USA, 2015. [CrossRef]
2. Bateman, L.; Bonilla, H.; Dempsey, T.; Felsenstein, D.; Komaroff, A.L.; Klimas, N.G.; Natelson, B.H.; Podell, R.; Ruhoy, I.; Vera-Nunez, M.; et al. US ME/CFS Clinician Coalition: Diagnosing and Treating ME/CFS. Available online: https://drive.google.com/file/d/1SG7hlJTCSDrDHqvioPMq-cX-rgRKXjfk/view (accessed on 30 November 2020).
3. WHO. *International Classification of Functioning, Disability and Health (ICF)*; WHO: Geneva, Switzerland, 2018.
4. McManimen, S.L.; McClellan, D.; Stootho, J.; Jason, L.A. Effects of unsupportive social interactions, stigma, and symptoms on patients with myalgic encephalomyelitis and chronic fatigue syndrome. *Community Psychol.* **2018**, *46*, 959–971. [CrossRef] [PubMed]
5. Valdez, A.R.; Hancock, E.E.; Adebayo, S.; Kiernicki, D.J.; Proskauer, D.; Attewell, J.R.; Bateman, L.; DeMaria, A.; Lapp, C.W.; Rowe, P.C.; et al. Estimating Prevalence, Demographics, and Costs of ME/CFS Using Large Scale Medical Claims Data and Machine Learning. *Front. Pediatr.* **2019**, *6*. [CrossRef]
6. Jason, L.A.; Mirin, A.A. Updating the National Academy of Medicine ME/CFS prevalence and economic impact figures to account for population growth and inflation. *Fatigue Biomed. Health Behav.* **2021**, 1–5. [CrossRef]
7. Chu, L.; Valencia, I.; Garvert, D.; Montoya, J. Onset Patterns and Course of Myalgic Encephalomyelitis/Chronic Fatigue Syndrome. *Front. Pediatr.* **2019**, *7*. [CrossRef]
8. Hickie, I.; Davenport, T.; Wakefield, D.; Vollmer-Conna, U.; Cameron, B.; Vernon, S.D.; Reeves, W.C.; Lloyd, A. Post-infective and chronic fatigue syndromes precipitated by viral and non-viral pathogens: Prospective cohort study. *BMJ* **2006**, *333*, 575. [CrossRef]
9. Lam, M.H.-B.; Wing, Y.-K.; Yu, M.W.-M.; Leung, C.-M.; Ma, R.C.W.; Kong, A.P.S.; So, W.Y.; Fong, S.Y.-Y.; Lam, S.-P. Mental Morbidities and Chronic Fatigue in Severe Acute Respiratory Syndrome Survivors: Long-term Follow-up. *Arch. Intern. Med.* **2009**, *169*, 2142–2147. [CrossRef] [PubMed]
10. Perrin, R.; Riste, L.; Hann, M.; Walther, A.; Mukherjee, A.; Heald, A. Into the looking glass: Post-viral syndrome post COVID-19. *Med. Hypotheses* **2020**, *144*. [CrossRef]
11. Hornig, M. What Does COVID-19 Portend for ME/CFS? Available online: https://solvecfs.org/wp-content/uploads/2020/04/COVID19-MECFS_Sci_Review.pdf (accessed on 30 November 2020).
12. Zimmer, K. News & Opinion: Could COVID-19 Trigger Chronic Disease in Some People? Available online: https://www.the-scientist.com/news-opinion/could-covid-19-trigger-chronic-disease-in-some-people-67749 (accessed on 30 November 2020).
13. Rubin, R. As Their Numbers Grow, COVID-19 "Long Haulers" Stump Experts. *JAMA* **2020**, *324*, 1381–1383. [CrossRef]
14. Komaroff, A.L.; Bateman, L. Will COVID-19 Lead to Myalgic Encephalomyelitis/Chronic Fatigue Syndrome? *Front. Med.* **2021**, *7*. [CrossRef]
15. Carruthers, B.M.; van de Sande, M.I.; De Meirleir, K.L.; Klimas, N.G.; Broderick, G.; Mitchell, T.; Staines, D.; Powles, A.C.P.; Speight, N.; Vallings, R.; et al. Myalgic encephalomyelitis: International Consensus Criteria. *J. Intern. Med.* **2011**, *270*, 327–338. [CrossRef]
16. Shepherd, C.; Chaudhuri, A. *ME/CFS/PVFS—An Exploration of the Clinical Issues*; The ME Association: Gawcott, UK, 2020.
17. Bringsli, G.J.; Gilje, A.; Getz Wold, B.K. The Norwegian ME Association National Survey Abridged English Version. Available online: https://huisartsvink.files.wordpress.com/2018/08/bringsli-2014-me-nat-norwegian-survey-abr-eng-ver-2.pdf (accessed on 30 November 2020).
18. Cairns, R.; Hotopf, M. A systematic review describing the prognosis of chronic fatigue syndrome. *Occup. Med.* **2005**, *55*, 20–31. [CrossRef]
19. Myhill, S.; Booth, N.E.; McLaren-Howard, J. Chronic fatigue syndrome and mitochondrial dysfunction. *Int. J. Clin. Exp. Med.* **2009**, *2*, 1–16. [PubMed]
20. Oldham, W.M.; Lewis, G.D.; Opotowsky, A.R.; Waxman, A.B.; Systrom, D.M. Unexplained Exertional Dyspnea Caused by Low Ventricular Filling Pressures: Results from Clinical Invasive Cardiopulmonary Exercise Testing. *Pulm. Circ.* **2016**, *6*, 55–62. [CrossRef] [PubMed]
21. Joseph, P.; Sanders, J.; Oaklander, A.L.; Rodriguez, T.C.A.A.; Oliveira, R.; Urbina, M.F.; Waxman, A.B.; Systrom, D.M. The Pathophysiology of Chronic Fatigue Syndrome: Results from an Invasive Cardiopulmonary Exercise Laboratory. In Proceedings of the American Thoracic Society 2019 International Conference, Dallas, TX, USA, 17–22 May 2019.
22. Oaklander, A.; Nolano, M. Scientific Advances in and Clinical Approaches to Small-Fiber Polyneuropathy: A Review. *JAMA Neurol.* **2019**. [CrossRef]
23. Geraghty, K.; Hann, M.; Kurtev, S. ME/CFS patients' reports of symptom changes following CBT, GET and Pacing Treatments: Analysis of a primary survey compared with secondary surveys. *J. Health Psychol.* **2017**. [CrossRef]
24. Roberts, E.; Wessely, S.; Chalder, T.; Chang, C.-K.; Hotopf, M. Mortality of people with chronic fatigue syndrome: A retrospective cohort study in England and Wales from the South London and Maudsley NHS Foundation Trust Biomedical Research Centre (SLaM BRC) Clinical Record Interactive Search (CRIS) Register. *Lancet* **2016**, *387*, 1638–1643. [CrossRef]
25. McManimen, S.L.; Devendorf, A.R.; Brown, A.A.; Moore, B.C.; Moore, J.H.; Jason, L.A. Mortality in Patients with Myalgic Encephalomyelitis and Chronic Fatigue Syndrome. *Fatigue* **2016**, *4*, 195–207. [CrossRef]

26. Johnson, M.L.; Cotler, J.; Terman, J.M.; Jason, L.A. Risk factors for suicide in chronic fatigue syndrome. *Death Stud.* **2020**, 1–7. [CrossRef]
27. NICE. Myalgic Encephalomyelitis (or Encephalopathy)/Chronic Fatigue Syndrome: Diagnosis and Management: Draft Guideline. Available online: https://www.nice.org.uk/guidance/gid-ng10091/documents/draft-guideline (accessed on 30 November 2020).
28. CDC. Myalgic Encephalomyelitis/Chronic Fatigue Syndrome. Available online: https://www.cdc.gov/me-cfs/index.html (accessed on 30 November 2020).
29. NIH. Estimates of Funding for Various Research, Condition, and Disease Categories (RCDC). Available online: https://report.nih.gov/categorical_spending.aspx (accessed on 30 November 2020).
30. Radford, G.; Chowdhury, S. ME/CFS Research Funding. Available online: https://meassociation.org.uk/wp-content/uploads/mecfs-research-funding-report-2016.pdf (accessed on 30 November 2020).
31. Dimmock, M.E.; Mirin, A.A.; Jason, L.A. Estimating the disease burden of ME/CFS in the United States and its relation to research funding. *J. Med. Ther.* **2016**, *1*. [CrossRef]
32. Mirin, A.A.; Dimmock, M.; Jason, L. Research update: The relation between ME/CFS disease burden and research funding in the USA. *Work* **2020**, *66*, 277–282. [CrossRef] [PubMed]
33. Mirin, A.A. Gender Disparity in the Funding of Diseases by the U.S. National Institutes of Health. *J. Women Health* **2020**. [CrossRef] [PubMed]
34. NIH. About ME/CFS. Available online: https://www.nih.gov/mecfs/about-mecfs (accessed on 30 November 2020).
35. CFSAC. CFSAC Meeting Summary. Available online: https://solvecfs.org/chronic-fatigue-syndrome-advisory-committee-cfsac-meeting-summary/ (accessed on 30 November 2020).
36. Kingdon, C.; Giotas, D.; Nacul, L.; Lacerda, E. Health Care Responsibility and Compassion-Visiting the Housebound Patient Severely Affected by ME/CFS. *Healthcare* **2020**, *8*, 197. [CrossRef] [PubMed]
37. Nath, A.; Smith, B. Neurological issues during COVID-19: An Overview. *Neurosci. Lett.* **2020**. [CrossRef]

 *healthcare*

Case Report

Severe ME in Children

Nigel Speight

Paediatrician, Southlands, Gilesgate, Durham DH1 1QN, UK; speight@doctors.org.uk

Received: 21 May 2020; Accepted: 9 July 2020; Published: 14 July 2020

Abstract: A current problem regarding Myalgic Encephalomyelitis/Chronic Fatigue Syndrome (ME/CFS) is the large proportion of doctors that are either not trained or refuse to recognize ME/CFS as a genuine clinical entity, and as a result do not diagnose it. An additional problem is that most of the clinical and research studies currently available on ME are focused on patients who are ambulant and able to attend clinics and there is very limited data on patients who are very severe (housebound or bedbound), despite the fact that they constitute an estimated 25% of all ME/CFS cases. This author has personal experience of managing and advising on numerous cases of severe paediatric ME, and offers a series of case reports of individual cases as a means of illustrating various points regarding clinical presentation, together with general principles of appropriate management.

Keywords: Myalgic Encephalomyelitis (ME); Chronic Fatigue Syndrome (CFS); Severe ME; Very Severe ME; Post-Exertional Malaise (PEM); Fabricated and Induced Illness; Cognitive Behavioural Therapy; Graded Exercise Therapy; Pervasive Refusal Syndrome; immunoglobulin

1. Definitions

Empirically, one can regard Severe Myalgic Encephalomyelitis/Chronic Fatigue Syndrome (ME/CFS) as affecting a patient who is housebound and bedbound most of the time, functioning at approximately 5–15% of normal capacity.

There are also even less fortunate patients who are totally bedbound, suffering continuous extremely unpleasant symptoms and in need of nursing care to meet all their needs for nutrition and personal hygiene. This can be termed Very Severe ME, and patients' functional capacity is less than 5%.

Paradoxically, the cardinal diagnostic feature of ME/CFS, Post-Exertional Malaise (PEM), is difficult to elicit in the very severely affected, as the patient is hardly capable of the slightest exertion. However, in common with all ME sufferers, there is usually evidence of deterioration after even minor forms of stress, such as light, sound, odour or motion. Hypersensitivities to various types of stimuli can cause "sensory overload" and can result in a "crash"—a period of immobilising mental or physical debility—even in patients who are not normally very severely affected [1].

Those in the group that are "only" severely affected *do* experience PEM to a marked degree, often in terms of increased intensity of pain.

2. Problems and Challenges for the Patient with Severe ME/CFS

Patients with ME have often suffered from disbelief on the part of their medical attendants, and even those severely affected are not protected from this disbelief.

This disbelief can operate at several different levels.

The first level is when the doctor disbelieves that ME/CFS even exists as an organic illness in the first place. This can lead to inappropriate referral to psychiatrists, and then to harmful management plans.

The next level is for the doctor to accept that ME/CFS exists, but to disbelieve that a particular patient has it.

A further level is to have an overoptimistic view on the efficacy of certain forms of "treatment", usually Cognitive Behavioural Therapy (CBT) and/or Graded Exercise Therapy (GET). This leads to the patient having the diagnosis of ME/CFS withdrawn when they fail to respond to this management strategy, or as so often happens when the patient's condition worsens. GET is especially likely to lead to such deterioration. Rather than drawing the conclusion that these "therapies" are misguided, the doctor often changes the diagnosis to one involving a psychological causation.

The final level is when the doctor diagnoses ME/CFS at a stage when it is still mild or moderate, and then the patient's condition worsens to the extent that it becomes severe. The doctor can then lose confidence in the original diagnosis, and change this to a psychological diagnosis such as Pervasive Refusal Syndrome (as in Case A below).

For the patient with Severe ME/CFS to be subjected to such disbelief is of course adding insult to injury.

The worst scenario is when the paediatrician, as a result of some of the above varieties of disbelief, makes a provisional diagnosis of Fabricated and Induced Illness (FII), and refers the patient to Social Services as a case needing safeguarding. This can lead to threats to remove the young person from their family. This is currently increasingly fashionable in the UK [2].

3. Challenges for the Doctor Faced with a Case of Severe ME/CFS

If the doctor has never seen a case of Severe ME/CFS before, as is often the case, there is a natural tendency to panic, and to worry that the patient might die, or that the diagnosis was wrong. In addition, the doctor may feel a sense of helplessness, with no obvious therapeutic options available. These emotions can lead to the doctor adopting a strategy of avoidance, refusing to accept that he/she has a duty of care to the patient, and regrettably many patients with Severe ME/CFS suffer near total neglect by their doctors. Alternatively, the doctor may try to insist on inappropriate "treatment" strategies, such as CBT or GET. Neither form of treatment has been studied in the severe or very severe population. Further, especially in the severe cases, rigidly enforced GET is nearly always seriously harmful. The current NICE Guidelines (2007) specifically do *not* recommend GET in Severe ME [3].

4. Possible Clinical Features of Severe and Very Severe ME/CFS

- Minimal energy levels, resulting in the patient being housebound or bedridden;
- Paralysis;
- Severe generalised continuous pain;
- Severe continuous headache;
- Hyperaesthesia/extreme sensitivity to touch;
- Abdominal pain, worse after food—this may be so severe as to interfere with nutrition;
- Sleep disturbance, possible hypersomnolence or difficulty sleeping on account of pain and headache;
- Major problems with cognition, concentration and short-term memory;
- Extreme sensitivity to light and sound;
- Multiple chemical sensitivities;
- Problems with eating and drinking—this can be due to either general weakness or actual dysphagia, and this may necessitate tube feeding;
- Aphonia (mechanism unclear);
- Myoclonic jerks;
- Incontinence.

5. General Points Regarding Impact of Symptoms on Quality of Life

It has been estimated that the sheer severity of suffering experienced by patients with the above symptoms can actually be worse than that suffered by patients with other chronic conditions such as multiple sclerosis and cancer [4].

The severity of the photosensitivity can be a further trigger to disbelief, as the doctor may find it difficult to accept that the patient not only has to lie in a darkened room but has to wear eye protection in addition.

The abdominal pain may be so severe as to interfere with nutrition, and some cases are due to an added complication, Mast Cell Activation Disorder [5]. This has probably been responsible for some actual fatalities. Specific treatment for MCAS includes oral cromoglycate and antihistamines.

Of course, simply having ME/CFS does not protect one from other causes of post-prandial abdominal pain, such as median arcuate ligament syndrome (MALS) [6], or Helicobacter pylori gastritis.

6. My First Severe Case

I will describe this in some detail as I made every mistake in the book.

When first seen, this 13-year-old girl was only moderately severe, being well enough to attend my clinic. A few months later she suffered marked deterioration, probably due to a virus infection. She was bedridden and suffered many of the symptoms listed above.

She had markedly raised antibodies to coxsackie B (1:520), which remained elevated at that level for over a year.

My first mistake was to panic, and to think that I might have been mistaken in my initial diagnosis. I accordingly asked an immunologist colleague in our regional hospital for a second opinion, as he had previously confirmed to me that he "believed" in ME/CFS. The girl and her parents were quite dubious about the referral as it involved a 20 mile journey, and they knew instinctively that this would be unpleasant for her and might well make her worse.

I got my colleague to promise that she would be seen the same day and then allowed straight home.

The immunologist made the same mistake as I had and asked for second and third opinions, with the result that she was kept in the regional hospital for three days, had numerous investigations, and was seen and examined by numerous doctors. One neurologist insisted on her having to get out of bed and walk for him. (One year later, aged 14, she wrote him a dignified letter of reproach, to which he had the grace to respond.)

No new diagnosis was proffered and she was allowed home, the entire exercise having been a negative one for her. The whole experience was so traumatic for her that it took her 2–3 years to forgive both me and her parents properly.

I continued to make similar mistakes, by asking in an experienced GP with a special interest in ME/CFS. I knew him as a fine and kind doctor but in the event, he upset her further by insisting on looking into her fundi and then talking about a similar patient he once had who had died.

Next to upset her was the district nurse, who bustled in, said she had read all about ME and all would be fine if she made an effort to walk every day,

The family doctor then convened a meeting, at which the Health Visitor (children's nurse) wondered whether it could all be due to father sexually abusing his daughter. This was duly considered, recorded in the minutes, and the minutes sent to the family, to their predictable outrage.

The main lesson I learnt from all this was how many ways it is possible to upset a severely ill ME patient, and that one needs to be proactively protective of one's patient, especially against other professionals, and disbelieving friends and relatives.

For professionals affected by "Furor Therapeuticus", i.e., the feeling that one has to do something, it is better to remember that there is no proven curative treatment, and that they should perhaps concentrate on the "Do No Harm" element in the Hippocratic Oath.

In this case, I gradually regained my patient's trust, with the help of a new family doctor. He and I shared the home visiting duties, going in alternately every 4 weeks or so.

Over the next six months, she continued to deteriorate and she found it increasingly difficult to eat and drink.

My next mistake was to give in for too long to her objections to nasogastric tube feeding. Once I finally persuaded her, life became much better for everyone.

Because of the evidence that immunoglobulin might be an effective therapy, she was treated with this by intramuscular injection, monthly, for 12 months. She made a very slow recovery, and tube feeding was stopped after 3 years. She continued to recover and 20 years later has a full-time job, and is operating at 95% of normal.

The final lesson to learn from this case is that virtually full recovery is possible even in severe cases, and one can use this fact to maintain hope for other similarly affected patients.

7. My Second Severe Case

This case is noteworthy as it demonstrates how much better things can go when the above lessons have been learnt. Again, the patient was a 13-year-old girl. Her presentation was unusual in that she presented acutely with severe loin pain, bad enough to be admitted with a presumptive diagnosis of renal colic. This diagnosis was excluded (I now think this illness started with a form of Bornholm Disease—convalescent titres against enterovirus were markedly elevated).

It soon transpired that she had multiple additional severe symptoms, including total prostration together with photophobia, generalised pain and hyperaesthesia. She was therefore given a presumptive diagnosis of ME/CFS. She was discharged home as soon as possible and, for the next year, needed round-the-clock nursing from her mother, who was herself a nurse. Tube feeding was started early to good effect, amitriptyline and carbamazepine were tried for her pain, and immunoglobulin was given.

After 12 months, she began to improve, and then her rate of improvement accelerated. By the end of two years, she had made a full recovery, and has never subsequently relapsed.

8. My Most Severe Case

Five years later, another 13-year-old girl presented. She was extremely ill, and I felt I had to get her seen by a colleague for a second opinion, and have a quick cranial MRI. Both occurred without the need for hospital admission. Over subsequent weeks, she lay motionless and in severe pain, and her breathing was so shallow that I was afraid she was going to die. Tube feeding and immunoglobulin were resorted to early on. In addition, I added clarithromycin, because of the work of Garth Nicolson from the US [7], in which he showed some cases of ME/CFS are due to atypical organisms including Mycoplasmas. She remained in this state for the next 9 months, with her mother providing total nursing care.

At nine months, I changed the clarithromycin to doxycycline, on the grounds that it was just possible that she might have a form of Lyme disease. From that time, she steadily improved, and within 12 months, she had recovered completely. At follow up 12 years later, she had suffered no relapses and was the healthy mother of a healthy child.

I am not sure one should draw too many conclusions from this case. However, it does suggest that total therapeutic nihilism is not the only approach, and that if one can think of therapeutic options that are probably harmless, one has a duty to consider them.

9. Overview and Discussion of These Cases

It is perhaps fortuitous that all these cases made good recoveries, and this may be a testament to the efficacy of immunoglobulin, for which further studies would seem to be indicated. However, many cases are not so fortunate, and remain severely or very severely affected for many years.

It is perhaps important to emphasise the importance of early diagnosis so that correct advice can be given from the outset. In other words, it is important to ignore those definitions which demand a duration of 6 months of symptoms before the patient can be given the diagnosis. Like so many "gold standard" definitions, these are more designed for those performing research rather than day-to-day clinical practice.

The correct advice is to rest in the early acute stage of the disease, rather than to attempt to fight one's way out of it.

Another factor that can lead to delays in diagnosis is the felt need on the part of the doctor to exclude a large number of other conditions. It is perhaps better to make an early *provisional* diagnosis of ME on the balance of strong probability, while keeping an open mind for the future. ME should be regarded as a positive clinical diagnosis based on a careful history rather than a diagnosis of exclusion.

10. The Value of Home Visiting by the Doctor

The magnitude of the challenge posed by a case of Severe ME can produce an avoidance reaction on the part of some doctors. Many patients are virtually abandoned by both their family doctors and local and regional specialists.

In the UK, the practice of doing home visits seems to have gone into decline, both with general practitioners and hospital consultants. This would seem to be regrettable, especially in the case of Severe ME. The above cases all seemed to benefit from the ongoing home visits they received, both for moral support and symptomatic treatment. Sadly, many cases of Severe ME lie at home without having seen a doctor for many years.

11. Symptomatic and Supportive Treatment

This is an area where the art of medicine comes into its own. Individual variation in response to treatment is common, and the best approach is a policy of a cautious therapeutic trial. Symptoms that deserve attempts to mitigate include:

- Pain—this is thought to be of neuropathic type, and drugs that can be tried include anticonvulsants, and tricyclic antidepressants such as amitriptyline (which can also help with sleep). Very severe cases deserve opiates.
- Sleep—here it is justified to use hypnotics and/or melatonin to try to increase the duration and quality of sleep, and reduce the wakeful hours of pain.
- Headache—here it is important to see whether some of the headache is migrainous in origin, in which case prophylactic drugs and or dietary change may help.
- Intercurrent illnesses—ME/CFS does not protect from these, and a supportive family doctor willing to do home visits is worth his/her weight in gold. As the patient is too unwell to attend the GP surgery, this is the only way episodes of intercurrent infections such as tonsillitis and otitis media will be treated.
- Depression—naturally, severely affected ME patients can suffer from depression as a *secondary* result of their condition (I am surprised how few actually do). Cautious use of antidepressants in selected cases may be worthwhile.

12. Caution Regarding Dosage

Most ME patients appear to be abnormally sensitive to a wide range of drugs, developing side effects at quite low doses. Accordingly, it is wise to start at very low doses and titrate upwards with all due caution.

Further guidance on lists of drugs and appropriate doses can be found in the Paediatric Primer on ME/CFS in children and young people [8].

13. The Role of Physiotherapy

Having already stressed the real dangers of active physiotherapy/GET, it must be stated that there is a role for gentle physiotherapy and massage, involving passive movements to reduce the risk of joint contractures and venous thrombosis.

14. The Risk of Osteoporosis

These patients are at risk of osteoporosis because of being bedbound for prolonged periods. Consideration should be given to use of bisphosphonates, without insisting on an attendance at hospital for a bone density scan.

15. Vitamin D Deficiency

Again prolonged bedrest in a darkened room puts Severe ME/CFS patients at risk of Vitamin D deficiency, which should accordingly be anticipated and prevented.

16. How Not to Manage Severe ME/CFS

It is sometimes true that one can learn more from a bad example than a good one. Here are presented a sequence of cases where the management was less than ideal.

16.1. Case A

Yet again, this case involves a 13-year-old girl. Like the first case, she initially presented as only moderately severe. The paediatrician diagnosed her as ME and was managing her appropriately. Her condition then deteriorated markedly and the paediatrician panicked. Unfortunately, the form her panic took was to refer her to the local Child Psychiatrist. This latter took an extremely dogmatic position, and promptly dismissed the original diagnosis of ME/CFS, replacing it with the alternative psychiatric diagnosis of Pervasive Refusal Syndrome.

The psychiatrist then took out an emergency court order to enforce admission to her psychiatric unit against the parents' and the patient's wishes. When visited a few days later, the patient presented a truly pitiable sight. She was already being tube fed. She was extremely sensitive to sound and was using ear defenders. However, the psychiatrist banned these on the grounds that she needed to be "desensitised" against this problem. The same approach was applied to her light sensitivity, so the nurses were told not to allow her to use eye shades. She was being nursed on the open ward close to the main ward door which kept opening and shutting. Every time it shut, the noise caused a convulsive myoclonic jerk which affected her whole body.

The diagnosis of Pervasive Refusal Syndrome was refuted by Dr Bryan Lask, the Child Psychiatrist who had first described the syndrome. He pointed out that as she was accepting tube feeding, she could hardly be said to have *pervasive* refusal. The court order was dropped and she was transferred to a private nursing home. Here, she received Tender Loving Care (TLC), and made a full spontaneous recovery.

16.2. Case B

A 16-year-old girl already had Severe ME when a new paediatrician took over her care. She (the paediatrician) could not believe that anyone could be so light sensitive as to have to lie in a darkened room and still wear eye shades. She applied to the court to get the girl admitted to her hospital ward for active physiotherapy, and assured the judge that with this treatment she would be back at school full time within 6 months. The judge granted the order and the girl was admitted to hospital. She received active physiotherapy for three months and her condition deteriorated even further. Before every session, she told the physiotherapist that she was in breach of her professional ethical guidelines.

The judge eventually lifted the court order and the girl was allowed home, in a significantly worse state than when she was admitted. Twenty years later, she remains severely affected, still being bedridden and needing tube feeding.

This case shows a combination of disbelief in the reality of Severe ME combined with a false belief in the efficacy of Graded Exercise Therapy.

16.3. Case C

This 29-year-old young woman had had Severe ME since childhood. For all this time, she had received excellent empathetic care from her family doctor and her physician. She was in severe pain requiring opiates, and was nursed on a ripple bed. She had a urinary catheter in place and received nutrition via a nasojejunal tube.

Despite the severity of her condition, with fierce intelligence and determination, she managed to write a book for doctors and families on how Severe ME should be managed [9].

I remember meeting her around this time and thinking that it only needed one extra stressor to tip her over the brink and cause her death.

Sadly, a few months later, she developed renal colic, which necessitated her admission to hospital. Her own physician was not allowed to manage her (for some reason of internal hospital politics). The physician assigned to her care freely admitted he knew nothing about ME/CFS, and although she was nursed in a side room, she was still subjected to sensory overload. The physician used to come to her bedside with his entourage and engage in long arguments with her. She died of her ME in hospital a few weeks later.

This case demonstrates that it is not only emotional upset that can result from maltreatment, but very real medical harm to an extent that can be life threatening.

17. Take Home Messages

- Severe ME constitutes a major challenge for both patient and doctor.
- Mismanagement in the form of "activation regimes" can result in permanent harm or even death of the patient.
- The patient deserves the total commitment of one doctor, who is willing to visit at home on a regular basis.
- Referral to a psychiatrist who does not believe in ME/CFS can be harmful.
- The patient should be protected from sensory overload.
- The doctor should resist the temptation to overinvestigate, or involve too many other professionals.
- Nursing at home is usually far preferable to admission to a busy general hospital.
- Tube feeding is indicated when the patient has problems with eating and drinking.
- Urinary catherization may be helpful in reducing the stress of having to micturate.
- Symptomatic treatment for pain and sleep problems is worthwhile.
- Full recovery is possible.
- The role of immunoglobulin deserves further study [10].
- There is a need to improve both undergraduate and postgraduate medical training in this area, and to provide greater resources for the patient population affected.

Funding: This research received no external funding.

Acknowledgments: The author thanks the following people for discussion and suggestions for this paper: Greg and Linda Crowhurst, Helen Brownlie and Helen Baxter (25% Group); Nina Muirhead, Paul Worthley, and Tony Crouch (TYMES Trust).

References

1. Carruthers, B.M.; Jain, A.K.; De Meirleir, K.L.; Peterson, D.L.; Klimas, N.G.; Lerner, A.M.; Bested, A.C.; Flor-Henry, P.; Joshi, P.; Powles, A.C.P.; et al. Myalgic Encephalomyelitis/Chronic Fatigue Syndrome; Clinical Working Case Definitions, Diagnostic and Treatment Protocols. *J. Chronic Fatigue Syndr.* **2003**, *11*, 7–115. [CrossRef]
2. Speight, N. Myalgic Encephalomyelitis in the young—Time to repent. *Acta Paediatr.* **2020**, *109*, 862. [CrossRef] [PubMed]

3.   Chronic Fatigue Syndrome/Myalgic Encephalomyelitis (or Encephalopathy): Diagnosis and Management. Available online: www.nice.org.uk/guidance/cg53 (accessed on 13 July 2020).

4.   Nacul, L.C.; Lacerda, E.M.; Campion, P.; Pheby, D.; de L Drachler, M.; Leite, J.C.; Poland, F.; Howe, A.; Fayyaz, S.; Molokhia, M. The functional status and wellbeing of people with myalgic encephalomyelitis/chronic fatigue syndrome and their carers. *BMC Public Health* **2011**, *11*, 402. [CrossRef] [PubMed]

5.   Weinstock, L.B.; Pace, L.A.; Rezaie, A.; Afrin, L.B.; Molderings, G.J. Mast Cell Activation Syndrome: A Primer for the Gastroenterologist. *Dig. Dis. Sci.* **2020**. [CrossRef] [PubMed]

6.   Mak, G.Z.; Speaker, C.; Anderson, K.; Stiles-Shields, C.; Lorenz, J.; Drossos, T.; Liu, D.C.; Skelly, C.L. Median arcuate ligament syndrome in the pediatric population. *J. Pediatr. Surg.* **2013**, *48*, 2261–2270. [CrossRef] [PubMed]

7.   Nicolson, G.L.; Nasralla, M.; Haier, J.; Nicolson, N.L. Diagnosis and treatment of Chronic Mycoplasmal Infections in Fibromyalgia and Chronic Fatigue Syndromes: Relation to Gulf War Illness. *Biomed. Ther.* **1998**, *16*, 216–271.

8.   Rowe, P.C.; Underhill, R.A.; Friedman, K.J.; Gurwitt, A.; Medow, M.S.; Schwartz, M.S.; Speight, N.; Stewart, J.M.; Vallings, R.; Rowe, K.S. Myalgic Encephalomyelitis/Chronic Fatigue Syndrome, Diagnosis and management in young people, A Primer. *Front. Paediatr.* 2017. [CrossRef]

9.   Collingridge, E.R. *Severe ME/CFS—A Guide to Living*; Action for ME: Bristol, UK, 2010.

10. Rowe, K.S. Double blind randomised control trial to assess the efficacy of intravenous gammaglobulin in the management of Chronic Fatigue Syndrome in adolescents. *J. Psychiatr. Res.* **1997**, *31*, 133–142. [CrossRef]

 *healthcare*

Commentary

Myalgic Encephalomyelitis/Chronic Fatigue Syndrome: When Suffering Is Multiplied

Anthony L. Komaroff

Division of General Medicine, Department of Medicine, Brigham & Women's Hospital, Harvard Medical School, 1620 Tremont, Boston, MA 02120, USA; komaroff@hms.harvard.edu

Abstract: Myalgic encephalomyelitis/chronic fatigue syndrome (ME/CFS) is an illness defined predominantly by symptoms. Routine laboratory test results often are normal, raising the question of whether there are any underlying objective abnormalities. In the past 20 years, however, new research technologies have uncovered a series of biological abnormalities in people with ME/CFS. Unfortunately, many physicians remain unaware of this, and some tell patients that "there is nothing wrong" with them. This skepticism delegitimizes, and thereby multiplies, the patients' suffering.

Keywords: myalgic encephalomyelitis/chronic fatigue syndrome; etiology; diagnostic testing

Citation: Komaroff, A.L. Myalgic Encephalomyelitis/Chronic Fatigue Syndrome: When Suffering Is Multiplied. *Healthcare* **2021**, *9*, 919. https://doi.org/10.3390/healthcare9070919

Academic Editors: Kenneth J. Friedman, Lucinda Bateman and Kenny Leo De Meirleir

Received: 15 June 2021
Accepted: 17 July 2021
Published: 20 July 2021

Publisher's Note: MDPI stays neutral with regard to jurisdictional claims in published maps and institutional affiliations.

The symptoms caused by any illness should be suffering enough. Yet, with some illnesses, the suffering often is multiplied by skepticism about the illness. That is the case with myalgic encephalomyelitis/chronic fatigue syndrome (ME/CFS).

In an article in *Healthcare*, Whitney Dafoe—who has been diagnosed with ME/CFS—describes his experience with an extremely severe form of the illness [1]. He describes the physical and mental crashes and the extreme sensitivity to any kind of sensory input. He also describes the isolation, the loss, the complete and sudden disruption in the life of a young adult, a life that was on the runway and cleared for takeoff.

Why have some physicians and biomedical scientists been skeptical about the "legitimacy" of ME/CFS? Primarily, it is because the illness has been defined largely by symptoms. Since it is difficult for symptoms to be confirmed objectively, physicians have sought objective laboratory evidence of underlying biological abnormalities—abnormalities that an individual cannot simply imagine, abnormalities that could explain the symptoms. Initially, that proved hard.

When interest in this condition was renewed in the mid-1980s, there was little such evidence: the "standard" laboratory tests ordered by physicians—typically, tests of red and white blood cells, a battery of about 20 chemistry tests, and a urinalysis—produced normal results. That posed a problem for the physicians. Their patients were suffering, and it was their job to make a diagnosis and prescribe a treatment, but the standard test results were normal: the physicians did not have a diagnosis.

At this point, the physicians had several options. First, they could have entertained some new hypotheses about what was causing the symptoms, and ordered new types of tests. Second, they could have said: "I just can't figure out what's making you sick, and don't know how to help you." Third, although they could not determine the diagnosis, they could have prescribed a treatment that might improve the symptoms even if they were not really sure what had caused the symptoms. That happens every day in the practice of medicine. For example, there is no diagnostic test for migraine headaches, yet doctors make that diagnosis every day based just on a combination of symptoms, and do not dispute the validity of the illness because there is no diagnostic test.

Unfortunately, the normal results of "standard" laboratory tests led some physicians to choose a fourth option: to conclude that there were no underlying biological abnormalities causing the symptoms. Even though the physicians knew that the "standard" tests they had ordered represented only a tiny fraction of all of the tests available to them, the normal results of that tiny fraction were enough for them to render a judgment. It was a harsh judgment: "There is nothing wrong with you."

For these physicians, it was an efficient solution: it transformed what had been their problem—the lack of a diagnosis they were expected to make—into their patient's problem. When the patients were told, implicitly or explicitly, that their symptoms were imaginary, it multiplied the suffering.

And then these skeptical physicians also conveyed their judgments, implicitly or explicitly, to the patients' families, friends and employers. The doctors' judgment led these people—the people who were most important in the patients' lives—to wonder whether the patients' suffering was legitimate. That further multiplied the suffering.

There was always an obvious alternative conclusion to the judgment that "there is nothing wrong with you": the standard laboratory tests might simply have been measuring the wrong things. Yet that alternative conclusion was ignored.

Since the resurgence of interest in ME/CFS 35 years ago, whole new technologies have become available that allow physicians and biomedical scientists to study human biology in ways that previously were not possible, e.g., noninvasive techniques for imaging the anatomy and physiology of the brain; polymerase chain reaction diagnostics; rapid nucleic acid sequencing; techniques for measuring gene expression; the ability to measure simultaneously thousands of molecules in a single sample (the "omics" revolution); metagenomic studies of the microbiome, and recognition of the impact of the microbiome on human health. In fact, these and other technologies have revealed things that the standard laboratory tests cannot—abnormalities that previously were invisible to doctors.

In 2015 the U.S. National Academy of Medicine (NAM) reviewed a literature of over 9000 publications on ME/CFS, and concluded that it was a "serious, chronic, complex systemic disease" [2]. The NAM estimated that in the U.S., alone, 836,000 to 2.5 million people suffer from ME/CFS [2], making it somewhat more common than multiple sclerosis [3].

A large literature now describes multiple underlying biological abnormalities in people with ME/CFS. Some of the evidence comes from tests that have been available for decades but are not part of the "standard" laboratory test battery [4], and some evidence comes from the new technologies mentioned above. Unfortunately, many physicians are unaware of the new discoveries about ME/CFS.

The abnormalities all converge on and can affect the brain, and fall into five categories. First, there are anatomic, physiologic and electrical abnormalities in the brain [5]. Second, various elements of the immune system are chronically activated and in some people those elements are exhausted—perhaps secondary to years of chronic activation [5]. This includes chronic activation of the brain's innate immune system—neuroinflammation [6]. It also includes evidence of autoimmunity, including autoantibodies directed at targets in the central and autonomic nervous system [7]. Third, there also is evidence of impaired energy metabolism: the person with ME/CFS feels he or she lacks "energy" because his or her cells have a reduced ability to generate energy molecules (adenosine triphosphate, or ATP) [8]. Along with the abnormalities in energy metabolism, there is associated oxidative stress, or redox imbalance [8]. Fourth, the autonomic nervous system is dysregulated, one consequence of which appears to be impaired blood flow to the brain [9]. Fifth, there are characteristic abnormalities of the gut microbiome [10], with increased numbers of pro-inflammatory bacterial species and decreased numbers of butyrate-producing anti-inflammatory species.

What remains unclear are the mechanistic details as to how the abnormalities in each of these five categories affect each other, and whether one of them is the initial and primary abnormality [5,8]. In this next decade, the growing community of global investigators who are studying ME/CFS should place a high priority on refining our understanding of each of these categories of abnormality, and an even higher priority on understanding how they are connected. This is essential for developing good diagnostic tests, and effective treatments.

Whitney Dafoe ends the description of his suffering by emphasizing the silver lining around the cloud that he has lived with for nearly 20 years. He says he has learned a great deal about what is important in life, and that "ME/CFS is the greatest teacher I've ever had."

I would like to think that ME/CFS will also prove to be a great teacher to the growing community of physicians and biomedical investigators involved in caring for and studying the illness. In particular, I speculate that the connections between the various abnormalities involving the central and autonomic nervous system, immune system, energy metabolism, redox imbalance, and the human microbiome that have been noted in ME/CFS will prove to be central also to the pathophysiology of many other diseases.

In particular, the COVID-19 pandemic appears to be producing millions of new cases of an ME/CFS-like condition [11], and NIH has allocated more than $1 billion to study this and other post-COVID chronic illnesses. Hopefully, this investment will produce more answers.

Of the personal lessons that I, as a physician, have learned from ME/CFS, perhaps the most important is that, if patients tell you they are suffering, your default assumption should be to believe them—even if you cannot find an answer with the diagnostic technology you first deploy. Above all, never succumb to the temptation to dismiss the patient's symptoms because you cannot explain them. That may ease your anxiety, but it only multiplies the patient's suffering.

Funding: The author received no support for writing this publication.

Institutional Review Board Statement: Not applicable.

Informed Consent Statement: Not applicable.

Data Availability Statement: Not applicable.

Conflicts of Interest: There was no Conflict of Interest for reported by the author.

References

1. Dafoe, W. Extremely severe ME/CFS-A personal account. *Healthcare* **2021**, *9*, 504. [CrossRef] [PubMed]
2. Institute of Medicine. *Beyond Myalgic Encephalomyelitis/Chronic Fatigue Syndrome: Redefining an Illness*; The National Academies Press: Washington, DC, USA, 2015.
3. Wallin, M.T.; Culpepper, W.J.; Campbell, J.D.; Nelson, L.M.; Langer-Gould, A.; Marrie, R.A.; Cutter, G.R.; Kaye, W.E.; Wagner, L.; Tremlett, H.; et al. The prevalence of MS in the United States. *Neurology* **2019**, *92*, e1029–e1040. [CrossRef] [PubMed]
4. Bates, D.W.; Buchwald, D.; Lee, J.; Kith, P.; Doolittle, T.; Rutherford, C.; Churchill, W.H.; Schur, P.; Wener, M.; Wybenga, D.; et al. Clinical laboratory test findings in patients with chronic fatigue syndrome. *Arch. Intern. Med.* **1995**, *155*, 97–103. [CrossRef] [PubMed]
5. Komaroff, A.L.; Lipkin, W.I. Insights from myalgic encephalomyelitis/chronic fatigue syndrome may help unravel the pathogenesis of post-acute COVID-19 syndrome. *Trends Mol. Med.* **2021**, in press. [CrossRef] [PubMed]
6. Nakatomi, Y.; Mizuno, K.; Ishii, A.; Wada, Y.; Tanaka, M.; Tazawa, S.; Onoe, K.; Fukuda, S.; Kawabe, J.; Takahashi, K.; et al. Neuroinflammation in patients with chronic fatigue syndrome/myalgic encephalomyelitis: An ^{11}C-(R)-PK11195 PET study. *J. Nucl. Med.* **2014**, *55*, 945–950. [CrossRef] [PubMed]
7. Sotzny, F.; Blanco, J.; Capelli, E.; Castro-Marrero, J.; Steiner, S.; Murovska, M.; Scheibenbogen, C. Myalgic encephalomyelitis/chronic fatigue syndrome—Evidence for an autoimmune disease. *Autoimmun. Rev.* **2018**, *17*, 601–609. [CrossRef] [PubMed]
8. Paul, B.D.; Lemle, M.D.; Komaroff, A.L.; Snyder, S.H. Redox imbalance links COVID-19 and myalgic encephalomyelitis/chronic fatigue syndrome. *Proc. Natl. Acad. Sci. USA* **2021**, *118*, e2024358118.
9. Van Campen, C.L.M.C.; Rowe, P.C.; Visser, F.C. Cerebral blood flow is reduced in severe myalgic encephalomyelitis/chronic fatigue syndrome patients during mild orthostatic stress testing: An exploratory study at 20 degrees of head-up tilt testing. *Healthcare* **2020**, *8*, 169. [CrossRef] [PubMed]
10. Giloteaux, L.; Goodrich, J.K.; Walters, W.A.; Levine, S.M.; Ley, R.E.; Hanson, M.R. Reduced diversity and altered composition of the gut microbiome in individuals with myalgic encephalomyelitis/chronic fatigue syndrome. *Microbiome* **2016**, *4*, 30. [CrossRef] [PubMed]
11. Komaroff, A.L.; Bateman, L. Will COVID-19 lead to myalgic encephalomyelitis/chronic fatigue syndrome? *Front. Med.* **2021**, *7*, 606824. [CrossRef] [PubMed]

Article

A Comprehensive Examination of Severely Ill ME/CFS Patients

Chia-Jung Chang [1,†], Li-Yuan Hung [2,†], Andreas M. Kogelnik [3], David Kaufman [1], Raeka S. Aiyar [1], Angela M. Chu [1], Julie Wilhelmy [1], Peng Li [2], Linda Tannenbaum [4], Wenzhong Xiao [1,2,*] and Ronald W. Davis [1,*]

[1] ME/CFS Collaborative Research Center at Stanford, Stanford Genome Technology Center, School of Medicine, Stanford University, Palo Alto, CA 94305, USA; chiajung@stanford.edu (C.-J.C.); david@centerforcomplexdiseases.com (D.K.); raeka.aiyar@gmail.com (R.S.A.); amchu@stanford.edu (A.M.C.); wilhelmy@stanford.edu (J.W.)
[2] ME/CFS Collaborative Research Center at Harvard, Massachusetts General Hospital, Boston, MA 02114, USA; LHUNG1@mgh.harvard.edu (L.-Y.H.); pli6@mgh.harvard.edu (P.L.)
[3] Basis Diagnostics, Newark, CA 94560, USA; andy@flexis.net
[4] Open Medicine Foundation, Agoura Hills, CA 91301, USA; ltannenbaum@omf.ngo
*   Correspondence: wenzhong.xiao@mgh.harvard.edu (W.X.); ron.davis@stanford.edu (R.W.D.)
†   These authors have contributed equally to this work and share first authorship.

Abstract: One in four myalgic encephalomyelitis/chronic fatigue syndrome (ME/CFS) patients are estimated to be severely affected by the disease, and these house-bound or bedbound patients are currently understudied. Here, we report a comprehensive examination of the symptoms and clinical laboratory tests of a cohort of severely ill patients and healthy controls. The greatly reduced quality of life of the patients was negatively correlated with clinical depression. The most troublesome symptoms included fatigue (85%), pain (65%), cognitive impairment (50%), orthostatic intolerance (45%), sleep disturbance (35%), post-exertional malaise (30%), and neurosensory disturbance (30%). Sleep profiles and cognitive tests revealed distinctive impairments. Lower morning cortisol level and alterations in its diurnal rhythm were observed in the patients, and antibody and antigen measurements showed no evidence for acute infections by common viral or bacterial pathogens. These results highlight the urgent need of developing molecular diagnostic tests for ME/CFS. In addition, there was a striking similarity in symptoms between long COVID and ME/CFS, suggesting that studies on the mechanism and treatment of ME/CFS may help prevent and treat long COVID and vice versa.

Keywords: severe ME/CFS; quality of life; clinical symptoms; sleep; cognitive tests; laboratory tests; viral infection; antibody and antigen; long COVID; post-acute sequelae SARS-CoV-2 infection (PASC)

Citation: Chang, C.-J.; Hung, L.-Y.; Kogelnik, A.M.; Kaufman, D.; Aiyar, R.S.; Chu, A.M.; Wilhelmy, J.; Li, P.; Tannenbaum, L.; Xiao, W.; et al. A Comprehensive Examination of Severely Ill ME/CFS Patients. *Healthcare* **2021**, *9*, 1290. https://doi.org/10.3390/healthcare9101290

Academic Editors: Kenneth J. Friedman, Lucinda Bateman and Kenny Leo De Meirleir

Received: 1 September 2021
Accepted: 22 September 2021
Published: 29 September 2021

Publisher's Note: MDPI stays neutral with regard to jurisdictional claims in published maps and institutional affiliations.

1. Introduction

Myalgic encephalomyelitis/chronic fatigue syndrome (ME/CFS) is a chronic complex disease characterized by unrelenting fatigue, post-exertional malaise, sleep problems, cognitive impairment, and orthostatic intolerance [1]. This debilitating illness is known to affect between 836,000 and 2.5 million people in the United States alone [1–3], and the majority of the patients remain undiagnosed [1,4]. Patients often report symptoms started with viral infection [1,2,5]. Patients of ME/CFS have been found to be more functionally impaired than those with major diseases such as cancers, heart disease, and rheumatoid arthritis [6], and their prognosis remains poor [7,8]. Despite the severity of the clinical symptoms, the etiology and pathophysiology of the disease remain unclear. To date, there is neither a validated biomarker for diagnosis nor an FDA-approved drug available for treatment.

An estimated 25% of patients with ME/CFS are unfortunately severely affected and physically confined to their homes or beds [1,9,10]. These severely affected patients suffer from extreme daily fatigue, grievous impairments, and other debilitating symptoms. They often require in-home assistance and support adjusted explicitly to their needs [11,12].

However, severely affected patients are rarely studied [10,11], partially due to the difficulties accessing clinical care facilities. The personal account of an extremely severe patient is presented in this Special Issue [13]. To reduce the significant gap between the needs of severe patients and the healthcare they receive, there is an urgent need to better characterize these patients' clinical conditions and discover the underlying biological abnormalities causing the symptoms [14]. In addition, as the condition worsens, the probability that biomarkers can be identified for the disease increases by studying severely ill patients.

Here we conducted a Severely Ill Patient Study (SIPS), which included a comprehensive examination of clinical symptoms and clinical lab tests of a cohort of severely ill patients and controls. First, questionnaires were administered to evaluate the patients' quality of life, health status, and symptoms. Second, the patients' daily activity, sleep profile, and cognitive capacity were monitored and examined to assess their symptoms objectively. Third, clinical laboratory testing and antigen & antibody tests against viral and bacterial pathogens were obtained. In addition, multiple omics studies are being conducted on the biological samples of these patients to identify molecular signatures of severe ME/CFS, and the results will be reported elsewhere. We have made the data and results available through a web-based data portal for the research community at https://endmecfs.stanford.edu.

2. Materials and Methods

2.1. Participants

Patients were identified for the study from an existing pool of homebound, and mostly bedbound, ME/CFS managed and diagnosed patients at the clinic sites of the investigators of the study and from those referred to the investigators to be eligible to participate in the study. ME/CFS clinicians at the study sites identified initially the potential subjects, who were most likely to be involved in this study, through the screening of medical records of these patients. Next, patients (age 18–70) were assessed for ME/CFS criteria online or by phone. They were consented if they met the International Consensus Criteria (ICC) for ME/CFS [15], were homebound (i.e., spending more than 14 h per day sedentary and in a reclined position as reported by patient or caregiver), and received a low score in physical status (i.e., SF-36 [16] physical functioning score and Karnofsky Performance Status Index [17] were both less than 70). They also must not fit the exclusion criteria. Consented patients were then provided with a FitBit device to confirm that they met the sedentary requirement. A complete blood count within the past 3–6 months was requested to verify anemia was not present (hematocrit > 34%). The detailed inclusion and exclusion criteria of the patients are listed in Section 2.1.1.

Healthy controls were evaluated for inclusion in the study based on meeting all inclusion criteria and not having any exclusion criteria (Section 2.1.2). These controls must be age 18 to 70, not carry a diagnosis of ME/CFS as defined by the ICC or active illness (acute or chronic), daily sedentary time $\leq$ 14 h, SF-36 physical functioning score $\geq$ 70, and without the conditions in the exclusion criteria.

All patient and control subjects were consented. Limited by available funding, 20 severely ill ME/CFS patients and 10 healthy controls were included in this study.

2.1.1. Severely Ill ME/CFS Patients Inclusion and Exclusion Criteria

Inclusion Criteria

1. Age 18–70, inclusive;
2. Must carry a diagnosis of ME/CFS as defined by the ICC criteria;
3. Subjects must be homebound and spend >14 h per day sedentary and in a reclined position (measured by FitBit and patient/family report);
4. SF-36 physical functioning score < 70; and
5. Be able to provide informed consent.

Exclusion Criteria

1. Patients, age < 18 years or > 70 years;
2. Women who are pregnant;
3. Unable to understand informed consent; or
4. Patients with known HCT < 34 mg/dL.

2.1.2. Healthy Control Inclusion and Exclusion Criteria

Inclusion Criteria

1. Age 18–70, inclusive;
2. Must not carry a diagnosis of ME/CFS as defined by the ICC criteria or active illness (acute or chronic);
3. Must be sedentary $\leq$ 14 h; and
4. SF-36 physical functioning score $\geq$ 70.

Exclusion Criteria

1. Patients, age < 18 years or > 70 years;
2. Women who are pregnant;
3. Unable to understand informed consent; or
4. Patients with a known HCT < 34 mg/dL.

2.2. Data Collection from Questionnaires

Questionnaires on Health Status and Quality of Life. The perceived health status and quality of life of the patients and the controls were evaluated by several sets of questionnaires, i.e., SF-36, Karnofsky Performance Status [17], Patient-Reported Outcomes Measurement Information System (PROMIS) instruments [18,19] (including Fatigue, Pain Behavior, Pain Interference, Sleep Disturbance, and Sleep-Related Impairment), Pittsburgh Sleep Quality Index (PSQI) [20], and a questionnaire on Restless Leg Syndrome (RLS) [21].

Evaluation of Common Symptoms in Patients. Patients were evaluated using a set of 7 symptoms-related questions which covered the common symptoms mentioned in ICC [15] and IOM [1]. The text-based answers were transformed to 79 numerical or categorical measurements, indicating if a subject had a particular impairment/symptom or quantifying the degree of the impairment/symptom. These were then grouped into 12 symptomatic categories, which represented the 5 core symptoms of ME/CFS in the 2015 IOM diagnostic criteria [1] (i.e., fatigue, post-exertional malaise, sleep disturbance, cognitive impairment, and orthostatic intolerance) and 7 additional common accompanying symptoms mentioned in the IOM or ICC criteria (i.e., pain, neurosensory disturbance, flu-like symptoms and/or susceptibility to viral infections, gastrointestinal tract impairment, loss of thermostatic stability and/or intolerance of extremes of temperature, respiratory impairments, and genitourinary impairments). In addition, the top 3 most troublesome symptoms of each patient were recorded.

2.3. Data Collection of Patient Activity, Sleep Monitoring, and Cognitive Tests

Activity Monitoring. Patients were provided with a Charge HR (FitBit, Inc., San Francisco, CA, USA) for two weeks. This device documented patient activity and continual heart rate to confirm that patients met the sedentary requirement. The measurements, including Active Minutes, Sleep Duration, Sleep Score, Sleep Time, Calories Burned, Distance, Floors, Steps and Resting Heart Rate, were retrieved with the R package fitbitScraper and summarized to the daily average.

Sleep Monitoring. Patients underwent an overnight sleep profiler study. The non-invasive sleep monitor was the Sleep Profiler [22] from Advanced Brain Monitoring (Carlsbad, CA, USA) and consisted of a 3-lead EEG, snore (audio) detector, activity/motion detector, and an eye movement detector. The overnight EEG and other signals were reviewed by the study staff. Thirty-five measurements on the sleep architecture & continuity (e.g., total sleep time, sleep efficiency and sleep latency) and cardio-respiratory signals (e.g., pulse rate and snoring) were analyzed and compared with the established normative ranges [23]. Sleep abnormalities were then identified and compared with sleep EEG biomarkers that were associated with chronic health conditions or neurological diseases [24].

Cognitive Tests and Extended EEG. WebNeuro Tests (Brain Resource Group, San Francisco, CA, USA) [25] were utilized to evaluate the cognitive performance of the patients and the controls. Four types of cognitive abilities (i.e., attention, maze, memory, and identifying emotions) were evaluated. The results were scored against a cohort of normative subjects in the Brain Resource International Database (BRID) [26]. The normalized scores (Z-scores) and the corresponding implications (e.g., Z-score £-2 implies clinical significance) were reported in WebNeuro Report (Version: WebNeuro Short 3.1.5). The clinical/research grade EEG device was a 24 channel Stat X24 also from Advanced Brain Monitoring. Twenty electrodes on the head were monitored in this study. Extended EEG monitoring was combined with the cognitive test for the patients and controls. Before or after the test, 15 min of EEG was monitored as the standard control. During the four tests: attention, maze, memory and emotion, EEG was monitored simultaneously.

2.4. Clinical Lab Tests

For clinical tests, a maximum of 160 ml of blood was collected from each subject for clinical tests. Blood samples were collected from all ME/CFS subjects when a research team visited the subject's home and performed the physical exam. Samples were collected from all healthy control subjects during their visit to the clinic. Urine over 24-h and saliva specimens were also collected from the subjects. To reduce the variability of the test results across the study population, all samples were collected on the same day during the patient's appointment. The samples were shipped to routine and specialty clinical labs. All clinical laboratories are CLIA approved.

The tests were chosen based on results from previous studies on ME/CFS (Table S1). These included complete blood count with differential, comprehensive metabolic panel, standard lipid panel, acylcarnitine profile, urinalysis of organic acids, hormones (including cortisol, thyroid-stimulating hormone/thyroid hormones (TSH/T3/T4), follicle-stimulating hormone and luteinizing hormone (FSH/LH), testosterone, estrogen, and arginine vasopressin), vitamins (B7/biotin, B12/folate, D, methylmalonic acid), selected chemistry analytes and disease biomarkers, lymphocyte subsets, and natural killer cell function. Salivary cortisol monitoring was tested for each subject at four time points of the day: 30 min after morning awakening, noon, afternoon, and night. All these tests were performed by Quest Diagnostics (Secaucus, NJ, USA).

2.5. Tests of Antibodies and Antigens against Pathogens

Also performed were tests on antibodies and antigens against viral and bacterial pathogens (Table S2). The tests of IgG and IgM antibodies against viruses were conducted at Quest Diagnostics, which included Herpes simplex virus 1 and 2 (HSV1/2, HHV1/2), Epstein-Barr Virus (EBV, HHV4), *Cytomegalovirus* (CMV, HHV5), Human Herpesvirus 6 and 7 (HHV6/7), and Primate *Erythroparvovirus* 1 (Parvovirus B19, B19).

Lyme disease antibody tests were performed at Quest Diagnostics, which included IgG and IgM antibody tests and the Western blot [27,28]. For the Western blot, *Borrelia burgdorferi* IgM was considered positive if two of the three bands were present; IgG was considered positive if five of the 10 bands were present [27]. In addition, Ceres Nanotrap antigen tests (Ceres Nanosciences, Manassas, VA, USA) were performed to detect *Borrelia* Outer surface protein A (OspA) antigen [29]. *Mycoplasma pneumoniae* IgG and IgM antibodies were tested by Quest Diagnostics. *Bartonella* tests were performed at Galaxy Diagnostics (Research Triangle Park, NC, USA), which included a PCR test of *Bartonella* species of the whole blood, serum and blood cultures at 8 days, 14 days, and 21 days. In addition, immunofluorescence assay (IFA) was used for the IgG of *Bartonella henselae* and *Bartonella quintana* and results with titers of $\geq$1:256 were considered to be positive for the analysis [30].

The biological samples collected were also archived for further omics studies of genes, proteins, metabolites, and microbes present in severely ill ME/CFS patients.

2.6. Data Analysis

To compare the quality of life and the patient-reported health status between SIPS patients and controls, Wilcoxon signed-rank test was used. To visualize the closeness/distance of SF-36 among SIPS samples, general CFS, and other related medical conditions, tSNE was utilized (implemented in the R package Rtsne) to project the SF-36 scores to two dimensions.

To quantify the severity and frequency of the 12 symptomatic categories in SIPS, we operationally defined a burden score for each category that could summarize the 95 symptomatic measurements from all questionnaires. We first unified the ranges and directions of the measurements. After the standardization, all the measurements ranged from 0 to 1, and the higher value indicated the worsening of the symptom. Specifically, for quantitative phenotypes, we used the formula $x-min(x)/max(x)-min(x)$ to re-scale the values of each measurement and reversed its direction if the average of healthy controls was larger than the average of SIPS patients. For binary phenotypes where 1 indicated having the symptoms, the values were weighted by the frequencies of the symptoms in the patients. We calculated the burden score of each symptomatic category by averaging the standardized measurements assigned to the category. The burden scores were visualized with the R package heatmaply, and the individuals were hierarchically clustered by their Euclidean distances.

For Fitbit measurements and cognitive test STEN (Standard Tens) scores, Student's *t*-test was performed to test if there was a significant difference between SIPS patients and controls. One-sided Fisher's exact test was performed to test if there is a significantly higher number of patients with a clinically significant low STEN score for the four types of cognitive abilities.

For each of the clinical tests, where the diverse raw values hardly followed a normal distribution, we performed Box-Cox transformation to fit the values from health controls into a normal distribution. A bootstrap *t*-test was also performed on the clinical tests to generate the *p*-values, and FDRs were also calculated.

The prevalence of the symptoms remaining after six months in long COVID reported in a recent study [31] was retrieved from the Appendix and Figure 11a of the article and compared with the correspondent symptoms in the SIPS patients.

All the analyses and visualization were performed with the R program.

3. Results

Results include Patient-reported health status and symptoms (Section 3.1), Activity, sleep monitoring, and cognitive tests (Section 3.2), Clinical laboratory testing (Section 3.3), and Antigen and antibody tests against viral and bacterial pathogens (Section 3.4). All results described below were based on data from the entire cohort unless otherwise indicated. The data and results are available through a web-based data portal at https://endmecfs.stanford.edu.

3.1. Patient-Reported Health Status and Symptoms of the Severely Ill

3.1.1. Demographics and Quality of Life of the Patients

The demographics of the subjects of the study are shown in Table 1. In the SIPS patients, the duration of the illness ranged from 2.4 years to 50 years, with a mean of 14.5 years. While all the patients were homebound, half of them required considerable assistance and frequent medical care, and 35% were disabled and needed special care and assistance, as indicated by the Karnofsky scale.

Table 1. Demographics and quality of life of severe myalgic encephalomyelitis/chronic fatigue syndrome (ME/CFS) patients and healthy controls.

	Patients (N = 20)	Controls (N = 10)	*p*-Value [1]
Age (years; mean ± s.d.)	47.4 ± 11.6	46.8 ± 9.2	0.552
Sex (% female)	65.0%	60.0%	0.813
BMI (kg/m^2; mean ± s.d.)	25.4 ± 6.8	22.0 ± 3.1	0.224
Duration of illness (years; mean ± s.d.)	14.5 ± 11.8	0.0 ± 0.0	
Karnofsky Performance status index (%)			<0.001
30: Severely disabled; hospital admission is indicated, although death is not imminent.	5.0%	0.0%	
40: Disabled; requires special care and assistance.	30.0%	0.0%	
50: Require considerable assistance and frequent medical care.	15.0%	0.0%	
60: Require occasional assistance, but is able to care for most personal needs.	50.0%	0.0%	
100. Normal; no complaints; no evidence of disease.	0.0%	100.0%	
Quality of life (SF-36 scores; mean ± s.d.)			
PF: Physical functioning	13.3 ± 12.8	99.0 ± 2.1	<0.001
RP: Role limitations due to physical health	1.9 ± 6.1	99.4 ± 2.0	<0.001
RE: Role limitations due to emotional problems	55.0 ± 45.9	94.2 ± 9.7	0.037
VT: Vitality/Energy/Fatigue	12.8 ± 19.3	80.0 ± 14.7	<0.001
MH: Mental health/Emotional well-being	56.0 ± 25.8	83.5 ± 16.0	0.005
SF: Social functioning	4.4 ± 12.4	92.5 ± 13.4	<0.001
BP: Body pain	33.4 ± 26.2	95.8 ± 7.6	<0.001
GH: General health	16.5 ± 7.3	83.5 ± 15.1	<0.001

[1] Wilcoxon signed-rank test.

The SF-36 results showed that SIPS patients had significantly lower scores in comparison with healthy controls (Table 1 and Figure 1a). In particular, scores on physical functioning (PF), role limitations due to physical health (RP), general health (GH), vitality/energy/fatigue (VT), and social functioning (SF) were extremely low, with each less than 20. As shown in Figure 1a, comparing to the scores of the general patients of ME/CFS [32], each of these five scales was further lowered significantly in the SIPS patients. This is also consistent with other published studies on the quality of life of ME/CFS. For example, in the phase 3 trial of rituximab (RituxME), the average PF score was >30 for the patients (35.2 ± 21.9 and 32.5 ± 19.1 for the treated and placebo groups, respectively) [33], while in this study, the PF score was <15 in the severely ill patients (13.3 ± 12.8). Our results suggested that severe illness had greatly reduced the quality of life of these severely ill patients, even further than the general ME/CFS patient population.

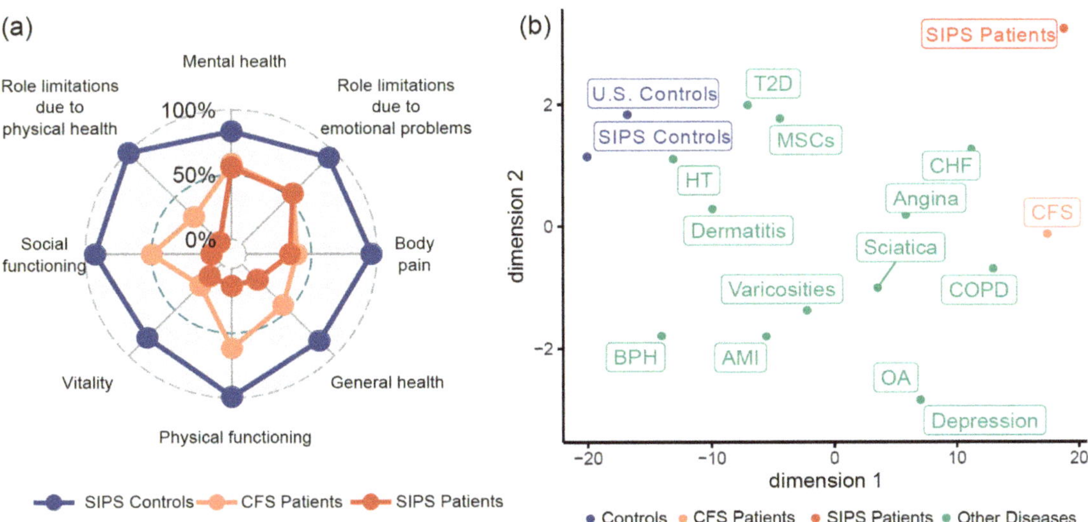

Figure 1. Comparison of the Quality of Life of Severely Ill myalgic encephalomyelitis/chronic fatigue syndrome (ME/CFS) and Other Major Diseases. (**a**) SF-36 scores of Severely Ill Patient Study (SIPS) patients, general CFS patients, and healthy controls. Compared with the general CFS patient population, scores on physical functioning (PF), role limitations due to physical health (RP), general health (GH), vitality/energy/fatigue (VT), and social functioning (SF) were significantly lower. (**b**) tSNE of SF-36 scores of SIPS, general CFS, and other medical conditions. T2D—type II diabetes, HT—hypertension, CHF—congestive heart failure, COPD—chronic obstructive pulmonary disease, MSCs—musculoskeletal complaints, BPH—benign prostatic hyperplasia, AMI—anterior myocardial infarction, and OA—Osteoarthritis. The quality-of-life scores of SIPS patients were clearly separated from that of controls, being most positively correlated with congestive heart failure (CHF) and most negatively correlated with clinical depression.

On the other hand, the role limitations due to emotional problems (RE) was much less impacted, followed by mental health/emotional well-being (MH), in SIPS patients, which appeared to be similar to general ME/CFS (Figure 1a).

We next compared the SF-36 scores of SIPS with that of other medical conditions in the USA [34]. The results showed that the SF-36 scores of SIPS were well separated from the general U.S. population as well as other medical conditions (Figure 1b). Compared to other major diseases, the severely ill ME/CFS patients had lower scores in six of the eight scales, except RE and MH (Figure S1a). In addition, among these medical conditions, the quality-of-life scores of the SIPS patients were most positively correlated with Congestive Heart Failure ($r = 0.63$) and most negatively correlated with Clinical depression ($r = -0.33$) (Figure S1b).

3.1.2. Patient-Reported Health Status

Several sets of questionnaires were administered to evaluate the health status of the patients. Five PROMIS instruments were utilized, which provided measures of physical, mental, and social well–being from the patient perspective. As shown in Table 2, comparing with the controls, the severely ill patients reported significant fatigue, sleep disturbance, sleep-related impairment, the experience of pain (pain behavior), and interference of pain on activities (pain interference).

Table 2. Comparison of patient-reported health status between severe ME/CFS patients and healthy controls.

	Patients	Controls	*p*-Value [1]
PROMIS Instruments (T-score; mean ± s.d.)			
Fatigue	75.2 ± 5.9	41.8 ± 9.6	<0.001
Sleep disturbance	64.5 ± 7.5	39.7 ± 7.4	<0.001
Sleep-related impairment	65.4 ± 7.4	37.5 ± 8.4	<0.001
Pain interference	67.0 ± 10.1	44.5 ± 4.8	0.003
Pain behavior	60.6 ± 8.9	42.4 ± 11.5	0.004
Pittsburgh Sleep Quality Index (mean ± s.d.)			
Sleep quality	2.1 ± 0.7	0.0 ± 0.0	<0.001
Sleep latency	2.1 ± 1.3	1.0 ± 0.8	0.093
Sleep duration	0.4 ± 0.8	0.3 ± 0.5	0.814
Habitual sleep efficiency	1.6 ± 1.3	0.0 ± 0.0	0.019
Sleep disturbances	1.9 ± 1.0	0.3 ± 0.5	0.009
Use of sleeping medications	2.2 ± 1.1	0.0 ± 0.0	<0.001
Daytime dysfunction	1.9 ± 1.3	0.8 ± 0.5	0.144
Global PSQI score	11.9 ± 3.4	2.3 ± 1.7	0.003
Restless Legs Syndrome (RLS; %)			
Probable RLS	23.5% (4/17)	0.0% (0/4)	

[1] Wilcoxon signed-rank test.

Similarly, the analysis of the results of the Pittsburg Sleep Quality Index (PSQI) showed significantly lower sleep quality, more sleep disturbances, and worse Global PSQI in the patients compared to the controls. In addition, 4 of the patients (20%) had probable Restless Leg Syndrome (RLS).

3.1.3. Evaluation of the Common Symptoms in the Patients

Data on specific symptoms known to be correlated with ME/CFS [1,15] were obtained using a standardized questionnaire. The results are shown in Figure S2a, indicating whether a patient or a control had a particular impairment or the degree of the impairment. Figure S2b shows a hierarchical clustering of these symptoms between the patients, where symptoms related to sleep disturbance and symptoms related to pain clustered together. These individual symptoms were then grouped into 12 symptomatic categories, which were mentioned in the IOM and ICC descriptions of ME/CFS. One of the extremely ill patients was not able to complete the questionnaire and was not included in the downstream analysis.

As shown in Figure 2a, all the patients had fatigue, sleep disturbance, and post-exertional malaise, and had either cognitive impairment (19/19 or 100%) or orthostatic intolerance (15/19 or 79%), or both. Therefore, all the patients met the IOM ME/CFS diagnosis criteria. Additional symptoms include pain, neurosensory disturbance, flu-like symptoms and/or susceptibility to viral infections, gastrointestinal tract impairments, Genitourinary impairment, and Respiratory impairment. Notably, 100% of the patients (19/19) suffer from the presence of significant pain and 89% (17/19) had sensitivity to light, noise, vibration, odor, taste, and touch (Figure S2a).

We next looked at the top 3 most troublesome symptoms of the severe patients (Figure 2b). The symptoms reported by the patients were fatigue (85%), pain (65%), cognitive impairment (50%), orthostatic intolerance (45%), sleep disturbance (35%), post-exertional malaise (30%), neurosensory disturbance (30%), GI tract impairment (30%), flu-like symptoms (15%), and loss of thermostatic stability (5%). Fatigue and post-exertional malaise were ranked most commonly as the top troublesome symptom by 50% and 20% of the patients, respectively.

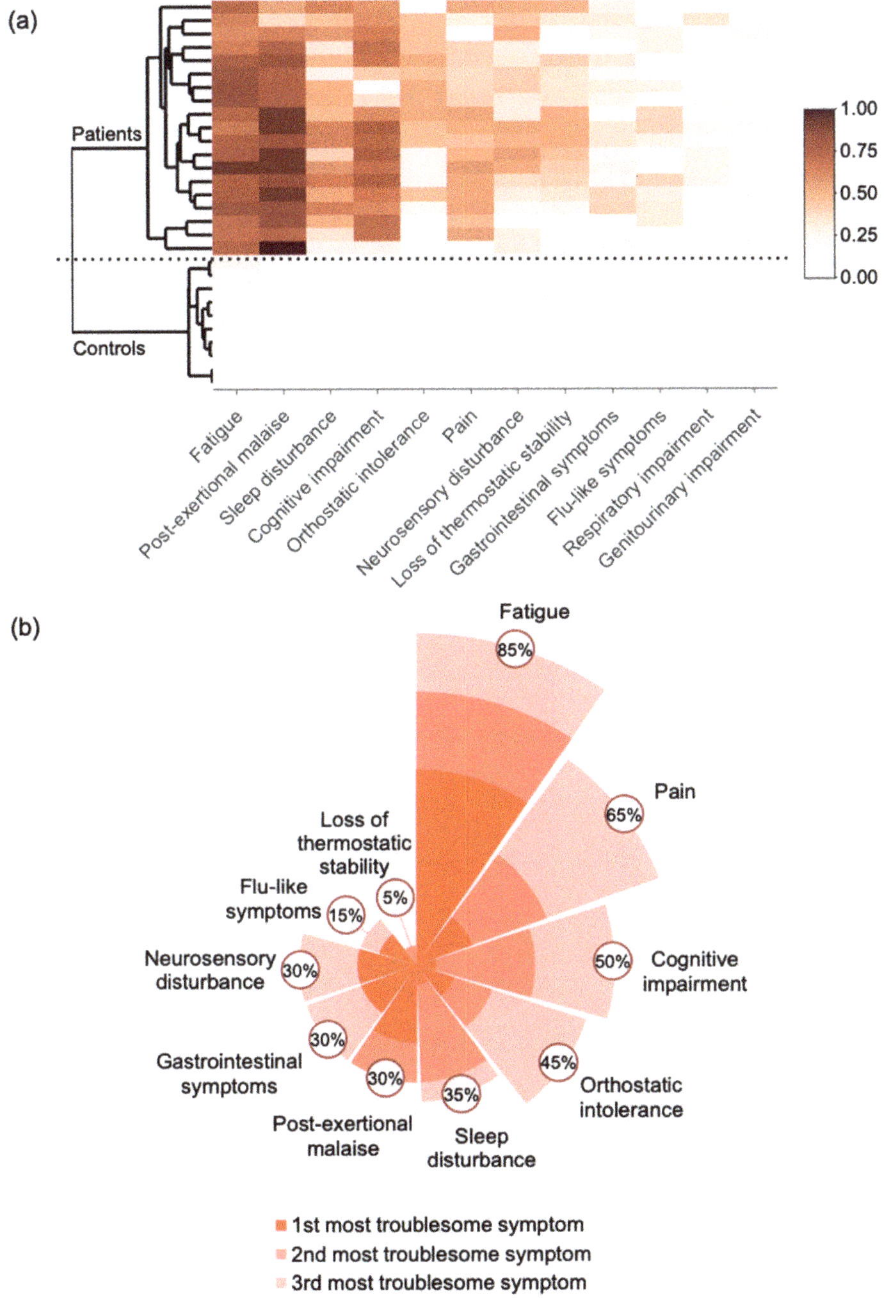

Figure 2. Common Symptoms in Severe ME/CFS Patients. (**a**) Similarity and variation of the symptoms of the SIPS patients and (**b**) the top three most troublesome symptoms of the SIPS Patients.

3.2. Activity, Sleep Monitoring, and Cognitive Tests of the Severe ME/CFS Patients

The severely ill patients in the study were homebound and spending more than 14 h per day sedentary and in a reclined position as reported by the patient or caregiver. To objectively monitor the physical activities of the severely ill patients, patients were provided with a FitBit device. The median daily steps taken by the SIPS patients was 912, which was significantly lower than that of the healthy controls as well as the reported values from previous studies of the U.S. population [35,36], and similar results were seen on the daily distance, the number of floors taken, and calories burned. These results confirmed that the mobility of the patients was severely limited by the disease.

Sleep-related problems, such as insomnia, sleep disturbances, and unrefreshing sleep, are among the core symptoms of ME/CFS [37,38]. Overnight sleep of the patients was monitored by a non-invasive Sleep Profiler (Advanced Brain Monitoring). Sleep time and efficiency, sleep architecture, latencies, and continuity, snoring, and cardio were reviewed by the study staff and analyzed comparing with the established normative ranges [23]. Five parameters were identified in the sleep profile where in more than 50% of the patients, the measurements were consistently out of the normal range, that is, either exclusively below the lower limit or above the higher limit of the normal range. Figure 3a shows these five parameters and the percentages of patients whose parameters fell out of the normal ranges. Among the severely ill patients, 75% had an abnormally higher number of awakenings (Awakening/hr $\geq$ 30 s), 65% had abnormally longer wake time after sleep onset (Wake after Sleep Onset), and 50% had sleep efficiency (Sleep Efficiency) below the normal range. Further, the EEG profile revealed that in 70% of the patients, the percentages of Stage R (REM) were below the normal range, and conversely, in 90% of the patients, the percentages of Stage N1 were above the normal range. The observed high percentages of Stage N1 and low percentages of Stage R were consistent with the frequent awakenings during the sleep observed in these patients [39].

Cognitive abnormalities are prevalent in ME/CFS, which include poor attention and concentration, slow information processing, and impaired memory registration and consolidation [40–42]. The cognitive performance of the patients and controls was evaluated using WebNeuro Tests (Brain Resource Group). Four types of cognitive abilities—attention, maze, memory, and identifying emotions—were evaluated and compared with established normal ranges [43]. When comparing the patients with controls, the most significant difference is the higher number of the SIPS patients who had issues in identifying emotions, where their scores were outside of the normal range (94% of the patients vs. 40% of the controls, p = 0.005). In particular, the reaction time of the patients was significantly longer than that of the controls for both happiness and anger (p = 0.015 and 0.007, respectively). In addition, the patients showed more attention problems than the controls (81% of the patients vs. 40% of the controls, p = 0.043). In contrast, the SIPS patients did not show a significant difference in the scores for memory and maze. Similarly, we did not identify any consistent difference between the patients and the controls in the EEG signal monitored taken during the cognitive tests, which potentially were due to the heterogeneity in the data acquired.

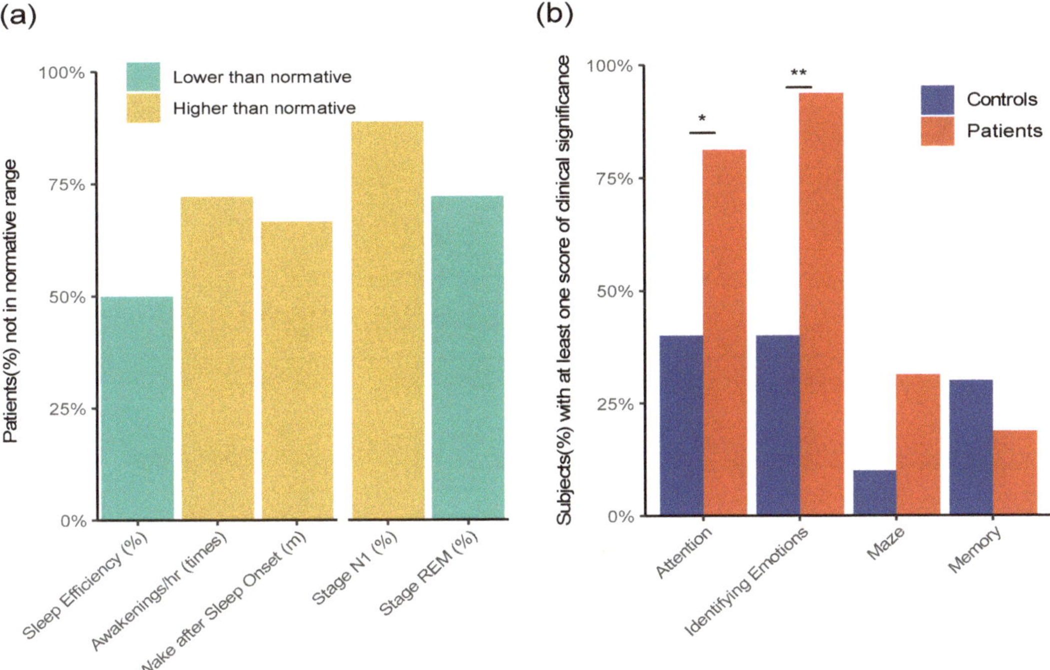

Figure 3. Sleep Monitoring and Cognitive Tests of the Severe ME/CFS Patients. (**a**) Five parameters in the overnight monitoring by Sleep Profiler where the values in ≥50% of the patients were consistently out of the normal ranges. These include lower sleep efficiency, more frequent awakenings per hour (>30 s), longer time of wake after sleep onset, a higher percentage of Stage N1, and a lower percentage of Stage R (REM). (**b**) Comparison between the patients and controls in each of the four sections of cognitive tests. The Y-axis represents the percentage of subjects that were identified as severe/deficit impairment. The patients compared with the controls showed significantly more problems in identifying emotions (94% of the patients vs. 40% of the controls, $p = 0.005$), as well as more attention problems (81% of the patients vs. 40% of the controls, $p = 0.046$).

3.3. Results of Clinical Laboratory Testing

To systematically evaluate whether clinically recognized biomarkers show the difference between severe ME/CFS and healthy controls, an extensive set of clinical laboratory tests were performed on the blood, urine, and saliva samples in this study.

The most significant difference between severe ME/CFS and the controls came from the 4-point salivary cortisol levels, which were tested upon wakening, at noon, afternoon, and night. In healthy individuals, the cortisol level increases upon wakening and steadily decreases throughout the day. As shown in Figure 4a, the severe patients showed significantly lower salivary cortisol concentrations in the morning, where the median levels were 0.20 mcg/dL and 0.45 mcg/dL in the patients and controls, respectively ($p = 0.002$). In addition, there was a significant reduction of the decrease in the cortisol level over the day in the patients compared to the controls: the mean coefficient (slope) of the cortisol level (in log scale) over time (in hours) was −0.059 in the patients and −0.156 in the controls ($p = 0.003$).

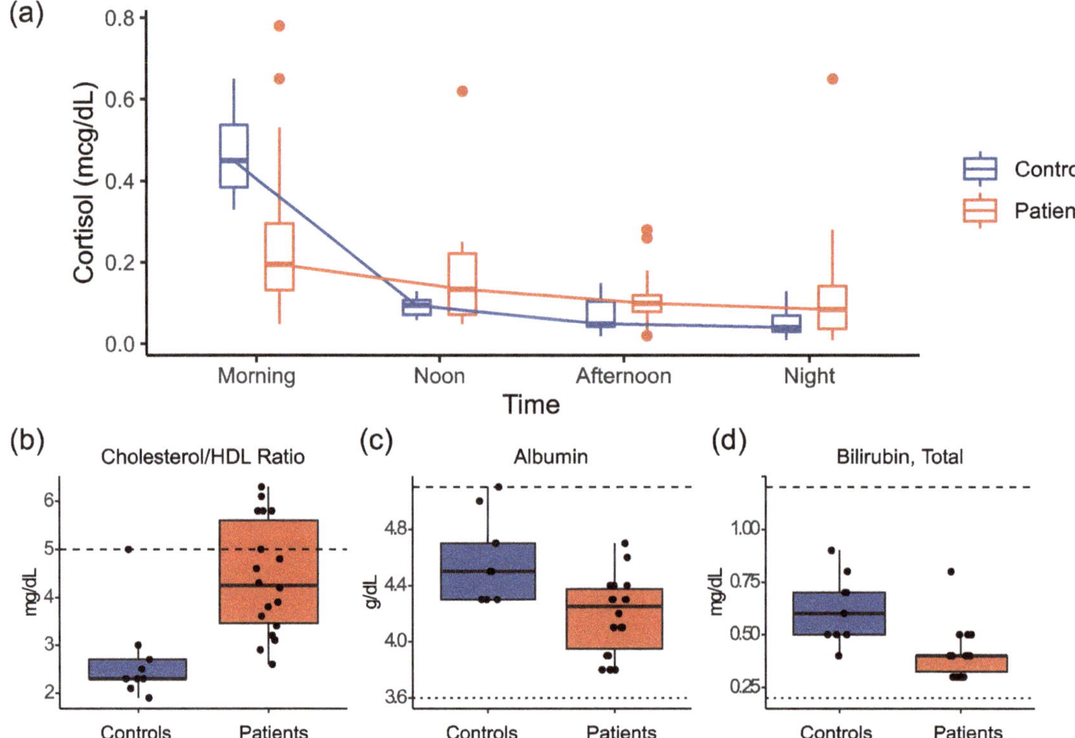

Figure 4. Clinical Lab Test Results Significantly Different between Severe ME/CFS and Controls. (**a**) Results of 4-point salivary cortisol upon wakening, at noon, afternoon, and night. The severe patients demonstrated significantly lower salivary cortisol concentrations in the morning and a significant flattening of the diurnal cortisol profile. (**b–d**) Results of a significantly higher level of cholesterol/HDL ratio (**b**), lower level of albumin, (**c**) and lower total bilirubin, (**d**) in the blood of the patients than of the controls.

Figure 4b–d show additional results significantly different between the severely ill patients and the controls (FDR < 0.1). These include a higher level of cholesterol/HDL ratio (b), lower level of albumin (c), and lower total bilirubin (d) in the blood of the patients than of the controls. On the other hand, no significant differences were observed in the rest of the lab tests, including CBC with DIFF/PLT, Lymphocyte Subsets, Natural Killer Cell function, Comprehensive Metabolic Panel, Standard Lipid Panel, Acylcarnitine Profile, Urinalysis of organic acids, hormones (TSH/T3/T4, FSH/LH, testosterone, estrogen, AVP)), vitamins (B7/biotin, B12/folate, D, Methylmalonic Acid), and selected chemistry analytes and disease biomarkers. The results are shown in Table S1.

3.4. Tests on Antigens and Antibodies against Viral and Bacterial Pathogens

Since ME/CFS patients often report symptoms started with a viral infection, we tested in the patients and the controls antibodies and antigens of a set of common pathogens. These included IgG and IgM antibodies against human herpesvirus 6 and 7 (HHV-6/7), herpes simplex virus 1 and 2 (HSV-1/2 or HHV-1/2), Epstein-Barr virus (EBV or HHV-4), *Cytomegalovirus* (CMV or HHV-5), and parvovirus B19. In addition, tests were performed to detect antigens and antibodies of *Borrelia burgdorferi*, *Bartonella* species, and *Mycoplasma pneumoniae*.

As shown in Table 3, there was no significant difference detected between the severely ill patients and the healthy controls in the tests performed. The percentages of samples identified as positive in each test were similar for each of the antibody and antigen tests of viral and bacterial pathogens. More detailed information can be found in Table S2.

Table 3. Tests on antibodies and antigens.

Viruses-Antibody Tests	Patients Positive/Total	Controls Positive/Total	*p*-Value [1]
Cytomegalovirus (IgG)	9/18	4/9	1
Cytomegalovirus (IgM)	1/18	0/9	1
Parvovirus B19 (IgG)	14/18	6/9	0.653
Parvovirus B19 (IgM)	0/18	0/9	1
Epstein-Barr Virus Early Antigen D (IgG)	2/18	2/9	0.582
Epstein-Barr Virus Viral Capsid Antigen (IgM)	0/18	0/9	1
Epstein-Barr Virus Viral Capsid Antigen (IgG)	17/18	9/9	1
Epstein-Barr Virus Nuclear Antigen (IgG)	16/19	8/8	0.532
Herpesvirus 6 (IgG)	19/19	9/9	1
Herpesvirus 6 (IgM)	1/18	0/9	1
Herpesvirus 7 (IgG)	0/19	0/9	1
Herpesvirus 7 (IgM)	0/18	0/9	1
Herpes Simplex Virus 1 (IgG)	6/18	2/9	0.676
Herpes Simplex Virus 2 (IgG)	6/18	1/9	0.363
Herpes Simplex Virus 1/2 (IgM)	2/18	1/9	1
Bacteria-Antigen and Antibody Tests	**Positive/Total**	**Positive/Total**	***p*-Value [1]**
Borrelia-Ceres Nanotrap Lyme Antigen Test	2/18	1/10	1
Lyme Disease Ab with Reflex to Blot (IgG)	0/18	0/9	1
Lyme Disease Ab with Reflex to Blot (IgM)	0/18	0/9	1
Borrelia burgdorferi (IgG)	0/18	0/9	1
Borrelia burgdorferi (IgM)	0/18	0/9	1
Mycoplasma pneumoniae (IgG)	13/18	6/7	0.637
Mycoplasma pneumoniae (IgM)	1/19	0/9	1
Bartonella DNA-(Blood, Serum, and Culture)	0/20	0/9	1
Bartonella henselae (IgG)	18/20	8/9	1
Bartonella quintana (IgG)	17/20	7/9	0.633
Immunoglobulin G Subclasses Panel	**Low/Total**	**Low/Total**	***p*-Value [1]**
Immunoglobulin G, subclass 1	1/19	0/9	1
Immunoglobulin G, subclass 2	0/19	2/9	0.095
Immunoglobulin G, subclass 3	3/19	0/9	0.530
Immunoglobulin G, subclass 4	1/19	2/9	0.234
Immunoglobulin G, serum	0/19	0/9	1

[1] Fisher's exact test.

IgM antibodies against the common viruses were either not detected or detected positive in very few of the patients and the controls at the same percentage. These include HHV-6 (1/18 in patients vs. 0/9 in controls), EBV (0/18 in patients vs. 0/9 in controls), B19 (0/18 in patients vs. 0/9 in controls), CMV (1/18 in patients vs. 0/9 in controls), and HSV-1/2 (2/17 in patients vs. 1/9 in controls). On the other hand, IgG antibodies were detected in large percentages of both the patients and the controls for these viruses, which included, in patients vs. in controls, HHV-6 (19/19 vs. 9/9), EBV (VCA: 17/18 vs. 9/9, and EBNA: 16/19 vs. 8/8), parvovirus B19 (14/18 vs. 6/9), CMV (9/18 vs. 4/9), and HSV-1 and HSV-2 (6/18 vs. 2/9 and 6/18 vs. 1/9, respectively).

Similarly, few bacterial antigen or IgM tests were positive in patients (0/18 for *Borrelia burgdorferi* IgM, 2/18 for *Borrelia* OspA, 1/19 for *Mycoplasma pneumoniae* IgM, and 0/20 for PCR of *Bartonella* DNA in blood, serum, and culture) without any significant difference comparing to the results of the controls (0/9 for *Borrelia burgdorferi* IgM, 1/10 for *Borrelia*

OspA, 0/10 for *Mycoplasma pneumoniae* IgM, and 0/9 for PCR of *Bartonella* Species in Blood, Serum, and Culture). In the same samples, IgG antibodies were detected at the same rate in patients vs. in controls (0/18 vs. 0/9 for *Borrelia Burgdorferi*, 13/18 vs. 6/7 for *Mycoplasma Pneumoniae*, 11/20 vs. 4/9 for *Bartonella Henselae*, and 7/20 vs. 4/9 for *Bartonella Quintana*).

4. Discussion

ME/CFS significantly reduces the quality of life of patients [6,44–46], and the severe cases studied here present a picture of a systematically debilitating disease. Severely affected patients who were homebound and mostly bedbound suffer from a greater reduction of their quality of life compared to other major chronic diseases as well as the general ME/CFS population. While physical functioning, energy/fatigue, and related functioning were extremely low in these patients, emotional well-being was clearly less impacted-a clear distinction from the frequent misdiagnosis of clinical depression in these patients.

The SIPS patients had all the core symptoms in the IOM criteria [1] and other symptoms such as pain and neurosensory disturbance, consistent with previous reports [1,11,12]. The most troublesome symptoms were fatigue (85%), pain (65%), cognitive impairment (50%), orthostatic intolerance (45%), sleep disturbance (35%), post-exertional malaise (30%), neurosensory disturbance (30%), GI tract impairment (30%). Pharmacological and non-pharmacological approaches to the relief of these symptoms could help individual patients manage this disease, since there are no treatments currently approved for ME/CFS [47,48].

Sleep disorders and cognitive impairments are core symptoms of ME/CFS [37,38,40–42]. Non-invasive sleep monitoring revealed that the majority of the severely ill patients had an abnormally high number of awakenings, abnormally long wake time after sleep onset, and sleep efficiency below the normal range, which are consistent with the high percentages of Stage N1 and low percentages of Stage R (REM) observed in their EEG profiles [39]. Cognitive tests showed significant differences in the severely ill patients in identifying emotions and having attention problems, while there was no difference in the maze and memory tests between the patients and the controls. Impairment of divided attention has been reported previously in ME/CFS patients [40,49], and our results are consistent with the hypothesis that the difficulty in divided attention may contribute significantly to the cognitive problems in ME/CFS. Further studies using sophisticated methodologies are essential to better characterize and understand the sleep and cognitive disorders in ME/CFS.

Currently, there is no diagnostic test for ME/CFS, and laboratory tests are primarily used in differential diagnosis to identify alternative conditions and comorbidities [14,47,50]. Here we evaluated an extensive set of clinical lab tests in blood, urine, and saliva samples. Between the severely ill patients and the controls, the most significant difference observed was lower salivary cortisol concentrations in the morning and the flattening of the daily cortisol profile in the patients, consistent with previously reported observations of the alterations in diurnal salivary cortisol rhythm in ME/CFS [51,52]. Other tests conducted did not show noticeable significance. While we did not perform all the recommended testing by the US ME/CFS Clinician Coalition [4,50], these lab results re-confirm the limitations of the standard laboratory test battery in ME/CFS and highlight the urgent need of developing new diagnostic tests for the disease [1,14]. For instance, lower-than-normal circulating blood volume could be associated with orthostatic intolerance seen in the severely ill patients, which would be worthwhile measuring [53].

Previous studies showed that in many ME/CFS patients, the 'sudden onset' of the disease appears to be a viral infection [2,5,54,55]. Therefore, we tested antibodies and antigens of a set of common viral and bacterial pathogens. The results showed no evidence for acute infections by the tested pathogens in the patients, while as expected, large percentages of both the patients and the controls had been exposed to some of these common viral or bacterial pathogens. Enteroviruses were proposed as a cause of ME/CFS [56,57] but were not tested in this study. Also, it is worth noting that certain pathogens are neurotropic and evidence of central nervous system (CNS) infection is not always revealed by serologic studies of blood. Further analysis of autoantibodies and detections of pathogens (e.g., by sequencing) in the relevant tissues, such as in the cerebrospinal fluid (CSF), will likely provide new insights into the link between pathogen exposure and ME/CFS.

The biological samples collected on the severely ill patients and the healthy controls are being further analyzed in multiple omics studies to identify signatures in genes, proteins, metabolites, heavy metals, and microbes of severe ME/CFS and the associated clinical symptoms.

Post-COVID conditions (long COVID, Post-Acute Sequelae of SARS-CoV-2 infection (PASC)), are affecting an increasingly large number of people worldwide, where patients suffer from prolonged fatigue and other symptoms [58–60]. A recent study [31] of 3762 confirmed or suspected COVID patients from 56 countries showed that the time to recovery in most patients exceeded 7 months, where the majority of the patients had multiple symptoms related to ME/CFS. Therefore, we compared the frequencies of the symptoms remaining after six months in the long COVID patients with those of the severely ill ME/CFS patients, and the results showed a striking similarity (Figure 5). This underscores the value of research to understand the mechanisms of ME/CFS for efforts to treat and prevent long COVID and other debilitating postviral conditions, which together affect millions in the United States alone [14,61].

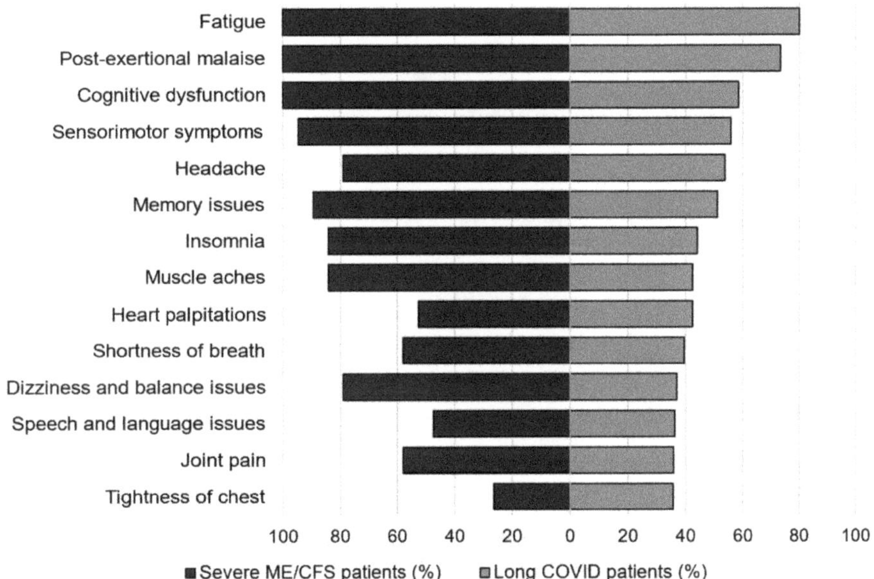

Figure 5. Comparison of the symptoms reported in the long COVID patients after 6 months with those in the severely ill ME/CFS patients. The symptoms are ranked based on the frequencies reported in the long COVID patients. The frequencies of these symptoms in the severely ill ME/CFS are similar to those reported in the long COVID.

Supplementary Materials: The following are available online at https://www.mdpi.com/article/10.3390/healthcare9101290/s1, Figure S1: Comparison of the Quality of Life of Severely Ill ME/CFS and Other Major Diseases, Figure S2: Similarities and Differences in Clinical Symptoms across 20 SIPS Patients. Table S1: List of Clinical Laboratory Tests. Table S2: List of Antibody and Antigen Tests of Viral and Bacterial Pathogens.

Author Contributions: Conceptualization, R.W.D., A.M.K. and W.X.; methodology, R.W.D., A.M.K., D.K., W.X.; investigation, A.M.K., D.K., C.-J.C., L.-Y.H., R.S.A., A.M.C., J.W., W.X.; analysis, C.-J.C., L.-Y.H., P.L., W.X.; data resources, C.-J.C., L.-Y.H., W.X.; writing, C.-J.C., L.-Y.H., R.S.A., W.X., R.W.D.;

project administration, R.W.D., A.M.K., L.T., R.S.A., W.X.; funding acquisition, R.W.D., L.T. All authors have read and agreed to the published version of the manuscript.

Funding: This research was funded by Open Medicine Foundation, and received no other external funding.

Institutional Review Board Statement: The study was conducted according to the guidelines of the Declaration of Helsinki and approved by the Western Institutional Review Board (protocol code 20152676 and 02/10/2016).

Informed Consent Statement: Informed consent was obtained from all subjects involved in the study.

Data Availability Statement: Deidentified data and results are available through a web-based data portal at https://endmecfs.stanford.edu.

Acknowledgments: We would also like to express our sincere gratitude to each of the severely ill patients who participated in this study. We are grateful to the Open Medicine Foundation and their generous donors for funding this research as part of the End ME/CFS Project.

Conflicts of Interest: The authors declare no conflict of interest. The funders had no role in the design of the study; in the collection, analyses, or interpretation of data; in the writing of the manuscript, or in the decision to publish the results.

References

1. Committee on the Diagnostic Criteria for Myalgic Encephalomyelitis/Chronic Fatigue Syndrome; Board on the Health of Select Populations; Institute of Medicine. *Beyond Myalgic Encephalomyelitis/Chronic Fatigue Syndrome: Redefining an Illness*; The National Academies Collection: Reports Funded by National Institutes of Health; National Academies Press (US): Washington, DC, USA, 2015; ISBN 978-0-309-31689-7.
2. Cortes Rivera, M.; Mastronardi, C.; Silva-Aldana, C.T.; Arcos-Burgos, M.; Lidbury, B.A. Myalgic Encephalomyelitis/Chronic Fatigue Syndrome: A Comprehensive Review. *Diagnostics* **2019**, *9*, 91. [CrossRef]
3. Valdez, A.R.; Hancock, E.E.; Adebayo, S.; Kiernicki, D.J.; Proskauer, D.; Attewell, J.R.; Bateman, L.; DeMaria, A.; Lapp, C.W.; Rowe, P.C.; et al. Estimating Prevalence, Demographics, and Costs of ME/CFS Using Large Scale Medical Claims Data and Machine Learning. *Front. Pediatr.* **2018**, *6*, 412. [CrossRef] [PubMed]
4. Bateman, L.; Bested, A.C.; Bonilla, H.F.; Chheda, B.V.; Chu, L.; Curtin, J.M.; Dempsey, T.T.; Dimmock, M.E.; Dowell, T.G.; Felsenstein, D.; et al. Myalgic Encephalomyelitis/Chronic Fatigue Syndrome: Essentials of Diagnosis and Management. In *Mayo Clinic Proceedings*; Elsevier: Amsterdam, The Netherlands, 2021. [CrossRef]
5. Rasa, S.; Nora-Krukle, Z.; Henning, N.; Eliassen, E.; Shikova, E.; Harrer, T.; Scheibenbogen, C.; Murovska, M.; Prusty, B.K.; European Network on ME/CFS (EUROMENE). Chronic Viral Infections in Myalgic Encephalomyelitis/Chronic Fatigue Syndrome (ME/CFS). *J. Transl. Med.* **2018**, *16*, 268. [CrossRef]
6. Falk Hvidberg, M.; Brinth, L.S.; Olesen, A.V.; Petersen, K.D.; Ehlers, L. The Health-Related Quality of Life for Patients with Myalgic Encephalomyelitis/Chronic Fatigue Syndrome (ME/CFS). *PLoS ONE* **2015**, *10*, e0132421. [CrossRef]
7. Bombardier, C.H.; Buchwald, D. Outcome and Prognosis of Patients with Chronic Fatigue vs Chronic Fatigue Syndrome. *Arch. Intern. Med.* **1995**, *155*, 2105–2110. [CrossRef]
8. Wilson, A.; Hickie, I.; Lloyd, A.; Hadzi-Pavlovic, D.; Boughton, C.; Dwyer, J.; Wakefield, D. Longitudinal Study of Outcome of Chronic Fatigue Syndrome. *BMJ* **1994**, *308*, 756–759. [CrossRef]
9. Fennell, P.A.; Jason, L.A.; Klein, S.M. Capturing the Different Phases of the CFS Illness. *CFIDS Chron.* **1998**, *11*, 13–16.
10. Pendergrast, T.; Brown, A.; Sunnquist, M.; Jantke, R.; Newton, J.L.; Strand, E.B.; Jason, L.A. Housebound versus Nonhousebound Patients with Myalgic Encephalomyelitis and Chronic Fatigue Syndrome. *Chronic Illn.* **2016**, *12*, 292–307. [CrossRef]
11. Wiborg, J.F.; van der Werf, S.; Prins, J.B.; Bleijenberg, G. Being Homebound with Chronic Fatigue Syndrome: A Multidimensional Comparison with Outpatients. *Psychiatry Res.* **2010**, *177*, 246–249. [CrossRef]
12. Centers for Disease Control and Prevention; National Center for Emerging and Zoonotic Infectious Diseases (NCEZID); Division of High-Consequence Pathogens and Pathology (DHCPP). Severely Affected Patients I Clinical Care of Patients I Healthcare Providers I Myalgic Encephalomyelitis/Chronic Fatigue Syndrome (ME/CFS) I CDC. Available online: https://www.cdc.gov/me-cfs/healthcare-providers/clinical-care-patients-mecfs/severely-affected-patients.html (accessed on 22 August 2021).
13. Dafoe, W. Extremely Severe ME/CFS-A Personal Account. *Healthc. Basel Switz.* **2021**, *9*, 504. [CrossRef]
14. Komaroff, A.L. Myalgic Encephalomyelitis/Chronic Fatigue Syndrome: When Suffering Is Multiplied. *Healthcare* **2021**, *9*, 919. [CrossRef]
15. Carruthers, B.M.; van de Sande, M.I.; De Meirleir, K.L.; Klimas, N.G.; Broderick, G.; Mitchell, T.; Staines, D.; Powles, A.C.P.; Speight, N.; Vallings, R.; et al. Myalgic Encephalomyelitis: International Consensus Criteria. *J. Intern. Med.* **2011**, *270*, 327–338. [CrossRef] [PubMed]
16. Ware, J.E.; New England Medical Center Hospital; Health Institute. *SF-36 Physical and Mental Health Summary Scales: A User's Manual*; Health Institute, New England Medical Center: Boston, MA, USA, 1994.

17. Karnofsky, D.A.; Abelmann, W.H.; Craver, L.F.; Burchenal, J.H. The Use of the Nitrogen Mustards in the Palliative Treatment of Carcinoma. With Particular Reference to Bronchogenic Carcinoma. *Cancer* **1948**, *1*, 634–656. [CrossRef]
18. PROMIS: Patient-Reported Outcomes Measurement Information System—Home Page. Available online: https://commonfund.nih.gov/promis/index (accessed on 27 April 2021).
19. Cella, D.; Yount, S.; Rothrock, N.; Gershon, R.; Cook, K.; Reeve, B.; Ader, D.; Fries, J.F.; Bruce, B.; Rose, M.; et al. The Patient-Reported Outcomes Measurement Information System (PROMIS): Progress of an NIH Roadmap Cooperative Group during Its First Two Years. *Med. Care* **2007**, *45*, S3–S11. [CrossRef]
20. Buysse, D.J.; Reynolds, C.F.; Monk, T.H.; Berman, S.R.; Kupfer, D.J. The Pittsburgh Sleep Quality Index: A New Instrument for Psychiatric Practice and Research. *Psychiatry Res.* **1989**, *28*, 193–213. [CrossRef]
21. Spencer, B.R.; Kleinman, S.; Wright, D.J.; Glynn, S.A.; Rye, D.B.; Kiss, J.E.; Mast, A.E.; Cable, R.G.; REDS-II RISE Analysis Group. Restless Legs Syndrome, Pica, and Iron Status in Blood Donors. *Transfusion (Paris)* **2013**, *53*, 1645–1652. [CrossRef]
22. Finan, P.H.; Richards, J.M.; Gamaldo, C.E.; Han, D.; Leoutsakos, J.M.; Salas, R.; Irwin, M.R.; Smith, M.T. Validation of a Wireless, Self-Application, Ambulatory Electroencephalographic Sleep Monitoring Device in Healthy Volunteers. *J. Clin. Sleep Med.* **2016**, *12*, 1443–1451. [CrossRef] [PubMed]
23. Walsleben, J.A.; Kapur, V.K.; Newman, A.B.; Shahar, E.; Bootzin, R.R.; Rosenberg, C.E.; O'Connor, G.; Nieto, F.J. Sleep and Reported Daytime Sleepiness in Normal Subjects: The Sleep Heart Health Study. *Sleep* **2004**, *27*, 293–298. [CrossRef]
24. Levendowski, D.J.; Ferini-Strambi, L.; Gamaldo, C.; Cetel, M.; Rosenberg, R.; Westbrook, P.R. The Accuracy, Night-to-Night Variability, and Stability of Frontopolar Sleep Electroencephalography Biomarkers. *J. Clin. Sleep Med.* **2017**, *13*, 791–803. [CrossRef] [PubMed]
25. Silverstein, S.M.; Berten, S.; Olson, P.; Paul, R.; Willams, L.M.; Cooper, N.; Gordon, E. Development and Validation of a World-Wide-Web-Based Neurocognitive Assessment Battery: WebNeuro. *Behav. Res. Methods* **2007**, *39*, 940–949. [CrossRef]
26. Gordon, E.; Cooper, N.; Rennie, C.; Hermens, D.; Williams, L.M. Integrative Neuroscience: The Role of a Standardized Database. *Clin. EEG Neurosci.* **2005**, *36*, 64–75. [CrossRef]
27. Centers for Disease Control and Prevention (CDC). Recommendations for Test Performance and Interpretation from the Second National Conference on Serologic Diagnosis of Lyme Disease. *MMWR Morb. Mortal. Wkly. Rep.* **1995**, *44*, 590–591.
28. Mead, P.; Petersen, J.; Hinckley, A. Updated CDC Recommendation for Serologic Diagnosis of Lyme Disease. *MMWR Morb. Mortal. Wkly. Rep.* **2019**, *68*, 703. [CrossRef] [PubMed]
29. Magni, R.; Espina, B.H.; Shah, K.; Lepene, B.; Mayuga, C.; Douglas, T.A.; Espina, V.; Rucker, S.; Dunlap, R.; Petricoin, E.F.I.; et al. Application of Nanotrap Technology for High Sensitivity Measurement of Urinary Outer Surface Protein A Carboxyl-Terminus Domain in Early Stage Lyme Borreliosis. *J. Transl. Med.* **2015**, *13*, 346. [CrossRef] [PubMed]
30. Theel, E.S.; Ross, T. Seasonality of Bartonella Henselae IgM and IgG Antibody Positivity Rates. *J. Clin. Microbiol.* **2019**, *57*, e01263-19. [CrossRef]
31. Davis, H.E.; Assaf, G.S.; McCorkell, L.; Wei, H.; Low, R.J.; Re'em, Y.; Redfield, S.; Austin, J.P.; Akrami, A. Characterizing Long COVID in an International Cohort: 7 Months of Symptoms and Their Impact. *EClinicalMedicine* **2021**, 101019. [CrossRef]
32. Rajeevan, M.S.; Dimulescu, I.; Murray, J.; Falkenberg, V.R.; Unger, E.R. Pathway-Focused Genetic Evaluation of Immune and Inflammation Related Genes with Chronic Fatigue Syndrome. *Hum. Immunol.* **2015**, *76*, 553–560. [CrossRef] [PubMed]
33. Fluge, Ø.; Rekeland, I.G.; Lien, K.; Thürmer, H.; Borchgrevink, P.C.; Schäfer, C.; Sørland, K.; Aßmus, J.; Ktoridou-Valen, I.; Herder, I.; et al. B-Lymphocyte Depletion in Patients With Myalgic Encephalomyelitis/Chronic Fatigue Syndrome: A Randomized, Double-Blind, Placebo-Controlled Trial. *Ann. Intern. Med.* **2019**, *170*, 585–593. [CrossRef]
34. Ware, J.E.; Snow, K.K.; Kosinski, M.; Gandek, B. *SF-36 Health Survey: MANUAL and Interpretation Guide*; Health Institute, New England Medical Center: Boston, MA, USA, 1993.
35. Tudor-Locke, C.E.; Myers, A.M. Methodological Considerations for Researchers and Practitioners Using Pedometers to Measure Physical (Ambulatory) Activity. *Res. Q. Exerc. Sport* **2001**, *72*, 1–12. [CrossRef]
36. Whitt, M.C.; DuBose, K.D.; Ainsworth, B.E.; Tudor-Locke, C. Walking Patterns in a Sample of African American, Native American, and Caucasian Women: The Cross-Cultural Activity Participation Study. *Health Educ. Behav. Off. Publ. Soc. Public Health Educ.* **2004**, *31*, 45S–56S. [CrossRef]
37. Jackson, M.L.; Bruck, D. Sleep Abnormalities in Chronic Fatigue Syndrome/Myalgic Encephalomyelitis: A Review. *J. Clin. Sleep Med.* **2012**, *8*, 719–728. [CrossRef]
38. Gotts, Z.M.; Newton, J.L.; Ellis, J.G.; Deary, V. The Experience of Sleep in Chronic Fatigue Syndrome: A Qualitative Interview Study with Patients. *Br. J. Health Psychol.* **2016**, *21*, 71–92. [CrossRef]
39. Shrivastava, D.; Jung, S.; Saadat, M.; Sirohi, R.; Crewson, K. How to Interpret the Results of a Sleep Study. *J. Community Hosp. Intern. Med. Perspect.* **2014**, *4*, 1–4. [CrossRef]
40. Teodoro, T.; Edwards, M.J.; Isaacs, J.D. A Unifying Theory for Cognitive Abnormalities in Functional Neurological Disorders, Fibromyalgia and Chronic Fatigue Syndrome: Systematic Review. *J. Neurol. Neurosurg. Psychiatry* **2018**, *89*, 1308–1319. [CrossRef]
41. Michiels, V.; de Gucht, V.; Cluydts, R.; Fischler, B. Attention and Information Processing Efficiency in Patients with Chronic Fatigue Syndrome. *J. Clin. Exp. Neuropsychol.* **1999**, *21*, 709–729. [CrossRef]
42. Schmaling, K.B.; Betterton, K.L. Neurocognitive Complaints and Functional Status among Patients with Chronic Fatigue Syndrome and Fibromyalgia. *Qual. Life Res.* **2016**, *25*, 1257–1263. [CrossRef]
43. OptumHealth; Brain Resource Company. *WebNeuro User Manual*; Brain Resource Company (BRC): San Francisco, CA, USA, 2008.

44. Nacul, L.C.; Lacerda, E.M.; Campion, P.; Pheby, D.; de Drachler, L.M.; Leite, J.C.; Poland, F.; Howe, A.; Fayyaz, S.; Molokhia, M. The Functional Status and Well Being of People with Myalgic Encephalomyelitis/Chronic Fatigue Syndrome and Their Carers. *BMC Public Health* **2011**, *11*, 402. [CrossRef] [PubMed]
45. Reeves, W.C.; Wagner, D.; Nisenbaum, R.; Jones, J.F.; Gurbaxani, B.; Solomon, L.; Papanicolaou, D.A.; Unger, E.R.; Vernon, S.D.; Heim, C. Chronic Fatigue Syndrome—A Clinically Empirical Approach to Its Definition and Study. *BMC Med.* **2005**, *3*, 19. [CrossRef] [PubMed]
46. Jason, L.; Brown, M.; Evans, M.; Anderson, V.; Lerch, A.; Brown, A.; Hunnell, J.; Porter, N. Measuring Substantial Reductions in Functioning in Patients with Chronic Fatigue Syndrome. *Disabil. Rehabil.* **2011**, *33*, 589–598. [CrossRef]
47. Nacul, L.; Authier, F.J.; Scheibenbogen, C.; Lorusso, L.; Helland, I.B.; Martin, J.A.; Sirbu, C.A.; Mengshoel, A.M.; Polo, O.; Behrends, U.; et al. European Network on Myalgic Encephalomyelitis/Chronic Fatigue Syndrome (EUROMENE): Expert Consensus on the Diagnosis, Service Provision, and Care of People with ME/CFS in Europe. *Medicina* **2021**, *57*, 510. [CrossRef]
48. US ME/CFS Clinician Coalition. *ME/CFS Treatment Recommendations*, Version 1; The US ME/CFS Clinician Coalition, 2021. Available online: https://mecfscliniciancoalition.org (accessed on 22 August 2021).
49. Ross, S.; Fantie, B.; Straus, S.F.; Grafman, J. Divided Attention Deficits in Patients with Chronic Fatigue Syndrome. *Appl. Neuropsychol.* **2001**, *8*, 4–11. [CrossRef]
50. US ME/CFS Clinician Coalition. *Testing Recommendations for Suspected ME/CFS*, Version 1; The US ME/CFS Clinician Coalition, 2021. Available online: https://mecfscliniciancoalition.org (accessed on 22 August 2021).
51. Nater, U.M.; Maloney, E.; Boneva, R.S.; Gurbaxani, B.M.; Lin, J.-M.; Jones, J.F.; Reeves, W.C.; Heim, C. Attenuated Morning Salivary Cortisol Concentrations in a Population-Based Study of Persons with Chronic Fatigue Syndrome and Well Controls. *J. Clin. Endocrinol. Metab.* **2008**, *93*, 703–709. [CrossRef]
52. Nater, U.M.; Youngblood, L.S.; Jones, J.F.; Unger, E.R.; Miller, A.H.; Reeves, W.C.; Heim, C. Alterations in Diurnal Salivary Cortisol Rhythm in a Population-Based Sample of Cases with Chronic Fatigue Syndrome. *Psychosom. Med.* **2008**, *70*, 298–305. [CrossRef]
53. van Campen, C.L.M.C.; Rowe, P.C.; Visser, F.C. Blood Volume Status in ME/CFS Correlates With the Presence or Absence of Orthostatic Symptoms: Preliminary Results. *Front. Pediatr.* **2018**, *6*, 352. [CrossRef]
54. Buchwald, D.S.; Rea, T.D.; Katon, W.J.; Russo, J.E.; Ashley, R.L. Acute Infectious Mononucleosis: Characteristics of Patients Who Report Failure to Recover. *Am. J. Med.* **2000**, *109*, 531–537. [CrossRef]
55. Katz, B.Z.; Shiraishi, Y.; Mears, C.J.; Binns, H.J.; Taylor, R. Chronic Fatigue Syndrome after Infectious Mononucleosis in Adolescents. *Pediatrics* **2009**, *124*, 189–193. [CrossRef] [PubMed]
56. Chia, J.; Chia, A.; Voeller, M.; Lee, T.; Chang, R. Acute Enterovirus Infection Followed by Myalgic Encephalomyelitis/Chronic Fatigue Syndrome (ME/CFS) and Viral Persistence. *J. Clin. Pathol.* **2010**, *63*, 165–168. [CrossRef] [PubMed]
57. O'Neal, A.J.; Hanson, M.R. The Enterovirus Theory of Disease Etiology in Myalgic Encephalomyelitis/Chronic Fatigue Syndrome: A Critical Review. *Front. Med.* **2021**, *8*, 688486. [CrossRef]
58. Nalbandian, A.; Sehgal, K.; Gupta, A.; Madhavan, M.V.; McGroder, C.; Stevens, J.S.; Cook, J.R.; Nordvig, A.S.; Shalev, D.; Sehrawat, T.S.; et al. Post-Acute COVID-19 Syndrome. *Nat. Med.* **2021**, *27*, 601–615. [CrossRef]
59. Carfi, A.; Bernabei, R.; Landi, F.; for the Gemelli Against COVID-19 Post-Acute Care Study Group. Persistent Symptoms in Patients After Acute COVID-19. *JAMA* **2020**, *324*, 603–605. [CrossRef]
60. Huang, C.; Huang, L.; Wang, Y.; Li, X.; Ren, L.; Gu, X.; Kang, L.; Guo, L.; Liu, M.; Zhou, X.; et al. 6-Month Consequences of COVID-19 in Patients Discharged from Hospital: A Cohort Study. *Lancet* **2021**, *397*, 220–232. [CrossRef]
61. Komaroff, A.L.; Bateman, L. Will COVID-19 Lead to Myalgic Encephalomyelitis/Chronic Fatigue Syndrome? *Front. Med.* **2021**, *7*, 1132. [CrossRef] [PubMed]

 *healthcare*

Article

Two-Day Cardiopulmonary Exercise Testing in Females with a Severe Grade of Myalgic Encephalomyelitis/Chronic Fatigue Syndrome: Comparison with Patients with Mild and Moderate Disease

C (Linda) M. C. van Campen [1,*], Peter C. Rowe [2] and Frans C. Visser [1]

[1] Stichting Cardio Zorg, 2132 HN Hoofddorp, The Netherlands; fransvisser@stichtingcardiozorg.nl
[2] Department of Paediatrics, John Hopkins University School of Medicine, John Hopkins University, Baltimore, MD 21218, USA; prowe@jhmi.edu
* Correspondence: info@stichtingcardiozorg.nl; Tel.: +31-208934125

Received: 28 April 2020; Accepted: 25 June 2020; Published: 30 June 2020

Abstract: Introduction: Effort intolerance along with a prolonged recovery from exercise and post-exertional exacerbation of symptoms are characteristic features of myalgic encephalomyelitis/chronic fatigue syndrome (ME/CFS). The gold standard to measure the degree of physical activity intolerance is cardiopulmonary exercise testing (CPET). Multiple studies have shown that peak oxygen consumption is reduced in the majority of ME/CFS patients, and that a 2-day CPET protocol further discriminates between ME/CFS patients and sedentary controls. Limited information is present on ME/CFS patients with a severe form of the disease. Therefore, the aim of this study was to compare the effects of a 2-day CPET protocol in female ME/CFS patients with a severe grade of the disease to mildly and moderately affected ME/CFS patients. Methods and results: We studied 82 female patients who had undergone a 2-day CPET protocol. Measures of oxygen consumption (VO_2), heart rate (HR) and workload both at peak exercise and at the ventilatory threshold (VT) were collected. ME/CFS disease severity was graded according to the International Consensus Criteria. Thirty-one patients were clinically graded as having mild disease, 31 with moderate and 20 with severe disease. Baseline characteristics did not differ between the 3 groups. Within each severity group, all analyzed CPET parameters (peak VO_2, VO_2 at VT, peak workload and the workload at VT) decreased significantly from day-1 to day-2 (*p*-Value between 0.003 and <0.0001). The magnitude of the change in CPET parameters from day-1 to day-2 was similar between mild, moderate, and severe groups, except for the difference in peak workload between mild and severe patients ($p = 0.019$). The peak workload decreases from day-1 to day-2 was largest in the severe ME/CFS group (−19 (11) %). Conclusion: This relatively large 2-day CPET protocol study confirms previous findings of the reduction of various exercise variables in ME/CFS patients on day-2 testing. This is the first study to demonstrate that disease severity negatively influences exercise capacity in female ME/CFS patients. Finally, this study shows that the deterioration in peak workload from day-1 to day-2 is largest in the severe ME/CFS patient group.

Keywords: chronic fatigue syndrome; cardiopulmonary exercise testing; oxygen consumption; VO_2 peak; ventilatory threshold; VO_2 VT; myalgic encephalitis; workload; ME/CFS severity grade

1. Introduction

Myalgic Encephalomyelitis/Chronic Fatigue Syndrome (ME/CFS) is a serious and potentially disabling chronic disease [1–4]. As in other diseases, ME/CFS severity can range from mild to severe. Some patients can perform their daily activities at the expense of extra resting, while others are bed-ridden and dependent on others for help with activities of daily living. Exercise intolerance along with a prolonged recovery from activity (physical as well as mental) and post-exertional exacerbation of symptoms [4], represent an important characteristic of ME/CFS termed post-exertional malaise (PEM) [5,6].

Cardiopulmonary exercise testing (CPET) is the gold standard for measuring the degree of physical activity intolerance [7–10]. Multiple studies have shown that peak oxygen consumption is reduced in the majority of ME/CFS patients [11–22]. However, studies have also shown that a single CPET test in ME/CFS patients may show peak VO_2 values that are similar to or only slightly lower than those of healthy sedentary controls [16]. To discriminate exercise capacity between ME/CFS patients and sedentary controls a 2-day CPET protocol has been proposed [19]. Studies using a 2-day CPET protocol, with two exercise tests separated by 24 h, have confirmed that ME/CFS patients have significantly lower VO_2 and workload parameters on day 2 than on day 1. In contrast, sedentary controls have unaltered or slightly improved peak VO_2 and workload [18–20,23,24]. We have recently confirmed the previous observations of a lower VO_2 and workload in ME/CFS patients in a large group of male and female ME/CFS patients [25,26].

As peak oxygen consumption differs between males and females [27–29], the available studies were analyzed according to gender. Four studies reported peak oxygen consumption data in females only [18–20,25]. One study reported peak oxygen consumption in males [26] and two studies reported on combined information on males and females [15,24].

Limited studies have been published on ME/CFS patients with a severe form of the disease. The aim of this study was to compare 2-day CPET results from severely affected and mildly and moderately affected female ME/CFS patients using the severity grading as proposed by Carruthers et al., in the International Consensus Criteria [2]. We furthermore explored disability for both testing days with Weber's disability metric [30].

2. Materials and Methods

Eligible participants were females with ME/CFS and exercise intolerance who had been referred to the Stichting Cardio Zorg, a cardiology clinic in the Netherlands that specializes in diagnosing and treating adults with ME/CFS. All patients underwent a detailed clinical history to establish the diagnosis of ME/CFS according to the ME criteria [2] and CFS criteria of Fukuda [1]. In all patients alternative diagnoses which could explain the fatigue and other symptoms were ruled out. The disease severity was scored according to the International Consensus Criteria. This was classified according to the paper as: "Symptom severity impact must result in a 50% or greater reduction in a patient's premorbid activity level for a diagnosis of ME. Mild: approximately 50% reduction in activity, moderate: mostly housebound, severe: mostly bedbound and very severe: bedbound and dependent on help for physical functions" [2].

We reviewed the clinical records of 93 females with a diagnosis of ME/CFS who had undergone a 2-day CPET protocol between June 2012 and November 2019. Six patients were excluded because the ventilatory threshold could not be accurately determined. Two patients were excluded because of heart rate or blood pressure lowering drugs. Three patients were excluded because of being identified as being outliers. They were removed from further analysis. No patients were excluded due to insufficient effort during either day of testing, as judged by the supervising cardiologist. This left 82 female patients with available data from a 2-day CPET protocol for analysis.

All patients gave informed consent to analyze their data. The use of clinical data for descriptive studies was approved by the ethics committee of the Slotervaart Hospital, the Netherlands (reference number P1736).

2.1. Cardiopulmonary Exercise Testing (CPET)

Patients underwent a symptom-limited exercise test on a cycle ergometer (Excalibur, Lode, Groningen, The Netherlands) according to a previously described protocol [26]. Briefly, a ramp workload protocol was used varying between 10–30 Watt/min. Oxygen consumption (VO_2), carbon dioxide release (VCO_2), and oxygen saturation were continuously measured (Cortex, Procare, The Netherlands), and displayed on screen using Metasoft software (Cortex, Biophysic Gmbh, Germany). An ECG was continuously recorded and blood pressures were measured using the Nexfin device (BMEYE, Amsterdam, The Netherlands) [31]. The metabolic measurement system (Cortex, Biophysic Gmbh, Germany) was calibrated before each test with ambient air, standard gases of known concentrations, and a 3-L calibration syringe. The ventilatory threshold (VT), a measure of the anaerobic threshold, was identified from expired gases using the V-Slope algorithm [32]. The same experienced cardiologist supervised the 2 tests and performed visual assessment and confirmation of the algorithm-derived VT. The same cardiologists also ensured that the tests were done with the maximal effort possible for each specific patient. The mean of the VO_2 measurements of the last 15 s before ending the exercise (peak VO_2) was taken. VO_2 at the peak and at the VT as well as the heart rate (HT) at peak exercise were expressed as a percentage of the normal values of a population study: %peak VO_2, %VT VO_2 [27]. Also, the mean respiratory exchange ratio (RER; VCO_2/VO_2) of the last 15 s was calculated by the software and presented in the results.

Maximum (or peak) oxygen uptake is used for the evaluation of cardiorespiratory endurance or aerobic fitness [7,8,33]. The anaerobic threshold (AT) was found to be an objective measure for aerobic work capacity [34–36]. Both invasive and non-invasive methods have been used for determining this value: invasive methods require blood lactate measurements (lactate threshold) and non-invasive methods rely on measuring respiratory gases, based on the relation between CO_2 expiration and O_2 inspiration (the respiratory exchange ratio or RER). Some studies have considered the AT to be at the point where the RER exceeds 1.0 [37–39]. As this has been considered inaccurate other methods have been proposed, like using the V-slope algorithm as described by Beaver et al. and used by others as in the current study [32]. In short, this method plots V_{O2} against V_{CO2}. During aerobic metabolism the slope is slightly less than 1. With the onset of anaerobic metabolism, the slope increases to a value greater than 1, reflecting the production of extra CO_2 resulting from HCO_3^- buffering of lactate being produced. This point from slope 1 to the steeper slope 2 is called the ventilatory threshold (or lactic acidosis threshold) and is considered equivalent to the anaerobic threshold [32].

2.2. Disability Metric

In the early 1980s Weber et al. described a disability metric in evaluating heart failure patients with cardiopulmonary exercise testing [30]. This disability metric termed the classes A, B, C and D to avoid confusion with the New York Heart Association classification. The disability metric classifies how much impairment in aerobic capacity is present (Table 1).

Table 1. Weber disability metric (30) [30].

Weber Class	Oxygen Consumption	Aerobic Capacity
Class A	>20 mL/min/kg	No impairment
Class B	16–20 mL/min/kg	Mild to moderate impairment
Class C	10–15 mL/min/kg	Moderate to severe impairment
Class D	<10 mL/min/kg	Severe impairment

2.3. Statistical Analysis

Data were analyzed using SPSS version 21 (IBM, Armonk, NY, USA). All continuous data were tested for normal distribution using the Shapiro-Wilk normality test, and presented as mean (SD) or as median (IQR), where appropriate. Nominal data (fibromyalgia and severity/disability) were compared using the chi-square test (a 3 × 2 table and a 4 × 3 table). For continuous data groups were compared using the paired or unpaired t-test where appropriate. Within group comparison was done by the ordinary one-way analysis of variance (ANOVA) or Kruskal-Wallis test where appropriate.

Where significant, results were then explored further using the post-hoc Tukey's test or Dunn's test where appropriate. Within group comparison was done by the two-way analysis of variance (ANOVA). Where significant, results were then explored further using the post-hoc Holm-Sidak test. Graphpad Prism version 8.4.2 (Graphpad software, La Jolla, CA, USA) was used for the graphical representation of data in the figures.

3. Results

3.1. Baseline Characteristics

Table 2 shows the baseline characteristics of the 82 female ME/CFS patients. According to the ICC criteria, 31 were graded as mild, 31 were graded as moderate and 20 were graded as severe. No significant differences were found with respect to age, height, weight, BSA, BMI and disease duration. In the mild ME/CFS group 14 (45%) patients were classified as having comorbid fibromyalgia, in the moderate ME/CFS group 18 (58%) were classified as having fibromyalgia and in the severe ME/CFS group 12 (60%) were classified as having fibromyalgia (chi-square analysis 3×2 table: $p = 0.48$).

Table 2. Baseline criteria for female ME/CFS patients.

	Group 1 Severe ($n = 20$)	Group 2 Moderate ($n = 31$)	Group 3 Mild ($n = 31$)	ANOVA/Kruskal-Wallis Test
Age (years)	39 (10)	41 (10)	42 (9)	$F_{(2, 79)} = 0.46$; $p = 0.63$
Height (cm)	171 (7)	171 (7)	169 (6)	$F_{(2, 79)} = 1.56$; $p = 0.22$
Weight (kg)	70 (61–77)	69 (63–80)	65 (60–72)	$X^2(2) = 1.144$; $p = 0.56$
BSA (m^2)	1.4 (1.2–1.8)	1.4 (1.3–1.5)	1.3 (1.2–1.4)	$X^2(2) = 1.690$; $p = 0.43$
BMI (kg/m^2)	22.7 (21.8–27.6)	23.9 (20.7–27.6)	23.4 (21.4–26.7)	$X^2(2) = 0.032$; $p = 0.98$
Disease duration (years)	15.9 (9.3)	13.3 (8.9)	13.5 (9.3)	$F_{(2, 79)} = 0.56$; $p = 0.57$

Data are presented as mean (SD) or median (IQR). BMI: body mass index (DuBois formula); BSA: body surface area.

3.2. Two-Day CPET Data for ME/CFS Female Patients with Severe, Moderate and Mild Disease

Figure 1 shows the peak oxygen consumption for both CPET-1 and CPET-2 for mild, moderate and severe ME/CFS. For the mild disease group there was a significant decrease in peak VO_2 of 2 mL/min/kg (−6%) (a change from 23 (5) to 21 (5) mL/min/kg; $p = 0.003$). For the moderate disease group there was a significant decrease in peak VO_2 of 2 mL/min/kg (−11%) (a change from 17 (3) to 16 (4) mL/min/kg: $p = 0.0001$). For the severe disease group there was a significant decrease in peak VO_2 of 2 mL/min/kg (−12%) (a change from 14 (3) to 12 (3) mL/min/kg: $p = 0.003$). Comparison of day 1 mild vs. moderate, mild vs. severe, and moderate vs. severe disease severity showed a significant difference between the groups (p ranging between 0.0001 and <0.0001). Comparison of day 2 mild vs. moderate, mild vs. severe, and moderate vs. severe disease severity showed a significant difference between the groups (all $p < 0.0001$).

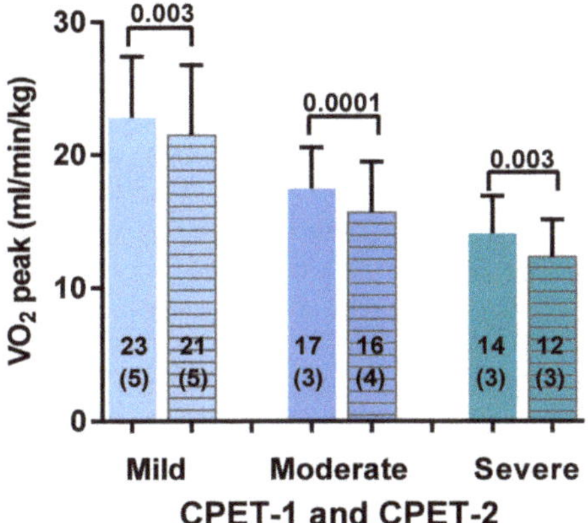

Figure 1. Peak oxygen consumption for both CPET-1 and CPET-2 for mild, moderate and severe ME/CFS. CPET: cardiopulmonary exercise test; VO_2: oxygen consumption. For between group comparisons see the result section. CPET-1 is presented by a solid bar and CPET-2 is represented by a lined bar.

Figure 2 shows the oxygen consumption at the ventilatory threshold for both CPET-1 and CPET-2 for mild, moderate and severe ME/CFS. For the mild disease group there was a significant decrease in VO_2 at the ventilatory threshold of 3 mL/min/kg (−21%) (a change from 14 (2) to 11 (2) mL/min/kg; $p < 0.0001$). For the moderate disease group there was a significant decrease in VO_2 at the ventilatory threshold of 3 mL/min/kg (−21%) (a change from 11 (2) to 9 (2) mL/min/kg: $p < 0.0001$). For the severe disease group there was a significant decrease in VO_2 at the ventilatory threshold of 2 mL/min/kg (−19%) (a change from 10 (2) to 8 (2) mL/min/kg: $p < 0.0001$). Comparison of day 1 mild vs. moderate, mild vs. severe, and moderate vs. severe disease severity showed a significant difference between the groups (p ranging between 0.008 and <0.0001). Comparison of day 2 moderate vs. severe disease severity was not significantly different ($p = 0.04$). Comparing mild vs. moderate and mild vs. severe disease severity showed a significant difference between the groups (p ranging between 0.0007 and <0.0001).

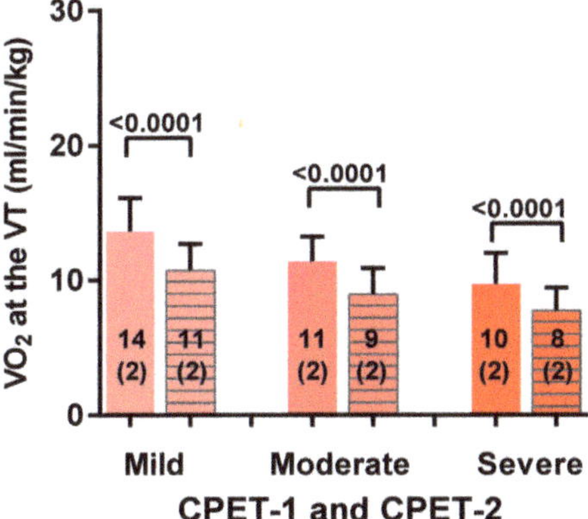

Figure 2. Oxygen consumption at the ventilatory threshold for both CPET-1 and CPET-2 for mild, moderate and severe ME/CFS patients. CPET: cardiopulmonary exercise test; VO_2: oxygen consumption; VT: ventilatory (or anaerobic) threshold. For between group comparisons see the result section. CPET-1 is presented by a solid bar and CPET-2 is represented by a lined bar.

Figure 3 shows the peak workload for both CPET-1 and CPET-2 for mild, moderate and severe ME/CFS. For the mild disease group there was a significant decrease in peak workload of 14 Watt (-10%) (a change from 144 (20) to 130 (22) Watt; $p < 0.0001$). For the moderate disease group there was a significant decrease in peak workload of 19 Watt (-16%) (a change from 117 (21) to 98 (24) Watt: $p < 0.0001$). For the severe disease group there was a significant decrease in peak workload of 17 Watt (-19%) (a change from 90 (26) to 73 (24) Watt: $p < 0.0001$). Comparison of day 1 mild vs. moderate, mild vs. severe, and moderate vs. severe disease severity showed a significant difference between the groups (p ranging between 0.0003 and <0.0001). Comparison of day 2 mild vs. moderate, mild vs. severe, and moderate vs. severe disease severity showed a significant difference between the groups (p ranging between 0.0003 and <0.0001).

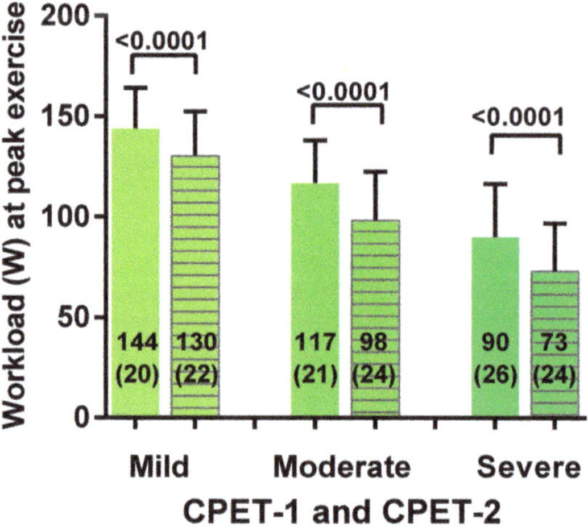

Figure 3. Peak workload for both CPET-1 and CPET-2 for mild, moderate and severe ME/CFS patients. CPET: cardiopulmonary exercise test. For between group comparisons see the result section. CPET-1 is presented by a solid bar and CPET-2 is represented by a lined bar.

Figure 4 shows the workload at the ventilatory threshold for both CPET-1 and CPET-2 for mild, moderate and severe ME/CFS. For the mild disease group there was a significant decrease in workload at the ventilatory threshold of 19 Watt (-26%) (a change from 69 (20) to 50 (19) Watt; $p < 0.0001$). For the moderate disease group there was a significant decrease in workload at the ventilatory threshold of 20 Watt (-31%) (a change from 61 (19) to 41 (15) Watt: $p < 0.0001$). For the severe disease group there was a significant decrease in workload at the ventilatory threshold of 18 Watt (-33%) (a change from 53 (19) to 36 (16) Watt: $p < 0.0001$). Comparison of day 1 mild vs. moderate and moderate vs. severe disease severity were not significantly different ($p = 0.15$ and 0.13 respectively. Comparison between mild vs. severe disease severity was significantly different ($p = 0.006$). Comparison of day 2 mild vs. moderate and moderate vs. severe disease severity were not significantly different ($p = 0.21$ and 0.05 respectively. Comparison between mild vs. severe disease severity was significantly different ($p = 0.006$).

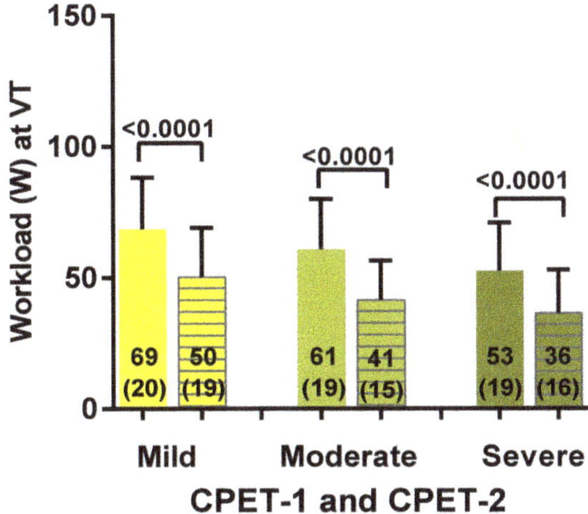

Figure 4. Workload at the ventilatory threshold for both CPET-1 and CPET-2 for mild, moderate and severe ME/CFS patients. CPET: cardiopulmonary exercise test; VT: ventilatory (or anaerobic) threshold. For between group comparisons see the result section. CPET-1 is presented by a solid bar and CPET-2 is represented by a lined bar.

3.3. Comparison of ME/CFS Patients with Severe, Moderate and Mild Disease for CPET Day-1 and Day-2 Variables

Table 3 shows the percent difference in CPET parameters from CPET-2 and CPET-1 for VO_2 VT, VO_2 peak, heart rate at the VT and peak exercise, and workload at the VT and at peak exercise for severe, moderate and mild ME/CFS patients. The post-hoc analysis showed that there was only a significantly higher decrease in the percent change in peak workload of severe patients compared to mild patients ($p = 0.019$).

Table 3. Percent differences from CPET-2 minus CPET-1 comparing female ME/CFS patients with severe, moderate and mild disease severity.

Percent Difference CPET-2 Minus CPET-1	Group 1 Severe ($n = 20$)	Group 2 Moderate ($n = 31$)	Group 3 Mild ($n = 31$)	ANOVA and Post-Hoc Tukey's Test
VO_2 peak (mL/min/kg)	−12 (14)	−11 (14)	−6 (11)	$F(2, 79) = 1.28; p = 0.28$
HR peak (bpm)	−7 (6)	−6 (7)	−3 (6)	$F(2, 79) = 2.08; p = 0.13$
Workload peak (Watts)	−19 (11)	−16 (15)	−10 (8)	$F(2, 79) = 4.37; p = 0.016$. Post-hoc tests: 1 vs. 2 $p = 0.083$; 1 vs. 3 $p = 0.019$ and 2 vs. 3 $p = 0.68$
VO_2 VT (mL/min/kg)	−19 (11)	−21 (12)	−21 (11)	$F(2, 79) = 0.26; p = 0.77$
HR VT (bpm)	−7 (5)	−9 (6)	−8 (7)	$F(2, 79) = 0.78; p = 0.46$
Workload VT (Watts)	−33 (20)	−31 (18)	−26 (18)	$F(2, 79) = 0.54; p = 0.58$
RER	−5 (7)	−2 (9)	−3 (7)	$F(2, 79) = 0.83; p = 0.44$

Data are presented as mean (SD). VT: ventilatory threshold; CPET: cardiopulmonary exercise test; HR: heart rate; VO_2: oxygen consumption; RER: respiratory exchange ratio.

Figure 5 shows the subdivision of mild, moderate and severe disease severity and Weber disability grades A-D for both CPET-1 and CPET-2 (panel A: mild ME/CFS CPET-1; panel B: mild ME/CFS CPET-2; panel C: moderate ME/CFS CPET-1; panel D: moderate ME/CFS CPET-2; panel E: severe ME/CFS CPET-1; panel F: severe ME/CFS CPET-2). A clear shift is visible for all three severity groups with more disability on day-2. Chi square testing was highly significantly different for both day-1 and day-2 between the three severity groups ($p < 0.0001$).

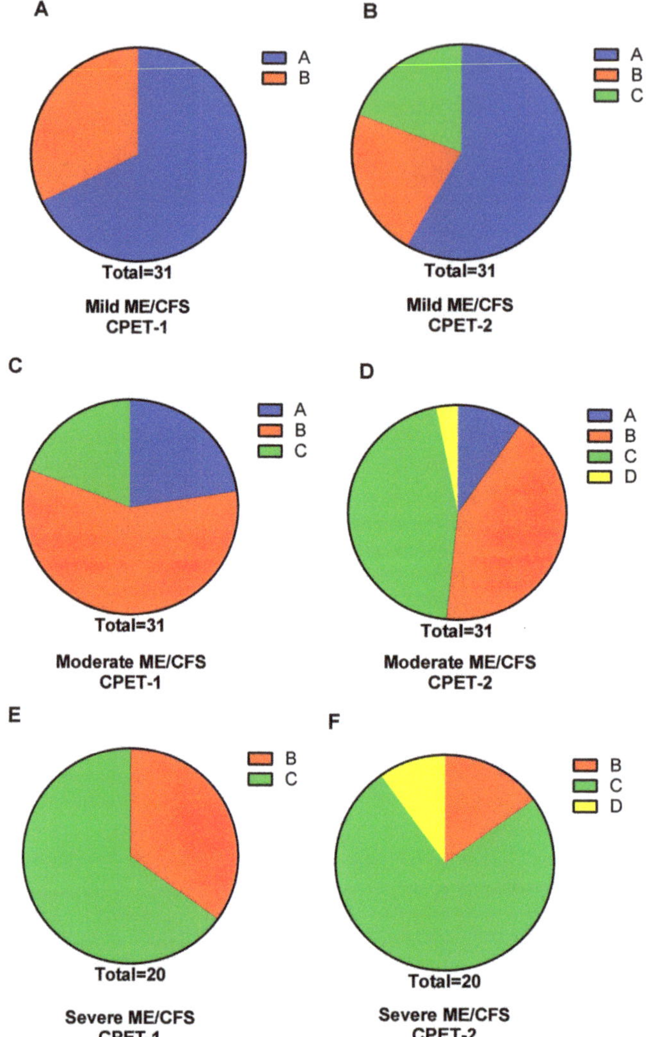

Figure 5. Disability grading according to Weber on both CPET-1 and CPET-2 for mild, moderate and severe ME/CFS patients. CPET: cardiopulmonary exercise test; Weber grading (**A**) (>20 mL/min/kg) (blue), (**B**) (16–20 mL/min/kg) (red), (**C**) (10–15 mL/min/kg) (green) and (**D**) (<10 mL/min/kg) (yellow). Panels (**A,C,E**) are representing CPET-1 for respectively mild, moderate and severe disease. Panels (**B,D,F**) are representing CPET-2 for respectively mild, moderate and severe disease [30].

Table 4 shows the RER results for each severity group by CPET day. A two-way analysis showed no significance for interaction between the 3 disease severity groups and either CPET day ($p = 0.59$).

Table 4. RER values for each severity group on each day.

Group 1 Severe ($n = 20$)	Group 2 Moderate ($n = 31$)	Group 3 Mild ($n = 31$)	ANOVA
CPET day-1			
1.08 (0.09)	1.09 (0.09)	1.13 (0.11)	$F(2.79) = 2.02; p = 0.14$
CPET day-2			
1.02 (0.11)	1.07 (0.11)	1.10 (0.11	$F(2.79) = 2.77; p = 0.07$

4. Discussion

The main finding of this study is that with a 2-day CPET protocol there is a consistent decrease from mildly affected to severely affected ME/CFS patients in peak oxygen consumption, oxygen consumption at the ventilatory threshold, peak workload and workload at the ventilatory threshold. We demonstrated a greater degree of disability in all disease severities when CPET-1 was compared to CPET-2. We believe this is the first study of 2-day CPET protocols in ME/CFS patients to include severity grading in the analysis [15,18–20,23–26].

4.1. Two-Day Cardiopulmonary Exercise Test Studies Reported in Literature

Eight studies have reported the results of 2-day CPETs in male and/or female ME/CFS patients. The relevant CPET parameters for female ME/CFS patients are presented in Table 5. All studies reported peak oxygen consumption and oxygen consumption at the ventilatory threshold [15,18–20,23–26]. Exact data could not be derived from one study, which was excluded in this overview [23]. Three studies reported percent peak oxygen consumption [15,25,26]. Two studies reported the percent oxygen consumption at the ventilatory threshold [25,26]. Six studies reported both the peak workload and workload at the ventilatory threshold [15,18–20,24–26].

Table 5. Ranges of CPET parameters (oxygen consumption at peak exercise and at the ventilatory threshold, workload at peak exercise and at the ventilatory threshold) in previous literature ranging from the minimal to the maximal value reported on day-1 and day-2 and the ranges from the CPET values from the present study ranging from the minimum value (severe disease) to the maximal value (mild disease) [15,18–20,24,25].

	Literature Day-1	Literature Day-2	Present Study Day-1	Present Study Day-2
Peak VO$_2$ (mL/min/kg)	19–26	17–21	14–23	12–21
VT VO$_2$ (mL/min/kg)	12–15	9–12	10–14	8–11
Peak Workload (Watt)	110–132	102–125	90–144	73–130
VT Workload (Watt)	50–62	22–54	53–69	36–50

VO$_2$: oxygen consumption; VT: ventilatory threshold.

Our lowest values were measured in severe ME/CFS patients, which are lower than results reported in the literature. It is therefore less likely that the more severely affected patients were included in these previous studies. Further studies of exercise intolerance from severely affected ME/CFS patients should improve our understandings of the true spectrum of disease severity.

Interestingly, with increasing severity, the percentage decrease in peak oxygen consumption from day-1 to day-2 was not significantly different between mild, moderate and severe patients, whereas the percentage decrease in peak workload from day-1 to day-2 was significantly different only between mild and severe patients. The larger percentage decrease in workload at day-2 in severe ME/CFS patients compared to mild and moderate ME/CFS patients, may indicate a more pronounced absence of recovery from day-1 as a measure of post-exertional malaise. This also needs to be confirmed in larger prospective studies.

The change in Weber's disability metric is an alternative validation of the clinical severity grading from ICC. The increase in disability grading on day-2 versus day-1 supports the severity classification. The concordance between the Weber disability metric and the ICC severity grading was not examined as part of this study.

We included patients who reached a maximal clinical effort as judged by the supervising cardiologist, even if the RER was below 1.1. Although an RER of less than 1.1 is often viewed as indicating inadequate effort, patients with severely impaired exercise tolerance can develop skeletal muscle exhaustion earlier than central hemodynamic and ventilatory factors become limiting. This in turn interrupts exercise at peak respiratory exchange ratio values lower than 1.00 [40]. Metabolic skeletal muscle abnormalities are present ME/CFS [6,41,42], and pain in those with co-morbid fibromyalgia can also limit maximal exercise performance. In support of including those with maximal clinical effort but an RER less than 1.1, we have recently shown that adults with ME/CFS with and without an RER of at least 1.1 did not differ with regard to correlations between other measures of exercise, including the physical activity subscale of the SF-36, peak oxygen consumption and the number of steps per day [43]. In those with a concomitant diagnosis of fibromyalgia the RER in that study was significantly lower than in the ME/CFS patients without fibromyalgia; exercise in this subgroup was terminated due to of muscle pain. The %peak VO_2, the number of steps and the physical functioning scale were not different between ME/CFS patients with and without fibromyalgia. The results of the current study are consistent with the earlier report, and argue for including in ME/CFS studies those who meet criteria for maximal effort as judged by a clinician even when the RER is less than 1.1. Excluding these patients has the potential to underestimate the severity of the activity limitations in ME/CFS.

4.2. Limitations

First, we did not include a group of sedentary controls for comparison. Second, this was not a prospective trial, as patients underwent the 2-day CPET protocol not only for clinical testing, but also for security claim reasons or to examine the hypothesis that deconditioning accounted for the results. Poor exercise results during testing and in daily life are often thought to be caused by deconditioning rather than being the result of a disease with prominent disabilities. This may have led to inclusion bias.

5. Conclusions

This large 2-day CPET protocol study confirms and extends previous findings of the reduction of various exercise variables in ME/CFS patients on day-2. This is the first study to demonstrate that disease severity negatively influences exercise capacity in female ME/CFS patients, confirming that deterioration in peak workload from day-1 to day-2 is largest in the severe ME/CFS patient group.

Author Contributions: C.M.C.v.C., P.C.R., and F.C.V. conceived the study, C.M.C.v.C. and F.C.V. collected the data, C.M.C.v.C. performed the primary data analysis and F.C.V., and P.C.R. performed secondary data analyses. All authors were involved in the drafting and review of the manuscript. All authors have read and agreed to the published version of the manuscript.

Funding: This research received no external funding.

Acknowledgments: In this section you can acknowledge any support given which is not covered by the author contribution or funding sections. This may include administrative and technical support, or donations in kind (e.g., materials used for experiments).

Conflicts of Interest: The authors declare no conflict of interest.

References

1. Fukuda, K. The Chronic Fatigue Syndrome: A Comprehensive Approach to Its Definition and Study. *Ann. Intern. Med.* **1994**, *121*, 953–959. [CrossRef] [PubMed]
2. Carruthers, B.M.; van de Sande, M.I.; de Meirleir, K.L.; Klimas, N.G.; Broderick, G.; Mitchell, T.; Staines, D.; Powles, A.C.P.; Speight, N.; Vallings, R.; et al. Myalgic encephalomyelitis: International consensus criteria. *J. Intern. Med.* **2011**, *270*, 327–338. [CrossRef] [PubMed]
3. Clayton, E.W. Beyond Myalgic Encephalomyelitis/Chronic Fatigue Syndrome. *JAMA* **2015**, *313*, 1101. [CrossRef] [PubMed]
4. Institute of Medicine (IOM). *Beyond Mayalgic Encephalomyelitis/Chronic Fatigue Syndrome: Redefining an Illness*; The National Academies Press: Washington, DC, USA, 2015.
5. Paul, L.; Wood, L.; Behan, W.M.; MacLaren, W.M. Demonstration of delayed recovery from fatiguing exercise in chronic fatigue syndrome. *Eur. J. Neurol.* **1999**, *6*, 63–69. [CrossRef]
6. Jones, D.E.J.; Hollingsworth, K.G.; Taylor, R.; Blamire, A.; Newton, J.L. Abnormalities in PH handling by peripheral muscle and potential regulation by the autonomic nervous system in chronic fatigue syndrome. *J. Intern. Med.* **2010**, *267*, 394–401. [CrossRef]
7. Mitchell, J.H.; Sproule, B.J.; Chapman, C.B. The physiological meaning of the maximal oxygen intake test. *J. Clin. Investig.* **1958**, *37*, 538–547. [CrossRef]
8. Saltin, B.; Astrand, P.O. Maximal oxygen uptake in athletes. *J. Appl. Physiol.* **1967**, *23*, 353–358. [CrossRef]
9. Mezzani, A.; Agostoni, P.; Cohen-Solal, A.; Corrà, U.; Jegier, A.; Kouidi, E.; Mazic, S.; Meurin, P.; Piepoli, M.; Simon, A.; et al. Standards for the use of cardiopulmonary exercise testing for the functional evaluation of cardiac patients: A report from the Exercise Physiology Section of the European Association for Cardiovascular Prevention and Rehabilitation. *Eur. J. Cardiovasc. Prev. Rehab.* **2009**, *16*, 249–267. [CrossRef]
10. Guazzi, M.; Adams, V.; Conraads, V.; Halle, M.; Mezzani, A.; Vanhees, L.; Arena, R.; Fletcher, G.F.; Forman, D.E.; Kitzman, D.W.; et al. EACPR/AHA Scientific Statement. Clinical recommendations for cardiopulmonary exercise testing data assessment in specific patient populations. *Circulation* **2012**, *126*, 2261–2274. [CrossRef]
11. De Becker, P.; Roeykens, J.; Reynders, M.; McGregor, N.; De Meirleir, K. Exercise capacity in chronic fatigue syndrome. *Arch. Intern. Med.* **2000**, *160*, 3270–3277. [CrossRef]
12. Fulcher, K.Y.; White, P.D. Strength and physiological response to exercise in patients with chronic fatigue syndrome. *J. Neurol. Neurosurg. Psychiatry* **2000**, *69*, 302–307. [CrossRef] [PubMed]
13. Hodges, L.; Nielsen, T.; Baken, D. Physiological measures in participants with chronic fatigue syndrome, multiple sclerosis and healthy controls following repeated exercise: A pilot study. *Clin. Physiol. Funct. Imaging* **2017**, *38*, 639–644. [CrossRef] [PubMed]
14. Jammes, Y.; Steinberg, J.G.; Mambrini, O.; Bregeon, F.; Delliaux, S. Chronic fatigue syndrome: Assessment of increased oxidative stress and altered muscle excitability in response to incremental exercise. *J. Intern. Med.* **2005**, *257*, 299–310. [CrossRef] [PubMed]
15. Keller, B.A.; Pryor, J.L.; Giloteaux, L. Inability of myalgic encephalomyelitis/chronic fatigue syndrome patients to reproduce VO2peak indicates functional impairment. *J. Transl. Med.* **2014**, *12*, 104. [CrossRef] [PubMed]
16. Sargent, C.; Scroop, G.C.; Nemeth, P.M.; Burnet, R.B.; Buckley, J.D. Maximal oxygen uptake and lactate metabolism are normal in chronic fatigue syndrome. *Med. Sci. Sports Exerc.* **2002**, *34*, 51–56. [CrossRef] [PubMed]
17. Sisto, S.A.; LaManca, J.; Cordero, D.L.; Bergen, M.T.; Ellis, S.P.; Drastal, S.; Boda, W.L.; Tapp, W.N.; Natelson, B.H. Metabolic and cardiovascular effects of a progressive exercise test in patients with chronic fatigue syndrome. *Am. J. Med.* **1996**, *100*, 634–640. [CrossRef]
18. Snell, C.R.; Stevens, S.R.; Davenport, T.E.; Van Ness, J.M. Discriminative validity of metabolic and workload measurements for identifying people with chronic fatigue syndrome. *Phys. Ther.* **2013**, *93*, 1484–1492. [CrossRef]
19. Vanness, J.M.; Snell, C.R.; Stevens, S.R. Diminished cardiopulmonary capacity during post-exertional malaise. *J. Chronic Fatigue Syndr.* **2007**, *14*, 77–85. [CrossRef]

20. Vermeulen, R.C.W.; Kurk, R.M.; Visser, F.C.; Sluiter, W.; Scholte, H.R. Patients with chronic fatigue syndrome performed worse than controls in a controlled repeated exercise study despite a normal oxidative phosphorylation capacity. *J. Transl. Med.* **2010**, *8*, 93. [CrossRef]

21. Vermeulen, R.C.W.; van Eck, I.W.V. Decreased oxygen extraction during cardiopulmonary exercise test in patients with chronic fatigue syndrome. *J. Transl. Med.* **2014**, *12*, 20. [CrossRef]

22. Wallman, K.; Morton, A.R.; Goodman, C.; Grove, J.R. Physiological responses during a submaximal cycle test in chronic fatigue syndrome. *Med. Sci. Sports Exerc.* **2004**, *36*, 1682–1688. [CrossRef] [PubMed]

23. Lien, K.; Johansen, B.; Veierød, M.B.; Haslestad, A.S.; Bøhn, S.K.; Melsom, M.N.; Kardel, K.R.; Iversen, P.O. Abnormal blood lactate accumulation during repeated exercise testing in myalgic encephalomyelitis/chronic fatigue syndrome. *Physiol. Rep.* **2019**, *7*, e14138. [CrossRef] [PubMed]

24. Nelson, M.J.; Buckley, J.D.; Thomson, R.L.; Clark, D.; Kwiatek, R.; Davison, K. Diagnostic sensitivity of 2-day cardiopulmonary exercise testing in Myalgic Encephalomyelitis/Chronic Fatigue Syndrome. *J. Transl. Med.* **2019**, *17*, 80. [CrossRef] [PubMed]

25. Van Campen, C.L.M.; Visser, F.C. Validity of 2-day cardiopulmonary exercise testing in female patients with myalgic encephalomyelitis/chronic fatigue syndrome. *Int. J. Curr. Res.* **2020**, *12*, 10436–10442.

26. Van Campen, C.L.M.; Rowe, P.C.; Visser, F.C. Validity of 2-day cardiopulmonary exercise testing in male patients with myalgic encephalomyelitis/Chronic fatigue syndrome. *Adv. Phys. Educ.* **2020**, *10*, 68–80. [CrossRef]

27. Glaser, S.; Koch, B.; Ittermann, T.; Schaper, C.; Dorr, M.; Felix, S.B.; Henry, V.; Ewert, R.; Hansen, J.E. Influence of age, sex, body size, smoking, and beta blockade on key gas exchange exercise parameters in an adult population. *Eur. J. Cardiovasc. Prev. Rehabil.* **2010**, *17*, 469–476.

28. Cureton, K.; Bishop, P.; Hutchinson, P.; Newland, H.; Vickery, S.; Zwiren, L. Sex difference in maximal oxygen uptake. Effect of equating haemoglobin concentration. *Eur. J. Appl. Physiol. Occup. Physiol.* **1986**, *54*, 656–660. [CrossRef]

29. Fletcher, G.F.; Balady, G.J.; Amsterdam, E.A.; Chaitman, B.; Eckel, R.; Fleg, J.; Froelicher, V.F.; Leon, A.S.; Piña, I.L.; Rodney, R.; et al. Exercise standards for testing and training: A statement for healthcare professionals from the American Heart Association. *Circulation* **2001**, *104*, 1694–1740. [CrossRef]

30. Weber, K.T.; Janicki, J.S. Cardiopulmonary exercise testing for evaluation of chronic cardiac failure. *Am. J. Cardiol.* **1985**, *55*, A22–A31. [CrossRef]

31. Martina, J.R.; Westerhof, B.E.; Van Goudoever, J.; De Beaumont, E.M.F.H.; Truijen, J.; Kim, Y.-S.; Immink, R.V.; Jöbsis, D.A.; Hollmann, M.W.; Lahpor, J.R.; et al. Noninvasive continuous arterial blood pressure monitoring with Nexfin®. *Anesthesiology* **2012**, *116*, 1092–1103. [CrossRef]

32. Beaver, W.L.; Wasserman, K.; Whipp, B.J. A new method for detecting anaerobic threshold by gas exchange. *J. Appl. Physiol.* **1986**, *60*, 2020–2027. [CrossRef] [PubMed]

33. Williams, C.G.; Wyndham, C.H.; Kok, R.; Rahden, M.J.E. Effect of training on maximum oxygen intake and on anaerobic metabolism in man. *Graefe's Arch. Clin. Exp. Ophthalmol.* **1967**, *24*, 18–23. [CrossRef]

34. Wasserman, K.; McIlroy, M.B. Detecting the threshold of anaerobic metabolism in cardiac patients during exercise. *Am. J. Cardiol.* **1964**, *14*, 844–852. [CrossRef]

35. Coen, B.; Urhausen, A.; Kindermann, W. Individual anaerobic threshold: Methodological aspects of its assessment in running. *Int. J. Sports Med.* **2001**, *22*, 8–16. [CrossRef] [PubMed]

36. Katz, S.D.; Berkowitz, R.; LeJemtel, T.H. Anaerobic threshold detection in patients with congestive heart failure. *Am. J. Cardiol.* **1992**, *69*, 1565–1569. [CrossRef]

37. Yeh, M.P.; Gardner, R.M.; Adams, T.D.; Yanowitz, F.G.; Crapo, R.O. "Anaerobic threshold": Problems of determination and validation. *J. Appl. Physiol.* **1983**, *55*, 1178–1186. [CrossRef]

38. Myers, J.; Ashley, E. Dangerous curves. A perspective on exercise, lactate, and the anaerobic threshold. *Chest* **1997**, *111*, 787–795. [CrossRef]

39. Solberg, G.; Robstad, B.; Skjønsberg, O.H.; Borchsenius, F. Respiratory gas exchange indices for estimating the anaerobic threshold. *J. Sports Sci. Med.* **2005**, *4*, 29–36.

40. Mezzani, A. Cardiopulmonary exercise testing: Basics of methodology and measurements. *Ann. Am. Thorac. Soc.* **2017**, *14*, S3–S11. [CrossRef]

41. Brown, A.; Jones, D.E.; Walker, M.; Newton, J.L. Abnormalities of AMPK activation and glucose uptake in cultured skeletal muscle cells from individuals with chronic fatigue syndrome. *PLoS ONE* **2015**, *10*, e0122982. [CrossRef]

42. Jones, D.E.J.; Hollingsworth, K.G.; Jakovljevic, D.G.; Fattakhova, G.; Pairman, J.; Blamire, A.; Trenell, M.; Newton, J.L. Loss of capacity to recover from acidosis on repeat exercise in chronic fatigue syndrome: A case-control study. *Eur. J. Clin. Investig.* **2011**, *42*, 186–194. [CrossRef] [PubMed]
43. Van Campen, C.M.C.; Rowe, P.C.; Verheugt, F.W.A.; Visser, F.C. Physical activity measures in patients with myalgic encephalomyalitis/chronic fatigue syndrome: Correlations between peak oxygen consumption, the physical functioning scale of the SF-36 scale, and the number of steps from an activity meter. *J. Transl. Med.* **2020**, *18*, 228–238. [CrossRef] [PubMed]

Article

Homebound versus Bedridden Status among Those with Myalgic Encephalomyelitis/Chronic Fatigue Syndrome

Karl Conroy *, Shaun Bhatia, Mohammed Islam and Leonard A. Jason

Center for Community Research, DePaul University, Chicago, IL 60614, USA; sbhatia3@depaul.edu (S.B.); mislam9@depaul.edu (M.I.); LJASON@depaul.edu (L.A.J.)
* Correspondence: kconro10@depaul.edu

Abstract: Persons living with myalgic encephalomyelitis/chronic fatigue syndrome (ME/CFS) vary widely in terms of the severity of their illness. It is estimated that of those living with ME/CFS in the United States, about 385,000 are homebound. There is a need to know more about different degrees of being homebound within this severely affected group. The current study examined an international sample of 2138 study participants with ME/CFS, of whom 549 were severely affected (operationalized as 'Homebound'). A subsample of 89 very severely affected participants (operationalized as 'Homebound-bedridden') was also examined. The findings showed a significant association between severely and very severely affected participants within the post-exertional malaise (PEM) symptom domain. The implications of these findings are discussed.

Keywords: ME/CFS; chronic fatigue syndrome; myalgic encephalomyelitis; illness severity; homebound; bedridden

Citation: Conroy, K.; Bhatia, S.; Islam, M.; Jason, L.A. Homebound versus Bedridden Status among Those with Myalgic Encephalomyelitis/ Chronic Fatigue Syndrome. *Healthcare* **2021**, *9*, 106. https://doi.org/ 10.3390/healthcare9020106

Academic Editor: Kenneth Friedman
Received: 29 December 2020
Accepted: 19 January 2021
Published: 20 January 2021

Publisher's Note: MDPI stays neutral with regard to jurisdictional claims in published maps and institutional affiliations.

1. Introduction

Myalgic encephalomyelitis/chronic fatigue syndrome (ME/CFS) is a debilitating chronic illness that affects about 0.42% of the adult population [1]. Persons with this disease experience diverse symptoms, including post-exertional malaise, cognitive impairment, and unrefreshing sleep [2–4]. When compared to adults with conditions such as cancer, stroke, schizophrenia, and renal failure, those with ME/CFS have reported a lower quality of life [5]. Research has also found that persons with ME/CFS have a poorer prognosis than those with a variety of other serious medical conditions [6].

Because those with ME/CFS vary significantly in their symptom presentation and functional status, study participants have been classified according to four categories of illness severity: (1) mild; (2) moderate; (3) severe; and (4) very severe [7–10]. Participants classified as 'mild' (Grade 1) can work and complete domestic tasks, but are restricted in leisure activities; participants classified as 'moderate' (Grade 2) are less mobile, restricted in daily activities, and have stopped working; participants classified at 'severe' (Grade 3) can perform only minimal self-care tasks (e.g., face washing, teeth cleaning) and are homebound; and finally, participants classified as 'very severe' (Grade 4) cannot complete daily tasks without assistance and are often bedridden [8]. It is also possible for participants to be classified as not homebound ('mild' or 'moderate'), homebound ('severe'), and bedridden ('very severe').

It has been estimated that between 10–25% of persons with ME/CFS might be homebound as they are severely or very severely affected [11,12]. A recent review article of 21 studies examined findings from severely and very severely affected study participants over the past two decades [13]. Most studies were limited by small samples of participants [14–18]. Out of the 21 studies identified by Strassheim and colleagues [13], only four included samples of more than 70 severely or very severely affected participants [8,19–21]. Since the review published by Strassheim and colleagues [13], several other studies exploring illness severity in ME/CFS have been published [22–24], but in each investigation, the number of participants classified as severely affected was also relatively small.

Among the studies on ME/CFS severity with larger samples, Cox and Findley [8] administered surveys to 72 inpatients who were severely or very severely affected in order to track their perceived activity level (at six-month intervals) and overall symptom duration (in years). However, the survey did not measure dimensions of symptomatology. Pheby and Faffron [19] surveyed 1104 study participants, of whom 124 were severely affected, for the purpose of associating severe ME/CFS to pre-illness risk factors such as occupation, personality type, and smoking or chemical exposure. Although several risk factors for severe ME/CFS were identified (e.g., being a homemaker, exposure to chemicals in the home), the instruments used to measure personality traits had not been validated for persons with ME/CFS, raising concern among other researchers [21]. Additionally, Friedberg and colleagues [20] enrolled 137 severely affected participants in a clinical trial to access the efficacy of fatigue self-management (as opposed to treatment at a clinic), and suggested that self-management might benefit those who are homebound.

Another study that examined a larger sample [21] compared severely affected participants ($n = 128$) to non-severely affected participants ($n = 409$) using a validated measure of ME/CFS symptomatology [25]. The study found that those who were severely affected (operationalized as homebound) reported significantly higher scores for 35 out of 54 symptoms. Furthermore, the study compared participants' functional status and found that homebound participants reported higher levels of bodily pain and lower levels of physical and social functioning than non-homebound participants.

Nonetheless, the findings presented by Pendergrast and colleagues [21] did not attempt to determine which symptoms were the most predictive of a participant being homebound. In addition, that study did not differentiate those who were homebound and bedridden from those who were homebound but not bedridden. The current study examined predictors of homebound versus not homebound status in participants with ME/CFS, and in a follow up analysis, examined predictors of participants being homebound but not bedridden versus homebound and bedridden.

2. Materials and Methods

2.1. Participants

The dataset for the current study was aggregated across several international samples as described below.

DePaul sample. An international convenience sample of adults who self-identified as having ME/CFS was collected by investigators at DePaul University. Eligible participants were those at least 18 years of age with a current self-reported and diagnosis of CFS or ME. The sample included 210 participants, of which 83.7% were female. The mean age of participants was 52.1 years ($SD = 11.2$). Most of the participants (74.2%) had completed at least a standard college degree.

BioBank 2016 sample. Collected by the Solve ME/CFS Initiative (https://solvecfs.org), the BioBank sample included participants recruited by physicians who specialized in diagnosing ME/CFS. Following exclusion due to missing data, the final sample consisted of 492 participants. In total, 77.3% of the sample was female with a mean age of 54.6 years ($SD = 12.0$). Seventy percent of participants had completed at least a standard college degree.

Newcastle sample. Participants from the Newcastle sample were those referred to the Newcastle-upon-Tyne Royal Victoria Infirmary clinic for a medical assessment due to a suspected diagnosis of CFS. Following exclusions due to missing data, the final sample consisted of 85 participants. The majority of the participants were female (80.0%) with a mean age of 45.9 years ($SD = 13.5$). Fifty percent of the sample had obtained at least a standard college degree.

Norway 1 sample. Participants from the Norway 1 sample were recruited from southern Norway, and were contacted via healthcare professionals, ME/CFS organizations, and the waiting list for a ME/CFS education program. Participants were required to be at least 18 years of age with a diagnosis of ME/CFS by a physician or medical specialist. Following exclusion due to incomplete data, the final sample consisted of 168 participants. Most participants (87.4%) were female with a mean age of 43.5 years (SD = 11.8). Just over half (50.6%) of the sample had completed at least a standard college degree.

Norway 2 sample. Participants from the Norway 2 sample were recruited from two sites: an inpatient medical ward for severely ill patients, and an outpatient clinic at a multidisciplinary ME/CFS center. Eligible participants were those between 18 and 65 years of age, who were able to read and write in Norwegian. Participants underwent a comprehensive medical history and examination conducted by an experienced physician and a psychologist. Following exclusion due to incomplete data, the final sample consisted of 51 participants. Most of the of participants (82.4%) were female with a mean age of 35.8 years (SD = 11.9). Approximately 39.2% of the sample had completed at least a standard college degree.

Norway 3 sample. Participants from the Norway 3 sample were recruited while attending a specialist ME/CFS tertiary care center. Eligible participants were those examined by an experienced physician and determined to meet the Canadian Consensus criteria for ME/CFS [3]. Participants were required to be between 18 and 65 years of age. Following exclusion due to incomplete data, the final sample consisted of 167 participants. The majority of the sample (82.0%) was female with a mean age of 38.7 years (SD = 11.2). Over half of participants (57.5%) had received at least a standard college degree.

Chronic Illness sample. The Chronic Illness respondents were from a convenience sample of adults living with chronic illnesses, including ME/CFS, collected by investigators at DePaul University [26]. Participants were recruited online using support groups, research forums, and social media platforms. Following the exclusion of participants due to missing data, the final sample consisted of 324 participants with a self-reported diagnosis of ME/CFS. Most of the sample (88.1%) was female with a mean age of 50.1 years (SD = 13.5). Most of the participants (70.9%) had completed at least a standard college degree.

Japan sample. Participants from the Japan sample were recruited from the ME Japan Association (https://mecfsjapan.com) and affiliated physician clinics specializing in ME/CFS. In total, 111 were included in the present study following exclusionary procedures due to incomplete data. Much of the sample (79.1%) were female with a mean age of 46.4 years (SD = 13.3). A little over half of the sample (52.7%) had completed at least a standard college degree.

Spain sample. Participants from the Spain sample were recruited from a tertiary referral center in Barcelona, Spain by a specialist physician with experience diagnosing ME/CFS. Eligible participants were required to be at least 18 years of age and meet the 1994 Fukuda case definition for CFS [2]. In total, 182 participants were included in the present study following exclusionary procedures due to incomplete data. Most of the sample (85.7%) was female with a mean age of 50.4 years (SD = 8.7), and 14.8% of participants had completed a least a standard college degree.

Amsterdam sample. Participants from the Amsterdam sample were selected from an outpatient clinic in the Netherlands (the CFS Medical Center in Amsterdam). Following exclusion due to incomplete data, the final sample consisted of 348 participants, all with physician report of ME/CFS diagnosis. Much of the sample (77.9%) was female with a mean age of 37.1 years (SD = 11.5). Under half of the participants (41.4%) had obtained at least a standard college degree.

2.2. Measures

The DePaul Symptom Questionnaire. Participants across all datasets completed the DePaul Symptom Questionnaire [25], a 54-item self-report measure of ME/CFS symptomatology. Participants were asked to rate the frequency of each symptom over the past six months on a five-point Likert scale with 0 = none of the time, 1 = a little of the time, 2 = about half the time, 3 = most of the time, and 4 = all of the time. Likewise, participants were asked to rate the severity of each symptom over the past six months on a similar scale with 0 = symptom not present, 1 = mild, 2 = moderate, 3 = severe, and 4 = very severe. All frequency and severity scores were standardized to a 100-point scale. Furthermore, the frequency and severity scores for each symptom were averaged to create a composite score, where higher scores indicated worse symptoms. These item composite scores were averaged, resulting in eight standardized symptom domain scores: (1) sleep dysfunction; (2) post-exertional malaise (PEM); (3) neurocognitive dysfunction; (4) immune dysfunction; (5) neuroendocrine dysfunction; (6) pain; (7) gastro-intestinal distress; and (8) orthostatic intolerance [25].

The DSQ-1 has shown good test-retest reliability among persons with ME/CFS and controls [27] and yielded valid, clinically useful results [28,29]. The DSQ-1 is available in the shared library of Research Electronic Data Capture (REDCap) [30,31] hosted at DePaul University. The full questionnaire can be viewed here: https://redcap.is.depaul.edu/surveys/?s=tRxytSPVVw.

Medical Outcomes Study 36-Item Short-Form Health Survey (SF-36 or RAND Questionnaire). Participants also completed the RAND 36-Item Short-Form Health Survey (SF-36) [32], a self-report measure assessing the impact of health outcomes on physical and mental functioning across eight domains: (1) physical functioning; (2) bodily pain; (3) role physical (limitations due to physical health problems); (4) role emotional (role limitations due to personal or emotional problems); (5) mental health; (6) social functioning; (7) vitality; and (8) general health. All domains are measured on 100-point scales, where higher scores indicate better health functioning.

The SF-36 has produced short- and long-term results that are psychometrically stable [33], has demonstrated strong internal consistency and good discriminant validity [34], and has shown utility across multiple illness groups [35], including fatiguing illnesses such as ME/CFS [36].

2.3. Illness Severity Status

Homebound versus Not Homebound. The DSQ-1 includes an item that asks participants to describe their fatigue/energy related illness over the past six months. Those who responded affirmatively to one of the following items were classified as 'Homebound:' "I am not able to work or do anything, and I am bedridden"; "I can walk around the house, but I cannot do light housework." Participants who responded affirmatively to any of the remaining items were classified as 'Not homebound': "I can do light housework, but I cannot work part-time"; "I can only work part-time at work or on some family responsibilities"; "I can work full-time, but I have no energy left for anything else"; and "I can work full-time and finish some family responsibilities, but I have no energy left for anything else" [25]. The group classified as 'Homebound' constituted 25.7% of the total sample (549 out of 2138). Although it is possible that some who indicated that they can do light housework but cannot work part-time are actually homebound, we decided to conservatively classify them as not homebound as it is at least conceivable that some within this group were able to engage in some activities outside their houses.

Bedridden versus Not Bedridden. Among those who were classified as 'Homebound,' we created two subcategories: 'Homebound-bedridden' and 'Homebound-not bedridden'. Participants who selected "I am not able to work or do anything, and I am bedridden" were classified as 'Homebound-bedridden,' whereas participants who selected "I can walk around the house, but I cannot do light housework" were classified as 'Homebound-not bedridden'. The group classified as 'Homebound-bedridden' comprised 16.2% of the original 'Homebound' group (89 out of 549) and the 'Homebound-not bedridden' group represented the remaining 83.8% (460 out of 549).

2.4. Statistical Procedure

Demographics. Chi-squared tests were conducted to determine if significant differences were present for demographic characteristics (gender, educational status, and marital status) across severity classifications. Independent samples *t*-tests were conducted to determine if significant differences in age were present. Variables which indicated the presence of demographic heterogeneity were included in base logistic regression models. These models were developed for two severity analyses ('Homebound' compared to 'Not homebound' and 'Homebound-bedridden' compared to 'Homebound-not bedridden').

Binary logistical regression. The criterion variable for our first analysis was 'Homebound' status compared to 'Not homebound' status, whereas the criterion variable for our second analysis was 'Homebound-bedridden' status compared to 'Homebound-not bedridden' status. Using a top-down analytic approach, we performed binary logistical regressions on the DSQ-1 and SF-36 domains, where each DSQ-1 domain represented a linear combination of individual symptom items [25]. To reduce chance findings with so many potential comparisons, we initially focused on DSQ-1 domains, and if significance was found within a domain, symptom items within those domains were examined in a later step. This multi-step process for variable selection [37] was chosen to facilitate an efficient analysis of the DSQ-1's extensive symptom inventory (54 items).

The demographic base models were used when testing the DSQ-1 and SF-36 domains, with each domain being tested individually (Step 1). Domains that were observed to be statistically significant in predicting severity status were entered into a forward stepwise selection procedure using likelihood ratio tests (Step 2).

If a statistically significant domain was identified in the forward stepwise procedure, we tested the domain's component symptom scores individually with the demographic base model and all statistically significant SF-36 domains (Step 3). In the development of the final models, those symptoms that were found to be statistically significant were entered into another forward stepwise selection procedure based on likelihood ratio tests (Step 4). IBM SPSS Statistics version 25 was used for all analyses [38].

In summary, rather than utilizing a statistical approach similar to Pendargrast and colleagues [21] to detect mean differences in specific symptoms in our ME/CFS severity groups, our iterative regression process specified two a priori group comparisons that were of interest (i.e., 'Homebound' versus 'Not homebound' and 'Homebound-bedridden' versus 'Homebound-not bedridden'). In preliminary work, we did find significant differences in symptoms and functionality between the three groups of participants, but our intent in the current study was to investigate two sets of comparisons among the illness severity groups.

3. Results

3.1. Demographics

Table 1 shows the demographic characteristics of the 'Homebound' versus 'Not homebound' groups. Statistical differences were observed in gender [χ^2 (1, 2, 112) = 13.07, $p < 0.001$] and educational status [χ^2 (1, 2, 106) = 15.71, $p < 0.001$]. The 'Homebound' group compared to the 'Not homebound' group had a smaller percentage of male participants (13.0% compared to 19.9%) and a smaller percentage of participants who had completed at least a standard college degree (48.9% compared to 58.7%). Based on these findings, subsequent regression analysis of the 'Homebound' group compared to the 'Not homebound' group was adjusted for gender and educational status.

Additionally, Table 1 shows the demographic characteristics for two sub-divisions of the 'Homebound' group: 'Homebound-not bedridden' and 'Homebound-bedridden'. Significant differences were observed in age [$t(526) = -6.79, p < 0.001$] and marital status [$\chi^2$ (1, 539) = 4.63, $p = 0.031$]. The 'Homebound-bedridden' group was significantly younger than the 'Homebound-not bedridden' group (mean ages were 37.5 and 48.3, respectively) and fewer participants were married (40.9% compared to 53.4%). Subsequent regression analysis of the 'Homebound-bedridden' group compared to the 'Homebound-not bedridden' group was adjusted for age and marital status.

Table 1. Demographic characteristics.

	Total Sample Size (*n* = 2138)			Homebound (*n* = 549)		
Characteristics	**Homebound**	**Not Homebound**	*p*	**Bedridden**	**Not Bedridden**	*p*
	(*n* = 549) *M (SD)*	(*n* = 1589) *M (SD)*		(*n* = 89) *M (SD)*	(*n* = 460) *M (SD)*	
Age	46.6 (14.0)	47.2 (13.5)	0.393	37.5 (12.1)	48.3 (13.7)	<0.001
	% (n)	*% (n)*		*% (n)*	*% (n)*	
Gender			<0.001			0.111
Male	13.0 (70)	19.9 (313)		18.2 (16)	11.9 (54)	
Female	87.0 (470)	80.1 (1259)		81.8 (72)	88.1 (398)	
Education			<0.001			0.811
At least a standard college degree	48.9 (265)	58.7 (918)		47.7 (42)	49.1 (223)	
Less than a standard college degree	51.1 (277)	41.3 (646)		52.3 (46)	50.9 (231)	
Marital			0.071			0.031
Married	51.4 (277)	55.9 (875)		40.9 (36)	53.4 (241)	
Not married	48.6 (262)	44.1 (691)		59.1 (52)	46.6 (210)	

3.2. Homebound Status

Table 2 (Step 1) shows the regression results for each DSQ-1 and SF-36 domain, tested individually and adjusted for gender and educational status. Every SF-36 and DSQ-1 domain was found to be a statistically significant predictor of 'Homebound' status. Table 2 (Step 2) shows the results of a forward stepwise selection procedure of statistically significant predictors from Step 1. Regarding the DSQ-1 domains, more severe scores in the PEM domain increased the odds of a participant being 'Homebound' [odds ratio (OR) = 1.034, 95% CI, (1.025, 1.044)]; no other DSQ-1 domains were found to be statistically significant. Regarding the SF-36 domains, higher levels of physical functioning and social functioning decreased the odds of a participant being 'Homebound' [OR = 0.957, 95% CI, (0.950, 0.965); OR = 0.980, 95% CI, (0.974, 0.987)].

Table 2 (Step 3) shows regression results for every symptom that constitutes the PEM domain, tested individually and adjusted for demographics, physical functioning, and social functioning. "Dead, heavy feeling after starting to exercise," "next day soreness or fatigue after non-strenuous, everyday activity," "mentally tired after the slightest effort," "minimum exercise makes you physically tired," and "physically drained or sick after mild activity" were found to be statistically significant predictors of 'Homebound' status; "muscle weakness" was not a significant predictor. Table 2 (Step 4) shows the results of a second forward stepwise selection of statistically significant predictors from Step 3, adjusted for gender, educational status, physical functioning, and social functioning. Regarding the symptoms, more severe scores for "next day soreness or fatigue after non-strenuous, everyday activity" and "physically drained or sick after mild activity" both increased the odds of a participant being 'Homebound' [OR = 1.016, 95% CI, (1.008, 1.025); OR = 1.024, 95% CI, (1.015, 1.032)]; no other symptoms were statistically significant.

3.3. Bedridden Status

Table 2 (Step 1) shows the regression results for each DSQ-1 and SF-36 domain, tested individually and adjusted for age and marital status. The PEM and neurocognitive dysfunction domains were the only statistically significant predictors of a participant being 'Homebound-bedridden'. Table 2 (Step 2) shows the results of a forward stepwise selection operation of statistically significant domain scores from Step 1 (PEM and neurocognitive dysfunction), adjusted for age and marital status. The only statistically significant domain was PEM, where more severe scores decreased the odds of a participant being 'Homebound-bedridden' (compared to 'Homebound-not bedridden') [OR = 0.974, 95% CI, (0.959, 0.989)].

Table 2. Binary logistic regressions predicting 'Homebound' status compared to 'Not homebound' status and 'Homebound-bedridden status compared to 'Homebound-not bedridden' status.

Iteration	Homebound [a] e [b] (95% CI)	Bedridden [b] e [b] (95% CI)
Step 1		
DSQ-1 domain		
Sleep	1.028 (1.022, 1.033)	1.005 (0.992, 1.017)
PEM	1.071 (1.063, 1.080)	0.974 (0.959, 0.989)
Neurocognitive	1.028 (1.022, 1.033)	0.985 (0.975, 0.997)
Immune	1.028 (1.021, 1.033)	0.998 (0.986, 1.010)
Neuroendocrine	1.015 (1.010, 1.019)	0.989 (0.978, 1.000)
Pain	1.018 (1.014, 1.022)	0.993 (0.984, 1.002)
Gastro-intestinal	1.012 (1.009, 1.016)	0.998 (0.989, 1.007)
Orthostatic	1.033 (1.027, 1.038)	0.993 (0.982, 1.005)
SF-36 domain		
Physical functioning	0.935 (0.929, 0.942)	1.000 (0.987, 1.014)
Role physical	0.965 (0.952, 0.978)	0.986 (0.951, 1.022)
Bodily pain	0.970 (0.964, 0.975)	1.001 (0.991, 1.012)
General health	0.967 (0.960, 0.974)	0.999 (0.982, 1.016)
Vitality	0.964 (0.957, 0.972)	1.006 (0.988, 1.026)
Social functioning	0.956 (0.950, 0.962)	1.004 (0.992, 1.017)
Role emotional	0.996 (0.994, 0.998)	1.003 (0.997, 1.008)
Mental health	0.988 (0.983, 0.993)	0.996 (0.985, 1.007)
Step 2		
Domain		
Age	-	0.937 (0.918, 0.956)
Marital status	-	0.802 (0.480, 1.339)
Gender	0.778 (0.612, 0.988)	-
Grade	0.841 (0.597, 1.186)	-
PEM	1.034 (1.025, 1.044)	0.974 (0.959, 0.989)
Physical functioning	0.957 (0.950, 0.965)	-
Social functioning	0.980 (0.974, 0.987)	-
Step 3		
DSQ-1 Symptom		
Heavy feeling	1.007 (1.002, 1.012)	0.988 (0.979, 0.997)
Soreness after activities	1.029 (1.021, 1.036)	0.979 (0.966, 0.992)
Mentally tired	1.013 (1.007, 1.018)	0.989 (0.978, 1.000)
Minimum exercise	1.027 (1.018, 1.034)	0.975 (0.962, 0.988)
Feeling drained	1.032 (1.024, 1.040)	0.977 (0.963, 0.989)
Muscle weakness	1.003 (0.998, 1.008)	0.997 (0.988, 1.006)
Step 4		
Variable		
Age	-	0.936 (0.917, 0.955)
Marital status	-	0.767 (0.453, 1.299)
Grade	0.813 (0.637, 1.038)	-
Gender	0.880 (0.620, 1.250)	-
Minimum exercise	-	0.974 (0.961, 0.988)
Soreness after activities	1.016 (1.008, 1.025)	-
Feeling drained	1.024 (1.015, 1.032)	-
Physical functioning	0.955 (0.948, 0.963)	-
Social functioning	0.985 (0.978, 0.992)	-

[a] 'Homebound' compared to 'Not homebound, [b] 'Homebound-bedridden' compared to 'Homebound-not bedridden'.

Table 2 (Step 3) shows regression results for every symptom that constitutes the PEM domain, tested individually and adjusted for age and marital status. "Dead, heavy feeling after starting to exercise," "next day soreness or fatigue after non-strenuous, everyday activity," "minimum exercise makes you physically tired," and "physically drained or sick after mild activity" were all significant predictors of a participant being 'Homebound-bedridden'; "mentally tired after the slightest effort" and "muscle weakness" were not significant. Table 2 (Step 4) shows the results of another forward stepwise selection operation of statistically significant predictors from Step 3, adjusted for age and marital status. Regarding the symptoms, more severe scores for "minimum exercise makes you physically tired" decreased the odds of a participant being 'Homebound-bedridden' (compared to 'Homebound-not bedridden') [OR = 0.974, 95% CI, (0.961, 0.988)]; no other symptoms were statistically significant.

4. Discussion

The findings of the current study indicated that PEM, social functioning, and physical functioning were significant predictors of a participant with ME/CFS being 'Homebound' (compared to 'Not homebound'). Among symptom items in the DSQ-1 PEM domain, "next day soreness or fatigue after non-strenuous, everyday activity" and "physically drained or sick after mild activity" were the strongest predictors of 'Homebound' status. These predictive results were consistent with the mean comparisons reported by Pendergrast and colleagues [21]. Moreover, the unique aspect of our study was subdividing the 'Homebound' group into two subgroups: 'Homebound-bedridden' and 'Homebound-not bedridden. We found that higher symptom scores in the PEM domain decreased the odds of a participant being 'Homebound-bedridden' (versus 'Homebound-not bedridden'). Among the PEM symptom items, "minimum exercise makes you physically tired" significantly decreased the odds of a participant being 'Homebound-bedridden.

Although several studies have mentioned the need for research that differentiates those with ME/CFS at varying levels of illness severity [3,8,39], existing research has focused on differences between participants who are severely and moderately affected [16–18,20–24] but not severely and very severely affected. While the illness severity gradation (i.e., mild, moderate, severe, very severe) proposed by Cox and Findley [8] describes differences between those who are severely affected (homebound) and those who are very severely affected (bedridden) in terms of symptom presentation (e.g., those who are bedridden might be sensitive to noise and light), their distinctions lacked an empirical foundation.

According to our findings, participants who were 'Homebound' (compared to participants who were 'Not homebound') were at increased odds of being less physically and socially functional, as well as exhibiting PEM symptomology, where "next day soreness or fatigue after non-strenuous, everyday activity" and "physically drained or sick after mild activity" had the strongest effect among the PEM items tested at Step 4. Inversely, participants who were 'Homebound-bedridden (compared to participants who were 'Homebound-not bedridden') were at decreased odds of exhibiting PEM symptomology, where "minimum exercise makes you physically tired" had the strongest effect at Step 4. These findings can be explained by the fact that participants who are bedridden (very severely affected) have fewer opportunities to engage in activities. For example, if a severely affected 'Homebound-not bedridden' participant with ME/CFS is expending significant amounts of their limited energy around their household, they may risk triggering more PEM symptoms than very severely affected 'Homebound-bedridden' participants who are less active. Indeed, when we compared the 'Homebound-bedridden' group to the 'Homebound-not bedridden' group at Step 3 using the component symptoms within the PEM domain, we found that symptoms involving activity such as "dead, heavy feeling after starting to exercise," "next day soreness or fatigue after non-strenuous, everyday activity," "minimum exercise makes you physically tired," and "physically drained or sick after mild activity" were statistically significant, whereas symptoms that did not explicitly involve activity, such as "mentally tired after the slightest effort" and "muscle weakness," were not statistically significant, which could mean that PEM triggered by mental exertion is experienced more evenly between the two groups. These findings suggest that treatment programs targeting persons who are homebound with ME/CFS should account for the heterogeneity of this population (i.e., bedridden and not bedridden). The symptomological differences identified in our study highlight the need for treatment programs that are tailored for two subgroups.

Of interest, the proportion of participants we classified as severely affected (25.7%, 549/2138) matched estimates offered by ME/CFS advocacy groups [11,12]. A recent study [40] estimated that 1.5 million persons suffer from ME/CFS in the United States. If 25.7% of those with ME/CFS are not able to leave their homes, there may be as many as 385,000 persons in the US who are homebound due to ME/CFS. Furthermore, our study found that 16.2% (89/549) of those who were homebound with ME/CFS were also bedridden, which equates to roughly 62,000 persons in the US. These estimates indicate a serious public health problem, as many who are homebound or bedridden due to ME/CFS may lack access to the healthcare system. Providing this group with adequate services will require attention and resources at many levels (e.g., research, treatment, and policy-making). We maintain that a crucial first step is to focus research on those who are severely and very severely affected, which will require methods that are sensitive to the needs of this population [39].

The current study had two limitations. First, the total sample ($n = 2138$) was aggregated from multiple sources, so there were inconsistencies in terms of participant recruitment and assessment. While a number of sources recruited participants who had complete a full medical review (e.g., Norway 1–3 samples, Spain sample), others allowed participants to self-report their medical diagnosis (e.g., DePaul sample, Chronic Illness sample). Second, our study used a variety of case ascertainment methods from recruitment using the internet to tertiary care settings, but such methods might have also increased the generalizability of the findings.

Our study found that participants who reported worse symptoms in the PEM domain [25] and less physical and social functioning [32] were at increased odds of being 'Homebound' (compared to 'Not homebound'). Among participants who were classified as 'Homebound,' those who reported worse symptoms in the PEM domain were at decreased odds of being 'Homebound-bedridden' (compared to 'Homebound-not bedridden'). We hypothesized that for participants who are 'Homebound,' those who are 'Homebound-bedridden' may experience less PEM symptomology because they are expending less energy. Based on the proportion of participants who were 'Homebound' in our study, we estimate that as many as 385,000 persons with ME/CFS are homebound in the United States. There is a pressing need to find ways of providing services to this under-resourced group.

Author Contributions: Conceptualization, K.C. and L.A.J.; methodology, K.C. and S.B.; formal analysis, K.C., S.B., and M.I.; investigation, K.C., S.B., and M.I.; data curation, K.C.; writing—original draft preparation, K.C. and S.B.; writing—review and editing, L.A.J. and M.I.; supervision, L.A.J. All authors have read and agreed to the published version of the manuscript.

Funding: This research received no external funding.

Institutional Review Board Statement: Not applicable.

Informed Consent Statement: Informed consent was obtained from all subjects involved in the study.

Data Availability Statement: The data presented in this study are available on request from the corresponding author.

Conflicts of Interest: The authors declare no conflict of interest.

References

1. Jason, L.A.; Richman, J.A.; Rademaker, A.W.; Jordan, K.M.; Plioplys, A.V.; Taylor, R.R.; McCready, W.; Huang, C.; Plioplys, S. A community-based study of chronic fatigue syndrome. *Arch. Intern. Med.* **1999**, *159*, 2129–2137. [CrossRef] [PubMed]
2. Fukuda, K.; Straus, S.E.; Hickie, I.; Sharpe, M.C.; Dobbins, J.G.; Komaroff, A.; Group ICFSS. The chronic fatigue syndrome: A comprehensive approach to its definition and study. *Ann. Intern. Med.* **1994**, *121*, 953–959. [CrossRef] [PubMed]

3.	Carruthers, B.M.; Jain, A.K.; De Meirleir, K.L.; Peterson, D.L.; Klimas, N.G.; Lerner, A.M.; Bested, A.C.; Flor-Henry, P.; Joshi, P.; Powles, A.C.P.; et al. Myalgic encephalomyelitis/chronic fatigue syndrome: Clinical working case definition, diagnostic and treatment protocols. *J. Chronic Fatigue Syndr.* **2003**, *11*, 7–115. [CrossRef]
4.	Institute of Medicine. *Beyond Myalgic Encephalomyelitis/Chronic Fatigue Syndrome: Redefining an Illness*; National Academic Press: Washington, DC, USA, 2015.
5.	Hvidberg, M.F.; Brinth, L.S.; Olesen, A.V.; Petersen, K.D.; Ehlers, L. The health-related quality of life for patients with myalgic encephalomyelitis/chronic fatigue syndrome (ME/CFS). *PLoS ONE* **2015**, *10*, e0132421. [CrossRef]
6.	Cairns, R.; Hotopf, M. A systematic review of describing the prognosis of chronic fatigue syndrome. *Occup. Med.* **2005**, *55*, 20–31. [CrossRef]
7.	Cox, D.L.; Findley, L.J. The management of chronic fatigue syndrome in an inpatient setting: Presentation of an approach and perceived outcome. *Br. J. Occup. Ther.* **1998**, *61*, 405–409. [CrossRef]
8.	Cox, D.L.; Findley, L.J. Severe and very severe patients with chronic fatigue syndrome: Perceived outcome following an inpatient programme. *J. Chronic Fatigue Syndr.* **2000**, *7*, 33–47. [CrossRef]
9.	Baker, R.; Shaw, E.J. Diagnosis and management of chronic fatigue syndrome or myalgic encephalomyelitis (or encephalopathy): Summary of NICE guidelines. *BMJ* **2007**, *335*, 446–448. [CrossRef]
10.	Carruthers, B.M.; van de Sande, M.I.; De Meirleir, K.L.; Klimas, N.G.; Broderick, G.B.; Mitchell, T.; Stains, D.; Powles, A.C.P.; Speight, N.; Vallings, R.; et al. Myalgic encephalomyelitis: International consensus criteria. *J. Intern. Med.* **2011**, *270*, 327–338. [CrossRef]
11.	ME Research UK. Severe ME—What Do We Know? Available online: https://www.meresearch.org.uk/severe-me-the-facts/ (accessed on 1 November 2020).
12.	25%, M.E.Group. What Is Severe M.E? Available online: https://25megroup.org/me (accessed on 1 November 2020).
13.	Strassheim, V.; Lambson, R.; Hackett, K.; Newton, J.L. What is known about severe and very severe chronic fatigue syndrome? A scoping review. *Fatigue* **2017**, *5*, 167–183. [CrossRef]
14.	Wernham, W.; Pheby, D.; Saffon, L. Risk factors for the development of severe ME/CFS—A pilot study. *J. Chronic Fatigue Syndr.* **2004**, *12*, 47–50. [CrossRef]
15.	Wiborg, J.F.; van der Werf, S.; Prins, J.B.; Bleijenberg, G. Being homebound with chronic fatigue syndrome: A multidimensional comparison with outpatients. *Psychiatry Res.* **2010**, *177*, 246–249. [CrossRef] [PubMed]
16.	Hardcastle, S.L.; Brenu, E.W.; Johnston, S.; Nguyen, T.; Huth, T.; Ramos, S.; Staines, D.; Marshall-Gradisnik, S. Longitudinal analysis of immune abnormalities in varying severities of chronic fatigue syndrome/myalgic encephalomyelitis patients. *J. Transl. Med.* **2015**, *13*, 299. [CrossRef] [PubMed]
17.	Hardcastle, S.L.; Brenu, E.W.; Johnston, S.; Nguyen, T.; Huth, T.; Ramos, S.; Staines, D.; Marshall-Gradisnik, S. Serum immune proteins in moderate and severe chronic fatigue syndrome/myalgic encephalomyelitis patients. *Int. J. Med. Sci.* **2015**, *12*, 764–772. [CrossRef]
18.	Hardcastle, S.L.; Brenu, E.W.; Johnston, S.; Nguyen, T.; Huth, T.; Wong, N.; Ramos, S.; Staines, D.; Marshall-Gradisnik, S. Characterisation of cell function and receptors in chronic fatigue syndrome/myalgic encephalomyelitis (CFS/ME). *BMC Immunol.* **2015**, *16*, 35. [CrossRef]
19.	Pheby, D.; Saffron, L. Risk factors for severe ME/CFS. *Biol. Med.* **2009**, *1*, 50–74.
20.	Friedberg, F.; Adamowicz, J.; Caikauskaite, I.; Seva, V.; Napoli, A. Efficacy of two delivery modes of behavioral self-management in severe chronic fatigue syndrome. *Fatigue* **2016**, *4*, 158–174. [CrossRef]
21.	Pendergrast, T.; Brown, A.; Sunnquist, M.; Jantke, R.; Newton, L.N.; Strand, E.B.; Jason, L.A. Housebound versus nonhousebound patients with myalgic encephalomyelitis and chronic fatigue syndrome. *Chronic Illn.* **2016**, *12*, 292–307. [CrossRef]
22.	Nacul, L.; de Barros, B.; Kingdon, C.C.; Cliff, J.M.; Clark, T.G.; Mudie, K.; Dockrell, H.M.; Lacerda, E.M. Evidence of clinical pathology abnormalities in people with myalgic encephalomyelitis/chronic fatigue syndrome (ME/CFS) from an analytic cross-sectional study. *Diagnostics* **2019**, *9*, 41. [CrossRef]
23.	Nacul, L.C.; Mudie, K.; Kingdon, C.C.; Lacerda, E.M. Hand grip strength as a clinical biomarker for ME/CFS and disease severity. *Front. Neurol.* **2018**, *9*, 992. [CrossRef]
24.	Tomas, C.; Elson, J.L.; Strassheim, V.; Newton, J.L.; Walker, M. The effect of myalgic encephalomyelitis/chronic fatigue syndrome severity on cellular bioenergetic function. *PLoS ONE* **2020**, *15*, e0231136. [CrossRef] [PubMed]
25.	Jason, L.A.; Sunnquist, M. The development of the DePaul Symptom Questionnaire: Original, expanded, brief, and pediatric versions. *Front. Pediatr.* **2018**, *6*, 330. [CrossRef] [PubMed]
26.	Jason, L.A.; Ohanian, D.; Brown, A.; Sunnquist, M.; McManimen, S.; Klebek, L.; Fox, P.; Sorenson, M. Differentiating multiple sclerosis from myalgic encephalomyelitis and chronic fatigue syndrome. *Insights Biomed.* **2017**, *2*, 11. [CrossRef] [PubMed]
27.	Jason, L.A.; So, S.; Brown, A.A.; Sunnquist, M.; Evans, M. Test–retest reliability of the DePaul Symptom Questionnaire. *Fatigue* **2015**, *3*, 16–32. [CrossRef]
28.	Jason, L.A.; Sunnquist, M.; Brown, A.; Evans, M.; Vernon, S.D.; Furst, J.; Simonis, V. Examining case definition criteria for chronic fatigue syndrome and myalgic encephalomyelitis. *Fatigue* **2014**, *2*, 40–56. [CrossRef]
29.	Murdock, K.W.; Wang, X.S.; Shi, Q.; Cleeland, C.S.; Fagundes, C.P.; Vernon, S.D. The utility of patient reported outcome measures among patients with myalgic encephalomyelitis/chronic fatigue syndrome. *Qual. Life Res.* **2017**, *26*, 913–921. [CrossRef]

30. Harris, P.A.; Taylor, R.; Thielke, R.; Payne, J.; Gonzalez, N.; Conde, J.G. Research electronic data capture (REDCap)—A metadata-driven methodology and workflow process for providing translational research informatics support. *J. Biomed. Inform.* **2009**, *42*, 377–381. [CrossRef]
31. Obeid, J.S.; McGraw, C.A.; Minor, B.L.; Conde, J.G.; Pawluk, R.; Lin, M.; Wang, J.; Banks, S.R.; Hemphill, S.A.; Taylor, R.; et al. Procurement of shared data instruments for research electronic data capture (REDCap). *J. Biomed. Inform.* **2013**, *46*, 259–265. [CrossRef]
32. Ware, J.E., Jr.; Sherbourne, C.D. The MOS 36-item short-form health survey (SF-36): I. Conceptual framework and item selection. *Med. Care* **1992**, *30*, 473–483. [CrossRef]
33. Stewart, A.L.; Hays, R.D.; Ware, J.E., Jr. The MOS short-form general health survey: Reliability and validity in a patient population. *Med. Care* **1988**, *26*, 724–735. [CrossRef]
34. McHorney, C.A.; Ware, J.E., Jr.; Lu, J.F.R.; Sherbourne, C.D. The MOS 36-item short-form health survey (SF-36): III. Tests of data quality, scaling assumptions, and reliability across diverse patient groups. *Med. Care* **1994**, *32*, 40–66. [CrossRef] [PubMed]
35. McHorney, C.A.; Ware, J.E., Jr.; Raczek, A.E. The MOS 36-item short-form health survey (SF-36): II. Psychometric and clinical tests of validity in measuring physical and mental health constructs. *Med. Care* **1993**, *31*, 247–263. [CrossRef] [PubMed]
36. Buchwald, D.; Pearlman, T.; Umali, J.; Schmaling, K.; Katon, W. Functional status in patients with chronic fatigue syndrome, other fatiguing illnesses, and healthy individuals. *Am. J. Med.* **1996**, *101*, 364–370. [CrossRef]
37. Guyon, I.; Elisseeff, A. An introduction to variable and feature selection. *J. Mach. Learn. Res.* **2003**, *3*, 1157–1182.
38. IBM Corps. *IBM SPSS Statistics for Windows, version 25.0*; IBM Corps: Armonk, NY, USA, 2017.
39. Strassheim, V.J.; Sunnquist, M.; Jason, L.A. Defining the prevalence and symptom burden of those with self-reported severe chronic fatigue syndrome/myalgic encephalomyelitis (CFS/ME): A two-phase community pilot study in the North East of England. *BMJ Open* **2018**, *8*, e02775. [CrossRef]
40. Jason, L.A.; Mirin, A.A. Updating the National Academy of Medicine ME/CFS prevalence and economic impact figures to account for population growth and inflation. **2020**, Manuscript under review.

 *healthcare*

Article

Reliability and Validity of the Modified Korean Version of the Chalder Fatigue Scale (mKCFQ11)

Yo-Chan Ahn [1], Jin-Seok Lee [2] and Chang-Gue Son [2,*]

[1] Department of Health Service Management, Daejeon University, Daejeon 300-716, Korea; ycahn@dju.kr
[2] Department of Korean Medicine, Institute of Bioscience and Integrative Medicine, Daejeon University, Daejeon 300-716, Korea; neptune@dju.ac.kr
* Correspondence: ckson@dju.ac.kr; Tel.: +82-42-257-6397; Fax: +82-42-257-6398

Received: 4 September 2020; Accepted: 22 October 2020; Published: 24 October 2020

Abstract: Fatigue can accompany various diseases; however, fatigue itself is a key symptom for patients with chronic fatigue syndrome (CFS). Due to the absence of biological parameters for the diagnosis and severity of CFS, the assessment tool for the degree of fatigue is very important. This study aims to verify the reliability and validity of the modified Korean version of the Chalder Fatigue Scale (mKCFQ11). This study was performed using data from 97 participants (Male: 37, Female: 60) enrolled in a clinical trial for an intervention of CFS. The analyses of the coefficient between the mKCFQ11 score and the Fatigue Severity Scale (FSS), the Visual Analogue Scale (VAS) or the 36-item Short-Form Health Survey (SF-36) at two time points (baseline and 12 weeks) as well as their changed values were conducted. The mKCFQ11 showed strong reliability, as evidenced by the Cronbach's alpha coefficient of 0.967 for the whole item and two subclasses (0.963 for physical and 0.958 for mental fatigue) along with the suitable validity of the mKCFQ11 structure shown by the principal component analysis. The mKCFQ11 scores also strongly correlated (higher than 0.7) with the VAS, FSS and SF-36 on all data from baseline and 12 weeks and changed values. This study demonstrated the clinical usefulness of the mKCFQ11 instrument, particularly in assessing the severity of fatigue and the evaluation of treatments for patients suffering from CFS.

Keywords: chronic fatigue syndrome; chalder fatigue scale; visual analogue scale; fatigue severity scale

1. Introduction

Fatigue is a subjective complaint commonly experienced by the general population during their lifetimes, with an approximate 30–50% point prevalence [1]. Unlike acute fatigue, which disappears after resting or treatment of the causative diseases, uncontrolled chronic fatigue substantially impairs the health-related quality of life [2]. In particular, chronic fatigue syndrome (CFS), a typical medically unexplained chronic fatigue, is a debilitating illness that results in the unemployment of half of patients with CFS and a risk of suicide approximately seven-fold higher than that of healthy controls [3,4].

Although many findings have been achieved from diverse aspects, including the nervous system, endocrine system, immune system, metabolomics, and gut microbiota, no universally accepted etiology, pathophysiology, diagnosis, or treatment for CFS exists [5]. Accordingly, both physicians and patients encounter many difficulties in the management of this disorder and communication with each other [6]. The diagnosis of certain disorders and the assessment of their severity are fundamental steps in treatment processes. For the diagnosis of CFS, physicians have adapted case definitions or diagnostic criteria such as the Fukuda definition in 1994 by the Centers for Disease Control and Prevention (CDC) or the criteria by Institute of Medicine (IOM) in 2015 [7,8]. These tools have been developed depending upon the clinical features.

Healthcare **2020**, *8*, 427; doi:10.3390/healthcare8040427　　　　　　www.mdpi.com/journal/healthcare

In general, objective measurement of fatigue is very important for patient management as well as assessment of the intervention efficacy for fatigue-related disorders [9]. Since the diagnosis of CFS and its categorization of illness status rely on self-reporting consultation, an accurate quantification of fatigue severity and its associated symptoms is vital, especially for patients with CFS [10]. To date, many severity scales have been developed based on patient-reported outcomes (PROs) for CFS patients, such as the Visual Analogue Scale (VAS), Fatigue Severity Scale (FSS), Multidimensional Fatigue Inventory (MFI), Chalder Fatigue Scale (CFQ), and Fibromyalgia and Chronic Fatigue Syndrome Rating Scale (FibroFatigue scale) [11–13]. Most of these instruments, however, have limitations, including a lack of specificity for CFS compared to other fatigue-inducing disorders, such as primary depression [14].

Among many fatigue-measuring scales, one of the most commonly used is the CFQ, which was developed in 1993 [12]. The CFQ consists of 11 easily applicable self-rating items for two domains of physical and mental fatigue, which could clearly discriminate patients with CFS from a healthy control [15]. This instrument has been well adapted in studies not only for clinical features of patients with CFS but also for evaluations of interventions, such as rehabilitative therapies [16,17]. The Korean version of the CFQ (K-CFQ) was also validated using healthy subjects, Korean graduate students [18]. The CFQ was initially developed as a four-point scale that compares to the "usual" status, and thus, it is difficult to measure the change in fatigue severity for certain periods. Therefore, we slightly modified it into a 10-point Likert scale (between normal and worst status) to describe their illness condition after treatment, called the Modified Korean Version of the Chalder Fatigue Scale (mKCFQ11). We have adapted the mKCFQ11 as a primary measurement in clinical trials using herb-derived therapeutics in both patients with idiopathic chronic fatigue (ICF) and CFS [19,20].

Although this modified scale has been well applied, its reliability and validity have not yet been assessed. The present study thus aims to verify the reliability and validity of the mKCFQ11.

2. Materials and Methods

2.1. Study Population and Data Source

The study population comprised 97 patients with CFS (37 males and 60 female) between the ages of 18 and 65 years (mean age 39.6 ± 10.0 years) who were enrolled in a phase 2 trial conducted in two hospitals (Daejeon Korean Medicine Hospital of Daejeon University and Daejeon St. Mary's Hospital of the Catholic University of South Korea) from December 2016 to November 2017 [20]. All participants met the 1994 Fukuda CFS definition, which requires clinically evaluated, unexplained, persistent, or relapsing chronic fatigue [7]. The exclusion criteria were subjects who suffered from other illnesses that induced chronic fatigue within the past 6 months, such as anemia, liver, kidney, and thyroid dysfunction, depression and anxiety disorders.

The data resource for this validation study was from the above phase 2, randomized, placebo-controlled trial of Myelophil (a standardized anti-fatigue herbal agent). This trial was designed to primarily determine the efficacy of Myelophil using the changes in the CFQ-based fatigue scores between baseline and 12 weeks of treatment. In addition, this trial had another purpose to verify the reliability and validity of the mKCFQ11 for use in trial 3; thus, we used the mKCFQ11 data at two time points and its changes regardless of the allocation of participants. Two well-known fatigue instruments, the VAS and the FSS, and the 36-item short-form health survey (SF-36) as an indicator of health-related quality of life (QoL) were used to calculate Pearson correlation coefficients for the mKCFQ11.

2.2. Ethics Statement

The trial was implemented in accordance with ethical and safety guidelines upon the approval of the Ministry of Food and Drug Safety (MFDS) in South Korea (Approval number 12354) and the Institutional Review Board in two hospitals (approval number DJDSKH-17-DR-03 in Daejeon Korean Medicine Hospital, DIRB-00139-3 in Daejeon St. Mary's Hospital). This trial is registered at Clinical

Research Information Service (CRIS) in Korea with identifier number KCT0002317, and an independent medical monitor (by MEDICAL excellence) ensured the trial procedure according to the protocol and maintaining the data. Data were independently analyzed by a medical statistics specialist.

2.3. Modified Korean Version of CFQ (mKCFQ11)

The original version of the CFQ consists of 11 items to determine the fatigue-related status by comparing to the "usual" condition: "Less than usual", "No more than usual", "More than usual", and "Much more than usual". However, the reference point ("usual") made it difficult for Korean patients to express their illness status, especially for patients with CFS due to the long-term duration of this condition, which could be over 10 years, or the very frequent childhood diagnosis. Furthermore, this "usual"-based comparison of illness condition at certain time points was not easily adapted to measure the changed score of fatigue severity in clinical trials of intervention. Therefore, we slightly modified it into a 10-point Likert scale as (0 = not at all to 9 = unbearably severe condition) for the same 11 questions (physical fatigue questions 1st–7th items, and mental fatigue 8th–11th items, total score range 0–99).

Briefly, the English version of the CFQ11 questionnaire was independently translated into Korean by two Korean specialists on CFS and a native English speaker proficient in Korean. Next, four specialists reviewed the differences and merged them and then examined the practical performance of many patients suffering from fatigue, including CFS. Based on the responder's comments, specialists discussed and completed the Korean version of the CFQ with 10-point Likert scale. After repeated tests of the patients complaining of chronic fatigue, including CFS, and a pilot clinical trial for ICF [19], the final mKCFQ11 (Table 1) was determined and used as a primary measurement for the above trial [20].

Table 1. The mKCFQ11 and its component analysis.

Korean Questionnaire	Varimax Rotation	
	Factor 1	Factor 2
1. 당신이 평소 느끼는 피로의 정도는 어떻습니까? (Do you have problems with tiredness?)	0.852	0.255
2. 당신은 어느 정도의 휴식이 필요합니까? (Do you need to rest more?)	0.859	0.294
3. 당신은 어느 정도의 졸음을 느끼십니까? (Do you feel sleepy or drowsy?)	0.854	0.360
4. 당신은 피로감 때문에 일을 시작할 때 힘이 듭니까? (Do you have problems starting things?)	0.795	0.423
5. 당신은 기운(기력)이 없다고 느끼십니까? (Do you lack energy?)	0.810	0.469
6. 당신은 근육의 힘이 약해졌다고 느끼십니까? (Do you have less strength in your muscles?)	0.727	0.473
7. 당신은 허약해졌다고 느끼십니까? (Do you feel weak?)	0.768	0.469
8. 당신은 일에 대한 집중력이 떨어졌습니까? (Do you have difficulties concentrating?)	0.551	0.748
9. 당신은 명료하게 생각하는 것에 어려움이 있습니까? (Do you make slips of the tongue when speaking?)	0.431	0.866
10. 당신은 말할 때 적절한 단어선택이 어려운 경우가 있습니까? (Do you find it more difficult to find the right word?)	0.258	0.908
11. 당신의 기억력 저하는 없습니까? (How is your memory?)	0.384	0.868

The modified Korean Version of the Chalder Fatigue Scale (mKCFQ11) is 10-point Likert scale (0 = 'not at all' to 9 = 'unbearably severe condition'), while Chalder Fatigue Scale (CFQ) is 4-point scale ('less than usual', 'no more than usual', 'more than usual' and 'much more than usual' for Q1 to Q10 and 'better than usual', 'no worse than usual', 'worse than usual', and 'much worse than usual' for Q10).

2.4. Fatigue Severity Scale (FSS)

The FSS is a 9-item self-report questionnaire to easily measure physical, social, or cognitive effects of fatigue. This scale was developed in 1989 as a seven-point Likert scale (1 indicating "Strongly disagree" to 7 representing "Strongly agree", total score range 7–63) and was translated into Korean previously and shown to be clinically useful for patients with fatigue [21,22].

2.5. Visual Analogue Scale (VAS)

The VAS was assessed by asking the participants to specify their level of overall discomfort from CFS by indicating a position along a continuous 100 mm line between two end points. The left end indicated "no exhaustion at all" while the right end indicated "complete exhaustion", and the value was then determined by measuring the length (mm) from the left end of the line [23].

2.6. The 36-Item Short-Form Health Survey (SF-36)

Health-related quality of life was measured using a 36-item short-form health survey (SF-36), whose usefulness was confirmed in patients with CFS [24] and translated into the Korean version [25]. The SF-36 consists of eight scaled scores that broadly reflect two domains of physical and mental health status. The total score range of each domain is a minimum of 0 (indicating "maximum disability") to a maximum of 100, representing "no disability".

2.7. Statistical Analysis

All statistical analyses were performed using Predictive Analytics SoftWare (PASW) Statistics (SPSS Inc., Chicago, IL, USA) for Windows. Two domains of mKCFQ11 (physical and mental) were assessed for their internal consistency by using Cronbach's alpha. Principal component analysis (PCA) was performed to examine the mKCFQ11 factor structure; factors with eigen values of >1 was extracted. The convergent validity of the total cognition score was tested using the Pearson correlation coefficients with the FSS, VAS and SF-36. Differences between the fatigue groups were tested using the *t*-test and one-way analysis of variance (ANOVA).

3. Discussion

The present study aimed to verify the reliability and validity of the mKCFQ11, a modified Korean version of the Chalder Fatigue Scale, altered from a four-point scale comparing "usual" status to a 10-point Likert scale (between normal and worst status) was created. This modification was performed because Korean patients with CFS described the difficulty of assessing their fatigue-associated severity using the CFQ instrument, especially in comparison to the "usual". In fact, most patients with CFS have been diagnosed with the disease for many years of disease with fluctuating symptoms [26]; thus, they frequently hesitated in answering. The mKCFQ11 has been adapted well by participants, in particular RCTs because they chose to describe their condition between "no fatigue" and "unbearable fatigue" at a certain period [20].

The reliability of the mKCFQ11 was strongly shown from the results using Cronbach's alpha coefficients. In general, Cronbach's alpha values ≥ 0.7 are considered satisfactory [27], and the values of mKCFQ11 were 0.967 for total fatigue and 0.963 and 0.958 for the subscales of physical and mental fatigue. These internal consistencies of the mKCFQ11 were higher than those of the original English version of the CFQ (total value of 0.92) using 361 patients with CFS in England [15] or the K-CFQ using Korean graduate students (total value was 0.88, 0.87 for physical and 0.73 for mental fatigue) [18]. There must be differences in not only the language structure between English and Korean but also cultural gaps, affecting the final results after translations of self-report measures [28]. When we translated the CFQ, we therefore strictly reflected the initial meaning of each question but tried to make them easy for the patients to understand via modification of the English phrases.

The principal component analysis using the varimax rotation model demonstrated the suitable validity of the mKCFQ1 structure composed of 11 question items with two subclasses of physical and mental fatigue (Figure 1 and Tables 1 and 2). In addition, the mKCFQ11 was significantly correlated with the VAS, FSS and SF-36 on three different data points from baseline and 12 weeks (Table 3). These correlations were stronger at 12 weeks than at baseline, as anticipated, because the intervention (Myelophil) showed positive effects on all scores of the mKCFQ11, FSS, VAS and SF-36 compared to the control [20]. Furthermore, these strong correlations were repeated for the altered values of the mKCFQ11 and others. The correlations between mKCFQ11 (total and subclasses of physical and mental fatigue) and the FSS or VAS were higher than 0.7, which indicated the very strong associations between the two fatigue scales [29]. FSS is a well-known fatigue measure that is specific for individuals with CFS compared to those with multiple sclerosis or primary depression [30]. Among PRO-based measurements, VAS is a simple technique to obtain continuous- and interval-level measurement data and to reduce response-style biases of Likert-type scales [31]. Moreover, the correlation with the SF-36 was relatively low compared to that with the FSS or VAS, which could be because the SF-36 is a non-disease specific generic scale to assess health-related quality of life [32]. The RCT for Myelophil measured the serum concentrations for oxidative and antioxidant parameters, and then the statistically significant correlations with mKCFQ11 were observed only for total glutathione (GSH) contents, tumor necrosis factor-α (TNF-α) and interferon-gamma (IFN-γ) on baseline and for the changed values of reactive oxygen species (ROS) after 12 weeks of treatment (Supplementary Table S1).

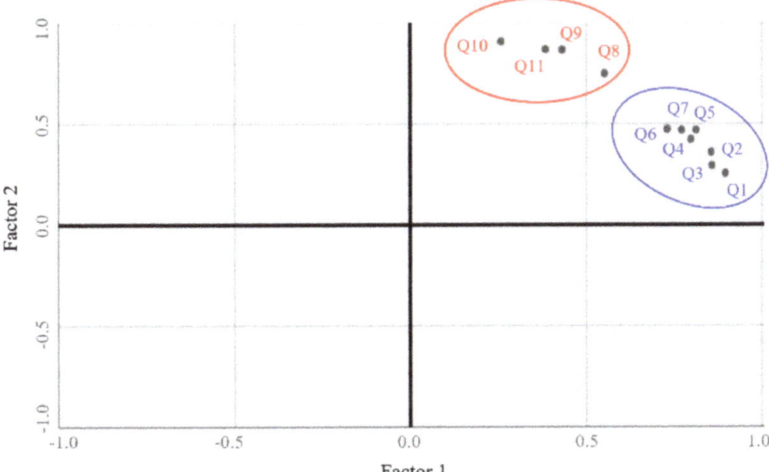

Figure 1. Principal component analysis (PCA) of the mCFQ11 structure. PCA was conducted for two-Figure 1. Q1 to Q7) and mental fatigue (factor 2, Q8 to Q11) based on covariance for a scale reliability of 11 questions.

Table 2. Principal component analysis after varimax rotation of the mKCFQ11.

N. of Subclass	Variance Explained			Extraction of Sums of Squared Leading		
	Eigenvalue	% Variance	% Cumulative	Factor	Subclass Factor 1	Subclass Factor 2
1	8.3126	75.597	75.597	Total	8.32	1.07
2	1.066	9.692	85.288			
3	0.387	3.514	88.803	% total of variance	75.6	9.7
4	0.261	2.369	91.172			
5	0.212	1.929	93.100	Questions	Q1	
6	0.189	1.716	94.816		Q2	Q8
7	0.171	1.551	96.368		Q3	Q9
					Q4	Q10
8	0.139	1.267	97.634		Q5	Q11
9	0.117	1.067	98.701		Q6	
					Q7	
10	0.083	0.751	99.453	Interpretation	Physical	Mental
11	0.060	0.547	100.000			

After varimax rotation of mKCFQ11, the number of subclasses was determined as initial Eigenvalue > 1.

Table 3. Correlation between the mKCFQ11 and the FSS, VAS or SF-36 on 0 and 12-week.

mKCFQ11		FSS	VAS	SF-36
0-week	Total	0.755 **	0.732 **	−0.383 **
	Physical	0.766 **	0.777 **	−0.403 **
	Mental	0.659 **	0.608 **	−0.311 **
12-week	Total	0.859 **	0.862 **	−0.698 **
	Physical	0.825 **	0.864 **	−0.672 **
	Mental	0.812 **	0.760 **	−0.672 **
Changes between 0 and 12-week	Total	0.757 **	0.860 **	−0.592 **
	Physical	0.769 **	0.887 **	−0.596 **
	Mental	0.698 **	0.761 **	−0.527 **

The mKCFQ11: Modified Korean Version of the Chalder Fatigue Scale, VAS: Visual Analogue Scale, FSS: Fatigue Severity Scale, SF-36: 36-item Short-Form Health Survey. The statistical significance of correlation was presented as ** $p < 0.001$.

In the assessment of subjective complaint disorders such as CFS or chronic pain, disease-specific PRO measures are the most important strategy to determine the treatment response because the patient is the most important judge of whether changes are important or meaningful [33,34]. The Multidimensional Fatigue Inventory (MFI-20) is another typical instrument used to assess fatigue severity in individuals with CFS, and the reliability and validity of its Korean version (MFI-K) was recently compared to the VAS and FSS [35]. The Cronbach's alpha coefficient of the MFI-K was 0.88, and the correlation coefficients with the VAS score (0.419) and the FSS score (0.635) were lower than that of the mKCFQ11. On the other hand, regarding the appropriate selection of the participants with CFS for especially RCTs, we would like to now recommend the combination of any diagnostic instrument such as CDC 1994 or IOM diagnostic criteria and cut-off scores of severity using mKCFQ11 or MFI.

4. Results

4.1. General Characteristics

A total of 97 participants (37 males and 60 females) who had a median age of 40 years (range 21 to 64 years) and a mean body mass index (BMI) of 22.6 ± 2.6 were included. All measurement scores for fatigue severity showed improvements at 12 weeks compared to at zero weeks, such as from 61.9 ± 15.5 to 37.7 ± 17.9 in the mKCFQ11, from 7.1 ± 1.7 to 4.3 ± 2.0 in the VAS, from 45.4 ± 9.8 to 32.3 ± 12.0 in the FSS, and 89.8 ± 15.8 to 101.2 ± 13.2 in the SF-36 (Table 4). These results were expected because nearly half of the participants (48 of 97) had taken an anti-fatigue herbal agent (Myelophil) for 12 weeks. There was no gender-related difference in score of mKCFQ11 (data not shown).

Table 4. Characteristics of the subjects and measurements.

Basic Characteristic	Male	Female	Total
Number of subjects (%)	37 (37.8)	60 (62.8)	97 (100)
Median age (year, range)	42 (25 to 63)	39 (21 to 64)	40 (21 to 64)
Mean value of BMI (kg/m^2)	24.1 ± 1.9	21.7 ± 2.6	22.6 ± 2.6
Measurement	**0-week**	**12-week**	**Change**
mKCFQ11: Total	61.9 ± 15.5	37.7 ± 17.9	24.2 ± 20.5
Physical	42.3 ± 9.1	26.0 ± 11.7	16.3 ± 13.0
Mental	19.9 ± 7.3	11.9 ± 6.8	8.0 ± 8.2
VAS	7.1 ± 1.7	4.3 ± 2.0	2.8 ± 2.4
FSS	45.4 ± 9.8	32.3 ± 12.0	13.2 ± 13.2
SF-36	89.8 ± 15.8	101.2 ± 13.2	−11.4 ± 17.2

BMI: Body Mass Index, mKCFQ11: Modified Korean Version of the Chalder Fatigue Scale, VAS: Visual Analogue Scale, FSS: Fatigue Severity Scale, SF-36: 36-item Short-Form Health Survey.

4.2. Internal Consistency

The Cronbach's alpha coefficient of mKCFQ11 was 0.967. The internal consistencies of the two subclasses were as follows: 0.963 for physical fatigue (Q1 to Q7) and 0.958 for mental fatigue (Q8 to Q11). In the absence condition for each item, Cronbach's alpha values were smaller than the values by the all-existence condition, which indicates the internal consistency of all question items.

4.3. Structural Validity of the mKCFQ11

The result of principal component analysis (with a varimax rotation) showed the structural validity of the mKCFQ11. Each question item was clustered together according to two subclasses and was distinct from the other item (Figure 1 and Table 1). Two subclass factors explained 85.3% of the total variance. Factor one explained 75.6% of the total variance and included all of the "Physical fatigue" items (Q1 to Q7). Factor two explained 9.7% of the total variance and included all "Mental fatigue" items (Q8 to Q11) (Table 2)

4.4. Convergent Validity with the VAS, FSS and SF-36

The mKCFQ11 had good convergent validity. The total mKCFQ11 score was significantly correlated with the VAS, FSS and SF-36 at both baseline (zero weeks, $p < 0.001$) and 12 weeks ($p < 0.001$). In addition, two subclasses (physical and mental fatigue) scores were also well correlated with the scores, with statistical significance at both timepoints (except between mental fatigue and mental SF-36 score at zero weeks, Table 3).

4.5. Convergent Validity of Changed Values with the VAS, FSS and SF-36

The changed values of the mKCFQ11 between zero weeks and 12 weeks were also significantly correlated with those of VAS, FSS and SF-36 ($p < 0.001$). Moreover, the changes in the two subclasses (physical and mental fatigue) scores were also significantly correlated with the VAS, FSS and SF-36 scores (physical and mental SF-36, Table 3).

5. Conclusions

This study demonstrated the clinical usefulness of the mKCFQ11, particularly in assessing the degree of fatigue and the changes in fatigue-related symptoms after treatment of patients with CFS. However, further studies are required, especially regarding subjects with other fatigue disorders and comparisons with an unmodified version of the CFQ.

Supplementary Materials: The following are available online at http://www.mdpi.com/2227-9032/8/4/427/s1, Table S1. Correlation between the mKCFQ11 biological parameters.

Author Contributions: Conceptualization, Y.-C.A. and C.-G.S.; methodology, validation, investigation, and data curation, Y.-C.A.; writing—original draft preparation, J.-S.L.; writing—review and editing, C.-G.S.; supervision, C.-G.S. All authors have read and agreed to the published version of the manuscript.

Funding: This research was supported by the Ministry of Education, Science and Technology (NRF-2018R1A6A1A03025221).

Conflicts of Interest: The authors declare no conflict of interest.

Ethical Statement: The study was conducted in accordance with ethical and safety guidelines upon the approval of Ministry of Food and Drug Safety (MFDS) in South Korea (Approval number 12354) and the Institutional Review Board in two hospitals (Approval number DJDSKH-17-DR-03 in Daejeon Korean Medicine Hospital, DIRB-00139-3 in Daejeon St. Mary's Hospital).

References

1. Van't Leven, M.; Zielhuis, G.A.; van der Meer, J.W.; Verbeek, A.L.; Bleijenberg, G. Fatigue and chronic fatigue syndrome-like complaints in the general population. *Eur. J. Public Health* **2010**, *20*, 251–257. [CrossRef]

2. Jorgensen, R. Chronic fatigue: An evolutionary concept analysis. *J. Adv. Nurs.* **2008**, *63*, 199–207. [CrossRef] [PubMed]

3. Castro-Marrero, J.; Faro, M.; Zaragoza, M.C.; Aliste, L.; de Sevilla, T.F.; Alegre, J. Unemployment and work disability in individuals with chronic fatigue syndrome/myalgic encephalomyelitis: A community-based cross-sectional study from Spain. *BMC Public Health* **2019**, *19*, 840. [CrossRef] [PubMed]

4. Kapur, N.; Webb, R. Suicide risk in people with chronic fatigue syndrome. *Lancet* **2016**, *387*, 1596–1597. [CrossRef]

5. Missailidis, D.; Annesley, S.J.; Fisher, P.R. Pathological mechanisms underlying myalgic encephalomyelitis/Chronic fatigue syndrome. *Diagnostics* **2019**, *9*, 80. [CrossRef] [PubMed]

6. Deale, A.; Wessely, S. Patients' perceptions of medical care in chronic fatigue syndrome. *Soc. Sci. Med.* **2001**, *52*, 1859–1864. [CrossRef]

7. Fukuda, K.; Straus, S.E.; Hickie, I.; Sharpe, M.C.; Dobbins, J.G.; Komaroff, A. The chronic fatigue syndrome: A comprehensive approach to its definition and study. International Chronic Fatigue Syndrome Study Group. *Ann. Intern. Med.* **1994**, *121*, 953–959. [CrossRef]

8. Beyond Myalgic Encephalomyelitis/Chronic Fatigue Syndrome: Redefining an Illness. *Mil. Med.* **2015**, *180*, 721–723. [CrossRef]

9. Walters, S.J.; Stern, C.; Stephenson, M. Fatigue and measurement of fatigue: A scoping review protocol. *JBI Evid. Synth.* **2019**, *17*, 261–266. [CrossRef]

10. Hardcastle, S.L.; Brenu, E.W.; Johnston, S.; Staines, D.; Marshall-Gradisnik, S. Severity scales for use in primary health care to assess chronic fatigue syndrome/Myalgic encephalomyelitis. *Health Care Women Int.* **2016**, *37*, 671–686. [CrossRef]

11. Smets, E.M.; Garssen, B.; Bonke, B.; De Haes, J.C. The Multidimensional Fatigue Inventory (MFI) psychometric qualities of an instrument to assess fatigue. *J. Psychosom. Res.* **1995**, *39*, 315–325. [CrossRef]

12. Chalder, T.; Berelowitz, G.; Pawlikowska, T.; Watts, L.; Wessely, S.; Wright, D.; Wallace, E.P. Development of a fatigue scale. *J. Psychosom. Res.* **1993**, *37*, 147–153. [CrossRef]

13. Zachrisson, O.; Regland, B.; Jahreskog, M.; Kron, M.; Gottfries, C.G. A rating scale for fibromyalgia and chronic fatigue syndrome (the FibroFatigue scale). *J. Psychosom. Res.* **2002**, *52*, 501–509. [CrossRef]

14. Jason, L.A.; Evans, M.; Brown, M.; Porter, N.; Brown, A.; Hunnell, J.; Anderson, V.; Lerch, A. Fatigue Scales and Chronic Fatigue Syndrome: Issues of Sensitivity and Specificity. *Disabil. Stud. Q.* **2011**, *31*, 1375. [CrossRef]

15. Cella, M.; Chalder, T. Measuring fatigue in clinical and community settings. *J. Psychosom. Res.* **2010**, *69*, 17–22. [CrossRef]

16. Slomko, J.; Newton, J.L.; Kujawski, S.; Tafil-Klawe, M.; Klawe, J.; Staines, D.; Marshall-Gradisnik, S.; Zalewski, P. Prevalence and characteristics of chronic fatigue syndrome/myalgic encephalomyelitis (CFS/ME) in Poland: A cross-sectional study. *BMJ Open* **2019**, *9*, e023955. [CrossRef]

17. Chalder, T.; Goldsmith, K.A.; White, P.D.; Sharpe, M.; Pickles, A.R. Rehabilitative therapies for chronic fatigue syndrome: A secondary mediation analysis of the PACE trial. *Lancet Psychiatry* **2015**, *2*, 141–152. [CrossRef]

18. Ha, H.; Jeong, D.; Hahm, B.J.; Shim, E.J. Cross-Cultural Validation of the Korean Version of the Chalder Fatigue Scale. *Int. J. Behav. Med.* **2018**, *25*, 351–361. [CrossRef]

19. Cho, J.H.; Cho, C.K.; Shin, J.W.; Son, J.Y.; Kang, W.; Son, C.G. Myelophil, an extract mix of Astragali Radix and Salviae Radix, ameliorates chronic fatigue: A randomised, double-blind, controlled pilot study. *Complement. Ther. Med.* **2009**, *17*, 141–146. [CrossRef]

20. Joung, J.Y.; Lee, J.S.; Cho, J.H.; Lee, D.S.; Ahn, Y.C.; Son, C.G. The Efficacy and Safety of Myelophil, an Ethanol Extract Mixture of Astragali Radix and Salviae Radix, for Chronic Fatigue Syndrome: A Randomized Clinical Trial. *Front. Pharmacol.* **2019**, *10*, 991. [CrossRef]

21. Krupp, L.B.; LaRocca, N.G.; Muir-Nash, J.; Steinberg, A.D. The fatigue severity scale. Application to patients with multiple sclerosis and systemic lupus erythematosus. *Arch. Neurol.* **1989**, *46*, 1121–1123. [CrossRef]

22. Chung, K.I.; Song, C.H. Clinical usefulness of fatigue severity scale for patients with fatigue, and anxiety or depression. *Korean J. Psychosom. Med.* **2001**, *9*, 164–173.

23. Khanna, D.; Pope, J.E.; Khanna, P.P.; Maloney, M.; Samedi, N.; Norrie, D.; Ouimet, G.; Hays, R.D. The minimally important difference for the fatigue visual analog scale in patients with rheumatoid arthritis followed in an academic clinical practice. *J. Rheumatol.* **2008**, *35*, 2339–2343. [CrossRef]

24. Myers, C.; Wilks, D. Comparison of Euroqol EQ-5D and SF-36 in patients with chronic fatigue syndrome. *Qual. Life Res.* **1999**, *8*, 9–16. [CrossRef]

25. Han, C.W.; Lee, E.J.; Iwaya, T.; Kataoka, H.; Kohzuki, M. Development of the Korean version of Short-Form 36-Item Health Survey: Health related QOL of healthy elderly people and elderly patients in Korea. *Tohoku J. Exp. Med.* **2004**, *203*, 189–194. [CrossRef] [PubMed]

26. Castro-Marrero, J.; Faro, M.; Aliste, L.; Saez-Francas, N.; Calvo, N.; Martinez-Martinez, A.; de Sevilla, T.F.; Alegre, J. Comorbidity in Chronic Fatigue Syndrome/Myalgic Encephalomyelitis: A Nationwide Population-Based Cohort Study. *Psychosomatics* **2017**, *58*, 533–543. [CrossRef]

27. Cronbach, L.J. Coefficient alpha and the internal structure of tests. *Psychometrika* **1951**, *16*, 297–334. [CrossRef]

28. Beaton, D.E.; Bombardier, C.; Guillemin, F.; Ferraz, M.B. Guidelines for the process of cross-cultural adaptation of self-report measures. *Spine* **2000**, *25*, 3186–3191. [CrossRef] [PubMed]

29. Cohen, J. *Statistical Power Analysis for the Behavioural Science*, 2nd ed.; Lawrence Erlbaum Associates: Hillsdale, NJ, USA, 1988; pp. 79–81.

30. Pepper, C.M.; Krupp, L.B.; Friedberg, F.; Doscher, C.; Coyle, P.K. A comparison of neuropsychiatric characteristics in chronic fatigue syndrome, multiple sclerosis, and major depression. *J. Neuropsychiatry Clin. Neurosci.* **1993**, *5*, 200–205. [CrossRef]

31. Sung, Y.T.; Wu, J.S. The Visual Analogue Scale for Rating, Ranking and Paired-Comparison (VAS-RRP): A new technique for psychological measurement. *Behav. Res. Methods* **2018**, *50*, 1694–1715. [CrossRef]

32. Stull, D.E.; Wasiak, R.; Kreif, N.; Raluy, M.; Colligs, A.; Seitz, C.; Gerlinger, C. Validation of the SF-36 in patients with endometriosis. *Qual. Life Res.* **2014**, *23*, 103–117. [CrossRef]

33. Acquadro, C.; Berzon, R.; Dubois, D.; Leidy, N.K.; Marquis, P.; Revicki, D.; Rothman, M.; Group, P.R.O.H. Incorporating the patient's perspective into drug development and communication: An ad hoc task force report of the Patient-Reported Outcomes (PRO) Harmonization Group meeting at the Food and Drug Administration, February 16, 2001. *Value Health* **2003**, *6*, 522–531. [CrossRef] [PubMed]

34. Turk, D.C.; Dworkin, R.H.; Burke, L.B.; Gershon, R.; Rothman, M.; Scott, J.; Allen, R.R.; Atkinson, J.H.; Chandler, J.; Cleeland, C.; et al. Developing patient-reported outcome measures for pain clinical trials: IMMPACT recommendations. *Pain* **2006**, *125*, 208–215. [CrossRef] [PubMed]

35. Song, S.W.; Kang, S.G.; Kim, K.S.; Kim, M.J.; Kim, K.M.; Cho, D.Y.; Kim, Y.S.; Joo, N.S.; Kim, K.N. Reliability and Validity of the Korean Version of the Multidimensional Fatigue Inventory (MFI-20): A Multicenter, Cross-Sectional Study. *Pain Res. Manag.* **2018**, *2018*, 3152142. [CrossRef] [PubMed]

Publisher's Note: MDPI stays neutral with regard to jurisdictional claims in published maps and institutional affiliations.

 *healthcare*

Communication

Cardiac Dimensions and Function are Not Altered among Females with the Myalgic Encephalomyelitis/Chronic Fatigue Syndrome

Per Ole Iversen [1,2,*], Thomas Gero von Lueder [3], Kristin Reimers Kardel [1] and Katarina Lien [1,4]

1   Department of Nutrition, Institute of Basic Medical Sciences, University of Oslo, 0317 Oslo, Norway; k.r.kardel@medisin.uio.no (K.R.K.); Katarina.Lien@follolms.no (K.L.)
2   Department of Hematology, Oslo University Hospital, 0424 Oslo, Norway
3   Department of Cardiology, Oslo University Hospital, 0424 Oslo, Norway; tomvonoslo@yahoo.com
4   CFS/ME Centre, Division of Medicine, Oslo University Hospital, 0424 Oslo, Norway
*   Correspondence: p.o.iversen@medisin.uio.no; Tel.: +47-22-85-13-91

Received: 24 September 2020; Accepted: 15 October 2020; Published: 16 October 2020

Abstract: *Background:* Myalgic encephalomyelitis/chronic fatigue syndrome (ME/CFS) is a debilitating condition associated with several negative health outcomes. A hallmark of ME/CFS is decreased exercise capacity and often profound exercise intolerance. The causes of ME/CSF and its related symptoms are unknown, but there are indications of a dysregulated metabolism with impaired glycolytic vs oxidative energy balance. In line with this, we recently demonstrated abnormal lactate accumulation among ME/CFS patients compared with healthy controls after exercise testing. Here we examined if cardiac dimensions and function were altered in ME/CFS, as this could lead to increased lactate production. *Methods:* We studied 16 female ME/CFS patients and 10 healthy controls with supine transthoracic echocardiography, and we assessed cardiac dimensions and function by conventional echocardiographic and Doppler analysis as well as novel tissue Doppler and strain variables. *Results:* A detailed analyses of key variables of cardiac dimensions and cardiac function revealed no significant differences between the two study groups. *Conclusion:* In this cohort of well-described ME/CFS patients, we found no significant differences in echocardiographic variables characterizing cardiac dimensions and function compared with healthy controls.

Keywords: cardiac function; echocardiography; myalgic encephalomyelitis/chronic fatigue syndrome

---

1. Introduction

Myalgic encephalomyelitis/chronic fatigue syndrome (ME/CFS) is characterized by fatigue, post-exertional malaise, pain, sleeping disturbances and exercise intolerance. Although its cause remains unknown, recent data point to dysregulated energy metabolism as a possible contributing factor [1,2]. In line with this, we recently demonstrated abnormal lactate accumulation and early gas exchange threshold among ME/CFS patients compared with healthy subjects during and after exercise testing [3]. Notably, the differences in lactate accumulation and gas exchange threshold between the patients and controls increased when they were re-tested after 24 h. Neither resting nor maximum heart rate at peak exercise differed significantly between the groups [3]. Such alterations in the oxidative vs. glycolytic energy balance might be due to intrinsic abnormalities in various energy-yielding metabolic pathways and/or reduced tissue oxygen supply [1,4].

Studies with magnetic resonance imaging (MRI) of the heart have suggested lower left ventricular dimensions and provided a plausible basis for reduced left ventricular function and exercise intolerance in ME/CSF [5]. In support of this, Miwa et al. have shown that ME patients often have small hearts as measured using chest roentgenograms and cardiac dysfunction evaluated with echocardiography [6].

Corroborating this notion of impaired cardiovascular capacity in ME/CFS are data from a systematic meta-analysis showing differences in various heart-rate indices in ME/CFS compared to healthy controls [4]. Notwithstanding these findings, detailed studies of cardiac function in rigorously defined ME/CFS cohorts are scarce. We therefore assessed conventional and advanced echocardiographic variables in a well-defined cohort of ME/CFS patients and in healthy controls to examine potential differences in cardiac dimensions and function.

2. Materials and Methods

2.1. Study Approvals and Participants

Approval was obtained from the Regional Committee for Medical and Health Research Ethics in Norway (no. 2012/571-1), and the original study [3] is registered with ClinicalTrials.org (ID NCT02970240). The original cohort consisted of 18 female, normotensive ME/CFS patients of moderate severity and 15 healthy, normotensive female controls. The patients did not use any drugs regularly, and they were diagnosed with ME/CFS > 2 years prior to the study. We limited the study population to females because (i) CFS/ME is more prominent among women compared with men and (ii) we could more consistently match patients and controls. Among these, two patients and three controls did not volunteer for the heart examination, and two other controls did not attend due to logistical reasons. Hence, 16 patients and 10 controls were available for the current study. The patients fulfilled The Canadian Consensus Criteria for ME/CFS [7]. Pregnant women, those who were completely bedridden or had comorbidities and those who used heart/lung medication were excluded. The enrolment procedure and characteristics of the two study groups have been reported [3].

2.2. Echocardiographic Measurements

We performed supine transthoracic echocardiography (GE Vingmed E9 scanner, Horten, Norway) after an overnight fast. Conventional echocardiographic, Doppler data and strain variables were analyzed offline by a trained specialist (TGvL) unaware of group affiliation, in accordance with established guidelines [8] and using commercially available software (Echopac vers. 201, GE Healthcare). The two study groups were matched for age and body mass index (BMI).

2.3. Statistical Analyses

All datasets showed a normal distribution as evidenced by Q–Q plots and Shapiro–Wilk's test. Thus, we used Student's t test to examine differences between the two study groups. Due to multiple comparisons between the two study groups, there was a possibility of familywise error rate (and thus Type I errors). We therefore adjusted the p-values using the Holm–Bonferroni sequential correction method [9]. This is a less strict method than the conventional Bonferroni correction, so the chances of Type II errors are reduced. Statistical significance was set at $p < 0.05$.

3. Results

3.1. Characteristics of the Study Participants

The mean (range) ages of the ME/CFS patients and the healthy controls were 40.1 (23.9–52.0) and 35.5 (25.0–44.3) years ($p = 0.13$), respectively, and the corresponding BMIs were 24.9 (18.6–31.3) and 23.7 (18.8–35.6) kg/m^2, respectively ($p = 0.54$). On the day of the examination, cardiorespiratory symptoms were not reported by any of the participants in either of the two study groups.

3.2. Echocardiographic Findings

Cardiac dimensions and function determined by echocardiography are summarized in Table 1. Here, no significant differences in systolic or diastolic left ventricular diameters were found between the ME/CFS patients and the healthy controls. Normalization of left ventricular dimensions to body surface area was also not significantly different between study groups. Moreover, the left atrial diameter was not different ($p > 0.05$) either. In line with these findings, no significant differences in systolic or diastolic blood pressure or in estimated systolic pulmonary artery pressure were found (data not shown). We next performed a wide range of assessments characterizing cardiac function in detail, both in diastole and systole. In addition to standard analysis of biplane left ventricular volumes and ejection fraction (ad modum Simpson), novel speckle-tracking echocardiography was performed. Global longitudinal left ventricular strain revealed similar results ($p > 0.05$) between the two study groups. Right ventricular function evidenced by tricuspid annulus plane systolic excursion (TAPSE) and peak systolic tissue Doppler velocity (not shown) was also comparable.

Table 1. Echocardiographic variables obtained in the ME/CFS patients and in the healthy controls.

Echocardiographic Variables	ME/CFS Patients ($n = 16$)	Controls ($n = 10$)	Crude-p	Adj.-p
Cardiac dimensions				
Left atrial area (cm^2)	17.0 (3.4)	16.9 (3.4)	0.60	>0.90
Left ventricular end-diastolic diameter (cm)	4.78 (0.38)	4.70 (0.32)	0.95	>0.90
Septal wall thickness (cm)	0.78 (0.16)	0.70 (0.10)	0.17	>0.90
Left ventricular systolic diameter (cm)	3.0 (0.29)	2.86 (0.27)	0.67	>0.90
Left ventricular end-diastolic volume (mL)	91.3 (20.4)	95.0 (22.4)	0.72	>0.90
Systolic and diastolic function				
Heart rate (beats/min)	73 (10)	68 (14)	0.29	>0.90
Ejection fraction EF (%)	60.1 (4.6)	56.5 (3.6)	0.044	0.62
Fractional shortening (%)	35.4 (5.9)	39.2 (3.8)	0.10	>0.90
Global longitudinal strain (%)	19.1 (2.0)	19.9 (1.0)	0.33	>0.90
Early transmitral flow (E; m/s)	0.67 (0.14)	0.66 (0.08)	0.86	>0.90
Atrial transmitral flow (A; m/s)	0.46 (0.12)	0.41 (0.16)	0.35	>0.90
Ratio of early and atrial transmitral flow (E/A)	1.55 (0.56)	1.85 (0.62)	0.21	>0.90
Pulsed tissue Doppler (e′; cm/s)	11.6 (1.9)	13.6 (2.4)	0.046	0.62
Ratio of early transmitral flow and e′ (E/e′)	5.96 (1.14)	5.06 (1.16)	0.10	>0.90
Tricuspid annular plane systolic excursion (cm)	5.0 (8.0)	2.4 (0.4)	0.31	>0.90
Stroke volume (mL)	59.5 (9.4)	62.3 (9.2)	0.49	>0.90
Stroke volume index (mL/m^2)	24.3 (15.1)	36.0 (4.7)	0.0094	0.14
Cardiac output (L/min)	4.3 (0.7)	4.2 (0.9)	0.74	>0.90
Cardiac index (L/min/m^2)	1.8 (1.1)	2.4 (0.4)	0.098	>0.90

Values are mean (SD). Crude-p, unadjusted p-values; Adj.-p, p-values adjusted according to the Holm–Bonferroni sequential method.

Impaired diastolic function is an established contributor to impaired exertional capacity. Here, early and late diastolic filling capacities were similar, as well as was early relaxation by tissue Doppler imaging and computed left ventricular filling pressures (i.e., E/é) ($p > 0.05$). In summary, we could not find any significant differences in any parameters of cardiac function between the ME/CFS patients and the healthy controls.

4. Discussion

In this study, we were not able to detect any significant differences between ME/CFS patients and healthy controls in a wide range of variables characterizing cardiac dimensions and function with the use of conventional and advanced echocardiography. These findings, therefore, do not support the hypothesis that reduced cardiac dimensions or function may contribute to early exertional lactate accumulation, early gas exchange threshold or the low exercise capacity in ME/CFS that we recently reported [3]. Our findings are similar to those reported by Montague et al. [10], but at variance with previous studies reporting smaller hearts and dysregulated autonomic regulation of cardiac function in ME/CFS [4–6]. Possible explanations for these discrepancies include differences in age, gender, diagnostic criteria for ME/CFS and methods for assessments of cardiac dimensions and function. For example, Miwa et al. applied the Fukuda criteria [11], whereas we used the Canadian Consensus Criteria for ME/CFS [7]. Importantly, the occurrence of post-exertional malaise is only mandatory in the latter criteria. Notably, whereas these previous reports did not correct for multiple statistical testing among MRI- and echocardiographic-derived variables, we included robust adjustments. Notwithstanding these contrasting results, those reported between ME/CFS and healthy controls in previous studies might be too small to fully explain exercise intolerance and the various metabolic abnormalities associated with ME/CSF [1–3].

Evidence for a reduced cardiac capacity in ME/CFS may not be evident in the resting state due to compensatory mechanisms, and absence of statistical differences in key echocardiographic parameters at rest do not necessarily mean that cardiac exertional capacities are similar. Moreover, in our previous study, we found similar blood concentrations of lactate among ME/CFS patients and controls at rest before the first exercise test, but during exercise the lactate concentration accumulated faster among the ME/CFS patients [3]. It would therefore be of interest to analyze cardiac function during exercise testing and examine whether cardiac variables are associated with markers of metabolic pathways such as enzymes and/or substrates for oxidative phosphorylation to generate ATP. For example, abnormal regulation of pyruvate dehydrogenase kinase has been associated with ME/CFS, and this could lead to an increased lactate accumulation during exercise [1].

Limitations of our study were the small sample size and that we only included female ME/CFS patients with moderate disease severity. Strengths include well-characterized participants, including a matched control group, and use of detailed, state-of-the-art echocardiography.

5. Conclusions

We found no significant differences in a wide range of conventional and novel echoardiographic variables characterizing cardiac dimensions and function when comparing ME/CFS patients with healthy controls. Further studies performed with exercising individuals, as well as directly linking cardiac function to biomarkers of metabolism, would be of interest.

Author Contributions: Conceptualization, P.O.I. and K.L.; methodology, T.G.v.L.; formal analysis, P.O.I., T.G.v.L., K.R.K. and K.L.; writing—original draft preparation, P.O.I.; writing—review and editing, P.O.I., T.G.v.L., K.R.K. and K.L.; project administration, P.O.I.; funding acquisition, P.O.I. All authors have read and agreed to the published version of the manuscript.

Funding: The study was sponsored in part by The Norwegian Extra Foundation and The Norwegian ME-Association.

Conflicts of Interest: The authors declare no conflict of interest. The funders had no role in the design of the study; in the collection, analyses, or interpretation of data; in the writing of the manuscript, or in the decision to publish the results.

References

1. Fluge, O.; Mella, O.; Bruland, O.; Risa, K.; Dyrstad, S.E.; Alme, K.; Rekeland, I.G.; Sapkota, D.; Røsland, G.V.; Fosså, A.; et al. Metabolic profiling indicates impaired pyruvate dehydrogenase function in myalgic encephalopathy/chronic fatigue syndrome. *JCI Insight* **2016**, *1*, e89376. [CrossRef] [PubMed]
2. Tomas, C.; Brown, A.; Strassheim, V.; Elson, J.L.; Newton, J.; Manning. P. Cellular bioenergetics is impaired in patients with chronic fatigue syndrome. *PLoS ONE* **2017**, *12*, e0186802. [CrossRef] [PubMed]
3. Lien, K.; Johansen, B.; Veierod, M.B.; Haslestad, A.S.; Bøhn, S.K.; Melsom, M.N.; Kardel, K.R.; Iversen, P.O. Abnormal blood lactate accumulation during repeated exercise testing in myalgic encephalomyelitis/chronic fatigue syndrome. *Physiol. Rep.* **2019**, *7*, e14138. [CrossRef] [PubMed]
4. Nelson, M.J.; Bahl, J.S.; Buckley, J.D.; Thomson, R.L.; Davison, K. Evidence of altered cardiac autonomic regulation in myalgic encephalomyelitis/chronic fatigue syndrome: A systematic review and meta-analysis. *Medicine* **2019**, *98*, e17600. [CrossRef] [PubMed]
5. Olimulder, M.A.; Galjee, M.A.; Wagenaar, L.J.; van Es, J.; van der Palen, J.; Visser, F.C.; Vermeulen, R.C.W.; von Birgelen, C. Chronic fatigue syndrome in women assessed with combined cardiac magnetic resonance imaging. *Neth. Heart J.* **2016**, *24*, 709–716. [CrossRef] [PubMed]
6. Miwa, K.; Fujita, M. Cardiovascular dysfunction with low cardiac output due to a small heart in patients with chronic fatigue syndrome. *Intern. Med.* **2009**, *48*, 1849–1854. [CrossRef] [PubMed]
7. Carruthers, B.; Jain, A.K.; De Meirleir, K.L.; Peterson, D.L.; Klimas, N.G.; Lerner, A.M.; Bested, A.C.; Flor-Henry, P.; Joshi, P.; Powles, P.; et al. Myalgic Encephalomyelitis/Chronic Fatigue Syndrome. *J. Chron. Fatigue Syndr.* **2003**, *11*, 7–115. [CrossRef]

8.	Lang, R.M.; Badano, L.P.; Mor-Avi, V.; Afilalo, J.; Armstrong, A.; Ernande, L.; Flachskampf, F.A.; Foster, E.; Goldstein, S.A.; Kuznetsova, T.; et al. Recommendations for cardiac chamber quantification by echocardiography in adults: An update from the American Society of Echocardiography and the European Association of Cardiovascular Imaging. *J. Am. Soc. Echocardiogr.* **2015**, *28*, 1–39. [CrossRef] [PubMed]

9.	Holm, S. A simple sequential rejective method procedure. *Scand. J. Stat.* **1979**, *6*, 65–70.

10.	Montague, T.J.; Marrie, T.J.; Klassen, G.A.; Bewick, D.J.; Horacek, B.M. Cardiac function at rest and with exercise in the chronic fatigue syndrome. *Chest* **1989**, *95*, 779–784. [CrossRef] [PubMed]

11.	Fukuda, K.; Straus, S.E.; Hickle, I.; Sharpe, M.C.; Dobbins, J.G.; Komaroff, A. The chronic fatigue syndrome: A comprehensive approach to its definition and study. International Chronic Fatigue Syndrome Study Group. *Ann. Intern. Med.* **1994**, *121*, 953–959. [CrossRef] [PubMed]

Publisher's Note: MDPI stays neutral with regard to jurisdictional claims in published maps and institutional affiliations.

 *healthcare*

Article

Cerebral Blood Flow Is Reduced in Severe Myalgic Encephalomyelitis/Chronic Fatigue Syndrome Patients During Mild Orthostatic Stress Testing: An Exploratory Study at 20 Degrees of Head-Up Tilt Testing

C (Linda) M.C. van Campen [1,*], Peter C. Rowe [2] and Frans C. Visser [1]

[1]   Stichting CardioZorg, 2132 HN Hoofddorp, The Netherlands; fransvisser@stichtingcardiozorg.nl

[2]   Department of Paediatrics, John Hopkins University School of Medicine, Baltimore, MD 21205, USA; prowe@jhmi.edu

*   Correspondence: info@stichtingcardiozorg.nl; Tel.: +31-208934125

Received: 5 May 2020; Accepted: 10 June 2020; Published: 13 June 2020

Abstract: Introduction: In a study of 429 adults with myalgic encephalomyelitis/chronic fatigue syndrome (ME/CFS), we demonstrated that 86% had symptoms of orthostatic intolerance in daily life. Using extracranial Doppler measurements of the internal carotid and vertebral arteries during a 30-min head-up tilt to 70 degrees, 90% had an abnormal reduction in cerebral blood flow (CBF). A standard head-up tilt test of this duration might not be tolerated by the most severely affected bed-ridden ME/CFS patients. This study examined whether a shorter 15-min test at a lower 20 degree tilt angle would be sufficient to provoke reductions in cerebral blood flow in severe ME/CFS patients. Methods and results: Nineteen severe ME/CFS patients with orthostatic intolerance complaints in daily life were studied: 18 females. The mean (SD) age was 35(14) years, body surface area (BSA) was 1.8(0.2) m^2 and BMI was 24.0(5.4) kg/m^2. The median disease duration was 14 (IQR 5–18) years. Heart rate increased, and stroke volume index and end-tidal CO_2 decreased significantly during the test (p ranging from <0.001 to <0.0001). The cardiac index decreased by 26(7)%: $p < 0.0001$. CBF decreased from 617(72) to 452(63) mL/min, a 27(5)% decline. All 19 severely affected ME/CFS patients met the criteria for an abnormal CBF reduction. Conclusions: Using a less demanding 20 degree tilt test for 15 min in severe ME/CFS patients resulted in a mean CBF decline of 27%. This is comparable to the mean 26% decline previously noted in less severely affected patients studied during a 30-min 70 degree head-up tilt. These observations have implications for the evaluation and treatment of severely affected individuals with ME/CFS.

Keywords: orthostatic intolerance; cerebral blood flow; 20 degree tilt table testing; myalgic encephalomyelitis; chronic fatigue syndrome; postural orthostatic tachycardia syndrome; stroke volume index; cardiac index

---

1. Introduction

Myalgic encephalomyelitis/chronic fatigue syndrome (ME/CFS) patients have a high prevalence of orthostatic intolerance [1,2]. In a study of 429 adult ME/CFS patients, we recently demonstrated that 86% had orthostatic intolerance symptoms during daily life. Moreover, during a 30-min head-up tilt table test, 90% had an abnormal cerebral blood flow (CBF) reduction as assessed by extracranial Doppler measurements [2,3]. This abnormal CBF reduction was not only present in ME/CFS patients with well-defined heart rate and blood pressure abnormalities during tilt testing, like orthostatic hypotension, postural orthostatic tachycardia syndrome (POTS) and syncope [4–6], but also in ME/CFS patients with a normal heart rate and blood pressure response to upright posture [2]. The mean CBF reduction of 26% in the entire study population with ME/CFS was significantly different from the 7% reduction observed in healthy controls in response to the same orthostatic stress.

Documenting an abnormal cerebral blood flow is helpful in guiding therapy for orthostatic intolerance (OI). As described in the IOM report: "Orthostatic intolerance is defined as a clinical condition in which symptoms worsen upon assuming and maintaining upright posture and are ameliorated (although not necessarily abolished) by recumbency" [1]. Symptoms of orthostatic intolerance sought in the history of patients "are those caused primarily by [1] cerebral underperfusion (such as light-headedness, near-syncope or syncope, impaired concentration, headaches, and dimming or blurring of vision), or [2] sympathetic nervous system activation (such as forceful beating of the heart, palpitations, tremulousness, and chest pain. Other common signs and symptoms of orthostatic intolerance are fatigue, a feeling of weakness, intolerance of low-impact exercise, nausea, abdominal pain, facial pallor, nervousness, and shortness of breath".

A limitation of the extracranial Doppler measurements is that image acquisition lasts between 2 and 7 min [2]. In our recent study, patients were excluded if they were unable to maintain the upright position during the acquisition period, and also if a rapid drop in heart rate and blood pressure prevented complete image acquisition. Moreover, patients can develop post-exertional malaise after conventional 60–90 degree head-up orthostatic stress testing [1]. Due to these problems—image acquisition time, the potential for post-exertional malaise and inability to stand long enough— conventional orthostatic testing may not be advisable in severe ME/CFS patients. Moreover, in previous work, we showed that 15 of 444 patients could not complete standing during the head-up tilt test [2]. Wyller et al. described a different method of orthostatic stress testing [7], which involved a low-grade head-up tilt test of 20 degrees in 27 ME/CFS adolescents. The rationale was that a 70 degree head-up tilt testing is associated with a high rate of false-positive results in adolescents. Using the 20 degree tilt over a period of 15 min, the authors showed that heart rate, blood pressure and stroke volume index changes were different in ME/CFS adolescents when compared with age- and gender-matched controls.

Assuming that severe ME/CFS patients cannot tolerate prolonged standing during tilt testing and may have more hemodynamic abnormalities including a rapid decline of blood pressure, the aim of the current study was to test the hypothesis that reduced cerebral blood flow and reduced stroke volume index/cardiac index could also be confirmed in severe ME/CFS patients during 15 min of low-grade (20 degree) head-up tilt testing.

2. Materials and Methods

2.1. Patients

From June 2019 to April 2020, 139 patients visited the outpatient clinic of the Stichting CardioZorg, Hoofddorp, the Netherlands, because of a suspicion of ME/CFS. This cardiology clinic specializes in diagnosing and treating adults with ME/CFS. All patients were evaluated by the same clinician (FVC). During the first visit, it was determined whether patients satisfied the criteria for CFS and ME, taking the exclusion criteria into account. Patients were classified as having CFS, chronic fatigue or no chronic fatigue as defined by Fukuda and colleagues [8] and as having ME or no ME as defined by Carruthers and colleagues [9]. Disease severity was scored by a clinician according to the ICC, with severity ranging between mild, moderate, severe and very severe. This was classified according to the paper as: "Symptom severity impact must result in a 50% or greater reduction in a patient's premorbid activity level for a diagnosis of ME. Mild: approximately 50% reduction in activity, moderate: mostly housebound, severe: mostly bedbound and very severe: bedbound and dependent on help for physical functions" [9]. The clinician ascertained for the presence of orthostatic intolerance symptoms in daily life like dizziness/light-headedness, prior (near)-syncope and nausea, among others, as well as triggering events like standing in a line. Over this 10 month period, 137 patients met the criteria for ME/CFS, 19 (14%) of whom met the criteria for severe ME/CFS. In those patients, an orthostatic stress test was performed at a low-grade head-up tilt angle of 20 degrees. The study was carried out in accordance with the Declaration of Helsinki. All ME/CFS patients gave informed, written consent. The study was approved by the medical ethics committee of the Slotervaart Hospital, Amsterdam, The Netherlands (reference number P1736).

2.2. Head-Up Tilt Test with Cerebral Blood Flow and Stroke Volume Measurements

Measurements were performed as described previously [3], with the main exception being that patients were positioned supine for 20 min before being tilted head-up to 20 degrees for a maximum of 15 min instead of the more classic approach of 70 degrees for 25–30 min. They were investigated in the morning, at least 3 h after a light breakfast or in the afternoon 3 h after a light lunch. No formal hydration protocol was applied, but subjects were asked to ingest an ample amount of fluid. If patients developed severe orthostatic symptoms, the test was stopped prematurely, but after upright image acquisition. The test was prematurely stopped in 6 patients due to an increase in symptoms. Heart rate and systolic and diastolic blood pressures were continuously recorded by finger plethysmography using the Nexfin device (BMeye, Amsterdam, The Netherlands). Heart rate and blood pressures were extracted from the device and imported into an Excel spreadsheet. End-tidal PCO_2 ($PetCO_2$) was monitored using a Lifesense device (Nonin Medical, Minneapolis, MN, USA).

2.3. Cerebral Blood Flow Determination by Doppler Echographic Measurements

Internal carotid artery and vertebral artery Doppler flow velocity frames were acquired by one operator (FCV), using a Vivid-I system (GE Healthcare, Hoevelaken, the Netherlands) equipped with a 6–13 MHz linear transducer. Flow data of the internal carotid artery (ICA) on the right and on the left side were obtained ~1.0–1.5 cm distal to the carotid bifurcation and of the vertebral artery (VA) on the right and on the left side, data were obtained at the C3–C5 level. Care was taken to ensure the insonation angle was less than 60 degrees, that the sample volume was positioned in the center of the vessel and that it covered the width of the vessel. High-resolution B mode images, color Doppler images and the Doppler velocity spectrum (pulsed wave mode) were recorded in one frame. The order of imaging was fixed: left internal carotid artery (ICA), left vertebral artery (VA), right internal carotid artery (ICA) and right vertebral artery (VA). At least two consecutive series of six frames per artery were recorded. The recording time intervals of the first and last imaged artery were noted and these times were corrected to the times of a radio clock, setting the start of tilt at 0 min. Heart rate and blood pressures of the echo recording time intervals were averaged. In the supine position, image acquisition started 8 (2) min prior to tilting (supine data) and during the upright position acquisition started at 10 (4) min. Based on data from healthy controls during a 30-min 70 degree head-up tilt, we defined an abnormal reduction in CBF as a >13% decline during the tilt compared to the supine values [2]. Analysis is described in the data-analysis section.

2.4. Stroke Volume Determination by Doppler Echocardiographic Measurements:

To determine stroke volume, velocity time integral (VTI) frames were obtained in the resting supine position and the upright position as previously described [10]. The aortic VTI was measured using a continuous wave Doppler pencil probe (GE P2D: 2 MHz) connected to a Vivid I machine (GE, Hoevelaken, NL, USA) with the transducer positioned in the suprasternal notch. A maximal Doppler signal was assumed to be the optimal flow alignment. At least 2 frames of 6 s were obtained. Echo Doppler recordings were stored digitally. From a transthoracic echocardiogram, the diameter of the left ventricular outflow tract (LVOT) was obtained. Analysis is described in the data-analysis section.

2.5. Data Analysis

The changes in heart rate and blood pressure during the head-up tilt test were classified according to the consensus statements [4,6]: normal heart rate and blood pressure response, classic orthostatic hypotension (a decrease of over 20 mmHg in systolic blood pressure and over 30 mmHg in the case of a systolic blood pressure over 140 mmHg, or a decrease of 10 mmHg in diastolic blood pressure from 1–3 min after tilt), delayed orthostatic hypotension (a decrease of over 20 mmHg in systolic blood pressure and over 30 mmHg in the case of a systolic blood pressure over 140 mmHg, or a decrease of 10 mmHg in diastolic blood pressure after 3 min post tilt), postural orthostatic tachycardia syndrome (POTS) (a sustained increase of at least 30 bpm within 10 min, without a significant decrease in BP) and syncope or near-syncope.

Blood flows of the internal carotid and vertebral arteries were calculated offline by an investigator (CMCvC) who was unaware of the patient severity status and unaware of the hemodynamic outcome of the head-up tilt test. Vessel diameters were manually traced by CMCvC on B-mode images, from the intima to the opposite intima. Surface area was calculated: the peak systolic and end diastolic diameters were measured, and the mean diameter was calculated as: mean diameter = (peak systolic diameter × 1/3) + (end diastolic diameter × 2/3) [11]. Blood flow in each vessel was calculated from the mean blood flow velocities times the vessel surface area and expressed in mL/min. Flow in the individual arteries was calculated in 3–6 cardiac cycles and data were averaged. Total cerebral blood flow was calculated by adding the flow of the four arteries. We previously demonstrated that this methodology had good intra- and inter-observer variability [3].

To determine stroke volumes, the velocity time integral was measured offline by manual tracing of at least 6 cardiac cycles, using the GE EchoPac post-processing software, by one operator (CMCvC). Stroke volumes were calculated from the corrected VTI and the LVOT cross-sectional area, as described previously [12,13]. Stroke volume index (SVI) was calculated by the equation: corrected LVOT cross-sectional area times the corrected aortic VTI, divided by the body surface area (BSA; DuBois formula) (mL/m^2). SVIs of the separate cycles were averaged. Cardiac index was calculated by the equation: SVI × heart rate/1000 (L/min/m^2).

2.6. Statistical Analysis

Data were analyzed using Graphpad Prism version 6.05 (Graphpad software, La Jolla, CA, USA). All continuous data were tested for normal distribution using the D'Agostino–Pearson omnibus normality test, and presented as mean (SD) or as median with the IQR, where appropriate. For continuous data, paired and non-paired *t*-tests were used for comparison, when appropriate. Linear regression analysis was performed correlating the percent cardiac index change with the percent cerebral blood flow change. A *p* value of <0.05 was considered significant.

3. Results

Patient Clinical and Echo Doppler Data

Nineteen patients with a severe grade of ME/CFS were studied (18 females). Baseline characteristics were as follows: mean age 35 (14) years, height 171 (4) cm, weight 70 (17) kg, BSA 1.8 (0.2) m^2 and BMI 24.0 (5.4) kg/m^2. The median disease duration was 14 (IQR 5–18) years. All patients met the Fukuda criteria for CFS, and all met the ICC criteria for ME. Daily life orthostatic intolerance symptoms were reported by all 19 ME/CFS patients. At the time of the test, no patients were being treated with drugs influencing heart rate or blood pressure and none were being treated with selective serotonin reuptake inhibitors (SSRI). Of the 19 ME/CFS patients, four met the criteria for POTS at 20 degrees. None of the patients developed orthostatic hypotension or vasovagal syncope at 20 degrees.

Table 1 shows the tilt table results of the study participants during the low-grade 20 degree head-up tilt testing. Heart rate increased from 83 (13) supine to 104 (23) bpm (*p* < 0.0001), and end-tidal CO_2 decreased from 38 (3) to 29 (6) mmHg (*p* < 0.001). Cerebral blood flow supine was 619 (68) mL and at the end of study was 453 (63) mL, a decline of 27 (5)%, with a decrease ranging between 21% and 37%. All 19 patients met criteria for an abnormal reduction in cerebral blood flow, being a more than 13% reduction. Stroke volume index decreased from 36 (6) to 25 (5) mL/m^2, a decline of 31 (8)%. Cardiac index fell from 2.9 (0.5) to 2.1 (0.4) L/min/m^2, a decline of 27 (7%).

Table 1. Tilt table test data of severe myalgic encephalomyelitis/chronic fatigue syndrome (ME/CFS) patients (n = 19).

	Supine	End of Study	*p*-Value
Heart rate (bpm)	83 (13)	104 (23)	<0.0001
Systolic blood pressure (mmHg)	134 (11)	138 (14)	0.10
Diastolic blood pressure (mmHg)	80 (9)	86 (9)	0.0006
End-tidal CO$_2$ (mmHg)	38 (3)	29 (6)	<0.0001
Cerebral blood flow (mL/min)	619 (68)	453 (63)	<0.0001
Stroke volume index (mL/m^2)	36 (6)	25 (5)	<0.0001
Cardiac index (L/min/m^2)	2.9 (0.6)	2.1 (0.4)	<0.0001
Cerebral blood flow %change		−27 (5)%	
Stroke volume index %change		−31 (8)%	
Cardiac index %change		−27 (7)%	

%change: percent change from supine data.

Figure 1 shows the graphical representation of the supine cerebral blood flow and the cerebral blood flow at the end of the low-grade head-up tilt test: the difference was highly statistically significant ($p < 0.0001$). Figure 2 shows the hemodynamic changes in stroke volume index supine and at the end of the test in panel A, and the hemodynamic changes in cardiac index supine and at the end of the test in panel B. Both stroke volume index and cardiac index declined significantly (both $p < 0.0001$). Figure 3 shows the relation between the change in cardiac index and the change in cerebral blood flow: this relation was significant ($p < 0.005$).

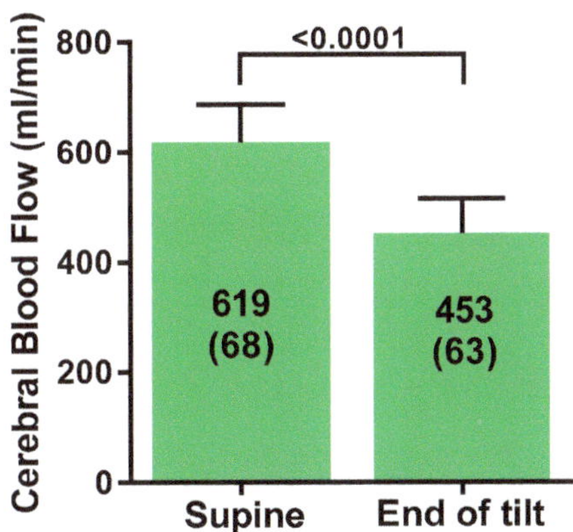

Figure 1. Cerebral blood flow in mL/min supine and at the end of a 20 degree tilt in severe ME/CFS patients.

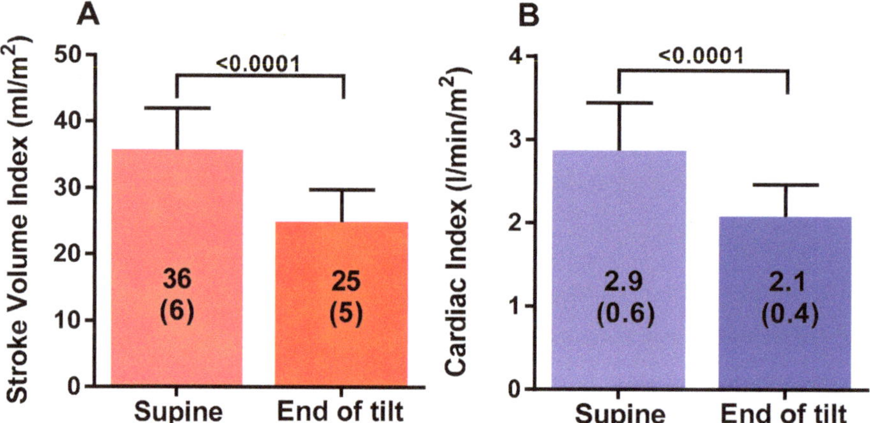

Figure 2. Stroke volume index in mL/m^2 supine and at the end of a 20 degree tilt in severe ME/CFS patients (panel **A**). Cardiac index in L/min/m^2 supine and at the end of a 20 degree tilt in severe ME/CFS patients (panel **B**).

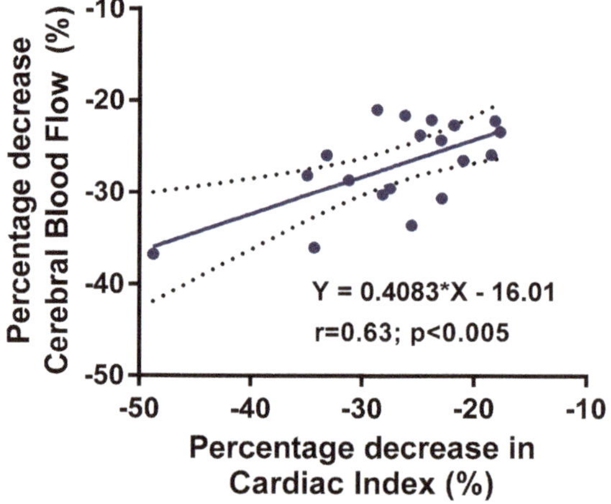

Figure 3. Correlation between the percent decrease in cardiac index and the percent decrease in cerebral blood flow.

Table 2 shows the comparison between ME/CFS patients with a normal heart rate and blood pressure response and those with POTS. The heart rate in POTS patients was significantly higher at the end of the study compared with the normal heart rate and blood pressure patients ($p < 0.05$). All other data were not significantly different between the two groups. Table 2 also shows the comparison per group of supine and end of study data. All values were statistically significantly different, except for diastolic blood pressure in the POTS group.

Table 2. Tilt table test data of severe ME/CFS patients: normal heart rate and blood pressure response and postural orthostatic tachycardia syndrome (POTS) response at 20 degrees tilting.

	Group 1 Norm BPHR n = 15	Group 2 POTS n = 4	*p*-Value Group 1 vs. Group 2	*p*-value Group 1 Supine vs. End Tilt	*p*-Value Group 2 Supine vs. End Tilt
HR supine (bpm)	80 (12)	91 (10)	0.14		
HR end-tilt (bpm)	98 (21)	126 (10)	0.03	0.001	0.0002
SBP supine (mmHg)	135 (11)	132 (7)	0.63		
SBP end-tilt (mmHg)	141 (13)	126 (3)	0.06	0.02	0.19
DBP supine (mmHg)	80 (10)	80 (4)	0.98		
DBP end-tilt (mmHg)	87 (10)	84 (3)	0.57	0.002	0.05
$PetCO_2$ supine (mmHg)	38 (3)	38 (3)	0.85		
$PetCO_2$ end-tilt (mmHg)	30 (6)	28 (6)	0.57	<0.0001	0.05
CBF supine (mL/min)	615 (71)	632 (71)	0.70		
CBF end-tilt (mL/min)	452 (57)	474 (67)	0.54	<0.0001	<0.0001
CBF end-tilt %change	−26 (5)%	−25 (2)	0.69		
SVI supine (mL/m^2)	36 (5)	36 (12)	0.95		
SVI end-tilt (mL/m^2)	25 (4)	23 (8)	0.54	<0.0001	0.01
SVI end tilt %change	−30 (8)%	−35 (7)%	0.27		
CI supine (L/min/m^2)	2.8 (0.5)	3.2(0.8)	0.17		
CI end tilt (L/min/m^2)	2.0 (0.3)	2.4 (0.6)	0.08	<0.0001	0.01
CI end tilt %change	−27 (8)%	−26 (6)	0.72		

DBP: diastolic blood pressure; CBF: cerebral blood flow; CI: cardiac index HR: heart rate; norm: normal; DBP: diastolic blood pressure; CBF: cerebral blood flow; CI: cardiac index; HR: heart rate; $P_{et}CO_2$: end-tidal CO_2 pressure; SBP: systolic blood pressure; %change: percent change from supine data; SVI: stroke volume index.

4. Discussion

The main finding of this exploratory study is that in severe ME/CFS patients, a significant reduction in cerebral blood flow can be provoked during a brief 20 degree head-up tilt test. The combination of low-grade head-up tilt testing and extracranial Doppler echography has not been described before. The 27% reduction in cerebral blood flow after 15 min compares to the 26% reduction observed after 30 min of 70 degree head-up tilt in a less severely affected population of ME/CFS patients [2]. We also observed a significant reduction in stroke volume index during this mild orthostatic stress, and found a significant correlation between the decrease in cardiac index and the decrease in cerebral blood flow. Finally, even with this low orthostatic stress, four patients fulfilled the heart rate criteria for POTS, while none of the patients had an orthostatic hypotension or (near) syncope. A milder orthostatic stress of 20 degrees also allowed the accurate measurement of CBF declines that might have been difficult to measure in those who have rapid drops in blood pressure associated with classical orthostatic hypotension or vasovagal syncope when tested at 70 degrees [2]. It remains to be determined whether a 20 degree tilt angle would be adequate for the diagnosis of orthostatic intolerance/significant cerebral blood flow reduction in less severely affected ME/CFS patients.

A previous study with 20 degree head-up tilt testing for 15 min in adolescent CFS patients showed a significantly higher heart rate and lower stroke volume index early after tilting (0.5–2.5 min after onset of tilt) in CFS patients compared with controls [7]. The present study in adults confirms the changes in stroke volume index during 20 degree tilt testing in CFS adolescents. Remarkably, supine heart rates were higher in our patient population (mean 35 years) than in the adolescent CFS population of Wyller et al. (mean 15 years) [7]. A large scale population study in healthy participants showed that resting heart rate in 15 year old adolescents normally is slightly higher than in adults over 20 years [14]. Our observation of a higher resting heart rate in the severe ME/CFS patient population might be related to disease severity, with a higher heart rate in more severe patients. However, this needs to be studied in a larger sample size.

It is generally assumed that part of the OI symptomatology is related to cerebral hypoperfusion [4,5,15,16]. One technique to study cerebral perfusion is transcranial Doppler. Using this technique, OH and POTS have been studied in different diseases and under different physiological conditions like aging, high-altitude, space flights and heat stress [17–32] However, it has been noted that OI symptoms during HUT may be present, even in the absence of abnormalities of heart rate or blood pressure [33–36]. Three recent studies used transcranial Doppler to investigate cerebral perfusion in patients with a normal HUT and without an abnormal HR and BP response like POTS or OH [33,34,36]. The three studies found that the blood flow velocity decrease in patients with a normal HUT but with OI symptoms was larger than in healthy volunteers and patients without OI, and similar to the patients with POTS or OH. These observations suggest that cerebral hypoperfusion is not only present in POTS and OH patients, but also in patients with OI symptoms without POTS and OH.

Several points about the study findings deserve emphasis. First, in our earlier study using a 70 degree tilt test, the decrease in stroke volume index was 31% at 15 min post tilt with no differences between mild, moderately and severely affected ME/CFS patients [10]. The stroke volume index decrease in the present study was 30%. This suggests that mild orthostatic stress testing in severe ME/CFS patients results in similar stroke volume index reductions when compared to 70 degree testing in ME/CFS patients of less severe disease. Further studies are needed to compare stroke volume changes at the different tilt angles in this patient population with varying degrees of disease severity.

Second, in a previous study, we found that cerebral blood flow reduction was 26% during a 70 degree tilt test [2]. In the present study, a cerebral blood flow decrease of 27% was observed. In contrast, studies using transcranial Doppler have shown no differences in cerebral blood flow velocities between CFS patients and healthy controls both at low-grade and high-grade tilt angles [37]. However, the authors showed that the end-tidal CO_2 values were lower in CFS patients compared with controls, at all tilt angles. Previous studies have shown that hypocapnia can reduce intracranial vessel diameters, thereby altering the relation between flow velocity changes and hemodynamic changes [38–40]. Therefore, the absence of a difference in cerebral blood flow velocities of patients versus controls using transcranial Doppler may be related to end-tidal CO_2-related vasoconstriction of the middle cerebral artery in CFS patients. Vasoconstriction leads to increases in cerebral flow velocities, resulting in non-significant differences in the TCD measurement between CFS patients and controls. The more direct measurement using extracranial Doppler identifies the global reduction in CBF that can be missed with transcranial Doppler. The similarity of cerebral blood flow reduction in the present study and our previous study despite the lower degree of orthostatic stress and the shorter tilt is most likely related to the more severe disease status of patients in the present study [2]. Consistent with our previous study [2], the results of this study again clearly demonstrate that reduced cerebral blood flow is a cardinal contributor to orthostatic intolerance symptoms in ME/CFS patients.

Third, as seen in our previous study and other recent studies using transcranial Doppler, we again demonstrated that an orthostatic intolerance/abnormal cerebral blood flow reduction may be present without heart rate and blood pressure abnormalities [2,33–36]. Patients with a normal heart rate and blood pressure response during a tilt test would have been misclassified as having no abnormalities. The present study suggests that cerebral blood flow measurements are needed in order to more accurately measure the prevalence of orthostatic intolerance in ME/CFS patients.

Fourth, the data of this study extend the observation that hypocapnia is significant in ME/CFS patients and is a likely contributor to reductions in cerebral blood flow [2,34,41]. The end-tidal CO_2 reductions also support the observation that 20 degrees is sufficient to provoke similar cerebral blood flow changes in severe ME/CFS patients compared with a less severely diseased group of ME/CFS. When hypocapnia is observed, a focus on respiration depth and speed has the potential to be one of the therapeutic guidance options for these patients to lessen orthostatic intolerance complaints. Another possible factor besides vasculature, autonomic nervous system and end-tidal CO_2 might be altered blood cell behavior, especially red blood cell [42,43].

Fifth, in a previous review on the relation between acute cardiac output changes and cerebral blood flow changes, Meng et al. found in healthy volunteers that a 30% reduction in cardiac output resulted in a 10% reduction in cerebral blood flow [44]. In the present study, we also found a significant relation between the reduction in cardiac index and cerebral blood flow. This provides further evidence for the validity of the extracranial Doppler measurements to determine cerebral blood flow. Furthermore, from the linear regression analysis in the present study, we calculated that a 30% reduction in cardiac index resulted in a 28% reduction in cerebral blood flow, which is much larger than the 10% in healthy volunteers in the review of Meng et al. [44]. Whether these differences are specific to ME/CFS are related to the use of transcranial Doppler vs. extracranial Doppler or are related to the significant reductions in end-tidal CO_2 in the present study will need to be determined in future and larger studies.

Finally, even during a brief 20 degree tilt, four (21%) patients developed POTS. This has not been described before and might be a reflection of the severity of the disease. It also implies that the diagnosis of POTS cannot be dismissed when patients have complaints suggestive of POTS in non-standing positions like sitting or lying down with a slight head-up position.

4.1. Clinical Implications

Patients are advised to lie down when they experience orthostatic intolerance complaints. Our findings of a clinically significant cerebral blood flow reduction at just 20 degrees suggest that a slight head-up position may not be adequate enough to resolve symptoms of orthostatic intolerance in some patients. Furthermore, the European Society of Cardiology syncope guidelines and other papers advocate the use of a nocturnal head-up position of more than 10 degrees to prevent nocturnal polyuria and the consequent circulatory underfilling [45–48]. In light of the presented results, this advice has the potential to be detrimental in some ME/CFS patients.

4.2. Limitations

This study only included ME/CFS patients who were bedbound, and we caution that the 20 degree head-up tilt angle needs further study before it can replace longer 70 degree tilt angles for assessing less severely impaired ME/CFS patients. Comparisons of the hemodynamic and cerebral blood flow abnormalities of 20 and 70 degrees of tilting are needed. We also did not include healthy controls for comparison. It is possible that healthy controls would have little or no perturbation in response to a 20 degree head-up angle, which would have the effect of widening the physiologic differences between ME/CFS patients and controls. Whether disease severity differences lead to differences in cerebral blood flow reduction needs to be studied in the future. Finally, while it is reasonable to expect that the 20 degree abbreviated tilt test would be less taxing than a longer 70 degree tilt test, and therefore less likely to provoke post-exertional malaise, this hypothesis remains to be tested.

5. Conclusions

This study demonstrates that a short 15-min tilt using a mild 20 degree head-up angle is sufficient to provoke a clinically significant reduction in cerebral blood flow in patients with severe ME/CFS. This method of orthostatic testing has the potential to improve the assessment of the prevalence of orthostatic intolerance in severely affected ME/CFS patients who are reluctant to undergo a 70 degree tilt. In this patient population, a milder orthostatic stress was able to confirm the following: cerebral blood flow abnormalities in the absence of heart rate and blood pressure abnormalities; POTS is a small subset; and associated reductions in end-tidal CO_2, stroke volume and cardiac index.

Author Contributions: C.(L.)M.C.v.C., P.C.R., and F.C.V. conceived the study, C.(L.)M.C.v.C. and F.C.V. collected the data, C.(L.)M.C.v.C. performed the primary data analysis and F.C.V. and P.C.R. performed secondary data analyses. All authors have read and agreed to the published version of the manuscript.

Funding: This research received no external funding.

Acknowledgments: Rowe is supported by the Sunshine Natural Wellbeing Foundation Professorship.

Conflicts of Interest: The authors declare no conflict of interest.

References

1. Institute of Medicine (IOM) (Ed.) *Beyond Mayalgic Encephalomyelitis/Chronic Fatigue Syndrome: Redefining an Illness*; The National Academies Press: Washington, DC, USA, 2015.
2. Van Campen, C. (L).M.C.; Verheugt, F.W.A.; Rowe, P.C.; Visser, F.C. Cerebral blood flow is reduced in ME/CFS during head-up tilt testing even in the absence of hypotension or tachycardia: A quantitative, controlled study using Doppler echography. *Clin. Neurophysiol. Pract.* **2020**, *5*, 50–58. [CrossRef] [PubMed]
3. Van Campen, C. (L).M.C.; Verheugt, F.W.A.; Visser, F.C. Cerebral blood flow changes during tilt table testing in healthy volunteers, as assessed by Doppler imaging of the carotid and vertebral arteries. *Clin. Neurophysiol. Pract.* **2018**, *3*, 91–95. [CrossRef] [PubMed]
4. Sheldon, R.S.; Grubb, B.P., II; Olshansky, B.; Shen, W.K.; Calkins, H.; Brignole, M.; Raj, S.R.; Krahn, A.D.; Morillo, C.A.; Stewart, J.M.; et al. 2015 heart rhythm society expert consensus statement on the diagnosis and treatment of postural tachycardia syndrome, inappropriate sinus tachycardia, and vasovagal syncope. *Heart Rhythm* **2015**, *12*, e41–e63. [CrossRef] [PubMed]
5. Shen, W.K.; Sheldon, R.S.; Benditt, D.G.; Cohen, M.I.; Forman, D.E.; Goldberger, Z.D.; Grubb, B.P.; Hamdan, M.H.; Krahn, A.D.; Link, M.S.; et al. 2017 ACC/AHA/HRS guideline for the evaluation and management of patients with syncope: Executive summary: A report of the American College of Cardiology/American Heart Association Task Force on Clinical Practice Guidelines and the Heart Rhythm Society. *J. Am. Coll. Cardiol.* **2017**, *70*, 620–663. [CrossRef] [PubMed]
6. Freeman, R.; Wieling, W.; Axelrod, F.B.; Benditt, D.G.; Benarroch, E.; Biaggioni, I.; Cheshire, W.P.; Chelimsky, T.; Cortelli, P.; Gibbons, C.H.; et al. Consensus statement on the definition of orthostatic hypotension, neurally mediated syncope and the postural tachycardia syndrome. *Auton. Neurosci.* **2011**, *161*, 46–48. [CrossRef] [PubMed]
7. Wyller, V.B.; Due, R.; Saul, J.P.; Amlie, J.P.; Thaulow, E. Usefulness of an abnormal cardiovascular response during low-grade head-up tilt-test for discriminating adolescents with chronic fatigue from healthy controls. *Am. J. Cardiol.* **2007**, *99*, 997–1001. [CrossRef]
8. Fukuda, K.; Straus, S.E.; Hickie, I.; Sharpe, M.C.; Dobbins, J.G.; Komaroff, A.; International Chronic Fatigue Syndrome Study Group. The chronic fatigue syndrome: A comprehensive approach to its definition and study. *Ann. Intern. Med.* **1994**, *121*, 953–959. [CrossRef]
9. Carruthers, B.M.; Van de Sande, M.I.; DE Meirleir, K.L.; Klimas, N.G.; Broderick, G.; Mitchell, T.; Staines, D.; Powles, A.C.P.; Speight, N.; Vallings, R.; et al. Myalgic encephalomyelitis: International consensus criteria. *J. Intern. Med.* **2011**, *270*, 327–338. [CrossRef]
10. Van Campen, C.M.C.; Visser, F.C. The abnormal Cardiac Index and Stroke Volume Index changes during a normal Tilt Table Test in ME/CFS patients compared to healthy volunteers, are not related to deconditioning. *J. Thromb. Circ.* **2018**. [CrossRef]
11. Sato, K.; Ogoh, S.; Hirasawa, A.; Oue, A.; Sadamoto, T. The distribution of blood flow in the carotid and vertebral arteries during dynamic exercise in humans. *J. Physiol.* **2011**, *589 Pt 11*, 2847–2856. [CrossRef]
12. Kusumoto, F.; Venet, T.; Schiller, N.B.; Sebastian, A.; Foster, E. Measurement of aortic blood flow by Doppler echocardiography: Temporal, technician, and reader variability in normal subjects and the application of generalizability theory in clinical research. *J. Am. Soc. Echocardiogr.* **1995**, *8 Pt 1*, 647–653. [CrossRef]
13. Van Campen, C. (L).M.C.; Visser, F.C.; De Cock, C.C.; Vos, H.S.; Kamp, O.; Visser, C.A. Comparison of the haemodynamics of different pacing sites in patients undergoing resynchronisation treatment: Need for individualisation of lead localisation. *Heart* **2006**, *92*, 1795–1800. [CrossRef] [PubMed]
14. Ostchega, Y.; Porter, K.S.; Hughes, J.; Dillon, C.F.; Nwankwo, T. *Resting Pulse Rate Reference Data for Children, Adolescents, and Adults: United States, 1999–2008*; National Health Stat Report; National Center for Health Statistics: Hyattsville, MD, USA, 2011; pp. 1–16.
15. Schondorf, R.; Benoit, J.; Stein, R. Cerebral autoregulation in orthostatic intolerance. *Ann. N. Y. Acad. Sci.* **2001**, *940*, 514–526. [CrossRef] [PubMed]

16. Moya, A.; Sutton, R.; Ammirati, F.; Blanc, J.J.; Brignole, M.; Dahm, J.B.; Deharo, J.C.; Gajek, J.; Gjesdal, K.; Krahn, A.; et al. Guidelines: Guidelines for the diagnosis and management of syncope (version 2009): The Task Force for the Diagnosis and Management of Syncope of the European Society of Cardiology (ESC). *Eur. Heart J.* **2009**, *30*, 2631–2671.

17. Haubrich, C.; Pies, K.; Dafotakis, M.; Block, F.; Kloetzsch, C.; Diehl, R.R. Transcranial Doppler monitoring in Parkinson's disease: Cerebrovascular compensation of orthostatic hypotension. *Ultrasound Med. Biol.* **2010**, *36*, 1581–1587. [CrossRef] [PubMed]

18. Mankovsky, B.N.; Piolot, R.; Mankovsky, O.L.; Ziegler, D. Impairment of cerebral autoregulation in diabetic patients with cardiovascular autonomic neuropathy and orthostatic hypotension. *Diabet. Med.* **2003**, *20*, 119–126. [CrossRef]

19. Purewal, T.S.; Watkins, P.J. Postural hypotension in diabetic autonomic neuropathy: A review. *Diabet. Med.* **1995**, *12*, 192–200. [CrossRef]

20. Fuente Mora, C.; Palma, J.A.; Kaufmann, H.; Norcliffe-Kaufmann, L. Cerebral autoregulation and symptoms of orthostatic hypotension in familial dysautonomia. *J. Cereb. Blood Flow Metab.* **2017**, *37*, 2414–2422. [CrossRef]

21. Bailey, D.M.; Jones, D.W.; Sinnott, A.; Brugniaux, J.V.; New, K.J.; Hodson, D.; Marley, C.J.; Smirl, J.D.; Ogoh, S.; Ainslie, P.N. Impaired cerebral haemodynamic function associated with chronic traumatic brain injury in professional boxers. *Clin. Sci. (Lond.)* **2013**, *124*, 177–189. [CrossRef]

22. Wong, B.J.; Sheriff, D.D. Role of splanchnic constriction in governing the hemodynamic responses to gravitational stress in conscious dogs. *J. Appl. Physiol.* **2011**, *111*, 40–47. [CrossRef]

23. Deegan, B.M.; Sorond, F.A.; Galica, A.; Lipsitz, L.A.; O'Laighin, G.; Serrador, J.M. Elderly women regulate brain blood flow better than men do. *Stroke* **2011**, *42*, 1988–1993. [CrossRef] [PubMed]

24. Razumovsky, A.Y.; DeBusk, K.; Calkins, H.; Snader, S.; Lucas, K.E.; Vyas, P.; Hanley, D.F.; Rowe, P.C. Cerebral and systemic hemodynamics changes during upright tilt in chronic fatigue syndrome. *J. Neuroimaging* **2003**, *13*, 57–67. [CrossRef] [PubMed]

25. Urbano, F.; Roux, F.; Schindler, J.; Mohsenin, V. Impaired cerebral autoregulation in obstructive sleep apnea. *J. Appl. Physiol.* **2008**, *105*, 1852–1857. [CrossRef] [PubMed]

26. Treger, I.; Shafir, O.; Keren, O.; Ring, H. Cerebral blood flow velocity during postural changes on tilt table in stroke patients. *Eur. Medicophys.* **2005**, *41*, 293–296.

27. Iwasaki, K.; Levine, B.D.; Zhang, R.; Zuckerman, J.H.; Pawelczyk, J.A.; Diedrich, A.; Ertl, A.; Cox, J.; Cooke, W.; Giller, C.; et al. Human cerebral autoregulation before, during and after spaceflight. *J. Physiol.* **2007**, *579 Pt 3*, 799–810. [CrossRef]

28. Thomas, K.N.; Burgess, K.R.; Basnyat, R.; Lucas, S.J.; Cotter, J.D.; Fan, J.L.; Peebles, K.C.; Lucas, R.A.I.; Ainslie, P.N. Initial orthostatic hypotension at high altitude. *High Alt. Med. Biol.* **2010**, *11*, 163–167. [CrossRef]

29. Van Beek, A.H.; Sijbesma, J.C.; Jansen, R.W.; Rikkert, M.G.; Claassen, J.A. Cortical oxygen supply during postural hypotension is further decreased in Alzheimer's disease, but unrelated to cholinesterase-inhibitor use. *J. Alzheimers Dis.* **2010**, *21*, 519–526. [CrossRef]

30. Lipsitz, L.A.; Mukai, S.; Hamner, J.; Gagnon, M.; Babikian, V. Dynamic regulation of middle cerebral artery blood flow velocity in aging and hypertension. *Stroke* **2000**, *31*, 1897–1903. [CrossRef]

31. Low, P.A.; Novak, V.; Spies, J.M.; Novak, P.; Petty, G.W. Cerebrovascular regulation in the postural orthostatic tachycardia syndrome (POTS). *Am. J. Med. Sci.* **1999**, *317*, 124–133. [CrossRef]

32. Morrison, S.A.; Ainslie, P.N.; Lucas, R.A.; Cheung, S.S.; Cotter, J.D. Compression garments do not alter cerebrovascular responses to orthostatic stress after mild passive heating. *Scand. J. Med. Sci. Sports* **2014**, *24*, 291–300. [CrossRef]

33. Novak, P. Hypocapnic cerebral hypoperfusion: A biomarker of orthostatic intolerance. *PLoS ONE* **2018**, *13*, e0204419. [CrossRef] [PubMed]

34. Bush, V.E.; Wight, V.L.; Brown, C.M.; Hainsworth, R. Vascular responses to orthostatic stress in patients with postural tachycardia syndrome (POTS), in patients with low orthostatic tolerance, and in asymptomatic controls. *Clin. Auton. Res.* **2000**, *10*, 279–284. [CrossRef] [PubMed]

35. Park, J.; Kim, H.T.; Park, K.M.; Ha, S.Y.; Kim, S.E.; Shin, K.J.; Kim, S.E.; Jang, W.; Kim, J.S.; Youn, J.; et al. Orthostatic dizziness in Parkinson's disease is attributed to cerebral hypoperfusion: A transcranial doppler study. *J. Clin. Ultrasound* **2017**, *45*, 337–342. [CrossRef]

36. Shin, K.J.; Kim, S.E.; Park, K.M.; Park, J.; Ha, S.Y.; Kim, S.E.; Kwon, O.Y. Cerebral hemodynamics in orthostatic intolerance with normal head-up tilt test. *Acta Neurol. Scand.* **2016**, *134*, 108–115. [CrossRef] [PubMed]

37. Ocon, A.J.; Messer, Z.R.; Medow, M.S.; Stewart, J.M. Increasing orthostatic stress impairs neurocognitive functioning in chronic fatigue syndrome with postural tachycardia syndrome. *Clin. Sci. (Lond.)* **2012**, *122*, 227–238. [CrossRef]

38. Verbree, J.; Bronzwaer, A.S.; Ghariq, E.; Versluis, M.J.; Daemen, M.J.; Van Buchem, M.A.; Dahan, A.; Van Lieshout, J.J.; Van Osch, M.J.P. Assessment of middle cerebral artery diameter during hypocapnia and hypercapnia in humans using ultra-high-field MRI. *J. Appl. Physiol.* **2014**, *117*, 1084–1089. [CrossRef]

39. Coverdale, N.S.; Gati, J.S.; Opalevych, O.; Perrotta, A.; Shoemaker, J.K. Cerebral blood flow velocity underestimates cerebral blood flow during modest hypercapnia and hypocapnia. *J. Appl. Physiol.* **2014**, *117*, 1090–1096. [CrossRef]

40. Al-Khazraji, B.K.; Shoemaker, L.N.; Gati, J.S.; Szekeres, T.; Shoemaker, J.K. Reactivity of larger intracranial arteries using 7 T MRI in young adults. *J. Cereb. Blood Flow Metab.* **2019**, *39*, 1204–1214. [CrossRef]

41. Immink, R.V.; Pott, F.C.; Secher, N.H.; Van Lieshout, J.J. Hyperventilation, cerebral perfusion, and syncope. *J. Appl. Physiol.* **2014**, *116*, 844–851. [CrossRef]

42. Huisjes, R.; Bogdanova, A.; Van Solinge, W.W.; Schiffelers, R.M.; Kaestner, L.; Van Wijk, R. Squeezing for Life—Properties of Red Blood Cell Deformability. *Front. Physiol.* **2018**, *9*, 656. [CrossRef]

43. Qiu, Y.; Myers, D.R.; Lam, W.A. The biophysics and mechanics of blood from a materials perspective. *Nat. Rev. Mater.* **2019**, *4*, 294–311. [CrossRef] [PubMed]

44. Meng, L.; Hou, W.; Chui, J.; Han, R.; Gelb, A.W. Cardiac output and cerebral blood flow: The integrated regulation of brain perfusion in adult humans. *Anesthesiology* **2015**, *123*, 1198–1208. [CrossRef] [PubMed]

45. Brignole, M.; Moya, A.; De Lange, F.J.; Deharo, J.C.; Elliott, P.M.; Fanciulli, A.; Fedorowski, A.; Furlan, R.; Kenny Rose, A.; Martín, A.; et al. 2018 ESC Guidelines for the diagnosis and management of syncope. *Kardiol. Pol.* **2018**, *76*, 1119–1198. [CrossRef] [PubMed]

46. Eschlbock, S.; Wenning, G.; Fanciulli, A. Evidence-based treatment of neurogenic orthostatic hypotension and related symptoms. *J. Neural. Transm. (Vienna)* **2017**, *124*, 1567–1605. [CrossRef]

47. Hainsworth, R.; Al-Shamma, Y.M. Cardiovascular responses to upright tilting in healthy subjects. *Clin. Sci. (Lond.)* **1988**, *74*, 17–22. [CrossRef]

48. Cooper, V.L.; Hainsworth, R. Head-up sleeping improves orthostatic tolerance in patients with syncope. *Clin. Auton. Res.* **2008**, *18*, 318–324. [CrossRef]

Systematic Review

Systematic Review of Sleep Characteristics in Myalgic Encephalomyelitis/Chronic Fatigue Syndrome

Rebekah Maksoud [1,2,*], Natalie Eaton-Fitch [1,2,3], Michael Matula [1,2], Hélène Cabanas [1,2], Donald Staines [1,2] and Sonya Marshall-Gradisnik [1,2]

[1] National Centre for Neuroimmunology and Emerging Diseases (NCNED), Menzies Health Institute Queensland, Griffith University, Gold Coast 4222, Australia; natalie.eaton-fitch@griffithuni.edu.au (N.E.-F.); m.matula@griffith.edu.au (M.M.); h.cabanas@griffith.edu.au (H.C.); d.staines@griffith.edu.au (D.S.); s.marshall-gradisnik@griffith.edu.au (S.M.-G.)

[2] Consortium Health International for Myalgic Encephalomyelitis, Griffith University, Gold Coast 4222, Australia

[3] School of Medical Science, Griffith University, Gold Coast 4222, Australia

* Correspondence: r.maksoud@griffith.edu.au

Citation: Maksoud, R.; Eaton-Fitch, N.; Matula, M.; Cabanas, H.; Staines, D.; Marshall-Gradisnik, S. Systematic Review of Sleep Characteristics in Myalgic Encephalomyelitis/Chronic Fatigue Syndrome. *Healthcare* **2021**, *9*, 568. https://doi.org/10.3390/healthcare9050568

Academic Editors: Kenneth J. Friedman, Lucinda Bateman and Kenny Leo De Meirleir

Received: 13 April 2021
Accepted: 7 May 2021
Published: 11 May 2021

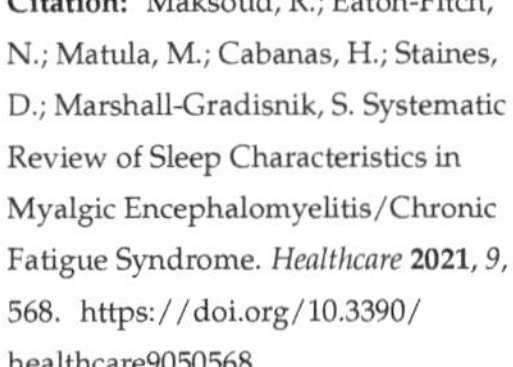

Abstract: (1) Background—Myalgic Encephalomyelitis/Chronic Fatigue Syndrome (ME/CFS) is a multifaceted illness characterized by profound and persistent fatigue unrelieved by rest along with a range of other debilitating symptoms. Experiences of unrefreshing and disturbed sleep are frequently described by ME/CFS patients. This is the first systematic review assessing sleep characteristics in ME/CFS. The aim of this review is to determine whether there are clinical characteristics of sleep in ME/CFS patients compared to healthy controls using objective measures such as polysomnography and multiple sleep latency testing. (2) Methods—the following databases—Pubmed, Embase, Medline (EBSCO host) and Web of Science, were systematically searched for journal articles published between January 1994 to 19 February 2021. Articles that referred to polysomnography or multiple sleep latency testing and ME/CFS patients were selected, and further refined through use of specific inclusion and exclusion criteria. Quality and bias were measured using the Joanna Briggs Institute checklist. (3) Results—twenty observational studies were included in this review. The studies investigated objective measures of sleep quality in ME/CFS. Subjective measures including perceived sleep quality and other quality of life factors were also described. (4) Conclusions—Many of the parameters measured including slow- wave sleep, apnea- hypopnea index, spectral activity and multiple sleep latency testing were inconsistent across the studies. The available research on sleep quality in ME/CFS was also limited by recruitment decisions, confounding factors, small sample sizes and non-replicated findings. Future well-designed studies are required to understand sleep quality in ME/CFS patients.

Keywords: Myalgic Encephalomyelitis; chronic fatigue syndrome; sleep; polysomnography; multiple sleep latency testing

1. Introduction

Myalgic Encephalomyelitis/Chronic Fatigue Syndrome (ME/CFS) is a medical condition characterised by non-restorative, incapacitating fatigue that is unrelieved by rest in combination with a plethora of other symptoms such as neurological, immune and endocrine disruption [1]. Unrefreshing or disturbed sleep is an almost universal symptom reported in about 91% of patients in the absence of a primary sleep disorder (PSD) [1,2]. The presentation and severity of these symptoms ranges between patients and results in considerable loss of quality of life [3]. There currently remains no diagnostic test nor targeted treatment for this condition. Diagnosis is instead dependent on the application of symptom-specific case criteria following the exclusion of any other potential medical cause [4].

There are three main criteria used in research and clinical practice to diagnose ME/CFS and include: (1) The Center for Disease Control's (CDC) Fukuda criteria (FC) (1994); (2) The Canadian Consensus Criteria (CCC) (2003) and (3), The International Consensus Criteria (ICC) (2011) [1,5,6]. Diagnosis with the FC is dependent on the presence of persistent fatigue that is unrelieved by rest in combination with four out of a potential eight additional symptoms including but not limited to unrefreshing sleep [5]. The revised CCC criteria builds upon the FC with emphasis on post-exertional malaise as a key symptom of ME/CFS. In this criteria, sleep disturbance was also described as a potential symptom of this disorder [6]. The ICC criteria divides sleep specific symptoms into two categories: disturbed sleep patterns and unrefreshing sleep and includes the most homogenous subset of patients [1]. The most recent institute of medicine criteria (IOMC) has unrefreshing sleep listed as one of the three required symptoms [4]. Unrefreshing or disturbed sleep can include the following sleep characteristics: reversed sleep rhythms and frequent awakenings [7].

Polysomnography (PSG) is the most common objective measure of sleep quality. PSG measures at various sleep phases including rapid eye movement (REM) and non-REM sleep. Non-REM sleep phases include: stage N1, N2 and N3/4 or Slow wave sleep (SWS). SWS is the deepest phase of non-REM sleep [8]. Other parameters including sleep onset latency (SOL) which is the time taken from being fully awake to fast asleep as well as apnoea- hypopnoea index (AHI) and microarousal Index (MAI) values [9]. AHI values are the number of apnoea and hypopnea events per hour of sleep. MAI values allows measurement of sleep fragmentation [10]. Multiple Sleep Latency Testing (MSLT) an objective measure to assess the ability to fall asleep under controlled conditions is at times used in combination with PSG [11].

This is the first systematic review to critically appraise primary studies that assess objective measures of sleep quality in ME/CFS patients using PSG and/or MSLT. Secondary to this, subjective measures including sleep quality and depression scores were also evaluated.

2. Methods

This study was conducted according to Cochrane reviews and Preferred Reporting Items for Systematic Reviews and Meta-analyses 2020 (PRISMA 2020) guidelines [12,13]. To ensure that international standards were maintained when reporting information in this systematic review these guidelines were used. Four electronic databases (Pubmed, Medline [EBSCOHost], Embase and Web of Science) were systematically searched. Articles containing the following medical Subject Headings (MeSH) terms "Syndrome, Chronic Fatigue" [Mesh] AND ('Multiple Sleep Latency Test*' OR 'Polysomnography' OR 'Polysomnograph*') were searched between January 1995 and 19th February 2021 (full list of terms can be found in Table S1). Terms were combined with the Boolean operators 'AND' in order to tie the disease of interest with objective measures of sleep quality and 'OR' to expand the search for all expressions of cases. Two identical literature searches were conducted separately by two different authors. Citation searching was completed, and no additional papers were selected. Searching for unpublished literature was not performed. No additional papers were identified in the final search or through alternative databases such as Griffith University institute library or Google Scholar.

2.1. Inclusion/ Exclusion Criteria

Studies were included in the review if they contained two or more of the key search terms in the abstract or title and adhered to the following inclusion criteria: (i) published in 1995 or later as the FC was established in December 1994; (ii) human participants who were aged 18 years or over; (iii) full- text articles written in English; (iv) were observational studies reporting on original research; (v) ME/CFS was defined according to the following case criteria: FC (1994), CCC (2003) or ICC (2011) and IOMC (2015); (vi) all studies investigated objective measures of sleep quality.

Articles were excluded from this review if they did not include at least two key search terms in the abstract or title or if they had any of the following exclusion criteria: (i) written prior to the introduction of the FC in 1994; (ii) conducted in participants that were under 18; (iii) articles not written in English or weren't available as full-text; (iv) were interventional based or reported on non-original data including: duplicate studies, case reports or review articles; (v) use of alternative case criteria; (vi) studies were not relevant to the scope of this review. (vii) Publications were also excluded if the ME/CFS cohort was compared with another patient group (e.g., fibromyalgia, depression etc.) and not compared with HC.

2.2. Selection of Studies

The referencing management software package Endnote X9 was used to screen, sort and store all articles from the databases. Duplicates were removed using Endnote's automatic feature. The title and abstract of each article were screened for selected keywords and those which did not contain at least one ME/CFS keyword and one sleep test keyword. The remaining articles that also adhered to inclusion and exclusion criteria were selected. This process was independently conducted by RM and MM. There were minor differences between the two authors, however, these were discussed, and a final list was compiled and approved by both authors. The final list was then reviewed and deemed accordant by all other authors.

2.3. Data Extraction

The following data was extracted from the included studies: (1) diagnostic criteria; (2) study design; (3) sample size; (4) age; (5) sex; (6) BMI; (7) total sleep duration; (8) method of analysis; (9) primary outcomes; (10) secondary outcomes.

2.4. Quality Analysis

All publications included in this systematic review were evaluated for quality and bias using the Joana Briggs Institute (JBI) Critical Appraisal Checklist for Case Control Studies (CACCCS) (File S1. JBI CACCCS and justification). This checklist was selected due to it being an internationally recognised and validated method of evaluating study quality and bias. Quality assessment was separately completed by two authors (RM and NEF). As Item four, five and nine were intervention based these items were excluded in all studies except one [14].

3. Results

Using the selected search terms, a total of 275 papers were identified using the following databases: Embase (108), Pubmed (50), Medline (61), and Web of Science (56). Following the screening process the total number of papers was 20. A detailed outline of the search process is presented in Figure 1. All included papers investigated objective measures of sleep quality in ME/CFS patients compared to HC.

3.1. Participant and Study Characteristics

Participant and study characteristics are presented in Table S2. Four (20%) out of the 20 articles included in this review were observational twin studies [15–18]. The remainder of the included articles (80%) were observational case-control studies [9,14,19–32]. Across all studies, the mean number of ME/CFS patients and HCs included was 26.6 and 24 respectively. Majority of the included participants were female (87%). The mean age across all studies was 41.5 for the ME/CFS group and 39.2 for the HC group [9,14–32]. Fifteen out of 20 papers reported a value for body mass index (BMI) [9,14,15,19–28,31,32]. The average BMI was 25.5 for ME/CFS patients and for 25.4 HC [9,14,15,19–28,31,32]. The FC was used in all studies to diagnose participants [9,14–32]. One study, however, used both the FC criteria and the CCC to diagnose [32]. Average total sleep time was 397.03 min for ME/CFS patients and 400.1 min for the HC group [9,14–32].

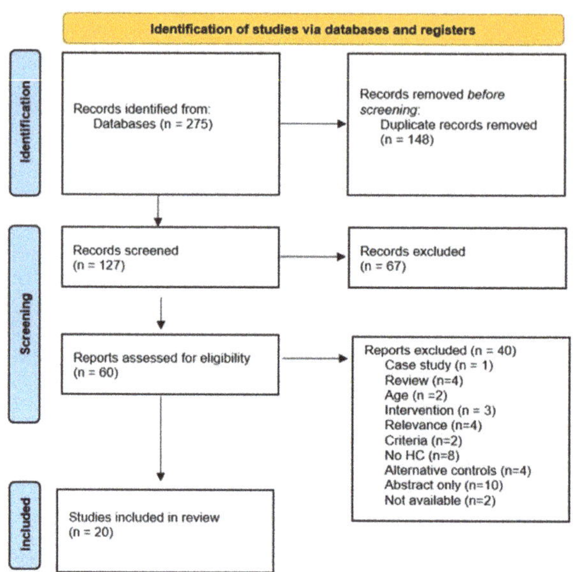

Figure 1. PRISMA 2020 flow diagram of literature search for included studies in this review of sleep and ME/CFS.

3.2. Literature Reporting Changes in Objective Sleep Measures

Objective sleep measures are presented in Table S3. Two twin studies identified an increase in REM sleep in ME/CFS patients compared to their healthy twin [16,17]. One article reported significantly reduced REM to non-REM sleep stage transitions [30]. Alteration of transition patterns resulting in greater relative transition frequency was also observed [30]. Sleep onset latency (SOL) was investigated in 13 articles [9,14,17,19,21–28,31]. All 13 papers reported no differences in SOL between the ME/CFS patients and the HC [9,14,17,19,21–28,31]. Non-REM sleep stages, (NREM) including stage 1- 4 sleep, % was investigated in 12 studies [9,14–17,19,21,22,25,26,29,31]. Two of the 12 studies reported increased stage 3 sleep, % [16,22]. All other findings were insignificant [9,14,15,17,19,21,25,26,29,31]. There were 11 studies that investigated slow-wave sleep (SWS) duration [9,14,15,21–25,28,29,31]. From these studies, only three found that SWS in ME/CFS was significantly longer in duration compared to HC [24,25,28]. The remaining studies reported no difference between the two groups [9,14,15,21–23,29,31].

When assessing sleep apnoea characteristics, five studies detected no differences in AHI [14,15,21–23]. Three studies detected differences in AHI [16,24,31]. MAI was measured in five studies [14,23–25,31]. An increase of MAI in ME/CFS patients was found in all the studies [14,23–25,31].

3.3. Literature Reporting Changes in Spectral Activity

Three articles investigated spectral activity during sleep [15,20,22]. A twin study found no significant differences in spectral power in any frequency band assessed: REM latency, delta-wave, fast frequency beta or alpha power between the twin with ME/CFS and the healthy twin [15]. Another study showed that there was diminished alpha power during stage 2, slow wave, and REM sleep in the ME/CFS cohorts compared to HC [20]. Delta power was found to be decreased during SWS but then was elevated during stage 1 and REM in the ME/CFS cohort. Theta, sigma and beta spectral power during stage 2, SWS and REM were significantly reduced in patients compared to HC [20]. One article found that ultra-slow delta power was significantly lower in ME/CFS patients compared to HC during N3 sleep while all other frequencies tested: theta, alpha, sigma and beta did not differ [22].

3.4. Literature Reporting Changes in MSLT

Changes in MSLT were investigated in six articles [18–20,23,24,26]. One study found reduced mean sleep latency on MSLT in ME/CFS patients compared with HC [23]. Another study found a negative correlation between individual Epworth Sleepiness Scale (ESS) and mean latency scores in both groups [18]. All other articles investigating MSLT identified no significant differences between ME/CFS patients and HC [14,19,26].

3.5. Literature Reporting Changes in Secondary Outcomes

Participant and study characteristics are presented in Table S4. Various secondary outcome measures were investigated in 14 out of 20 included studies [9,14,17–19,22–25,27–29,31,32]. Additionally, different tools were used to measure the same outcomes. Subjective sleep quality or sleepiness was measured in 13 of the studies [9,14,17–19,22–25,28,29,31,32]. All these studies reported significant differences in sleep quality or perceived sleepiness in ME/CFS patients compared with HC. Depression scores were significantly higher in all six studies that included values [14,24,25,27,31,32]. In the five studies that measured anxiety, the ME/CFS scores were significantly different from HC in all but one study [14,23,24,31,32]. Insomnia was investigated in two studies and was found to be significantly higher in ME/CFS patients compared with HC [18,19]. Fatigue levels were also significantly greater in ME/CFS patients in all seven studies that measured this variable [14,21,23–25,28,31]. One study investigated emotional awareness in ME/CFS patients compared with HC [32]. Significant differences in some emotional awareness parameters including TAS-20, TAS total and LEAS-self were found and these correlated with number of awakenings in ME/CFS patients [32].

3.6. Quality Assessment

The Joanna Briggs Institute (JBI) Critical Appraisal Checklist for Case Control Studies (CACCCS) was used to review the selected articles quality and bias. Justification can be found in file S1. Item 4, 5 and 9 were excluded in all studies except Neu 2014B [14]. The study included an exposure to a cognitive test. The authors successfully measured the effect of the exposure for an appropriate duration in a standard, valid, and reliable way across patients and HC [14]. Item 8 was most frequently addressed where 100% of the studies assessed outcomes in a standard, valid and reliable way [9,14–32]. Nineteen out of 20 studies successfully identified confounding factors [9,14–31]. The confounding factors that were addressed were effectively mitigated in 17 of the studies [14–26,28–31]. Sixteen studies had appropriately matched patients and HC [14–20,22–24,26–29,31,32]. Nineteen articles utilised consistent criteria to identify ME/CFS patients and HC [9,14–23,25–32]. Item 2 was the least addressed item where only seven studies appropriately defined and matched source population for ME/CFS patients and HC [9,15–20]. Thirteen of the articles included appropriate statistical analysis [9,14,19–22,24,25,27,28,30–32].

4. Discussion

ME/CFS patients report a significant number of sleep complaints [9,14–32]. The aim of this systematic review was to investigate primary studies that assess objective measures of sleep quality in ME/CFS patients using PSG and/or MSLT compared with HC. Subjective scores including depression, anxiety and QOL scores were also measured. Variable results from these studies were found.

This is the first systematic review assessing objective measures of sleep quality in ME/CFS patients with respect to HC. This method allows the inclusion of all relevant articles. A review of sleep in ME/CFS patients however was undertaken by Jackson et al. [7]. The major findings reported in this publication include: objective and subjective contrasts in sleep quality as well as early evidence suggesting differences in sleep stage transitions, sleep instability and heart rate variability in ME/CFS patients compared with

HC [7]. This review was published in 2012, therefore, a significant amount of time has passed since its publication [7]. Additional studies, in comparison to Jackson et al. have also been identified through this systematic review process [7,9,14,17,22–25,27,32]. A subset of sleep studies was also included in review in a neuroimaging paper by Maksoud et al. [33]. This current systematic review is important as it brings a complete and up-to-date picture of sleep and ME/CFS.

The average age of patients in the included studies of this systematic review was 41.5 years. Approximately 87% of the patients were female. This is consistent with literature showing that ME/CFS is most frequently reported in females aged between 29–35 years [34,35]. This current systematic review selected for participants over the age of 18 due to age-related differences in sleep [15,27,30]. The included studies had a maximum age cut-off for the same reason. Some studies (15%) only recruited females to account for sex- specific differences in sleep as well as to reduce patient pool heterogeneity [15,27,30]. Six of the studies included information on race or ethnicity where majority of the participants were Caucasian [16–20,26]. There was no significant difference in total sleep time between ME/CFS patients and HC. Selected studies restricted outliers of total sleep time in either group to control for potential sleep-related morbidities.

Four of the included studies were twin-based [15–18]. Recruitment of twins assists in moderating differences in genetic and environmental factors. The genetic contribution and potential familial vulnerability of ME/CFS on the unaffected twin is not currently known [15–18]. Ball et al. reported sleep disruption in both ME/CFS patients and their unaffected twin [16]. Therefore, future considerations may involve comparative studies with closely-matched non-relative controls to ensure that there is no genetic contribution to sleep disruption in the selected HC [16].

Paediatric and adolescent sleep characteristics have not been captured in this sleep review due to potentially significant age-related differences. Presentation of illness may also differ between adults and children [36]. Case criteria have also described unrefreshing sleep as a hallmark symptom [1,4]. One study was identified during the screening process that investigates sleep in adolescent ME/CFS patients [37]. This study found that there were significantly higher levels of sleep disruption in adolescents with ME/CFS, and includes brief and longer awakenings [37]. Further investigation of sleep disruption in paediatric and adolescent ME/CFS populations is required.

All of the included studies utilised the FC to classify ME/CFS patients [9,14–32]. One study used both FC and CCC [32]. Compared to the later definitions, the FC is considered too broad and often presents with a heterogenous subset of patients [4]. Consideration of future studies may include representation of patients diagnosed with more stringent definitions [1,4,6]. The more recent case definitions incorporate ME/CFS specific symptoms such as post-exertional malaise that allows a more representative subset of ME/CFS patients to be included [1,4,6].

A limitation to this systematic review is that it was restricted to articles that had PSG and MSLT in the abstract or title [9,14–32]. These terms were selected on the basis of being the primary objective measure of sleep used. Other measures that may describe sleep quality include actigraphy, observation, bed sensors, eyelid movement- and non-invasive arm sensors [38]. Reports on the use of actigraphy for measures investigated in this paper including sleep-wake cycles are controversial. These terms were also excluded due to their broad nature, although this may have resulted in potentially relevant articles not being captured. Some studies also utilised components of polysomnography including EEG and discussed features of sleep but did not undergo the whole polysomnography process [39]. Additionally, two studies by Neu et al. were not included in this review due to not containing any key words in the abstract or title [40,41]. These papers followed most of our selection criteria. One used PSG to assess cognitive impairment in ME/CFS [40]. ME/CFS performance in almost all cognitive tasks was lower compared with HC. EEG theta power was also significantly higher in ME/CFS patients. The other paper investigated sleep parameters in ME/CFS compared with HC and primary sleep disorders [41]. ME/CFS showed higher slow-wave sleep, however this is an inconsistent parameter across studies included in this review. In order to avoid selection bias this paper could not be handpicked to include in our study based on recommendations of Cochrane guidelines handbook [12,40,41].

Existing comorbid disorders may also play a role on sleep disruption in ME/CFS patients. Fibromyalgia syndrome (FMS), migraine and irritable bowel syndrome (IBS) all commonly occur in ME/CFS patients and have known implications on sleep efficiency. PSG studies of FMS patients reported poorer sleep quality as well as higher number of awakenings, higher arousal index, greater AHI and lower N1 sleep in FMS patients compared to HC. Sleep disturbance also exacerbates symptom severity in FMS [42,43]. One included study separated patients with ME/CFS alone or comorbid ME/CFS and FMS. There was a higher number of cases of sleep disorders among those diagnosed with IBS, further analysis is required, however, to understand this relationship [44]. All of the studies did not include patients who had a Diagnostic and Statistical Manual of Mental Disorders, 4th Edition (DSM-IV) disorder. Therefore, the sleep patterns that are observed cannot be attributed to major depression episodes or other associated conditions [9,14–32].

Care needs to be considered to ensure that all sleep characteristics are related to ME/CFS specifically, not other associated disorders. Some studies recruited ME/CFS patients without comorbidities to confirm the results observed were representative of ME/CFS [21]. In these studies, a minority of ME/CFS patients exhibited abnormalities in PSG data. Some studies even further classified ME/CFS patients in of groups into less sleepy and sleepier groups; this was conducted in two of the studies [27,28].

Confounding factors including consumption of alcohol and caffeine, medication, strenuous exercise, or a change in time zones may have contributed to varied results observed. Nine of the studies accounted for alcohol and/or caffeine [14–18,25,27,28,30]. Three of the studies also ensured that participants were not travelling from conflicting time zones within a certain timeframe of the study or adjusted the sleep schedule according to their place of residence [9,15,22]. In three of the studies participants, in particular HC were requested to refrain from strenuous exercise in the daytime prior to being assessed at night [27,28,30]. Nine of the studies controlled for medication [9,15–20,22,26]. These confounding factors may have influenced changes in sleep scheduling or temporarily impair the participants ability to sleep. Therefore, to ensure consistency across the studies, controlling for these confounding factors is a necessary consideration for future studies.

In the study conducted by Bileviciute-Ljungar et al. HC were included to measure emotional awareness parameters, however, they used previously recorded HC data for PSG comparisons and only conducted PSG recordings on patients [32]. It is important to include well-defined and matched controls for each study to ensure that there is consistency between groups and that all other experimental variables are appropriately controlled for [32,45].

Eleven out of 20 studies accounted for first night effects. Considerations included recording over consecutive days [9,14–18,20–23,26]. In a study examining the impact of first night effects in four groups of participants: sleep-related breathing disorders, insomnia, movement and behavioural disorders and HC, it was found that in all groups there was a significant first night effect [46]. Additionally, Le Bon et al. also investigated first night effects in ME/CFS patients and found clinically significant differences in PSG recordings including SPT, TST, Sleep Efficiency and REM Sleep that can be attributed to first night effects [47]. Recommendations from these studies included measuring participants sleep parameters for at least two consecutive nights to ensure that first night sleep effects are accounted for [46,47]. This is an important consideration for all sleep physiology studies.

Two of the studies used a take-home PSG kit [9,29]. Using this method means that conditions are not controlled for including light exposure and sleep disruptions that may come from an uncontrolled setting. Use of take-home polysomnography kits allows participation of a greater proportion of ME/CFS patients that are housebound, bedbound, or otherwise unable to attend a research site. As approximately 25% of ME/CFS patients have more severe symptoms this is an important consideration [3]. Eighteen studies required participants to attend a sleep clinic [14–28,30–32]. Those who spent overnight in a sleep clinic will have more appropriately monitored process, however, change in sleep setting may also affect results.

Investigations into other factors influencing sleep quality, including melatonin and other hormone levels, do not fall within the scope of this review as no interventional studies were analysed. Melatonin levels influence multiple physiological processes including immune cell pathways [48]. As the most consistent immunological feature of ME/CFS is reduced natural killer (NK) cell cytotoxicity, this area will benefit from additional research [49]. Dysregulation of 2-5A synthetase/RNase L antiviral pathway has been previously linked with sleep disruption in particular changes to alpha delta sleep, however, investigations by Van Hoof et al. did not support associations [50]. Van Hoof et al. was not included in our analysis because that study did not have a HC group [50]. Changes in other hormone profiles including the hypothalamic-pituitary axis (HPA) has also been implicated in ME/CFS pathogenesis. Dysregulation of HPA also has known implications on sleep [7].

As mentioned previously, intervention studies were not included in the scope of this review. Majority of the intervention studies that were captured by the search terms focused on implementing exercise or alternative sleep scheduling such as a four-hour sleep delay on ME/CFS patients [51,52]. Introducing these interventions at even a moderate capacity in ME/CFS patients may result in the exacerbation of symptoms including post-exertional malaise (PEM). Therefore careful study design to ensure patient safety must be incorporated [53]. A review of currently available literature on these intervention studies is yet to be conducted.

Variable results were found for sleep apnoea scores in ME/CFS patients compared with HC. Le Bon et al. suggested that the percentage of patients with obstructive sleep apnoea may be influenced by the cut-off selected [21]. Some ME/CFS patients with comorbid sleep disorders have found benefits using a continuous positive airway pressure (CPAP) machine. This includes cognitive and daytime sleepiness. This machine, however, does not remediate the underlying fatigue [7]. A study conducted by Libman et al. has suggested that sleep apnoea-hypopnea syndrome should not be an exclusion criterion for ME/CFS; it instead should be considered a potential comorbidity [54]. Including participants with comorbid primary sleep disorders, however, makes distinguishing sleep patterns in ME/CFS patients difficult [21].

One study although finding no significant changes in PSG recordings reported higher fractal scaling index $\alpha 1$, a measure of heart rate variability during nonrapid eye movement (non-REM) sleep (Stages 1, 2, and 3 sleep) in the a.m. sleepier ME/CFS group compared with HC [27]. This suggests contribution of RR interval dynamics, an electrocardiogram parameter or autonomic nervous system activity during non-REM sleep to disrupted sleep in ME/CFS patients [27]. Additional studies have shown the potential role of cardiovascular regulation in the pathomechanism of ME/CFS [27,55]. ME/CFS patients presented with increased heart rate, and reduced heart rate variability. Orthostatic intolerance also promoted increased symptom severity [27]. These changes may suggest that there is dysregulation of the autonomic nervous system in ME/CFS pathology. These findings also demonstrate the importance of addressing whether unrefreshing sleep is a consequence of another underlying pathology in ME/CFS patients. Due to this feature observed in ME/CFS patients, it may be an important future consideration to further stratify patients on the basis of having postural tachycardia syndrome (POTS) or any other form of orthostatic intolerance [56]. This may further assist in understanding their contribution to sleep quality in ME/CFS patients.

A report made throughout the studies was an increase in slow wave sleep. Ball et al., made an association of this finding with immunological changes in ME/CFS patients [16]. It was suggested that this feature may be related to the release of cytokines [16]. However, there is insufficient evidence on the role of cytokines in ME/CFS pathomechanism [57]. Some studies also showed that there were no differences in SWS in ME/CFS patients or that there were only changes following sleep challenge [51,58]. These studies, however, did not follow inclusion criteria and were not selected for review.

The 2012 study by Le Bon et al. found that there was decreased ultra-slow delta power in ME/CFS patients compared with HC [22]. This result emphasised the importance of looking beyond conventional EEG bands and to exercise caution when categorising sleep EEG into discrete stages alone as some trends may be overlooked [22,33].

MSLT results were inconsistent across the studies. One out of six studies that used MSLT reported significant disruptions in ME/CFS patients compared with HC [23]. It has been suggested that the presence of a comorbid sleep disorder in addition to ME/CFS may contribute to excessive daytime sleepiness [23].

A common trend in these sleep studies is that there is a discrepancy between subjective sleep measures and objective sleep measures. This misperception was further investigated by Shan et al., who identified that there were structural changes in the medial prefrontal cortex that correlates with unrefreshing sleep in ME/CFS patients [59]. Approximately 91% of ME/CFS patients exhibit symptoms of unrefreshing sleep [59]. This finding shows the importance of using alternative neuroimaging techniques available to address sleep quality impairment in ME/CFS [59]. Additionally, sleep disruption can also be explained by additional abnormalities that have been described including brainstem reticular activation system connectivity deficits [59,60]. A majority of the studies utilise well-established sleep scoring tools, however, validation of some of these tools in ME/CFS populations is required. Additionally, the use of these tools may be affected by self-report bias [61]. Further research on the discrepancy between subjective and objective measures of sleep quality is required.

Quality Assessment

There were variable quality levels across the studies. Standard measures for clinical evaluation were used across all studies as PSG as well as MSLT in selected studies were employed. All studies included information on ME/CFS selection criteria, however, in some studies HC selection criteria were not provided. Item one was successfully addressed if two or more forms of patient and HC matching is employed including age, sex and BMI/weight-matching. A greater proportion of studies identified confounding variables and provided methods to mitigate them. Item two which assesses whether socio-demographic characteristics between ME/CFS patients and HC were appropriately matched was the least addressed item. Recommendations for future studies include reporting and matching of patient socio-demographics.

5. Conclusions

In the five studies that investigated MAI, all studies showed an increase in this parameter. SOL and NREM were not significantly different between ME/CFS patients throughout the studies. Slow- wave sleep, AHI, spectral activity, and MSLT were inconsistent across the studies. These results require validation in future well-designed studies. Numerous considerations for future experiments have been recommended including recruitment of participants with more stringent ME/CFS criteria and controlling for first night effects. Effective control of confounding variables of sleep quality including medications, change in time zones or strenuous exercise can also be implemented to improve overall study design. Replication of these studies in larger well-matched populations is also required.

Supplementary Materials: The following are available online at https://www.mdpi.com/article/10.3390/healthcare9050568/s1, Table S1: Search code, Table S2: Study and Patient Characteristics, Table S3: Summary of Primary Outcome Measures, Table S4: Summary of Secondary Outcome Measures, File S1: JBI CACCCS and justification.

Author Contributions: Conceptualization, S.M.-G., D.S., R.M. and M.M.; methodology, R.M. and M.M.; software, R.M.; validation, M.M. and N.E.-F.; formal analysis, R.M.; investigation; writing—original draft preparation, R.M.; writing—review and editing, N.E.-F. and H.C.; All authors have read and agreed to the published version of the manuscript.

Funding: This study was supported by the Stafford Fox Medical Research Foundation, McCusker Charitable Foundation, Douglas Stutt, Alison Hunter Memorial Foundation, Buxton Foundation, Blake Beckett Trust, Henty Donation, Mason Foundation, and the Change for ME Charity. The funders had no role in study design, data collection and analysis, decision to publish, or preparation of the manuscript.

Institutional Review Board Statement: Not applicable.

Informed Consent Statement: Not applicable.

Conflicts of Interest: The authors declare no conflict of interest.

References

1. Carruthers, B.M.; Van de Sande, M.I.; De Meirleir, K.L.; Klimas, N.G.; Broderick, G.; Mitchell, T.; Staines, D.; Powles, A.C.P.; Speight, N.; Vallings, R.; et al. Myalgic Encephalomyelitis: International Consensus Criteria. *J. Intern. Med.* **2011**, *270*, 327–338. [CrossRef] [PubMed]
2. Nisenbaum, R.; Jones, J.F.; Unger, E.R.; Reyes, M.; Reeves, W.C. A Population-Based Study of the Clinical Course of Chronic Fatigue Syndrome. *Health Qual. Life Outcomes* **2003**, *1*, 49. [CrossRef] [PubMed]
3. Eaton-Fitch, N.; Johnston, S.C.; Zalewski, P.; Staines, D.; Marshall-Gradisnik, S. Health-Related Quality of Life in Patients with Myalgic Encephalomyelitis/Chronic Fatigue Syndrome: An Australian Cross-Sectional Study. *Qual. Life Res.* **2020**, *29*, 1521–1531. [CrossRef] [PubMed]
4. Committee on the Diagnostic Criteria for Myalgic Encephalomyelitis/Chronic Fatigue Syndrome; Board on the Health of Select Populations; Institute of Medicine. *Beyond Myalgic Encephalomyelitis/Chronic Fatigue Syndrome: Redefining an Illness*; The National Academies Collection: Reports funded by National Institutes of Health; National Academies Press (US): Washington, DC, USA, 2015; ISBN 978-0-309-31689-7.
5. Fukuda, K.; Straus, S.E.; Hickie, I.; Sharpe, M.C.; Dobbins, J.G.; Komaroff, A. The Chronic Fatigue Syndrome: A Comprehensive Approach to Its Definition and Study. International Chronic Fatigue Syndrome Study Group. *Ann. Intern. Med.* **1994**, *121*, 953–959. [CrossRef]
6. Carruthers, B.M.; Jain, A.K.; Meirleir, K.L.D.; Peterson, D.L.; Klimas, N.G.; Lerner, A.M.; Bested, A.C.; Flor-Henry, P.; Joshi, P.; Powles, A.C.P.; et al. Myalgic Encephalomyelitis/Chronic Fatigue Syndrome. *J. Chronic Fatigue Syndr.* **2003**, *11*, 7–115. [CrossRef]
7. Jackson, M.L.; Bruck, D. Sleep Abnormalities in Chronic Fatigue Syndrome/Myalgic Encephalomyelitis: A Review. *J. Clin. Sleep Med. JCSM Off. Publ. Am. Acad. Sleep Med.* **2012**, *8*, 719–728. [CrossRef]
8. Purves, D.; Augustine, G.J.; Fitzpatrick, D.; Katz, L.C.; LaMantia, A.-S.; McNamara, J.O.; Williams, S.M. Stages of Sleep. Available online: https://www.ncbi.nlm.nih.gov/books/NBK10996/ (accessed on 11 May 2021).
9. Gotts, Z.M.; Deary, V.; Newton, J.L.; Ellis, J.G. A Comparative Polysomnography Analysis of Sleep in Healthy Controls and Patients with Chronic Fatigue Syndrome. *Fatigue Biomed. Health Behav.* **2016**, *4*, 80–93. [CrossRef]
10. Shrivastava, D.; Jung, S.; Saadat, M.; Sirohi, R.; Crewson, K. How to Interpret the Results of a Sleep Study. *J. Community Hosp. Intern. Med. Perspect.* **2014**, *4*. [CrossRef]
11. Arand, D.L.; Bonnet, M.H. Chapter 26—The multiple sleep latency test. In *Handbook of Clinical Neurology*; Levin, K.H., Chauvel, P., Eds.; Clinical Neurophysiology: Basis and Technical Aspects; Elsevier: Amsterdam, The Netherlands, 2019; Volume 160, pp. 393–403.
12. Chapter 4: Searching for and Selecting Studies. Available online: /handbook/current/chapter-04 (accessed on 14 December 2020).
13. Page, M.J.; McKenzie, J.E.; Bossuyt, P.M.; Boutron, I.; Hoffmann, T.C.; Mulrow, C.D.; Shamseer, L.; Tetzlaff, J.M.; Akl, E.A.; Brennan, S.E.; et al. The PRISMA 2020 Statement: An Updated Guideline for Reporting Systematic Reviews. *BMJ* **2021**, *372*, n71. [CrossRef]
14. Neu, D.; Mairesse, O.; Montana, X.; Gilson, M.; Corazza, F.; Lefevre, N.; Linkowski, P.; Le Bon, O.; Verbanck, P. Dimensions of Pure Chronic Fatigue: Psychophysical, Cognitive and Biological Correlates in the Chronic Fatigue Syndrome. *Eur. J. Appl. Physiol.* **2014**, *114*, 1841–1851. [CrossRef]
15. Armitage, R.; Landis, C.; Hoffmann, R.; Lentz, M.; Watson, N.; Goldberg, J.; Buchwald, D. Power Spectral Analysis of Sleep EEG in Twins Discordant for Chronic Fatigue Syndrome. *J. Psychosom. Res.* **2009**, *66*, 51–57. [CrossRef] [PubMed]
16. Ball, N.; Buchwald, D.S.; Schmidt, D.; Goldberg, J.; Ashton, S.; Armitage, R. Monozygotic Twins Discordant for Chronic Fatigue Syndrome: Objective Measures of Sleep. *J. Psychosom. Res.* **2004**, *56*, 207–212. [CrossRef]
17. Watson, N.F.; Kapur, V.; Arguelles, L.M.; Goldberg, J.; Schmidt, D.F.; Armitage, R.; Buchwald, D. Comparison of Subjective and Objective Measures of Insomnia in Monozygotic Twins Discordant for Chronic Fatigue Syndrome. *Sleep* **2003**, *26*, 324–328. [CrossRef] [PubMed]
18. Watson, N.F.; Jacobsen, C.; Goldberg, J.; Kapur, V.; Buchwald, D. Subjective and Objective Sleepiness in Monozygotic Twins Discordant for Chronic Fatigue Syndrome. *Sleep* **2004**, *27*, 973–977. [CrossRef]

19. Majer, M.; Jones, J.F.; Unger, E.R.; Youngblood, L.S.; Decker, M.J.; Gurbaxani, B.; Heim, C.; Reeves, W.C. Perception versus Polysomnographic Assessment of Sleep in CFS and Non-Fatigued Control Subjects: Results from a Population-Based Study. *BMC Neurol.* **2007**, *7*, 40. [CrossRef]
20. Decker, M.J.; Tabassum, H.; Lin, J.-M.S.; Reeves, W.C. Electroencephalographic Correlates of Chronic Fatigue Syndrome. *Behav. Brain Funct. BBF* **2009**, *5*, 43. [CrossRef]
21. Le Bon, O.; Neu, D.; Valente, F.; Linkowski, P. Paradoxical NREMS Distribution in "Pure" Chronic Fatigue Patients a Comparison with Sleep Apnea-Hypopnea Patients and Healthy Control Subjects. *J. Chronic Fatigue Syndr.* **2008**, *14*, 45–59. [CrossRef]
22. Le Bon, O.; Neu, D.; Berquin, Y.; Lanquart, J.-P.; Hoffmann, R.; Mairesse, O.; Armitage, R. Ultra-Slow Delta Power in Chronic Fatigue Syndrome. *Psychiatry Res.* **2012**, *200*, 742–747. [CrossRef]
23. Neu, D.; Hoffmann, G.; Moutrier, R.; Verbanck, P.; Linkowski, P.; Le Bon, O. Are Patients with Chronic Fatigue Syndrome Just "tired" or Also "Sleepy"? *J. Sleep Res.* **2008**, *17*, 427–431. [CrossRef]
24. Neu, D.; Mairesse, O.; Verbanck, P.; Linkowski, P.; Le Bon, O. Non-REM Sleep EEG Power Distribution in Fatigue and Sleepiness. *J. Psychosom. Res.* **2014**, *76*, 286–291. [CrossRef]
25. Neu, D.; Mairesse, O.; Verbanck, P.; Le Bon, O. Slow Wave Sleep in the Chronically Fatigued: Power Spectra Distribution Patterns in Chronic Fatigue Syndrome and Primary Insomnia. *Clin. Neurophysiol. Off. J. Int. Fed. Clin. Neurophysiol.* **2015**, *126*, 1926–1933. [CrossRef] [PubMed]
26. Reeves, W.C.; Heim, C.; Maloney, E.M.; Youngblood, L.S.; Unger, E.R.; Decker, M.J.; Jones, J.F.; Rye, D.B. Sleep Characteristics of Persons with Chronic Fatigue Syndrome and Non-Fatigued Controls: Results from a Population-Based Study. *BMC Neurol.* **2006**, *6*, 41. [CrossRef] [PubMed]
27. Togo, F.; Natelson, B.H. Heart Rate Variability during Sleep and Subsequent Sleepiness in Patients with Chronic Fatigue Syndrome. *Auton. Neurosci. Basic Clin.* **2013**, *176*, 85–90. [CrossRef] [PubMed]
28. Togo, F.; Natelson, B.H.; Cherniack, N.S.; FitzGibbons, J.; Garcon, C.; Rapoport, D.M. Sleep Structure and Sleepiness in Chronic Fatigue Syndrome with or without Coexisting Fibromyalgia. *Arthritis Res. Ther.* **2008**, *10*, R56. [CrossRef]
29. Sharpley, A.; Clements, A.; Hawton, K.; Sharpe, M. Do Patients with "Pure" Chronic Fatigue Syndrome (Neurasthenia) Have Abnormal Sleep? *Psychosom. Med.* **1997**, *59*, 592–596. [CrossRef]
30. Kishi, A.; Struzik, Z.R.; Natelson, B.H.; Togo, F.; Yamamoto, Y. Dynamics of Sleep Stage Transitions in Healthy Humans and Patients with Chronic Fatigue Syndrome. *Am. J. Physiol. Regul. Integr. Comp. Physiol.* **2008**, *294*, R1980–R1987. [CrossRef]
31. Neu, D.; Mairesse, O.; Hoffmann, G.; Dris, A.; Lambrecht, L.J.; Linkowski, P.; Verbanck, P.; Le Bon, O. Sleep Quality Perception in the Chronic Fatigue Syndrome: Correlations with Sleep Efficiency, Affective Symptoms and Intensity of Fatigue. *Neuropsychobiology* **2007**, *56*, 40–46. [CrossRef]
32. Bileviciute-Ljungar, I.; Friberg, D. Emotional Awareness Correlated With Number of Awakenings From Polysomnography in Patients With Myalgic Encephalomyelitis/Chronic Fatigue Syndrome—A Pilot Study. *Front. Psychiatry* **2020**, *11*. [CrossRef]
33. Maksoud, R.; Preez, S.d.; Eaton-Fitch, N.; Thapaliya, K.; Barnden, L.; Cabanas, H.; Staines, D.; Marshall-Gradisnik, S. A Systematic Review of Neurological Impairments in Myalgic Encephalomyelitis/ Chronic Fatigue Syndrome Using Neuroimaging Techniques. *PLoS ONE* **2020**, *15*, e0232475. [CrossRef]
34. Prins, J.B.; van der Meer, J.W.; Bleijenberg, G. Chronic Fatigue Syndrome. *Lancet* **2006**, *367*, 346–355. [CrossRef]
35. Chu, L.; Valencia, I.J.; Garvert, D.W.; Montoya, J.G. Onset Patterns and Course of Myalgic Encephalomyelitis/Chronic Fatigue Syndrome. *Front. Pediatr.* **2019**, *7*. [CrossRef] [PubMed]
36. Collin, S.M.; Nuevo, R.; Putte, E.M.v.d.; Nijhof, S.L.; Crawley, E. Chronic Fatigue Syndrome (CFS) or Myalgic Encephalomyelitis (ME) Is Different in Children Compared to in Adults: A Study of UK and Dutch Clinical Cohorts. *BMJ Open* **2015**, *5*, e008830. [CrossRef] [PubMed]
37. Stores, G.; Fry, A.; Crawford, C. Sleep Abnormalities Demonstrated by Home Polysomnography in Teenagers with Chronic Fatigue Syndrome. *J. Psychosom. Res.* **1998**, *45*, 85–91. [CrossRef]
38. Van de Water, A.T.M.; Holmes, A.; Hurley, D.A. Objective Measurements of Sleep for Non-Laboratory Settings as Alternatives to Polysomnography–a Systematic Review. *J. Sleep Res.* **2011**, *20*, 183–200. [CrossRef] [PubMed]
39. Fatt, S.J.; Beilharz, J.E.; Joubert, M.; Wilson, C.; Lloyd, A.R.; Vollmer-Conna, U.; Cvejic, E. Parasympathetic Activity Is Reduced during Slow-Wave Sleep, but Not Resting Wakefulness, in Patients with Chronic Fatigue Syndrome. *J. Clin. Sleep Med. JCSM Off. Publ. Am. Acad. Sleep Med.* **2020**, *16*, 19–28. [CrossRef]
40. Neu, D.; Kajosch, H.; Peigneux, P.; Verbanck, P.; Linkowski, P.; Le Bon, O. Cognitive Impairment in Fatigue and Sleepiness Associated Conditions. *Psychiatry Res.* **2011**, *189*, 128–134. [CrossRef]
41. Neu, D.; Cappeliez, B.; Hoffmann, G.; Verbanck, P.; Linkowski, P.; Le Bon, O. High Slow-Wave Sleep and Low-Light Sleep: Chronic Fatigue Syndrome Is Not Likely to Be a Primary Sleep Disorder. *J. Clin. Neurophysiol. Off. Publ. Am. Electroencephalogr. Soc.* **2009**, *26*, 207–212. [CrossRef]
42. Çetin, B.; Sünbül, E.A.; Toktaş, H.; Karaca, M.; Ulutaş, Ö.; Güleç, H. Comparison of Sleep Structure in Patients with Fibromyalgia and Healthy Controls. *Sleep Breath. Schlaf Atm.* **2020**. [CrossRef]
43. Bigatti, S.M.; Hernandez, A.M.; Cronan, T.A.; Rand, K.L. Sleep Disturbances in Fibromyalgia Syndrome: Relationship to Pain and Depression. *Arthritis Rheum.* **2008**, *59*, 961–967. [CrossRef]
44. Wang, B.; Duan, R.; Duan, L. Prevalence of Sleep Disorder in Irritable Bowel Syndrome: A Systematic Review with Meta-Analysis. *Saudi J. Gastroenterol. Off. J. Saudi Gastroenterol. Assoc.* **2018**, *24*, 141–150. [CrossRef]

45. Ghorbani, S.; Nejad, A.; Law, D.; Chua, K.S.; Amichai, M.M.; Pimentel, M. Healthy Control Subjects Are Poorly Defined in Case-Control Studies of Irritable Bowel Syndrome. *Ann. Gastroenterol. Q. Publ. Hell. Soc. Gastroenterol.* **2015**, *28*, 87–93.
46. Newell, J.; Mairesse, O.; Verbanck, P.; Neu, D. Is a One-Night Stay in the Lab Really Enough to Conclude? First-Night Effect and Night-to-Night Variability in Polysomnographic Recordings among Different Clinical Population Samples. *Psychiatry Res.* **2012**, *200*, 795–801. [CrossRef]
47. Le Bon, O.; Minner, P.; Van Moorsel, C.; Hoffmann, G.; Gallego, S.; Lambrecht, L.; Pelc, I.; Linkowski, P. First-Night Effect in the Chronic Fatigue Syndrome. *Psychiatry Res.* **2003**, *120*, 191–199. [CrossRef]
48. Miller, S.C.; Pandi, P.S.R.; Esquifino, A.I.; Cardinali, D.P.; Maestroni, G.J.M. The Role of Melatonin in Immuno-Enhancement: Potential Application in Cancer. *Int. J. Exp. Pathol.* **2006**, *87*, 81–87. [CrossRef]
49. Eaton-Fitch, N.; Du Preez, S.; Cabanas, H.; Staines, D.; Marshall-Gradisnik, S. A Systematic Review of Natural Killer Cells Profile and Cytotoxic Function in Myalgic Encephalomyelitis/Chronic Fatigue Syndrome. *Syst. Rev.* **2019**, *8*. [CrossRef] [PubMed]
50. Van Hoof, E.; De Becker, P.; Lapp, C.; Cluydts, R.; De Meirleir, K. Defining the Occurrence and Influence of Alpha-Delta Sleep in Chronic Fatigue Syndrome. *Am. J. Med. Sci.* **2007**, *333*, 78–84. [CrossRef] [PubMed]
51. Armitage, R.; Landis, C.; Hoffmann, R.; Lentz, M.; Watson, N.F.; Goldberg, J.; Buchwald, D. The Impact of a 4-Hour Sleep Delay on Slow Wave Activity in Twins Discordant for Chronic Fatigue Syndrome. *Sleep* **2007**, *30*, 657–662. [CrossRef] [PubMed]
52. Kishi, A.; Togo, F.; Cook, D.B.; Klapholz, M.; Yamamoto, Y.; Rapoport, D.M.; Natelson, B.H. The Effects of Exercise on Dynamic Sleep Morphology in Healthy Controls and Patients with Chronic Fatigue Syndrome. *Physiol. Rep.* **2013**, *1*, e00152. [CrossRef]
53. Light, A.R.; Bateman, L.; Jo, D.; Hughen, R.W.; Vanhaitsma, T.A.; White, A.T.; Light, K.C. Gene Expression Alterations at Baseline and Following Moderate Exercise in Patients with Chronic Fatigue Syndrome and Fibromyalgia Syndrome. *J. Intern. Med.* **2012**, *271*, 64–81. [CrossRef]
54. Libman, E.; Creti, L.; Baltzan, M.; Rizzo, D.; Fichten, C.S.; Bailes, S. Sleep Apnea and Psychological Functioning in Chronic Fatigue Syndrome. *J. Health Psychol.* **2009**, *14*, 1251–1267. [CrossRef]
55. Bozzini, S.; Albergati, A.; Capelli, E.; Lorusso, L.; Gazzaruso, C.; Pelissero, G.; Falcone, C. Cardiovascular Characteristics of Chronic Fatigue Syndrome. *Biomed. Rep.* **2018**, *8*, 26–30. [CrossRef]
56. Garner, R.; Baraniuk, J.N. Orthostatic Intolerance in Chronic Fatigue Syndrome. *J. Transl. Med.* **2019**, *17*. [CrossRef] [PubMed]
57. Corbitt, M.; Eaton-Fitch, N.; Staines, D.; Cabanas, H.; Marshall-Gradisnik, S. A Systematic Review of Cytokines in Chronic Fatigue Syndrome/Myalgic Encephalomyelitis/Systemic Exertion Intolerance Disease (CFS/ME/SEID). *BMC Neurol.* **2019**, *19*, 207. [CrossRef] [PubMed]
58. Fischler, B.; Le Bon, O.; Hoffmann, G.; Cluydts, R.; Kaufman, L.; De Meirleir, K. Sleep Anomalies in the Chronic Fatigue Syndrome. A Comorbidity Study. *Neuropsychobiology* **1997**, *35*, 115–122. [CrossRef] [PubMed]
59. Shan, Z.Y.; Kwiatek, R.; Burnet, R.; Del Fante, P.; Staines, D.R.; Marshall-Gradisnik, S.M.; Barnden, L.R. Medial Prefrontal Cortex Deficits Correlate with Unrefreshing Sleep in Patients with Chronic Fatigue Syndrome. *NMR Biomed.* **2017**, *30*. [CrossRef] [PubMed]
60. Barnden, L.R.; Shan, Z.Y.; Staines, D.R.; Marshall-Gradisnik, S.; Finegan, K.; Ireland, T.; Bhuta, S. Intra Brainstem Connectivity Is Impaired in Chronic Fatigue Syndrome. *NeuroImage Clin.* **2019**, *24*, 102045. [CrossRef] [PubMed]
61. Althubaiti, A. Information Bias in Health Research: Definition, Pitfalls, and Adjustment Methods. *J. Multidiscip. Healthc.* **2016**, *9*, 211–217. [CrossRef]

 *healthcare*

Article

Validation of the Severity of Myalgic Encephalomyelitis/Chronic Fatigue Syndrome by Other Measures than History: Activity Bracelet, Cardiopulmonary Exercise Testing and a Validated Activity Questionnaire: SF-36

C. (Linda) M. C. van Campen [1,*], Peter C. Rowe [2] and Frans C. Visser [1]

[1] Stichting CardioZorg, 2132 HN Hoofddorp, The Netherlands; fransvisser@stichtingcardiozorg.nl

[2] Department of Paediatrics, John Hopkins University School of Medicine, Baltimore, MD 21205, USA; prowe@jhmi.edu

* Correspondence: info@stichtingcardiozorg.nl; Tel.: +31-208-934-125

Received: 27 June 2020; Accepted: 13 August 2020; Published: 14 August 2020

Abstract: Introduction: Myalgic encephalomyelitis/chronic fatigue syndrome (ME/CFS) is a severe and disabling chronic disease. Grading patient's symptom and disease severity for comparison and therapeutic decision-making is necessary. Clinical grading that depends on patient self-report is subject to inter-individual variability. Having more objective measures to grade and confirm clinical grading would be desirable. Therefore, the aim of this study was to validate the clinical severity grading that has been proposed by the authors of the ME International Consensus Criteria (ICC) using more standardized measures like questionnaires, and objective measures such as physical activity tracking and cardiopulmonary exercise testing. Methods and results: The clinical database of a subspecialty ME/CFS clinic was searched for patients who had completed the SF 36 questionnaire, worn a Sensewear[TM] armband for five days, and undergone a cardiopulmonary exercise test. Only patients who completed all three investigations within 3 months from each other—to improve the likelihood of stable disease—were included in the analysis. Two-hundred-eighty-nine patients were analyzed: 121 were graded as mild, 98 as moderate and 70 as having severe disease. The mean (SD) physical activity subscale of the SF-36 was 70 (11) for mild, 43 (8) for moderate and 15 (10) for severe ME/CFS patients. The mean (SD) number of steps per day was 8235 (1004) for mild, 5195 (1231) for moderate and 2031 (824) for severe disease. The mean (SD) percent predicted oxygen consumption at the ventilatory threshold was 47 (11)% for mild, 38 (7)% for moderate and 30 (7)% for severe disease. The percent peak oxygen consumption was 90 (14)% for mild, 64 (8)% for moderate and 48 (9)% for severe disease. All comparisons were $p < 0.0001$. Conclusion: This study confirms the validity of the ICC severity grading. Grading assigned by clinicians on the basis of patient self-report created groups that differed significantly on measures of activity using the SF-36 physical function subscale and objective measures of steps per day and exercise capacity. There was variability in function within severity grading groups, so grading based on self-report can be strengthened by the use of these supplementary measures.

Keywords: Sensewear[TM] armband; chronic fatigue syndrome; cardiopulmonary exercise testing; peak VO_2; VO_2 at the ventilatory threshold; physical activity subscale; SF 36 questionnaire; disease severity; steps

---

1. Introduction

Chronic fatigue syndrome (CFS) is a potentially severe and disabling chronic disease [1–3]. The pathophysiology has not been established but there is considerable evidence that CFS is associated with multi-systemic neuropathology, metabolic, and immunological abnormalities [4–16]. In this light, the name "myalgic encephalomyelitis" was suggested by Carruthers and colleagues, a name also more consistent with the neurological classification of this disease in the World Health Organization's International Classification of Diseases (ICD G93.3) [1].

Symptom and disease severity is discussed in the clinical application section of the International Consensus Criteria (ICC): "For a diagnosis of ME, symptom severity must result in a significant reduction in a patient's premorbid activity level. Mild: approximately 50% reduction in activity, moderate: mostly housebound, severe: mostly bedbound, and very severe: bedbound and dependent on help for physical functions" [1]. One of the ICC's considerations was to classify patient disease severity to increase patient group homogeneity in research. For example, we previously showed that the use of curcumin was only favorable in less severely ill ME/CFS patients and was not effective in severely ill ME/CFS patients [17].

History taking and clinical severity grading might be challenging because of its dependence on the interpretation of symptomatology by the patient. Patient expression of symptoms can be influenced by age, gender, education level, disease duration and co-morbidities like fibromyalgia, attention deficit disorder, and depression. Adding more objective measures to confirm symptom severity and clinical grading may be helpful. Severity grading is rarely reported in ME/CFS research and limited information is available on objective measures linked to disease severity. Also, the proposed grading of the ICC, which mainly focusses on the ability to perform physical activities, has not been validated.

Therefore, the aim of this study was to validate the clinical severity grading in ME/CFS, as suggested by the ICC [1], using both a standardized questionnaire (physical activity subscale of the SF-36 questionnaire) and two objective activity measures: a cardiopulmonary exercise test and a physical activity tracker.

2. Materials and Methods

2.1. Patients

From October 2012 to January 2018, 714 patients visited the outpatient clinic of the Stichting CardioZorg, Hoofddorp, the Netherlands, because of the suspicion of ME/CFS. This cardiology clinic specializes in diagnosing and treating adults with ME/CFS. All patients were evaluated by the same clinician (FVC). During the first visit, it was determined whether patients satisfied the criteria for CFS and ME, taking the exclusion criteria into account. Patients were classified as having chronic fatigue, or no chronic fatigue as defined by Fukuda and colleagues [3] and as having ME or no ME as defined by Carruthers and colleagues [1]. Disease severity was scored according to the ICC, with severity scored as mild, moderate, severe and very severe [1]. Very severe ME/CFS patients were not included in this analysis as none of these patients were able to undergo a cardiopulmonary exercise test (CPET).

Of the initial 714 patients, 675 were diagnosed with ME/CFS, fulfilling criteria of both ME and CFS. Of the remaining 39 patients, 26 did not fulfill the criteria of ME and/or CFS and 13 had an alternative diagnosis. The records of these 675 patients were searched for the availability of a completed SF-36 questionnaire, the availability of having worn a Sensewear™ armband for five days and the availability of results of a cardiopulmonary exercise test. Only patients who completed all three investigations within a maximum interval of 3 months were included in the analysis. This interval was chosen to minimize variation in disease severity. Stability of disease was confirmed by review of the patient charts. Four hundred six patients had all three investigations. The interval for performance of the three measures exceeded 3 months in 75 patients, who were excluded. There were no differences in demographic data between ME/CFS patients who were included or excluded from the study due to the interval of the three measures (data not shown). Another 42 patients were excluded: because of a BMI > 37 ($n = 7$), because of gross under or overactivity compared to their average daily activity ($n = 15$), inadequate wearing armband time ($n = 12$), or motion artifacts ($n = 8$) (see below). We therefore included 289 patients in the analysis.

The study was carried out in accordance with the Declaration of Helsinki. All ME/CFS patients gave informed, written consent. The use of clinical data for descriptive studies was approved by the ethics committee of the Slotervaart Hospital (reference number P1450).

2.2. Sensewear™ Activity Armband

To track activity and help determine grading of disability, the Sensewear™ armband (BodyMedia, Pittsburgh, PA, USA) was used. Patients wore the armband for approximately 5 days and were advised to take off the armband only during showering or bathing. Furthermore they were instructed to wear the armband if possible for a minimum of 23 h of the day. For assessing disability to guide treatment, patients were instructed to wear the armband in an on average reasonable period of their life, without excesses to be expected during the wearing time. From the armband data, the number of steps were recorded and normalized to 24 h to account for information of steps from only a part of the day. After wearing the armband patients were asked if the 5 days were "average" days. If there was a gross over- or under-activity due to mandatory physical activities or due to severe post-exertional malaise or current other illnesses (e.g., a viral infection), patients were excluded from the analysis. These exclusions were decided by the clinician and reported in the patient chart. Furthermore, patients who wore the armband less than 23 h. per day were excluded. Motion artifacts, due to horseback riding or motor bike riding, resulting in overestimation of steps, led to exclusion.

2.3. SF-36 Questionnaire: Physical Activity Subscale

SF-36 physical activity subscale asks whether the respondent's health limits activities ranging from washing/clothing to walking shorter and longer distances and even strenuous running, performed during a typical day, ranging from limited a lot, limited a little, or not limited at all. The Dutch version of the SF-36 for physical activity [18] was used. The scores of the 10 items of the questionnaire were transformed into a scale ranging from 0–100%, where a higher score represents better physical condition and the lower scores worse conditions. Patients were instructed to complete questionnaires on average days, similar to the activity tracker, in order to avoid analysis on information acquired on a day with postexertional malaise or a good day.

2.4. Cardiopulmonary Exercise Testing

Patients underwent a symptom-limited maximal exercise test on a cycle ergometer (Excalibur, Lode, Groningen, The Netherlands) according to a previously described protocol [19]. A more detailed description can be found in Appendix A.

2.5. Statistical Analysis

Data were analyzed using Graphpad Prism version 8.4.2 (Graphpad software, La Jolla, CA, USA) and using SPSS version 21 (IBM USA). All continuous data were tested for normal distribution using the D'Agostino–Pearson omnibus normality test. Data are presented as the mean (SD) or as median and interquartile range (IQR), where appropriate. Groups were compared using the paired or unpaired *t*-test where appropriate. Categorical and distribution data were tested by Chi-square analysis (3×2 table). Receiver operating curve (ROC) analysis was performed on the rand physical activity subscale of the SF-36, the number of steps on an activity meter, on the %VT VO_2 of the cardiopulmonary exercise test and on the %peak VO_2 of the cardiopulmonary exercise test to determine optimal cut-off values discriminating between mild and moderate and moderate and severe disease. For the analysis of the ROC curve Graphpad Prism was used. The sensitivities and specificities for different values of the different test measures were tabulated. This was performed separately for the mild versus moderate disease category as well as for the moderate versus severe disease category. This analysis resulted in an area under the curve (AUC), by a graphical representation of sensitivity% on the y-axis and 100%- specificity%. The most optimal discriminative value was obtained by the highest value of the multiplication of sensitivity with specificity [20,21]. Kappa's with 95% confidence intervals were also calculated to determine agreement between clinical severity grading and the other measures as physical activity subscale of the SF-36, the number of steps per day, the %VT VO_2 of the cardiopulmonary exercise test and on the %peak VO_2 of the cardiopulmonary exercise test. The 95% confidence intervals (CI) of the Kappa values that overlap are considered equally discriminative. A Cohen's kappa was calculated comparing the severity grading of two clinicians to determine reliability of clinical grading between the two. For this purpose a clinician (CMCvC) reviewed the charts of the first 162 patients. Within group comparison was done by the ordinary one way variance of analysis (ANOVA). Where significant, results were then explored further using the post-hoc Tukey's test. Nominal data were compared using the Chi-square test (in a 3×2 table). Within group comparison with two different categorical independent variables on one continuous dependent variable was done by the two-way mixed analysis of variance (ANOVA). Where significant, results were then explored further using the post-hoc Tukey test. A p-value of <0.05 was considered to be statistically significant.

3. Results

Patient Clinical Data

The studied group consisted of 289 ME/CFS patients and included 51 males (17.6%) and 238 females (82.4%). The mean (SD) age was 40 (11) years, the median BMI 23.4 (20.8–27.1) kg/m^2 and the mean disease duration 12 (9) years. The mean physical activity subscale score from the SF-36 was 48 (24). The mean number of steps per day was 5701 (2670). The oxygen consumption at the ventilatory threshold was 11 (3) mL/kg/min, and the mean percentage predicted oxygen consumption at the ventilatory threshold was 40 (11)%. The peak oxygen consumption was 21 (7) mL/kg/min and the percentage predicted peak oxygen consumption was 71 (20)%. Using the clinician-assigned ICC severity category, 121 (42%) were scored as having mild disease, 98 patients (34%) were scored as having moderate disease and 70 patients (24%) were scored as having severe disease. The calculated Cohen's kappa of the agreement in severity grading between the two clinicians was 0.86.

Table 1 shows the comparison of patient data between male and female ME/CFS patients. Only the peak VO_2 data were significantly higher in male ME/CFS patients compared to female ME/CFS patients ($p = 0.005$).

Table 1. Baseline characteristics all ME/CFS patients ($n = 289$) and for male ($n = 51$) and female ME/CFS patients ($n = 238$).

	Males ($n = 51$)	Females ($n = 238$)	p-Value
Age (years)	42 (11)	39 (11)	0.14
BMI (kg/m^2)	24.9 (3.9)	24.5 (5.3)	0.64
Disease duration (years)	11 (8)	12 (9)	0.69
Disease severity: mild/moderate/severe (%) *	23/15/13 (45/29/26%)	98/83/57 (41/29/20%)	Chi-square 0.75
Heart rate at rest (bpm)	89 (16)	89 (19)	0.84
SBP at rest (mmHg)	125 (16)	130 (16)	0.18
DBP at rest (mmHg)	85 (20)	84 (11)	0.85
SF-36 PAS	48 (23)	48 (24)	0.96
Number of steps/day	5768 (2511)	5687 (2713)	0.84

Table 1. *Cont.*

	Males (n = 51)	Females (n = 238)	p-Value
VT VO$_2$ (ml/kg/min)	12 (4)	11 (3)	0.09
%VT VO$_2$	38 (14)	40 (11)	0.33
peak VO$_2$ (ml/kg/min)	23 (8)	20 (6)	0.005
%peak VO$_2$	72 (24)	71 (19)	0.79

BMI: body mass index; SBP: systolic blood pressure; DBP: diastolic blood pressure, HR: heart rate; PAS: physical activity subscale of the SF-36; peak VO$_2$: oxygen consumption at peak exercise; %peak VO$_2$: oxygen consumption at peak exercise as percentile of a reference population; VT VO$_2$: oxygen consumption at the ventilatory threshold; %VT VO$_2$: oxygen consumption at the ventilatory threshold as percentile of a reference population. * Disease severity was classified by the clinician using the ICC categories [1].

Table 2 shows the baseline criteria and the results of the three tests for the different disease severity groups. Because the peak VO$_2$ was significantly higher in male ME/CFS patients and percent VT VO$_2$ and percent peak VO$_2$ were not different between female and male ME/CFS patients, for the comparison of the severity grading, the percent VO$_2$ data were used. No differences were found in baseline characteristics as age or disease duration. No significant difference was found between the ratio male/female patients. The physical activity subscale of the SF-36 was 70 (11) for mild, 43 (8) for moderate and 15 (10) for severe ME/CFS patients. The mean number of steps per day was 8235 (1004) for mild, 5195 (1231) for moderate and 2031 (824) for severe disease. The percent predicted oxygen consumption at the ventilatory threshold was 47 (11%) for mild, 38 (7%) for moderate and 30 (7%) for severe disease. The percent peak oxygen consumption was 90 (14%) for mild, 64 (8%) for moderate and 48 (9%) for severe disease. All comparisons were highly significantly different (all p < 0.0001). Figure 1 shows the graphical representation of those data.

Table 2. ME/CFS patient characteristics divided by ICC disease severity grading.

	Group 1: Mild	Group 2: Moderate	Group 3: Severe	One-Way ANOVA and Post-Hoc Tukey's Test
Number	121	98	70	
Male/female	23/98	15/83	13/57	Chi-square 0.75 (3 × 2 table)
Age (years)	43 (11)	38 (11)	35(11)	F (2, 286) = 12.1; p < 0.0001. Post-hoc tests: 1 vs. 2 p = 0.008; 1 vs. 3 p < 0.0001 and 2 vs. 3 p = 0.12
Disease duration (years)	12 (9)	12 (8)	12 (8)	F (2, 286) = 0.06; p = 0.94
SF-36 PAS	70 (11)	43 (8)	15 (10)	F (2, 286) = 717.6; p < 0.0001. Post-hoc tests: 1 vs. 2 p < 0.0001; 1 vs. 3 p < 0.0001 and 2 vs. 3 p < 0.0001
Number of steps/day	8235 (1004)	5195 (1231)	2031 (824)	F (2, 286) = 792.4; p < 0.0001. Post-hoc tests: 1 vs. 2 p < 0.0001; 1 vs. 3 p < 0.0001 and 2 vs. 3 p < 0.0001
%VT VO$_2$	47 (11)	38 (7)	30 (7)	F (2, 286) = 85.7; p < 0.0001. Post-hoc tests: 1 vs. 2 p < 0.0001; 1 vs. 3 p < 0.0001 and 2 vs. 3 p < 0.0001
%peak VO$_2$	90 (14)	64 (8)	48 (9)	F (2, 286) = 355.0; p < 0.0001. Post-hoc tests: 1 vs. 2 p < 0.0001; 1 vs. 3 p < 0.0001 and 2 vs. 3 p < 0.0001

BMI: body mass index; PAS: physical activity subscale of the SF-36; %peak VO$_2$: oxygen consumption at peak exercise as percentile of a reference population; %VT VO$_2$: oxygen consumption at the ventilatory threshold as percentile of a reference population.

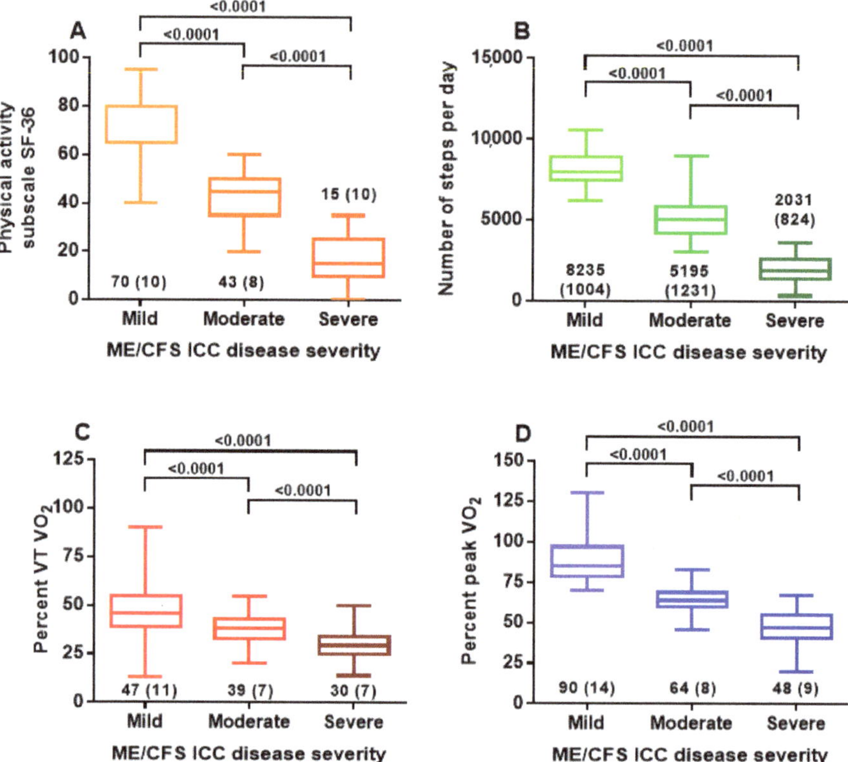

Figure 1. Physical activity subscale of SF-36 (panel **A**), number of steps per day (panel **B**), percent oxygen consumption at the ventilatory threshold (panel **C**) and percent peak oxygen consumption (panel **D**) in mild, moderate and severe ME/CFS according to clinical grading from International Consensus Criteria. Legend Figure 1 %peak VO_2: oxygen consumption at peak exercise as percentile of a reference population; %VT VO_2: oxygen consumption at the ventilatory threshold as percentile of a reference population.

To compare the overall ICC severity grading with the other measures of activity, we divided the groups based on the results of Receiving Operator Characteristic (ROC) curves that generated the best cut-off values between mild and moderate and between moderate and severe disease. For the physical activity subscale of the SF-36, the best cut-off values were <30, from 30-to 60 and >60 to optimally discriminate between severe, moderate and mild disease. The area under the curve between moderate and severe was 0.984 and between mild and moderate 0.981. For the number of steps the best cut-off values were <3500, from 3500 to 6250 and >6250 steps to optimally discriminate between severe, moderate and mild disease. The area under the curve between moderate and severe was 0.997 and between mild and moderate 0.962. For the percent predicted oxygen consumption at the ventilatory threshold the best cut-off values were <30%, from 30% to 45% and >45%. The area under the curve between moderate and severe was 0.794 and between mild and moderate 0.760. For the percent predicted peak oxygen consumption the best cut-off value were ≤57%, from 58% to 72% and >72%. The area under the curve between moderate and severe was 0.925 and between mild and moderate 0.973.

Table 3 shows the percentage of patients who are included in the range of the three predetermined cut-off values. Figure 2 shows the graphical representation of these results.

Table 3. Subdivision of cut-off values/measures of agreement as determined by ROC analysis for the four studied measures in relation to ICC clinical severity grading.

	Group 1 Mild	Group 2 Moderate	Group 3 Severe	Measure of Agreement (Kappa)
Physical Activity subscale of the SF-36 (PAS) (panel A Figure 2)				
>60	93 (77%)	0 (0%)	0 (0%)	Kappa 0.80 (95%CI: 0.742–0.859)
30–60	28 (23%)	93 (95%)	5 (7%)	
<30	0 (0%)	5 (5%)	65 (93%)	
Number of steps (panel B Figure 2)				
>6250	120 (99%)	15 (15%)	0 (0%)	Kappa 0.89 (95% CI: 0.848–0.938).
3500–6250	1 (1%)	82 (84%)	3 (4%)	
<3500	0 (0%)	1 (1%)	67 (96%)	
%predicted oxygen consumption at the ventilatory threshold (panel C Figure 2)				
>56%	70 (58%)	15 (15%)	2 (3%)	Kappa 0.39 (95% CI: 0.303–0.473).
30–45%	46 (38%)	68 (69%)	32 (46%)	
<30%	5 (4%)	15 (16%)	36 (51%)	
%predicted peak oxygen consumption (panel D Figure 2)				
>72%	117 (97%)	15 (15%)	0 (0%)	Kappa 0.78 (95% CI: 0.721–0.843).
58–72%	4 (3%)	70 (71%)	8 (11%)	
≤57%	0 (0%)	13 (14%)	62 (89%)	

CI: confidence intervals.

From the measures of agreement and confidence intervals, it follows that using the %VT VO_2 for severity grading has the lowest kappa. Grading using the physical activity subscale of the SF-36, the number of steps per day and the %peak VO_2 are equal with respect to the measured kappa, but the number of steps per day is superior to the %peak VO_2 with respect to severity grading, as the 95% CI do not overlap.

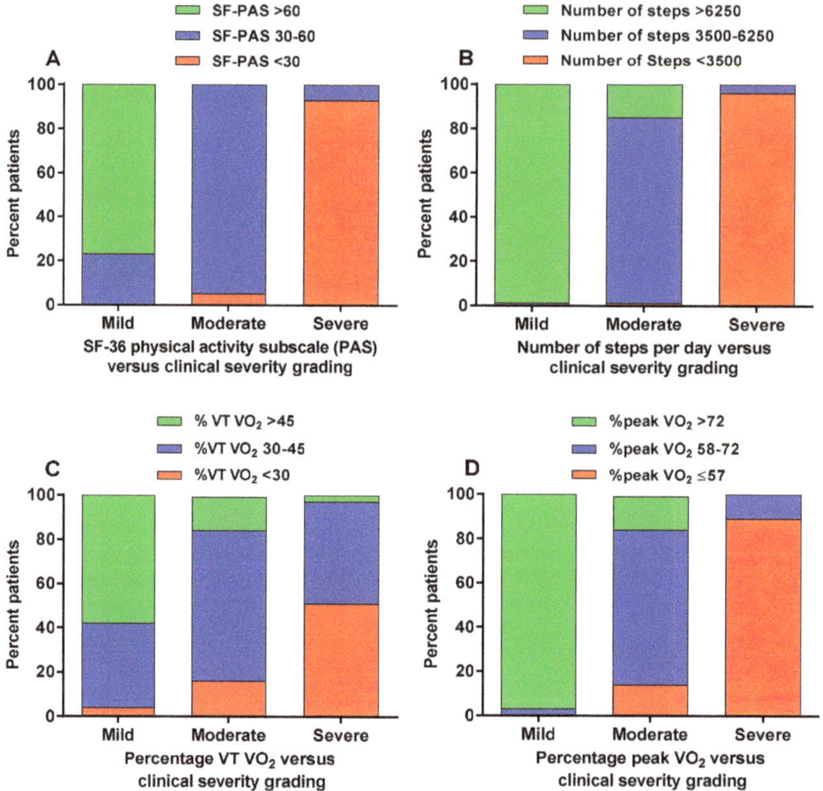

Figure 2. Stacked columns for physical activity subscale SF-36 (panel **A**) and number of steps per day (panel **B**), for percent oxygen consumption at the ventilatory threshold (panel **C**) and percent peak oxygen consumption (panel **D**) versus ICC clinical severity grading. Legend Figure 2 PAS: physical activity subscale of the SF-36 questionnaire; %peak VO_2: oxygen consumption at peak exercise as percentile of a reference population; %VT VO_2: oxygen consumption at the ventilatory threshold as percentile of a reference population.

Table 4 shows the results of a subgroup analysis: the objective results of female and male ME/CFS patients. As in the overall population, differences between mild, moderate and severe are all highly significantly different both in males and females. The 2-way ANOVA showed no significant differences in the severity groups and showed no significant interaction effect between the three disease severity groups and gender for the number of steps, the %VT VO_2 and the %peak VO_2. The physical activity subscale of the SF-36 showed a significant interaction between gender and disease severity ($p = 0.03$). Post hoc results are shown in the table.

Table 4. Comparison of female and male ME/CFS patients on validated questionnaire and more objective measures.

	Group 1: Mild	Group 2: Moderate	Group 3: Severe	Two-Way Mixed ANOVA with Post Hoc Tukey Test
	Female/Male ME/CFS patients			
Number of males/females	23/98	15/83	13/57	
Male SF-36 PAS	67 (13)	43 (7)	20 (11)	$F(2, 283) = 3.55$; $p = 0.030$. Post-hoc tests: female patients 1 vs. 2 $p < 0.0001$; 1 vs. 3 $p < 0.0001$ and 2 vs. 3 $p < 0.0001$; male patients 1 vs. 2 $p < 0.0001$; 1 vs. 3 $p < 0.0001$ and 2 vs. 3 $p < 0.0001$
Female SF-36 PAS	71 (10)	44 (9)	14 (10)	
Male Number of steps/day	8067 (1152)	5056 (892)	2523 (860)	$F(2, 283) = 2.51$; $p = 0.083$
Female Number of steps/day	8274 (969)	5220 (1286)	1919 (780)	
Male %VT VO$_2$	48 (15)	33 (5)	26 (6)	$F(2, 283) = 2.11$; $p = 0.12$.
Female %VT VO$_2$	47 (10)	39 (7)	30 (8)	
Male %peak VO$_2$	93 (15)	60 (9)	46 (9)	$F(2, 283) = 2.46$; $p = 0.088$
Female %peak VO$_2$	89 (13)	65 (7)	48 (9)	

PAS: physical activity subscale of the SF-36; %peak VO$_2$: oxygen consumption at peak exercise as percentile of a reference population; %VT VO$_2$: oxygen consumption at the ventilatory threshold as a percentile of a reference population.

4. Discussion

The ICC ME criteria proposed a severity classification including mild, moderate, severe and very severe disease [1]. The ICC's considerations to classify patient disease severity were to increase patient group homogeneity in the research, to help orient and monitor treatment, and to determine total disease burden. Thus far, we are not aware of any studies that have implemented this classification, nor are we aware of validation studies. The main finding of this study is that the physical function subscale of the SF-36 questionnaire, and objective measures such as the number of steps per day on an activity meter and measures of percent oxygen consumption at ventilatory threshold and at peak exercise, showed a clear distinction between mild, moderate and severe ME/CFS patients. The physical activity subscale of the SF-36, the number of steps per day and oxygen consumption data all decreased significantly with increasing ICC severity as is shown in Figure 1. Secondly, in the present study, differences in the SF-36 physical activity subscale, in the number of steps and in the oxygen consumption at the ventilatory threshold and at peak exercise of mild, moderate and severe patients were comparable in female and male ME/CFS patients, indicating the validity of the severity grading in both women and men as is shown in Table 1. These data are in contrast to the study of Faro et al. who demonstrated in a large patient population that the clinical phenotype of male ME/CFS patients differed from that of female ME/CFS patients [22]. In the study of Faro et al., the physical activity subscales of the SF-36 were significantly higher in men than in women. However, the authors did not analyze the physical functioning subscale in relation to the clinical severity degree, as no subdivision was made in patients based on a clinical grading as in the present study.

While the four measures were significantly different between mild, moderate and severe disease, Table 3 and Figure 2 show discrepancies between the four measures. The oxygen consumption at the ventilatory threshold has only a fair agreement (low Kappa value) with the clinical grading and is therefore not likely to be helpful. The reasons for the low diagnostic capacity of the ventilatory threshold are unknown and need to be explored in future studies. In a previous study, we showed highly significant correlations between the physical functioning scale, the number of steps/day and the percent peak oxygen consumption in female ME/CFS patients [23]. Despite the highly significant correlations, a large variation between the three measures was found in that study. Activity is partially determined by age, race, menopausal status, educational level, body mass index, depressive symptoms, smoking, chronic medical conditions, and pain [24]. Furthermore, physical activity can also vary due to social circumstances (taking care of parents and children, marriage status), age, and previous physical fitness. The peak VO_2 is influenced by genetics, gender, age, training status, exercise mode, bedrest, altitude, body composition, medication, the capacity of the respiratory and circulatory systems to take up and transport oxygen, and the capacity of the working muscles to receive and use oxygen. In ME/CFS patients the degree of fatigue/exhaustion, post-exertional malaise, underlying metabolic abnormalities, fibromyalgic pain, kinesiophobia and the use of medication may further influence physical activities. Moreover, the classification based on history taking may be difficult, due to a difference of interpretation between clinicians as the history can differ due to co-morbidities, cognitive dysfunction, gender, age, disease duration, social status, personality, dependence on social security funding and, importantly, on the presence or absence of post-exertional malaise during history taking. Thus, due to the large number of influencing factors, the cut-off values for the physical functioning scale, number of steps and peak VO_2 in relation to the severity of the disease cannot be taken as absolute values, but merely support the proper severity grading.

4.1. Physical Activity Questionnaires in Previous ME/CFS Studies:

A large number of studies have examined the validity of the SF-36 questionnaire, showing that the physical activity subscale discriminates between various diseases and healthy controls [25–33]. In ME/CFS patients Jason et al. reported the ability of the different subscales of the SF-36 questionnaire to discriminate CFS patients from healthy controls [34]. The authors found that the physical activity subscale was not very optimal to discriminate between patients and healthy controls, using an area-under-the-curve (AUC) cut-off value of >0.90 for optimal discrimination. In the community-based sample, the AUC of the physical activity scale was 0.84 and in the tertiary care sample 0.87. On the contrary, another study found an AUC for assessing decrease in the physical activity scale of 0.91 [35], suggesting that the use of the physical activity score is valid with an acceptable discriminative value. The current study did not compare patients with healthy controls, instead patients with mild, moderate and severe disease were compared. We found excellent areas under the curves of 0.981 for the distinction between mild and moderate disease and 0.984 for the distinction between moderate and severe disease. However, when interpreting results, care must be taken that completion of the questionnaire is not done in a period of post-exertional malaise, as underreporting can be expected, or in a period of revitalization where over reporting can be expected.

4.2. Number of Steps in Previous ME/CFS Studies

The Sensewear^{TM} armband is a triaxial accelerometer, which measures steps, determines upright and lying position, characterizes sleep and estimates energy expenditure. Two studies have compared the number of steps of the Sensewear^{TM} armband to other activity trackers [36,37]. Both studies found a high mean absolute percentage error for the Sensewear^{TM} armband, relative to other activity trackers, it was considered to be in the middle range of reliability. The mean absolute percentage error (MAPE) reflects the accuracy of a device tested against a gold standard. Both studies used counted steps as the gold standard against measured steps from the device [36,37]. However, the study of Wahl et al. studied sporting conditions and not daily life circumstances. The study of An et al. studied daily life circumstances besides comparison of devices on a treadmill. Both studies showed the Sensewear^{TM} underestimated the number of steps. In the study of An et al., subjects did not wear the device for the whole 24 h, but only at documented activity periods. In the current study, patients wore the device for 5 complete days, which is a completely different study protocol, making results less comparable. Nevertheless, the Sensewear^{TM} device has successfully been used in patients with rheumatoid arthritis [38] in hemodialysis [39], and chronic obstructive pulmonary disease patients [40], spinal cord injury patients [41], in obese patients [42] and in children [43].

The exercise intolerance in CFS patients, using a variety of accelerometers, has been demonstrated in a number of studies [44–50]. None of these studies explored the discriminative value of the number of steps for symptom severity. Our data, with areas under the curve of the distinction between mild and moderate disease of 0.962 and of the distinction between moderate and severe disease of 0.997 suggests that activity tracking can be used as a diagnostic criterion (see Table 2 and Figure 1). However, as with the questionnaire, care must be taken that patients wear the tracker on average days. In fact, 15 patients were excluded from analysis due to under/over-activity.

4.3. Cardiopulmonary Exercise Test in Previous ME/CFS Studies

Finally, cardiopulmonary exercise testing is considered the most objective way to characterize exercise performance in ME/CFS patients. Multiple studies have shown that peak oxygen consumption is reduced in the majority of ME/CFS patients [51–63]. However, only one study determined the relation between the peak VO_2 and accelerometer data in female ME/CFS patients: higher peak VO_2 values were related to a higher physical activity time, physical activity energy expenditure, and a mean energy expenditure [46]. In line with the study of Ickmans et al., our study showed that in more severely affected ME/CFS patients, activity as expressed by the number of steps is associated with a lower percent predicted peak oxygen consumption as well as with the percent predicted oxygen consumption at the ventilatory threshold as is shown in Tables 2 and 3. Studies have shown that cardiopulmonary exercise test values of males and females differ due to a variety of factors, including weight, height, total body fat, total muscle mass, hemoglobin, cardiac volumes, and lung volumes [64–69]. In the present study we also found a difference between male and female ME/CFS patients with respect to the absolute peak oxygen consumption as is shown in Table 1 However, when percent predicted values were used, no significant difference was found between male and female patients. This is due to the fact that results are normalized to a reference population in which reference values of males are higher of females. Whether the same cut-off values of the physical activity subscale of the SF-36, the number of steps per day and the percent parameters of the cardiopulmonary exercise test can be used for men and women to discriminate between mild, moderate and severe ME/CFS needs to be determined in a larger male patient sample.

4.4. Summary of Previous ICC Clinical Severity Category Grading

We have explored the ICC disease grading and reported this in several recent papers. In 99 female patients, we correlated the physical activity subscale of the SF-36 with more objective measures as the number of steps per day on an activity tracker and the percent predicted peak oxygen consumption. Subgroup analysis of a RER over or under 1.1, the presence or absence of fibromyalgia, the use or non-use of pain-medication and a subdivision above and below a BMI of 30 had no impact on the correlations [23].

Furthermore, we reported on 82 female ME/CFS patients undergoing a 2-day CPET protocol, including 31 patients with mild disease, 31 patients with moderate disease and 20 patients with severe disease according to the ICC criteria. With increasing disease severity, cardiopulmonary exercise variables like oxygen consumption at the ventilatory threshold and peak exercise and the workload at the ventilatory threshold and at peak exercise declined significantly with increasing disease severity [70].

4.5. Limitations

This was a retrospective study taking data from patients with a maximum interval of 3 month between the three different measurements. This was a retrospective study and data can be used as a guideline for prospective studies. Ensuring a stable patient population was confirmed by checking of the patient charts: no major changes in symptomatology were found over this period of time in all patients. Only 289 out of 675 fulfilled all the requirements for analysis. The main reason for data acquisition in these patients were social security claims. In 269 patients, not all three methods were obtained. This may have led to inclusion bias. We did not include a control population. A prospective study is needed to evaluate the variability in measurements over time. Although the

Sensewear^{TM} activity meter probably underestimates the number of steps per day, a clear distinction in the three patient severity groups was shown. In addition, the Sensewear^{TM} activity meter is not available anymore, but the present commercial actographs and smart watches have step measurements included. ME/CFS is a heterogeneous disease with complex phenotyping such as differences in onset, differences in symptom combinations and specific comorbidities. This heterogeneity might influence results/outcomes and needs to be studied in future.

5. Conclusions

Disease severity grading as suggested by Carruthers in the ICC on ME/CFS is validated by using questionnaires and more objective measures as the number of steps or cardiopulmonary exercise test parameters. We showed a well-defined difference between ICC severity categories and physical activity on the SF-36 questionnaire, as well as in the number of steps per day and the percentage predicted oxygen consumption at the ventilatory threshold and peak exercise. With this information, the history of the patients reported outcomes on the disease can be confirmed and be more comprehensible for the patient and his/her caretakers, treating physicians and authorities. Moreover, it increases patient group homogeneity in ME/CFS research.

Author Contributions: C.M.C.v.C., P.C.R., and F.C.V. conceived the study, C.M.C.v.C. and F.C.V. collected the data, C.M.C.v.C. performed the primary data analysis and F.C.V., and P.C.R. performed secondary data analyses. All authors were involved in the drafting and review of the manuscript. All authors have read and agreed to the published version of the manuscript.

Funding: This research received no external funding.

Acknowledgments: Rowe is supported by the Sunshine Natural Wellbeing Foundation Professorship.

Conflicts of Interest: The authors declare no conflict of interest.

Appendix A

A ramp workload protocol was used varying between 10–30 Watt/min increases, depending on sex, age, and expected exercise intolerance. Symptoms limiting the test were exhaustion in all patients, muscle pain in 80% as one of the symptoms at the end of the test, and dizziness in 8%. None showed any signs of cardiac ischemia. Oxygen consumption (VO_2), carbon dioxide release (VCO_2), and oxygen saturation were continuously measured (Cortex, Procare, Amsterdam, The Netherlands), and displayed on screen using Metasoft software (Cortex, Biophysic Gmbh, Leipzig, Germany). An ECG was continuously recorded and blood pressures were measured using the Nexfin device (BMEYE, Amsterdam, The Netherlands) [71]. The metabolic measurement system (Cortex, Biophysic Gmbh, Leipzig, Germany) was calibrated before each test with ambient air, standard gases of known concentrations, and a 3-L calibration syringe. The ventilatory threshold (VT), a measure of the anaerobic threshold, was identified from expired gases using the V-Slope algorithm [72]. An experienced cardiologist supervised the test and performed visual assessment and confirmation of the algorithm-derived VT. The mean of the VO_2 measurements of the last 15 s before ending the exercise (peak VO_2) was taken. VO_2 at the peak and at the VT were expressed as a percentage of the normal values of a population study of males and females: %peak VO_2 and %VT VO_2 [73]. Also the mean respiratory exchange ratio (RER; VCO_2/VO_2) of the last 15 s was calculated. Immediately after the test, the attending cardiologist noted the primary reason for termination of the exercise. We excluded patients with a body mass index of >37 as well as patients younger than 25 years, because the reference values of the population-based study of Glaser et al. reported on data of healthy controls older than 25 years and with a BMI less than 37 kg/m^2 [73].

References

1. Carruthers, B.M.; van de Sande, M.I.; DE Meirleir, K.L.; Klimas, N.G.; Broderick, G.; Mitchell, T.; Staines, D.; Powles, A.C.; Speight, N.; Vallings, R.; et al. Myalgic encephalomyelitis: International Consensus Criteria. *J. Intern. Med.* **2011**, *270*, 327–338. [CrossRef] [PubMed]
2. Clayton, E.W. Beyond myalgic encephalomyelitis/chronic fatigue syndrome: An IOM report on redefining an illness. *JAMA* **2015**, *313*, 1101–1102. [CrossRef] [PubMed]
3. Fukuda, K.; Straus, S.E.; Hickie, I.; Sharpe, M.C.; Dobbins, J.G.; Komaroff, A. The chronic fatigue syndrome: A comprehensive approach to its definition and study. International Chronic Fatigue Syndrome Study Group. *Ann. Intern. Med.* **1994**, *121*, 953–959. [CrossRef] [PubMed]
4. Arnett, S.V.; Alleva, L.M.; Korossy-Horwood, R.; Clark, I.A. Chronic fatigue syndrome—A neuroimmunological model. *Med. Hypotheses* **2011**, *77*, 77–83. [CrossRef] [PubMed]
5. Fulle, S.; Pietrangelo, T.; Mancinelli, R.; Saggini, R.; Fano, G. Specific correlations between muscle oxidative stress and chronic fatigue syndrome: A working hypothesis. *J. Muscle Res. Cell Motil.* **2007**, *28*, 355–362. [CrossRef] [PubMed]
6. Gerrity, T.R.; Papanicolaou, D.A.; Amsterdam, J.D.; Bingham, S.; Grossman, A.; Hedrick, T.; Herberman, R.B.; Krueger, G.; Levine, S.; Mohagheghpour, N.; et al. Immunologic aspects of chronic fatigue syndrome. Report on a Research Symposium convened by The CFIDS Association of America and co-sponsored by the US Centers for Disease Control and Prevention and the National Institutes of Health. *Neuroimmunomodulation* **2004**, *11*, 351–357. [CrossRef] [PubMed]
7. Gur, A.; Oktayoglu, P. Central nervous system abnormalities in fibromyalgia and chronic fatigue syndrome: New concepts in treatment. *Curr. Pharm. Des.* **2008**, *14*, 1274–1294. [CrossRef]
8. Jones, D.E.; Hollingsworth, K.G.; Taylor, R.; Blamire, A.M.; Newton, J.L. Abnormalities in pH handling by peripheral muscle and potential regulation by the autonomic nervous system in chronic fatigue syndrome. *J. Intern. Med.* **2010**, *267*, 394–401. [CrossRef] [PubMed]
9. Klimas, N.G.; Salvato, F.R.; Morgan, R.; Fletcher, M.A. Immunologic abnormalities in chronic fatigue syndrome. *J. Clin. Microbiol.* **1990**, *28*, 1403–1410. [CrossRef]
10. Komaroff, A.L.; Cho, T.A. Role of infection and neurologic dysfunction in chronic fatigue syndrome. *Semin. Neurol.* **2011**, *31*, 325–337. [CrossRef]
11. McCully, K.K.; Smith, S.; Rajaei, S.; Leigh, J.S., Jr.; Natelson, B.H. Blood flow and muscle metabolism in chronic fatigue syndrome. *Clin. Sci. (Lond.)* **2003**, *104*, 641–647. [CrossRef] [PubMed]
12. Naess, H.; Sundal, E.; Myhr, K.M.; Nyland, H.I. Postinfectious and chronic fatigue syndromes: Clinical experience from a tertiary-referral centre in Norway. *In Vivo* **2010**, *24*, 185–188. [PubMed]
13. Okamoto, L.E.; Raj, S.R.; Peltier, A.; Gamboa, A.; Shibao, C.; Diedrich, A.; Black, B.K.; Robertson, D.; Biaggioni, I. Neurohumoral and haemodynamic profile in postural tachycardia and chronic fatigue syndromes. *Clin. Sci. (Lond.)* **2012**, *122*, 183–192. [CrossRef] [PubMed]
14. Ortega-Hernandez, O.D.; Shoenfeld, Y. Infection, vaccination, and autoantibodies in chronic fatigue syndrome, cause or coincidence? *Ann. N. Y. Acad. Sci.* **2009**, *1173*, 600–609. [CrossRef]
15. Stewart, J.M. Autonomic nervous system dysfunction in adolescents with postural orthostatic tachycardia syndrome and chronic fatigue syndrome is characterized by attenuated vagal baroreflex and potentiated sympathetic vasomotion. *Pediatr Res.* **2000**, *48*, 218–226. [CrossRef]
16. Wong, R.; Lopaschuk, G.; Zhu, G.; Walker, D.; Catellier, D.; Burton, D.; Teo, K.; Collins-Nakai, R.; Montague, T. Skeletal muscle metabolism in the chronic fatigue syndrome. In vivo assessment by 31P nuclear magnetic resonance spectroscopy. *Chest* **1992**, *102*, 1716–1722. [CrossRef]
17. van Campen, C.L.M.C.; Visser, F.C. The Effect of Curcumin in Patients with Chronic Fatigue Syndrome/Myalgic Encephalomyelitis Disparate Responses in Different Disease Severities. *Pharmacovigil. Pharmacoepidemiol.* **2019**, *2*, 22–27. [CrossRef]
18. Viehoff, P.B.; van Genderen, F.R.; Wittink, H. Upper limb lymphedema 27 (ULL27): Dutch translation and validation of an illness-specific health-related quality of life questionnaire for patients with upper limb lymphedema. *Lymphology* **2008**, *41*, 131–138.
19. van Campen, C.L.M.C.; Visser, F.C. Validity of 2-day cardiopulmonary exercise testing in female patients with myalgic encephalomyelitis/chronic fatigue syndrome. *Int. J. Curr. Res.* **2020**, *12*, 10436–10442. [CrossRef]
20. Liu, X. Classification accuracy and cut point selection. *Stat. Med.* **2012**, *31*, 2676–2686. [CrossRef]

21. Hoo, Z.H.; Candlish, J.; Teare, D. What is an ROC curve? *Emerg. Med. J.* **2017**, *34*, 357–359. [CrossRef] [PubMed]

22. Faro, M.; Saez-Francas, N.; Castro-Marrero, J.; Aliste, L.; Fernandez de Sevilla, T.; Alegre, J. Gender differences in chronic fatigue syndrome. *Reum. Clin.* **2016**, *12*, 72–77. [CrossRef] [PubMed]

23. van Campen, C.L.M.C.; Rowe, P.C.; Verheugt, F.W.A.; Visser, F.C. Physical activity measures in patients with myalgic encephalomyalitis/chronic fatigue syndrome: Correlations between peak oxygen consumption, the physical functioning scale of the SF-36 scale, and the number of steps from an activity meter. *J. Transl. Med.* **2020**, *18*, 228–238. [CrossRef] [PubMed]

24. Dugan, S.A.; Everson-Rose, S.A.; Karavolos, K.; Sternfeld, B.; Wesley, D.; Powell, L.H. The impact of physical activity level on SF-36 role-physical and bodily pain indices in midlife women. *J. Phys. Act. Health* **2009**, *6*, 33–42. [CrossRef]

25. McHorney, C.A.; Ware, J.E., Jr.; Raczek, A.E. The MOS 36-Item Short-Form Health Survey (SF-36): II. Psychometric and clinical tests of validity in measuring physical and mental health constructs. *Med. Care* **1993**, *31*, 247–263. [CrossRef]

26. Matcham, F.; Scott, I.C.; Rayner, L.; Hotopf, M.; Kingsley, G.H.; Norton, S.; Scott, D.L.; Steer, S. The impact of rheumatoid arthritis on quality-of-life assessed using the SF-36: A systematic review and meta-analysis. *Semin. Arthritis Rheum.* **2014**, *44*, 123–130. [CrossRef]

27. Gu, M.; Cheng, Q.; Wang, X.; Yuan, F.; Sam, N.B.; Pan, H.; Li, B.; Ye, D. The impact of SLE on health-related quality of life assessed with SF-36: A systemic review and meta-analysis. *Lupus* **2019**, *28*, 371–382. [CrossRef]

28. Wang, Y.; Zhao, R.; Gu, C.; Gu, Z.; Li, L.; Li, Z.; Dong, C.; Zhu, J.; Fu, T.; Gao, J. The impact of systemic lupus erythematosus on health-related quality of life assessed using the SF-36: A systematic review and meta-analysis. *Psychol. Health Med.* **2019**, *24*, 978–991. [CrossRef]

29. Yang, X.; Fan, D.; Xia, Q.; Wang, M.; Zhang, X.; Li, X.; Cai, G.; Wang, L.; Xin, L.; Xu, S.; et al. The health-related quality of life of ankylosing spondylitis patients assessed by SF-36: A systematic review and meta-analysis. *Qual. Life Res.* **2016**, *25*, 2711–2723. [CrossRef]

30. Li, L.; Cui, Y.; Chen, S.; Zhao, Q.; Fu, T.; Ji, J.; Li, L.; Gu, Z. The impact of systemic sclerosis on health-related quality of life assessed by SF-36: A systematic review and meta-analysis. *Int. J. Rheum. Dis.* **2018**, *21*, 1884–1893. [CrossRef]

31. Wyld, M.; Morton, R.L.; Hayen, A.; Howard, K.; Webster, A.C. A systematic review and meta-analysis of utility-based quality of life in chronic kidney disease treatments. *PLoS Med.* **2012**, *9*, e1001307. [CrossRef] [PubMed]

32. Fuke, R.; Hifumi, T.; Kondo, Y.; Hatakeyama, J.; Takei, T.; Yamakawa, K.; Inoue, S.; Nishida, O. Early rehabilitation to prevent postintensive care syndrome in patients with critical illness: A systematic review and meta-analysis. *BMJ Open* **2018**, *8*, e019998. [CrossRef] [PubMed]

33. Filbay, S.; Pandya, T.; Thomas, B.; McKay, C.; Adams, J.; Arden, N. Quality of Life and Life Satisfaction in Former Athletes: A Systematic Review and Meta-Analysis. *Sports Med.* **2019**, *49*, 1723–1738. [CrossRef] [PubMed]

34. Jason, L.; Brown, M.; Evans, M.; Anderson, V.; Lerch, A.; Brown, A.; Hunnell, J.; Porter, N. Measuring substantial reductions in functioning in patients with chronic fatigue syndrome. *Disabil. Rehabil.* **2011**, *33*, 589–598. [CrossRef] [PubMed]

35. Thorpe, T.; McManimen, S.; Gleason, K.; Stoothoff, J.; Newton, J.L.; Strand, E.B.; Jason, L.A. Assessing current functioning as a measure of significant reduction in activity level. *Fatigue* **2016**, *4*, 175–188. [CrossRef]

36. An, H.S.; Jones, G.C.; Kang, S.K.; Welk, G.J.; Lee, J.M. How valid are wearable physical activity trackers for measuring steps? *Eur. J. Sport Sci.* **2017**, *17*, 360–368. [CrossRef]

37. Wahl, Y.; Duking, P.; Droszez, A.; Wahl, P.; Mester, J. Criterion-Validity of Commercially Available Physical Activity Tracker to Estimate Step Count, Covered Distance and Energy Expenditure during Sports Conditions. *Front. Physiol.* **2017**, *8*, 725. [CrossRef]

38. Almeida, G.J.; Wasko, M.C.; Jeong, K.; Moore, C.G.; Piva, S.R. Physical activity measured by the SenseWear Armband in women with rheumatoid arthritis. *Phys. Ther.* **2011**, *91*, 1367–1376. [CrossRef]

39. Avesani, C.M.; Trolonge, S.; Deleaval, P.; Baria, F.; Mafra, D.; Faxen-Irving, G.; Chauveau, P.; Teta, D.; Kamimura, M.A.; Cuppari, L.; et al. Physical activity and energy expenditure in haemodialysis patients: An international survey. *Nephrol. Dial. Transplant.* **2012**, *27*, 2430–2434. [CrossRef]

40. Hill, K.; Dolmage, T.E.; Woon, L.; Goldstein, R.; Brooks, D. Measurement properties of the SenseWear armband in adults with chronic obstructive pulmonary disease. *Thorax* **2010**, *65*, 486–491. [CrossRef]

41. Tanhoffer, R.A.; Tanhoffer, A.I.; Raymond, J.; Hills, A.P.; Davis, G.M. Comparison of methods to assess energy expenditure and physical activity in people with spinal cord injury. *J. Spinal Cord Med.* **2012**, *35*, 35–45. [CrossRef] [PubMed]

42. Monteiro, F.; Camillo, C.A.; Vitorasso, R.; Sant'anna, T.; Hernandes, N.A.; Probst, V.S.; Pitta, F. Obesity and Physical Activity in the Daily Life of Patients with COPD. *Lung* **2012**, *190*, 403–410. [CrossRef]

43. Calabro, M.A.; Welk, G.J.; Eisenmann, J.C. Validation of the SenseWear Pro Armband algorithms in children. *Med. Sci. Sports Exerc.* **2009**, *41*, 1714–1720. [CrossRef] [PubMed]

44. Brown, M.; Khorana, N.; Jason, L.A. The role of changes in activity as a function of perceived available and expended energy in nonpharmacological treatment outcomes for ME/CFS. *J. Clin. Psychol.* **2011**, *67*, 253–260. [CrossRef] [PubMed]

45. Evering, R.M.; Tonis, T.M.; Vollenbroek-Hutten, M.M. Deviations in daily physical activity patterns in patients with the chronic fatigue syndrome: A case control study. *J. Psychosom. Res.* **2011**, *71*, 129–135. [CrossRef] [PubMed]

46. Ickmans, K.; Clarys, P.; Nijs, J.; Meeus, M.; Aerenhouts, D.; Zinzen, E.; Aelbrecht, S.; Meersdom, G.; Lambrecht, L.; Pattyn, N. Association between cognitive performance, physical fitness, and physical activity level in women with chronic fatigue syndrome. *J. Rehabil. Res. Dev.* **2013**, *50*, 795–810. [CrossRef]

47. Jason, L.A.; King, C.P.; Frankenberry, E.L.; Jordan, K.M.; Tryon, W.W.; Rademaker, F.; Huang, C.F. Chronic fatigue syndrome: Assessing symptoms and activity level. *J. Clin. Psychol.* **1999**, *55*, 411–424. [CrossRef]

48. Kop, W.J.; Lyden, A.; Berlin, A.A.; Ambrose, K.; Olsen, C.; Gracely, R.H.; Williams, D.A.; Clauw, D.J. Ambulatory monitoring of physical activity and symptoms in fibromyalgia and chronic fatigue syndrome. *Arthritis Rheum.* **2005**, *52*, 296–303. [CrossRef]

49. Solomon-Moore, E.; Jago, R.; Beasant, L.; Brigden, A.; Crawley, E. Physical activity patterns among children and adolescents with mild-to-moderate chronic fatigue syndrome/myalgic encephalomyelitis. *BMJ Paediatr. Open* **2019**, *3*, e000425. [CrossRef]

50. Wyller, V.B.; Vitelli, V.; Sulheim, D.; Fagermoen, E.; Winger, A.; Godang, K.; Bollerslev, J. Altered neuroendocrine control and association to clinical symptoms in adolescent chronic fatigue syndrome: A cross-sectional study. *J. Transl. Med.* **2016**, *14*, 121. [CrossRef]

51. De Becker, P.; Roeykens, J.; Reynders, M.; McGregor, N.; De Meirleir, K. Exercise capacity in chronic fatigue syndrome. *Arch. Intern. Med.* **2000**, *160*, 3270–3277. [CrossRef] [PubMed]

52. Fulcher, K.Y.; White, P.D. Strength and physiological response to exercise in patients with chronic fatigue syndrome. *J. Neurol. Neurosurg. Psychiatry* **2000**, *69*, 302–307. [CrossRef] [PubMed]

53. Hodges, L.D.; Nielsen, T.; Baken, D. Physiological measures in participants with chronic fatigue syndrome, multiple sclerosis and healthy controls following repeated exercise: A pilot study. *Clin. Physiol. Funct. Imaging* **2018**, *38*, 639–644. [CrossRef] [PubMed]

54. Jammes, Y.; Steinberg, J.G.; Mambrini, O.; Bregeon, F.; Delliaux, S. Chronic fatigue syndrome: Assessment of increased oxidative stress and altered muscle excitability in response to incremental exercise. *J. Intern. Med.* **2005**, *257*, 299–310. [CrossRef]

55. Keller, B.A.; Pryor, J.L.; Giloteaux, L. Inability of myalgic encephalomyelitis/chronic fatigue syndrome patients to reproduce VO(2)peak indicates functional impairment. *J. Transl. Med.* **2014**, *12*, 104. [CrossRef]

56. Sargent, C.; Scroop, G.C.; Nemeth, P.M.; Burnet, R.B.; Buckley, J.D. Maximal oxygen uptake and lactate metabolism are normal in chronic fatigue syndrome. *Med. Sci. Sports Exerc.* **2002**, *34*, 51–56. [CrossRef]

57. Sisto, S.A.; LaManca, J.; Cordero, D.L.; Bergen, M.T.; Ellis, S.P.; Drastal, S.; Boda, W.L.; Tapp, W.N.; Natelson, B.H. Metabolic and cardiovascular effects of a progressive exercise test in patients with chronic fatigue syndrome. *Am. J. Med.* **1996**, *100*, 634–640. [CrossRef]

58. Snell, C.R.; Stevens, S.R.; Davenport, T.E.; Van Ness, J.M. Discriminative Validity of Metabolic and Workload Measurements to Identify Individuals With Chronic Fatigue Syndrome. *Phys. Ther.* **2013**. [CrossRef]

59. Vanness, J.M.; Snell, C.R.; Stevens, S.R. Diminished cardiopulmonary capacity during post-exertional malaise. *J. Chronic Fatigue Syndr.* **2007**, *14*, 77–85. [CrossRef]

60. Vermeulen, R.C.; Kurk, R.M.; Visser, F.C.; Sluiter, W.; Scholte, H.R. Patients with chronic fatigue syndrome performed worse than controls in a controlled repeated exercise study despite a normal oxidative phosphorylation capacity. *J. Transl. Med.* **2010**, *8*, 93. [CrossRef]

61. Vermeulen, R.C.; Vermeulen van Eck, I.W. Decreased oxygen extraction during cardiopulmonary exercise test in patients with chronic fatigue syndrome. *J. Transl. Med.* **2014**, *12*, 20. [CrossRef] [PubMed]
62. Wallman, K.E.; Morton, A.R.; Goodman, C.; Grove, R. Physiological responses during a submaximal cycle test in chronic fatigue syndrome. *Med. Sci. Sports Exerc.* **2004**, *36*, 1682–1688. [CrossRef] [PubMed]
63. Franklin, J.D.; Atkinson, G.; Atkinson, J.M.; Batterham, A.M. Peak Oxygen Uptake in Chronic Fatigue Syndrome/Myalgic Encephalomyelitis: A Meta-Analysis. *Int. J. Sports Med.* **2019**, *40*, 77–87. [CrossRef] [PubMed]
64. Cureton, K.; Bishop, P.; Hutchinson, P.; Newland, H.; Vickery, S.; Zwiren, L. Sex difference in maximal oxygen uptake. Effect of equating haemoglobin concentration. *Eur. J. Appl. Physiol. Occup. Physiol.* **1986**, *54*, 656–660. [CrossRef] [PubMed]
65. Fletcher, G.F.; Balady, G.J.; Amsterdam, E.A.; Chaitman, B.; Eckel, R.; Fleg, J.; Froelicher, V.F.; Leon, A.S.; Pina, I.L.; Rodney, R.; et al. Exercise standards for testing and training: A statement for healthcare professionals from the American Heart Association. *Circulation* **2001**, *104*, 1694–1740. [CrossRef]
66. Fomin, A.; Ahlstrand, M.; Schill, H.G.; Lund, L.H.; Stahlberg, M.; Manouras, A.; Gabrielsen, A. Sex differences in response to maximal exercise stress test in trained adolescents. *BMC Pediatr.* **2012**, *12*, 127. [CrossRef]
67. Higginbotham, M.B.; Morris, K.G.; Coleman, R.E.; Cobb, F.R. Sex-related differences in the normal cardiac response to upright exercise. *Circulation* **1984**, *70*, 357–366. [CrossRef]
68. Sharma, H.B.; Kailashiya, J. Gender Difference in Aerobic Capacity and the Contribution by Body Composition and Haemoglobin Concentration: A Study in Young Indian National Hockey Players. *J. Clin. Diagn. Res.* **2016**, *10*, CC09–CC13. [CrossRef]
69. Wheatley, C.M.; Snyder, E.M.; Johnson, B.D.; Olson, T.P. Sex differences in cardiovascular function during submaximal exercise in humans. *Springerplus* **2014**, *3*, 445. [CrossRef]
70. van Campen, C.L.M.C.; Rowe, P.C.; Visser, F.C. Two-day cardiopulmonary exercise testing in females with a severe grade of myalgic encephalomylitis/chornic fatigue syndrome: Comparison with patients with a mild and moderate disease. *Healthcare* **2020**, *8*, 192. [CrossRef]
71. Martina, J.R.; Westerhof, B.E.; van Goudoever, J.; de Beaumont, E.M.; Truijen, J.; Kim, Y.S.; Immink, R.V.; Jobsis, D.A.; Hollmann, M.W.; Lahpor, J.R.; et al. Noninvasive continuous arterial blood pressure monitoring with Nexfin(R). *Anesthesiology* **2012**, *116*, 1092–1103. [CrossRef] [PubMed]
72. Beaver, W.L.; Wasserman, K.; Whipp, B.J. A new method for detecting anaerobic threshold by gas exchange. *J. Appl. Physiol. (1985)* **1986**, *60*, 2020–2027. [CrossRef] [PubMed]
73. Glaser, S.; Koch, B.; Ittermann, T.; Schaper, C.; Dorr, M.; Felix, S.B.; Volzke, H.; Ewert, R.; Hansen, J.E. Influence of age, sex, body size, smoking, and beta blockade on key gas exchange exercise parameters in an adult population. *Eur. J. Cardiovasc. Prev. Rehabil.* **2010**, *17*, 469–476. [CrossRef] [PubMed]

Article

Reductions in Cerebral Blood Flow Can Be Provoked by Sitting in Severe Myalgic Encephalomyelitis/Chronic Fatigue Syndrome Patients

C (Linda) MC van Campen [1,*], Peter C. Rowe [2] and Frans C Visser [1]

[1] Stichting CardioZorg, 2132 HN Hoofddorp, The Netherlands; fransvisser@stichtingcardiozorg.nl
[2] Department of Paediatrics, John Hopkins University School of Medicine, Baltimore, MD 21205, USA; prowe@jhmi.edu
* Correspondence: info@stichtingcardiozorg.nl; Tel.: +0031-208934125

Received: 7 September 2020; Accepted: 9 October 2020; Published: 11 October 2020

Abstract: Introduction: In a large study with myalgic encephalomyelitis/chronic fatigue syndrome (ME/CFS) patients, we showed that 86% had symptoms of orthostatic intolerance in daily life and that 90% had an abnormal reduction in cerebral blood flow (CBF) during a standard tilt test. A standard head-up tilt test might not be tolerated by the most severely affected bed-ridden ME/CFS patients. Sitting upright is a milder orthostatic stress. The present study examined whether a sitting test, measuring cerebral blood flow by extracranial Doppler, would be sufficient to provoke abnormal reductions in cerebral blood flow in severe ME/CFS patients. Methods and results: 100 severe ME/CFS patients were studied, (88 females) and were compared with 15 healthy controls (HC) (13 females). CBF was measured first while seated for at least one hour, followed by a CBF measurement in the supine position. Fibromyalgia was present in 37 patients. Demographic data as well as supine heart rate and blood pressures were not different between ME/CFS patients and HC. Heart rate and blood pressure did not change significantly between supine and sitting both in patients and HC. Supine CBF was not different between patients and HC. In contrast, absolute CBF during sitting was lower in patients compared to HC: 474 (96) mL/min in patients and 627 (89) mL/min in HC; $p < 0.0001$. As a result, percent CBF reduction while seated was −24.5 (9.4)% in severe ME/CFS patients and −0.4 (1.2)% in HC ($p < 0.0001$). In the ten patients who had no orthostatic intolerance complaints in daily life, the CBF reduction was −2.7 (2.1)%, which was not significantly different from HC ($p = 0.58$). The remaining 90 patients with orthostatic intolerance complaints had a −26.9 (6.2)% CBF reduction. No difference in CBF parameters was found in patients with and without fibromyalgia. Patients with a previous diagnosis of postural orthostatic tachycardia syndrome (POTS) had a significantly larger CBF reduction compared with those without POTS: 28.8 (7.2)% vs. 22.3 (9.7)% ($p = 0.0008$). Conclusions: A sitting test in severe ME/CFS patients was sufficient to provoke a clinically and statistically significant mean CBF decline of 24.5%. Patients with a previous diagnosis of POTS had a larger CBF reduction while seated, compared to patients without POTS. The magnitude of these CBF reductions is similar to the results in less severely affected ME/CFS patients during head-up tilt, suggesting that a sitting test is adequate for the diagnosis of orthostatic intolerance in severely affected patients.

Keywords: orthostatic intolerance; cerebral blood flow; sitting; myalgic encephalomyelitis; chronic fatigue syndrome; severe disease; ME/CFS

---

1. Introduction

Orthostatic intolerance is defined as a clinical condition in which symptoms worsen upon assuming and maintaining upright posture and are ameliorated (although not necessarily abolished) by recumbency [1]. Myalgic encephalomyelitis/chronic fatigue syndrome (ME/CFS) patients have a high prevalence of orthostatic intolerance [1,2]. We reported in a study of 429 ME/CFS patients that 86% had orthostatic intolerance symptoms during daily life [2]. Furthermore, during a 30-min head-up tilt table test, 90% had an abnormal cerebral blood flow (CBF) reduction as assessed by extracranial Doppler measurements. This abnormal CBF reduction was not only present in ME/CFS patients with postural orthostatic tachycardia syndrome (POTS) and delayed orthostatic hypotension, but also in patients with a normal heart rate and blood pressure response during tilt testing. The mean CBF reduction of 26% in the entire study population with ME/CFS was significantly different from the 7% reduction observed in healthy controls in response to the same orthostatic stress.

Extracranial Doppler measurements may take between 2 and 7 min, and in our recent study, 15 patients were excluded because they were unable to maintain the upright position during the acquisition period [2]. Moreover, patients can develop post-exertional malaise after conventional 60–90 degree head-up orthostatic stress testing [1]. Due to these two restrictions, conventional orthostatic testing may not be feasible in severe ME/CFS patients. Miwa et al. described in a study of 44 ME/CFS patients that during a 10 min active sitting, test 30 patients (69%) reported orthostatic intolerance [3]. Moreover, in that study, orthostatic intolerance symptoms while sitting were more prevalent in patients with increased functional impairment.

Assuming that severe ME/CFS patients cannot tolerate prolonged standing during tilt testing and also that they have orthostatic symptoms during sitting, the aim of the current study was to test the hypothesis that reduced cerebral blood flow was present in severe ME/CFS patients during sitting.

2. Materials and Methods

2.1. Patients and Healthy Controls

From June 2011 to January 2018, 801 patients with the suspicion of ME/CFS visited the outpatient clinic of the Stichting CardioZorg. All patients were evaluated by the same clinician (FVC). During the first visit, it was determined whether patients satisfied the criteria for CFS and ME, taking the exclusion criteria into account [4,5]. Furthermore, in case of ME/CFS, the severity of the disease was classified. Disease severity was scored according to the International Consensus Criteria (ICC), with severity scored as mild, moderate, severe, and very severe [4]. The clinician also ascertained the presence or absence of orthostatic intolerance symptoms in daily life like dizziness/light-headedness, prior syncope and prior near-syncope, nausea, etc., as well as triggering events like standing in a line. In the present study, we analyzed 100 severe ME/CFS patients who had cerebral blood flow measurements while seated. A sitting test was chosen over a standing test or tilt-table test because patients were not able to stand for more than several minutes and we wanted to avoid the prolonged post-exertional malaise that can occur after tilt table testing. Furthermore, we recorded whether patients had a previous diagnosis of postural orthostatic tachycardia syndrome (POTS) or fibromyalgia. If those with POTS were being treated with cardio-active medications, patients were instructed to stop those drugs for the 3 days before testing. Patients taking pain medications and selective serotonin reuptake inhibitors continued taking these medications as usual. For comparison, 15 healthy controls were studied. None used heart rate or blood pressure lowering medications. The study was carried out in accordance with the Declaration of Helsinki. All ME/CFS patients gave informed, written consent. The study was approved by the medical ethics committee of the Slotervaart Hospital, Amsterdam, the Netherlands (reference number P1736 and P1450).

2.2. Sitting Test

Testing was first performed in the seated position. Subjects had been seated for at least one hour before testing: see Table 1. Subjects were positioned on a special examination table with a whole trunk inclination of 80 degrees. The upper limbs rested comfortably at their sides and the lower limbs had an inclination of −25 degrees (see Figure 1). The study protocol started with measurements of heart rate and blood pressure, followed by extracranial Doppler echography (see below). After completion of the measurements, patients were placed in the supine position for 10–15 min, after which the measurements were repeated.

Table 1. Summary of the sitting test procedure.

A minimum of 1 h sitting was required (including travel, time spent in the waiting room, and history taking)
Transfer to the research table and start sitting test
Measurement of heart rate and blood pressure in sitting position as shown in Figure 1
Cerebral blood flow measurements in sitting position
Lying down for 10–15 min
Measurement of heart rate and blood pressuring in supine position
Cerebral blood flow measurements in supine position
End of the sitting test

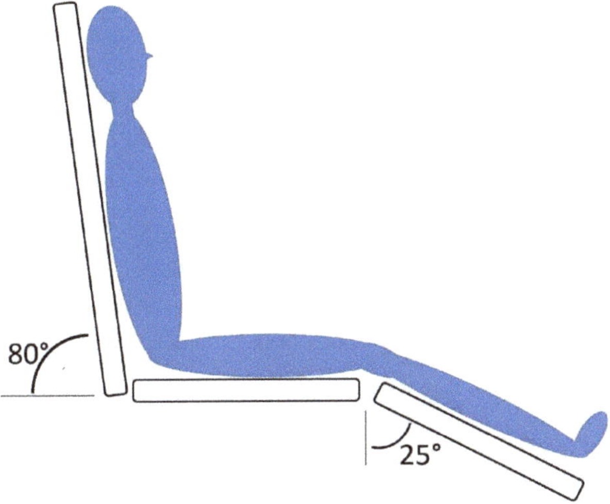

Figure 1. Research setup.

2.3. Cerebral Blood Flow Determination by Extracranial Doppler Echography

Cerebral blood flow measurements were performed with the same protocol as described in previous studies [2,6]. A more detailed description can be found in Appendix A.

2.4. Statistical Analysis

Data were analyzed using Graphpad Prism version 8.2.4 (Graphpad software, La Jolla, CA, USA). All continuous data were tested for normal distribution using the D'Agostino & Pearson omnibus normality test, and presented as mean (SD) or as median with the IQR, where appropriate. Nominal data were compared with a Chi-square analysis. For continuous data, a paired and non-paired t-test/Mann–Whitney was used for comparison, where appropriate. Groups were compared using the ordinary one-way analysis of variance (ANOVA), with post hoc Tukey's test in the event of a significant result. A *p* value of <0.01 was considered significant.

3. Results

Patient Clinical and Echo Doppler Data

Between June 2011 and January 2018, 801 patients were evaluated. Of those, 53 patients did not fulfill the ME/CFS criteria, leaving 748 patients diagnosed with ME/CFS. Of the 748 ME/CFS patients, 177 (24%) were classified as having severe ME/CFS. From this group, 38 patients were excluded because they underwent a passive standing test, 12 patients were excluded from the analysis because of the use of blood pressure and/or heart rate influencing medication, 12 patients refused orthostatic stress testing, and 15 patients had incomplete or insufficient image acquisitions, leaving 100 patients with a severe grade of ME/CFS studied (88 females/12 males). The estimated mean duration of sitting before the start of the sitting test was 1 h 55 min (SD 58 min). The maximal duration of sitting before starting the test was 3.5 h. Baseline characteristics are presented in Table 2. For the patient group, median disease duration was 14 (IQR 8–22) years. Daily life orthostatic intolerance symptoms were reported by 90 of the 100 ME/CFS patients. At the time of the test, nine patients were being treated with selective serotonin reuptake inhibitors (SSRIs). Baseline and clinical data in patients with and without SSRIs did not differ significantly (data not shown). Fibromyalgia was present in 37/100 patients. Of the 37 ME/CFS patients with fibromyalgia, 28 were being treated with pain-medication or benzodiazepines for sleeping disorders. Of the remaining 63 patients, 4 were being treated with benzodiazepines for sleeping disorders. A diagnosis of POTS was reported by 34 patients. Table 2 also shows supine and sitting test results for both healthy controls and severe ME/CFS patients. No statistically significant differences were present during supine testing. Heart rate and blood pressure was similar between groups during the sitting position, but there was a significant difference in the cerebral blood flow during sitting: 474 (96) mL/min in ME/CFS patients vs. 627 (89) mL/min in healthy controls ($p < 0.0001$). The percent decline in cerebral blood flow in the sitting position compared to the supine position was −24.5 (9.4)% in ME/CFS patients vs. −0.4 (1.2)% in healthy controls ($p < 0.0001$). No patients developed POTS or delayed orthostatic hypotension during the test.

Table 2. Demographic features and orthostatic stress test results in healthy controls and severe ME/CFS patients.

	HC (*n* = 15)	Severe ME/CFS (*n* = 100)	*p*-Value
Male/female	2/13	12/88	0.88
Age (years)	38 (15)	38 (12)	0.94
Height (cm)	172 (9)	174 (8)	0.44
Weight (kg)	68 (15)	73 (16)	0.27
BMI (kg/m^2)	22.9 (3.4)	24.3 (5.5)	0.33
Heart rate supine (bpm)	70 (14)	75 (15)	0.34
Heart rate sitting (bpm)	75 (10)	87 (19)	0.08
Systolic blood pressure supine (mmHg)	119 (14)	124 (17)	0.40
Systolic blood pressure sitting (mmHg)	122 (18)	134 (18)	0.05
Diastolic blood pressure supine (mmHg)	75 (12)	80 (11)	0.20
Diastolic blood pressure sitting (mmHg)	83 (10)	88 (11)	0.22
Cerebral blood flow supine (mL/min)	630 (85)	631 (121)	0.96
Cerebral blood flow sitting (mL/min)	627 (89)	474 (96)	<0.0001
Cerebral blood flow % change sitting vs. supine	−0.4 (1.2)	−24.5 (9.4)	<0.0001

HC: healthy controls; ME/CFS: myalgic encephalomyelitis/chronic fatigue syndrome.

Figure 2 shows the percent decrease in cerebral blood flow from supine to sitting in healthy controls and in severe ME/CFS patients with and without orthostatic intolerance in daily life. Only 10 (10%) severe ME/CFS patients had no orthostatic symptoms in daily life. Cerebral blood flow reductions in patients without daily life orthostatic intolerance symptoms did not differ significantly from those found in healthy controls.

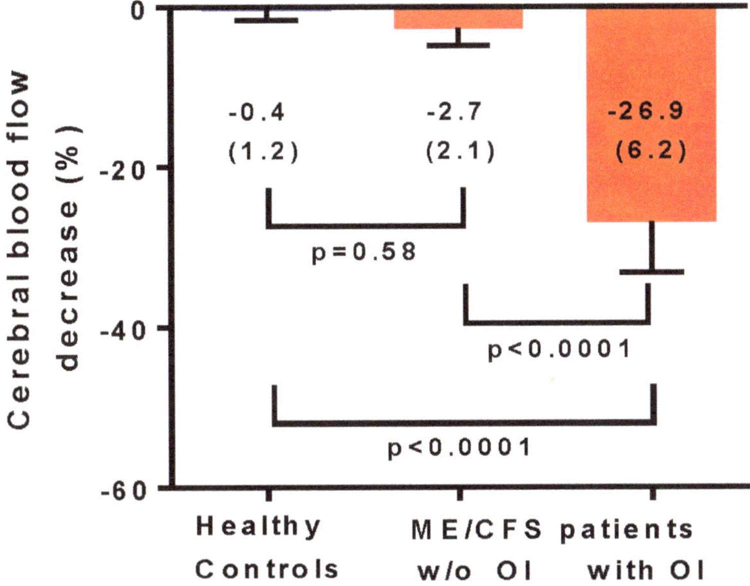

Figure 2. Cerebral blood flow decrease in percentage reduction between the supine position and the sitting position in healthy controls ($n = 15$), in severe ME/CFS patients without orthostatic intolerance in daily life ($n = 10$), and in severe ME/CFS patients with orthostatic intolerance in daily life ($n = 90$). OI: orthostatic intolerance; w/o: without; ME/CFS: myalgic encephalomyelitis/chronic fatigue syndrome; *p*-values shown are the result from an ordinary one-way analysis of variance (ANOVA), with post hoc Tukey's test. Results are presented as percentage cerebral blood flow decrease (standard deviation).

Table 3 shows the baseline, supine, and sitting test results for severe ME/CFS patients with ($n = 37$) and without ($n = 63$) fibromyalgia. Patients with fibromyalgia were significantly older and had a higher BMI. Other variables did not differ significantly. The percent decline in cerebral blood flow was −23.9 (9.4)% and −24.9 (9.5)% for patients with and without fibromyalgia respectively ($p = 0.51$) Comparing these data in patients with and without pain-medication and/or benzodiazepines showed similar results (data not shown).

Table 4 shows the baseline and supine and sitting test results for severe ME/CFS patients with ($n = 34$) and without ($n = 66$) POTS in their history. Patients with POTS in their history were significantly younger. Other variables did not differ significantly. The percent decline in cerebral blood flow was −22.3 (9.7)% and −28.8 (7.2)% for patients without and with POTS in their history respectively ($p < 0.001$).

Table 3. Baseline and orthostatic stress test results in severe ME/CFS patients with fibromyalgia ($n = 37$) and in severe ME/CFS patients without fibromyalgia ($n = 63$).

	FM Plus ($n = 37$)	FM Minus ($n = 63$)	*p*-Value
Male/female	2/35	10/53	0.12
Age (years)	43 (13)	36 (11)	<0.01
Disease duration (years) [#]	15 (8.5–23.5)	13 (8–19)	0.46
Height (cm)	172 (7)	174 (8)	0.15
Weight (kg)	78 (17)	70 (15)	0.02
BMI (kg/m^2)	26.3 (5.5)	23.1 (5.2)	<0.01
Heart rate supine (bpm)	71 (10)	78 (17)	0.02
Heart rate sitting (bpm)	81 (12)	91 (21)	0.02
Systolic blood pressure supine (mmHg)	125 (19)	124 (16)	0.95
Systolic blood pressure sitting (mmHg)	138 (20)	132 (16)	0.15
Diastolic blood pressure supine (mmHg)	82 (11)	79 (11)	0.24
Diastolic blood pressure sitting (mmHg)	90 (12)	86 (10)	0.09
Cerebral blood flow supine (mL/min)	639 (131)	627 (116)	0.62
Cerebral blood flow sitting (mL/min)	481 (88)	470 (101)	0.56
Cerebral blood flow % change sitting vs. supine	−23.9 (9.4)	−24.9 (9.5)	0.60

FM: fibromyalgia; [#] Data with median (IQR). ME/CFS: myalgic encephalomyelitis/chronic fatigue syndrome.

Table 4. Demographic and orthostatic stress test results in severe ME/CFS patients with and without a history of POTS.

	With POTS ($n = 34$)	Without POTS ($n = 66$)	*p*-Value
Male/female	2/32	10/56	0.18
Age (years)	31 (9)	42 (12)	<0.0001
Disease duration (years) [#]	13 (8–18.5)	14.5 (–23)	0.47
Orthostatic intolerance in daily life	34 (100%)	56 (85%)	0.02 [#]
Height (cm)	174 (7)	174 (8)	0.98
Weight (kg)	74 (18)	72 (15)	0.63
BMI (kg/m^2)	24.7 (6.3)	24.1 (5.1)	0.59
Heart rate supine (bpm)	77 (18)	74 (13)	0.30
Heart rate sitting (bpm)	89 (21)	86 (17)	0.48
Systolic blood pressure supine (mmHg)	126 (16)	124 (18)	0.67
Systolic blood pressure sitting (mmHg)	136 (15)	134 (19)	0.64
Diastolic blood pressure supine (mmHg)	82 (12)	79 (10)	0.30
Diastolic blood pressure sitting (mmHg)	89 (11)	87 (10)	0.64
Cerebral blood flow supine (mL/min)	656 (121)	619 (121)	0.15
Cerebral blood flow sitting (mL/min)	466 (96)	478 (96)	0.57
Cerebral blood flow % change sitting vs. supine	−28.8 (7.2)	−22.3 (9.7)	<0.001

POTS: postural orthostatic tachycardia syndrome; [#] Data with median (IQR). ME/CFS: myalgic encephalomyelitis/chronic fatigue syndrome. [#] Chi-square testing (2 × 2 table).

Figure 3 illustrates the percent reduction in cerebral blood flow between supine and sitting in severe ME/CFS patients based on the presence or absence of a history of POTS. Cerebral blood flow reductions in patients without POTS were significantly lower compared to those who had POTS ($p = 0.0008$).

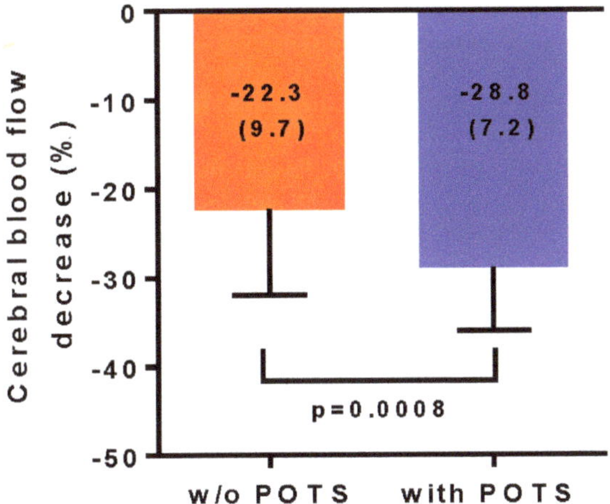

Figure 3. Cerebral blood flow reductions in percent reduction between the supine position and the sitting position in severe ME/CFS patients with ($n = 34$) and without ($n = 66$) a history of Postural Orthostatic Tachycardia Syndrome (POTS). A history of POTS could result from self-reporting, from the diagnosis with a standing test or a tilt-table test elsewhere in the past, or by tilt-table testing in our clinic in the past. POTS: postural orthostatic tachycardia syndrome; w/o: without. ME/CFS: myalgic encephalomyelitis/chronic fatigue syndrome; *p*-values shown are the result from an unpaired t-test. Results are presented as percentage cerebral blood flow decrease (standard deviation).

4. Discussion

This is the first study showing that in severe ME/CFS patients, a significant reduction in cerebral blood flow (mean reduction −24.5%) can be provoked by sitting. In contrast, in healthy controls, only minimal change in cerebral blood flow reduction was observed (mean reduction −0.4%). Studying sitting as an orthostatic stressor with extracranial Doppler echography has not been described before. However, not all patients showed this abnormal response. In patients without daily life orthostatic intolerance complaints, a minimal reduction in cerebral blood flow of −2.7% was found, which did not differ significantly from healthy controls. In contrast, in patients with orthostatic intolerance complaints in daily life, sitting resulted in a 26.9% reduction in cerebral blood flow. This is similar to the 27% reduction in cerebral blood flow in an exploratory study with 20 degrees tilting during 15 min in severe ME/CFS patients [7]. It also is comparable to the 26% reduction observed after 30 min of 70 degree head-up tilt in a less severely affected population of ME/CFS patients [2]. We observed no difference in cerebral blood flow results whether patients had concomitant fibromyalgia or not. In patients with a previous diagnosis of POTS, cerebral blood flow reductions while seated were significantly larger than in patients who did not have POTS. It remains to be determined whether a sitting test would be adequate for the diagnosis of orthostatic intolerance/significant cerebral blood flow reduction in less severely affected ME/CFS patients. Furthermore, it remains to be determined whether a standard head-up tilt test shows more abnormal cerebral blood flow reduction than a sitting test in severe ME/CFS patients without orthostatic intolerance symptoms in daily life.

In the present study, a difference close to a significance of 0.05 was found between the sitting systolic blood pressure between healthy controls and severe ME/CFS patients: 122 (18) mmHg for healthy controls and 134 (18) mm Hg for ME/CFS patients. This difference might be explained by the fact that in severe ME/CFS patients, a decrease in blood volume may be present [8–11]. Moreover, we showed that cardiac output reductions during tilt-table testing were significantly more pronounced in ME/CFS patients compared to controls [12]. Both findings may result in a reduced venous return, and subsequent vasoconstriction and elevation of the measured, sitting, and systolic blood pressure. However, this remains to be proven. Also, it is unknown whether the duration of sitting before the test influences the reduction in venous return, cardiac output, and cerebral blood flow reduction in a ME/CFS patient population. In the present study, we did not find a correlation between the estimated duration of sitting before the start of the test and the percent reduction in cerebral blood flow (data not shown).

Miwa et al. described in a study of 44 ME/CFS patients (11 males and 33 females) that during an active standing test 40 (91%) reported orthostatic intolerance symptoms, and 30 (69%) reported orthostatic symptoms during a 10 min active sitting test [3]. Sitting as trigger for the development of orthostatic intolerance complaints was reported with increasing severity of the disease.

There are limited data on the effects of sitting using transcranial Doppler (TCD). Carter et al. studied the effects of 4-h sitting with and without exercise breaks in 10 healthy males [13]. Although they did not perform a statistical analysis of the difference of the change of 4 h sitting to supine afterwards, the data showed that middle cerebral artery velocity increased from 53.8 (1.6) cm/sec after 4 h of sitting to 55.5 (2.1) cm/sec in the supine position afterwards, without intervening exercise breaks. Their raw data imply that sitting middle cerebral artery velocity was 3% lower than supine middle cerebral artery velocity. Wheeler et al. demonstrated that prolonged sitting without exercise and/or breaks resulted in a significant reduction of the cerebral blood flow velocities from the middle cerebral arteries by transcranial Doppler over an 8 h period of sitting observation in 12 elderly obese subjects [14]. Perdomo et al. also observed a significant drop in middle cerebral artery velocities after 3 h and 40 min sitting in 25 subjects with pre-/stage 1 hypertension, with a return to baseline values at the end of the second working period [15]. No studies have reported transcranial Doppler measurements during sitting in ME/CFS patients. Unlike extracranial Doppler measurements, which calculate total CBF inflow, TCD can only measure CBF velocity, which is affected by vessel diameter, and in the presence of vasoconstriction can underestimate changes in CBF.

In the present study, a significantly higher reduction in cerebral blood flow was found in severe ME/CFS patients with POTS compared to severe ME/CFS patients without a history of POTS. The results are consistent with our previous study where POTS patients had a significant higher reduction in CBF at the end of 30 min of head-up tilt compared to patients with a normal heart rate and blood pressure response: -29 (6)% vs. -24 (10)%; $p < 0.0005$ [2]. Several previous studies have described a higher decline in cerebral blood flow as measured by TCD, but compared the subjects only to healthy controls and not to other ME/CFS patients with different hemodynamic tilt test results attributable to POTS [16,17]. Consistent with earlier studies, our ME/CFS patients with a history of POTS were younger than those without POTS [18,19]

4.1. Clinical Implications

Patients are advised to rest when they experience orthostatic intolerance complaints. Our findings of a clinically significant cerebral blood flow reduction while seated suggest that sitting may not be adequate enough to resolve symptoms of orthostatic intolerance in some patients. Importantly, none of the patients developed POTS or delayed orthostatic hypotension while sitting. This suggests that this form of stress testing may be less taxing than tilt-table testing or standing.

4.2. Limitations

This study only included severe ME/CFS patients. Sitting cerebral blood flow needs to be measured in less severely affected ME/CFS patients and compared to the blood flow measured in active standing and standard 70 degree angle tilt-testing. Whether differences in disease severity lead to differences in seated cerebral blood flow reduction needs to be studied in the future. While sitting was sufficient to provoke a clinically and statistically significant reduction in cerebral blood flow, we do not know whether it would have dropped further during head-up tilt. The measurement of supine cerebral blood flow was performed 10–15 min after patients had been sitting upright for at least an hour. We cannot exclude the possibility that there was a delayed recovery from the orthostatic stress, and that the supine measurement might have been different had patients not been sitting beforehand. However, supine measures did not differ between ME/CFS patients and controls, so we think that the supine measurements were not likely to have been affected by a carryover effect. Finally, while it is reasonable to expect that the sitting test would be less taxing than a standard tilt test, and therefore is less likely to provoke post-exertional malaise, this hypothesis remains to be tested.

5. Conclusions

This study demonstrates that a sitting test can result in a clinically significant reduction in cerebral blood flow in patients with severe ME/CFS and that sitting, therefore, should be considered as an important orthostatic stressor for the severe ME/CFS patient group. This method of orthostatic testing has the potential to improve the assessment of the prevalence of orthostatic intolerance in severely affected ME/CFS patients who are reluctant or cannot tolerate a standard 70 degree tilt test. In this patient population, a milder orthostatic stress was able to confirm cerebral blood flow abnormalities in the absence of heart rate and blood pressure abnormalities.

Author Contributions: C.M.C.v.C., P.C.R., and F.C.V. conceived the study, C.M.C.v.C. and F.C.V. collected the data, C.M.C.v.C. performed the primary data analysis and F.C.V., and P.C.R. performed secondary data analyses. All authors were involved in the drafting and review of the manuscript. All authors have read and agreed to the published version of the manuscript.

Funding: This research received no external funding.

Acknowledgments: Rowe is supported by the Sunshine Natural Wellbeing Foundation Professorship.

Conflicts of Interest: The authors declare no conflict of interest.

Appendix A

Internal carotid artery and vertebral artery Doppler flow velocity frames were acquired by one operator, using a Vivid-I system (GE Healthcare, Hoevelaken, the Netherlands) equipped with a 6–13 MHz linear transducer. Flow data of the internal carotid artery on the right and on the left side were obtained ~1.0–1.5 cm distal to the carotid bifurcation. The vertebral artery data were obtained at the C3–C5 level. Care was taken to ensure the insonation angle was less than 60 degrees, that the sample volume was positioned in the center of the vessel, and that it covered the width of the vessel. High resolution B mode images, color Doppler images, and the Doppler velocity spectrum (pulsed wave mode) were recorded in one frame. The order of imaging was fixed: left internal carotid artery, left vertebral artery, right internal carotid artery, and right vertebral artery. At least two consecutive series of six frames per artery were recorded.

Blood flows of the internal carotid and vertebral arteries were calculated offline by a second investigator who was unaware of the patient severity status. Vessel diameters were manually traced on B-mode images, from the intima to the opposite intima. The surface area was calculated as follows: the peak systolic and end diastolic diameters were measured, and the mean diameter was calculated as: mean diameter = (peak systolic diameter $\times$ 1/3) + (end diastolic diameter $\times$ 2/3) [20]. The surface area = π * (diameter/2)^2. Blood flow in each vessel was calculated from the mean blood flow velocities times the vessel surface area and expressed in ml/min. Flow in the individual arteries was calculated in 3–6 cardiac cycles and the data were averaged. Total cerebral blood flow was calculated by adding the flow of the four arteries. We previously demonstrated that this methodology had good intra- and inter-observer variability [6].

References

1. Institute of Medicine (IOM) (Ed.) *Beyond Mayalgic Cncephalomyelitis/Chronic Fatigue Syndrome: Redefining an Illness*; The National Academies Press: Washington, DC, USA, 2015.
2. van Campen, C.M.C.; Verheugt, F.W.A.; Rowe, P.C.; Visser, F.C. Cerebral blood flow is reduced in ME/CFS during head-up tilt testing even in the absence of hypotension or tachycardia: A quantitative, controlled study using Doppler echography. *Clin. Neurophysiol. Pract.* **2020**, *5*, 50–58. [CrossRef] [PubMed]
3. Miwa, K.; Inoue, Y. The etiologic relation between disequilibrium and orthostatic intolerance in patients with myalgic encephalomyelitis (chronic fatigue syndrome). *J. Cardiol.* **2018**, *72*, 261–264. [CrossRef] [PubMed]
4. Carruthers, B.M.; van de Sande, M.I.; De Meirleir, K.L.; Klimas, N.G.; Broderick, G.; Mitchell, T.; Staines, D.; Powles, A.C.P.; Speight, N.; Vallings, R.; et al. Myalgic encephalomyelitis: International Consensus Criteria. *J. Intern. Med.* **2011**, *270*, 327–338. [CrossRef] [PubMed]
5. Fukuda, K.; Straus, S.E.; Hickie, I.; Sharpe, M.C.; Dobbins, J.G.; Komaroff, A. The chronic Fatigue syndrome: A comprehensive approach to its definition and study. *Ann. Intern. Med.* **1994**, *121*, 953–959. [CrossRef] [PubMed]
6. van Campen, C.M.C.; Verheugt, F.W.A.; Visser, F.C. Cerebral blood flow changes during tilt table testing in healthy volunteers, as assessed by Doppler imaging of the carotid and vertebral arteries. *Clin. Neurophysiol. Pract.* **2018**, *3*, 91–95. [CrossRef] [PubMed]
7. van Campen, C.M.C.; Rowe, P.C.; Visser, F.C. Cerebral Blood Flow Is Reduced in Severe Myalgic Encephalomyelitis/Chronic Fatigue Syndrome Patients During Mild Orthostatic Stress Testing: An Exploratory Study at 20 Degrees of Head-Up Tilt Testing. *Healthcare* **2020**, *8*, 169. [CrossRef] [PubMed]
8. van Campen, C.M.C.; Rowe, P.C.; Visser, F.C. Blood Volume Status in ME/CFS Correlates With the Presence or Absence of Orthostatic Symptoms: Preliminary Results. *Front. Pediatr.* **2018**, *6*, 4. [CrossRef] [PubMed]
9. Streeten, D.H.P.; Bell, D.S. Circulating Blood Volume in Chronic Fatigue Syndrome. *J. Chronic Fatigue Syndr.* **1998**, *4*, 3–11. [CrossRef]
10. Lin, Y.; Liu, H.L.; Fang, J.; Yu, C.H.; Xiong, Y.K.; Yuan, K. Anti-fatigue and vasoprotective effects of quercetin-3-O-gentiobiose on oxidative stress and vascular endothelial dysfunction induced by endurance swimming in rats. *Food Chem. Toxicol.* **2014**, *68*, 290–296. [CrossRef] [PubMed]
11. Hurwitz, B.E.; Coryell, V.T.; Parker, M.; Martín, P.; Laperriere, A.; Klimas, N.G.; Sfakianakis, G.N.; Bilsker, M.S. Chronic fatigue syndrome: Illness severity, sedentary lifestyle, blood volume and evidence of diminished cardiac function. *Clin. Sci.* **2010**, *118*, 125–135. [CrossRef] [PubMed]
12. van Campen, C.M.C.; Visser, F.C. The abnormal Cardiac Index and Stroke Volume Index changes during a normal Tilt Table Test in ME/CFS patients compared to healthy volunteers, are not related to deconditioning. *J. Thromb. Circ.* **2018**, *2018*, 1–8.
13. Carter, S.E.; Draijer, R.; Holder, S.M.; Brown, L.; Thijssen, D.H.J.; Hopkins, N.D. Regular walking breaks prevent the decline in cerebral blood flow associated with prolonged sitting. *J. Appl. Physiol.* **2018**, *125*, 790–798. [CrossRef] [PubMed]
14. Wheeler, M.J.; Dunstan, D.W.; Smith, B.; Smith, K.J.; Scheer, A.; Lewis, J.; Naylor, L.H.; Heinonen, I.; Ellis, K.A.; Cerin, E.; et al. Morning exercise mitigates the impact of prolonged sitting on cerebral blood flow in older adults. *J. Appl. Physiol.* **2019**, *126*, 1049–1055. [CrossRef] [PubMed]
15. Perdomo, S.J.; Gibbs, B.B.; Kowalsky, R.J.; Taormina, J.M.; Balzer, J.R. Effects of Alternating Standing and Sitting Compared to Prolonged Sitting on Cerebrovascular Hemodynamics. *Sport Sci. Health* **2019**, *15*, 375–383. [CrossRef] [PubMed]
16. Ocon, A.J.; Medow, M.S.; Taneja, I.; Clarke, D.; Stewart, J.M. Decreased upright cerebral blood flow and cerebral autoregulation in normocapnic postural tachycardia syndrome. *Am. J. Physiol. Heart Circ. Physiol.* **2009**, *297*, H664–H673. [CrossRef] [PubMed]
17. Stewart, J.M.; Medow, M.S.; Messer, Z.R.; Baugham, I.L.; Terilli, C.; Ocon, A.J. Postural neurocognitive and neuronal activated cerebral blood flow deficits in young chronic fatigue syndrome patients with postural tachycardia syndrome. *Am. J. Physiol. -Heart Circ. Physiol.* **2012**, *302*, H1185–H1194. [CrossRef] [PubMed]
18. Lewis, I.; Pairman, J.; Spickett, G.; Newton, J.L. Clinical characteristics of a novel subgroup of chronic fatigue syndrome patients with postural orthostatic tachycardia syndrome. *J. Intern. Med.* **2013**, *273*, 501–510. [CrossRef] [PubMed]

19. Reynolds, G.K.; Lewis, D.P.; Richardson, A.M.; Lidbury, B.A. Comorbidity of postural orthostatic tachycardia syndrome and chronic fatigue syndrome in an Australian cohort. *J. Intern. Med.* **2014**, *275*, 409–417. [CrossRef] [PubMed]

20. Sato, K.; Ogoh, S.; Hirasawa, A.; Oue, A.; Sadamoto, T. The distribution of blood flow in the carotid and vertebral arteries during dynamic exercise in humans. *J. Physiol.* **2011**, *589 (Pt 11)*, 2847–2856. [CrossRef]

Review

Back to the Future? Immunoglobulin Therapy for Myalgic Encephalomyelitis/Chronic Fatigue Syndrome

Helen Brownlie [1] and Nigel Speight [2,*]

1   Independent Researcher and Former Local Government Officer, Social Policy and Research, Glasgow G2 4P, UK; hm.brownlie@virginmedia.com
2   Paediatrician and Independent Researcher, Durham DH1 1QN, UK
*   Correspondence: speight@doctors.org.uk

Abstract: The findings of controlled trials on use of intravenous immunoglobulin G (IV IgG) to treat myalgic encephalomyelitis/chronic fatigue syndrome (ME/CFS) are generally viewed as representing mixed results. On detailed review, a clearer picture emerges, which suggests that the potential therapeutic value of this intervention has been underestimated. Our analysis is consistent with the propositions that: (1) IgG is highly effective for a proportion of patients with severe and well-characterised ME/CFS; (2) responders can be predicted with a high degree of accuracy based on markers of immune dysfunction. Rigorous steps were taken in the research trials to record adverse events, with transient symptom exacerbation commonly experienced in both intervention and placebo control groups, suggesting that this reflected the impact of participation on people with an illness characterised by post-exertional symptom exacerbation. Worsening of certain specific symptoms, notably headache, did occur more commonly with IgG and may have been concomitant to effective treatment, being associated with clinical improvement. The findings emerging from this review are supported by clinical observations relating to treatment of patients with severe and very severe ME/CFS, for whom intramuscular and subcutaneous administration provide alternative options. We conclude that: (1) there is a strong case for this area of research to be revived; (2) pending further research, clinicians would be justified in offering a course of IgG to selected ME/CFS patients at the more severe end of the spectrum. As the majority of trial participants had experienced an acute viral or viral-like onset, we further suggest that IgG treatment may be pertinent to the care of some patients who remain ill following infection with SARS-CoV-2 virus.

Keywords: immunoglobulin; myalgic encephalomyelitis; chronic fatigue syndrome; viral onset; cell-mediated immunity; post-acute sequelae of COVID-19; long-COVID

Citation: Brownlie, H.; Speight, N. Back to the Future? Immunoglobulin Therapy for Myalgic Encephalomyelitis/Chronic Fatigue Syndrome. *Healthcare* **2021**, *9*, 1546. https://doi.org/10.3390/healthcare9111546

Academic Editors: Kenneth J. Friedman, Lucinda Bateman and Kenny Leo De Meirleir

Received: 23 August 2021
Accepted: 20 October 2021
Published: 12 November 2021

Publisher's Note: MDPI stays neutral with regard to jurisdictional claims in published maps and institutional affiliations.

1. Introduction

Interest in researching immunoglobulin (IgG) therapy in patients with myalgic encephalomyelitis/chronic fatigue syndrome (ME/CFS) began in the mid-1980s, prompted by emerging evidence of immunoregulatory defects including disordered cell-mediated immunity (CMI) [1–9] and immunoglobulin subclass deficiencies [1,10–12]. A documented viral onset in some patients [2,13,14], the demonstration of enteroviral RNA in muscle tissue [15] and the presence of enteroviral antigen in serum [16] were viewed as supporting the hypothesis that ME/CFS may develop as a consequence of failed clearance of viral or other antigens [13,17] and strengthened the case for investigating an intervention predicated on the presence of immunologic dysfunction. High-dose intravenous (IV) IgG was known to ameliorate other disorders of immune regulation [1,18–20] and positive findings in a crossover study of intramuscular IgG therapy in patients with 'chronic mononucleosis syndrome' [21], in tandem with numerous individual case reports of beneficial outcome following IV IgG in patients with ME/CFS [18], provided further encouragement for this line of enquiry.

Four double-blind randomised controlled trials (RCTs) of IV IgG followed. Reports of effectiveness diverged [13,14,22–24], with findings collectively assessed as *"mixed"* [25] and *"inconclusive"* [26]. Following publication in 1997 of the last of these, [24] asserting an *"ineffective"* conclusion, interest in researching this treatment in patients with ME/CFS waned.

Recently, there have been renewed indications of research interest and the potential relevance of IgG in the treatment of patients with ME/CFS.

A detailed investigation of the effects of immunoglobulins on adrenergic receptors and immune function in ME/CFS was published in 2020 [27]. Additionally, in 2020, pilot work revealed that an autoimmune-associated small-fibre polyneuropathy (aaSFPN)—a condition known to respond to IV IgG—may occur in a substantial proportion of patients, leading the researcher to conclude: *"it may be important to identify ME/CFS patients who present with comorbid aaSFPN"* [28]. In March 2021, a paper reporting on immune-related pathogenesis revealed that several immunoglobulin genes are significantly increased in ME/CFS patients compared with controls [29]. On presenting at the 2020 International Association for Chronic Fatigue Syndrome/Myalgic Encephalomyelitis (IACFS/ME) Conference, the lead author expressed the clear opinion that findings indicate that therapeutic immunoglobulin studies are warranted [30]. Advice on the administration of IgG features in treatment recommendations produced by the US ME/CFS Clinician Coalition in February 2021 [31], while IgG is included in a summary of treatment approaches in an August 2021 paper from the Mayo Foundation for Medical Education and Research, setting out consensus recommendations for diagnosis and management [32].

Against this background, and with the recent lifting of a longstanding ban on use of UK-sourced blood plasma in the manufacture of immunoglobulins poised to ease global shortages of this product [33] and renewed interest in medium- to long-term immune dysfunctions following viral infection as post-acute sequelae of COVID-19 (PASC) [34,35], a fresh appraisal of this research, supported by published observations from clinical practice, is timely.

2. Nomenclature and Case Definitions

The research trials reviewed in this paper were published between 1990 and 1999 and refer to the disorder studied as 'chronic fatigue syndrome' or 'the chronic fatigue syndrome' (CFS). At that time, CFS was a recent entry to the medical lexicon, having been introduced in 1988 by the United States Centers for Disease Control (US CDC) with the stated intention of replacing the terms that were then current, i.e., 'chronic Epstein–Barr virus syndrome' or 'chronic mononucleosis', and variations [36]. 'Myalgic encephalomyelitis' (ME) was not utilised in the US but had been in common usage in the United Kingdom (UK), as was 'post-viral fatigue syndrome' (PVFS).

Hence, all of these identifiers are variously used to refer to the disorder studied in the respective papers cited from this period. The present paper will refer to 'ME/CFS', in keeping with current usage.

Amidst the range of possible descriptive terms, case definition for selection of patients remains the most crucial consideration with regard to assessing research: implications will be discussed.

3. The Immunoglobulin Trials

Four double-blind randomised placebo-controlled trials (RCTs) on the use of IV IgG to treat patients suffering from ME/CFS have been conducted. Results from the first two were published together in the American Journal of Medicine in November 1990 [13,14]. The respective authors had reached quite different conclusions. One study reported that immunoglobulin is effective in a *"significant number of patients"*, the other that IV IgG *"is unlikely to be of clinical benefit in CFS"*. This pattern was later repeated. In January 1997, an RCT reported a beneficial effect of IV IgG on adolescent patients [22]; later that year, a further, and to date final, trial reported that IV IgG *"is ineffective"* [24].

It is this apparently opaque and divergent picture that the present review seeks to illuminate. Scrutiny of recorded outcomes reveals that the findings of these trials were not mutually incompatible. To establish this, it is essential not to confine consideration to the most prominently reported conclusions, but to take into account features of the respective studies and their specific findings.

For simplicity, we shall refer to these four trials as Studies 1 [13], 2 [14], 3 [22] and 4 [24]. Tables 1 and 2 illustrate trial characteristics and participants, respectively.

Table 1. IV immunoglobulin trial characteristics.

Trial Characteristic	Study 1 Lloyd et al., 1990 [13]	Study 2 Peterson et al., 1990 [14]	Study 3 Rowe 1997 [22]	Study 4 Vollmer-Conna et al., 1997 [24]
Immunoglobulin G (IgG) dosage	2 g per kg	1 g per kg	1 g per kg, to a maximum of 60 g	2 g, 1 g, or 0.5 g per kg
IgG Brand	Intragram™	Gammagard®	Intragram™	Intragram™
Placebo infusion	unclear	1% albumin solution	1% albumin in a 10% weight/volume maltose solution [1]	1% albumin in a 10% weight/volume maltose solution [1]
Number and frequency of infusions	3 at monthly intervals	6 at 30 day intervals	3 at monthly intervals	3 at monthly intervals
Participants recruited	*n* = 49 23 IgG 26 Placebo	*n* = 30 15 IgG 15 placebo	*n* = 71 36 IgG 35 Placebo	*n* = 99 23 at 2 g IgG per kg 28 at 1 g IgG per kg 22 at 0.5 g IgG per kg 26 placebo
Drop outs	2 from the IgG group	1 from each group	none 1 lost to follow-up	3 IgG recipients (dosage not specified)
Findings based on	*n* = 49 'intention to treat'	*n* = 28	*n* = 70	*n* = 99 'intention to treat'
Timing of final follow-up	3 months after final infusion + 'responders' reviewed at 6 and 12 months	within 48 h of final infusion	6 months after final infusion [1]	3 months after final infusion

[1] The IgG brand used, Intragram™, is likewise prepared in a 10% weight/volume maltose solution. A report including feedback on participant experiences over the five years following participation was published in 1999 [23]; however, it was not possible to carry out a placebo vs. intervention group comparison at this time as many placebo group patients were subsequently treated with IgG.

Table 2. IV immunoglobulin trial participant characteristics.

Participant Characteristic	Study 1 Lloyd et al., 1990 [13]	Study 2 Peterson et al., 1990 [14]	Study 3 Rowe 1997 [22]	Study 4 Vollmer-Conna et al., 1997 [24]
Age range	16–63	Eldest 74, youngest not stated	11–18	16–73
Duration of illness	Range 1–15 years median 47 months	Mean 3.8 years	mean in months: IgG 19.2 placebo 16.9	mean in years: IgG 2 g per kg 5 IgG 1 g per kg 7 IgG 0.5 g per kg 6 placebo 7
Illness onset	Acute viral-like onset in 37 (76%)	All but one began with episode of an acute viral-like illness	Over 85% developed the illness secondary to an infection	Acute viral-like onset in 75 (76%)
Severity at baseline	*"well-characterised, severe, and long-standing CFS"*	Not directly reported, but apparently less severe that was the norm for clinic (see text for details)	*"by selection at the more severe end of the spectrum"*	77% able to work 28% working more than half time; average 5 h of non-sedentary activity daily

Table 2. *Cont.*

Participant Characteristic	Study 1 Lloyd et al., 1990 [13]	Study 2 Peterson et al., 1990 [14]	Study 3 Rowe 1997 [22]	Study 4 Vollmer-Conna et al., 1997 [24]
Recruitment criteria	Australian 1988 [37] (Lloyd et al.), in which marked exercise-induced muscle fatigue with prolonged recovery time is essential	Centers for Disease Control (CDC) 1988 [36] (Holmes et al.)	CDC 1994 [38] (Fukada et al.) Also: identifiable time of onset; exacerbated by minor exercise	Lloyd et al., 1990 [37] and Schluderberg et al., 1992 [38] [1]
Cohort illness characteristics	All chronic and persisting; none relapsing and remitting	All but one reported prolonged post-exertional fatigue In addition to fatigue, had been experiencing an average of 8.8 of the CFS symptoms listed by the CDC [36]		*"probably, more heterogeneous than that enrolled in our earlier study"*

[1] An account of the National Institutes of Health workshop which initiated the development of the CDC criteria, subsequently published in 1994 [39].

The measures employed to record the nature and severity of symptoms experienced and their impact on daily life varied from study to study, comprising a mix of established protocols [40–49], modifications of established protocols [50], and rating methods devised by the authors. These are listed in Appendix A.

All trials obtained measures of cell-mediated immunity and, with the exception of Study 2, these were considered at baseline and at follow-up stage; Study 2 focused on humoral immunity (IgG subclass levels) when assessing outcomes. The tests of cell-mediated immunity performed in the respective studies are listed at Appendix B.

All trials were double blinded; however, details of exactly how the blinding of investigators was conducted are not provided. Studies 1 and 3 record that the follow-up assessment was conducted by a physician who was unaware of the treatment regimen.

All four trials took rigorous steps to record and respond to any adverse experiences. This aspect is summarised and discussed at Section 5.1 below.

3.1. Study 1 (Australia) Lloyd et al., 1990

This trial involved 49 participants aged 16–63, randomised to receive a total of three infusions of IV IgG at 2 g per kilogram of bodyweight ($n = 23$) or placebo ($n = 26$), administered at monthly intervals [13].

3.1.1. Cell-Mediated Immunity at Baseline

Two-thirds of participants (33) had reduced cutaneous delayed-type hypersensitivity (DTH) responses, while 40 (82%) had abnormal CMI in this form and/or as evidenced by T-cell lymphopenia. CD8 cell count was below the normal reference range for the authors' laboratory in 18 participants (37%) and CD4 in nine (18%), while the absolute T-cell count was below the reference range in 21 (43%).

3.1.2. Summary of Reported Findings

Three months after the final infusion, 10 immunoglobulin recipients (43%) and three placebo recipients (12%) were designated as 'responders' on blinded evaluation by physician following participant interviews where specific details of employment, social and leisure activities were obtained. The higher incidence of 'responders' in the IV IgG group was statistically significant.

On this basis, the authors conclude: *"The results of this study demonstrate that a significant proportion of patients (43%) with well-characterized, severe, and long standing CFS responded to high-dose intravenous immunoglobulin therapy."*

It is notable that this categorisation of participants as 'responders' or 'non-responders' was not a matter of fine judgment. All 'responders' had improved considerably: *"This response was characterized by recommencement of employment, leisure, and social activities, as well as by a significant reduction in physical and psychologic morbidity and by an improvement in cell-mediated immunity."* (see Box 1 below).

Box 1. Employment, Leisure and Social Activities—three months after final infusion, Lloyd et al., 1990 [13]; IgG 'responders' *n* = 10; placebo 'responders' *n* = 3.

Employment and Household Duties:
Six had resumed pre-morbid employment status in full-time occupation or housework; three had been working on a part-time basis and were now able to work full-time; one had been unemployed before contracting ME/CFS and remained so, but was now able to resume his leisure activity of bushwalking. *One had been in full-time work before trial commencement and remained so; two moved from part-time to full-time work.*
Leisure Activities:
None had been involved in any leisure activities prior to the trial; all but one were now participating in leisure activities including tennis, surfing and gardening; the person who did not resume any leisure activities was now working full-time as a labourer.
Two resumed tennis and swimming, respectively; the third person, described as a 'housewife', did not participate in leisure activities before or after the trial.
Social Activities:
All had increased participation in social activities; seven were now involved in social events at least once a week, three who had previously managed no social activity were now participating in social events less often than once a week.
All had increased participation, one to at least once a week.

In contrast, the 'non-responders' had experienced little or no change, a pattern that emerged consistently across the range of outcome measures: *"The distribution of the change in the self-report measures of physical and psychologic morbidity fell in a dichotomous (i.e., almost "all or nothing") pattern, consistent with the categorized "response" or "no response" assessment from the physician's interviews."*

Every 'responder'—whether in the IgG or placebo group—who underwent immunologic follow-up was found to have experienced a significant improvement in cell-mediated immunity (CMI). These twelve patients (one placebo 'responder' was unable to undergo immunological testing at follow up) recorded a significantly greater improvement in CD4 (helper T) cell count when compared with the 36 participants designated as 'non-responders', with the group average up 37% versus down 3% ($p < 0.01$). Reduced DTH responses, present in eight at entry, had resolved to normal values for all but one.

The authors observe: *"The association between recovery from CFS and resolution of abnormalities in cell-mediated immunity is strengthened by the demonstration of immunologic improvement in both the placebo responders tested at follow-up".*

Among the 23 IgG recipients, the percentage change in score on a modified Quality of Life visual analogue scale (see Appendix A) was positively correlated with improvement in CD4 count ($r = 0.4$, $p < 0.05$) and with improved DTH ($r = 0.3$, $p = 0.08$).

Most of those who benefitted from IgG had improved promptly: *"in 8 of the 10 IgG 'responders', improvement in symptoms and function was noted within 3 weeks of the first infusion and tended to increase incrementally after subsequent infusions".*

Further therapy was able to reverse any subsequent decline. Clinical follow-up identified that, by one year after the final infusion, eight of the ten IgG 'responders' *"had substantial return of symptoms and disability".* These patients received further IgG therapy *"that was associated with an identical remission of symptoms in improvement in function in each case".*

3.1.3. Discussion of Study 1

PARTICIPANT CHARACTERISTICS The assertion that participants were patients whose illness was *"well-characterized, severe and longstanding"* is borne out by the recruitment criteria employed [51]. All were required to experience *"marked exercise-aggravated muscle fatigue, with abnormally prolonged recovery time"*, ensuring that a core defining feature of ME/CFS was present.

OUTCOMES Described as *"almost 'all or nothing'"*, the distribution of outcomes *"prevented significant differences in physical, psychologic, or immune measures from being demonstrated between the immunoglobulin and placebo recipients when analyzed as a group"*. Therefore, confining analysis to a comparison of intervention and control group means on conclusion of the trial—a widely accepted method of judging efficacy—would have concealed the significantly greater likelihood of the patient experiencing clear improvement if treated with IgG.

PREDICTORS OF OUTCOME Predictors of improvement were found to lie in immunologic measures recorded at entry, particularly a lowed CD4 lymphocyte count: *"supporting the concept that cellular immunity is important in both the pathogenesis and response to treatment in patients with CFS."* The analysis conducted implies that ME/CFS patients who have scope to benefit from IgG may be predicted in advance with a high degree of accuracy. A separate analysis based on age, sex and disease duration did not significantly predict response.

3.2. Study 2 (USA) Peterson et al., 1990

This trial involved 30 participants aged up to 74, with the IV IgG group ($n = 15$) receiving half the dose used in Study 1, at 1 g per kilogram of bodyweight. A total of six infusions were administered, at 30 day intervals [14].

3.2.1. Cell-Mediated Immunity at Baseline

Five participants (18%) had reduced numbers of T3 (pan T) cells; five had reduced numbers of CD4 cells. CD8 cells were low in three participants (11%). This was the only study to assess B cells: none had a reduced number of B lymphocytes; however, an abnormally high number was found in two patients (7%).

IgG subclass levels—reflecting humoral immunity—were also assessed in this trial. Low levels of IgG1 were found in 12 participants (43%) and low levels of IgG3 in 18 (64%).

3.2.2. Summary of Reported Findings

The final outcome point involved ratings of the period between penultimate and final infusions. Comparison of intervention and control group means on various measures indicated that the only statistically significant difference between the IgG and placebo control groups over the course of the trial had been *"a slight relative improvement in social functioning of the placebo group"*. However, significant changes had occurred for the respective groups. An improvement in health perceptions was found in the IgG group, where mean score on the relevant scale of the Medical Outcome Study short-form (MOS SF) [42,43] had risen from 8.5 to 20.5 (on a scale of 1–100), an increase of 12 points. Group mean score on physical function, also measured by MOS SF, had significantly deteriorated in the placebo group.

Regarding impact on IgG levels, when assessed within 48 h of the final infusion, all participants recorded IgG 1 levels within the normal range, the deficiency recorded in 12 patients at baseline having resolved. On the other hand, IgG3 had normalised in only six of the 18 patients where it had previously been abnormal. The authors note that this is *"consistent with the shorter half-life of IgG3 (7 days) versus IgG1 (22 days)"*, continuing: *"If a deficiency of IgG3 is of pathogenic importance, this could provide a pharmacodynamic explanation for treatment failure."*

Cell-mediated immunity was not assessed after baseline.

The authors observe that findings *"do not, of course, rule out a potential benefit of other immunoglobulin preparations, dosage regimens, or longer courses of therapy"*. Nonetheless, their prominent conclusion was that *"IV IgG is unlikely to be of clinical benefit in CFS."*

3.2.3. Discussion of Study 2

SAMPLE SIZE Based on the hypothesis of symptomatic benefit of IgG in 67% and placebo in 25%, the authors had calculated that a trial size of 30 would have sufficient power to detect a statistically significant difference. However, the marked improvements observed in Study 1 had occurred in a lower proportion: 43% of IgG and 12% of placebo recipients. An analysis involving a trial size of 28 (one participant was lost from each group) therefore lacked the power to detect change occurring at the level observed in Study 1.

RESULTS It is remarkable that the significant improvement in health perceptions unique to the IV IgG group did not temper the negative conclusion presented. The authors describe all recorded changes as of a *"small magnitude"* and therefore of questionable clinical importance. It is far from apparent that the scale of this change—from a mean of 8.5 to 20.5 i.e., +141% —was of a *"small magnitude"*. Furthermore, the standard deviation was substantial (25.0), implying that some of the IgG group recorded an improved level of health perception following treatment that was far in excess of the group mean.

There was no comparable analysis to that carried out in Study 1, which identified a distinct pattern of change—i.e., participants improved greatly or scarcely at all—a change that was masked when group means were compared.

PARTICIPANT CHARACTERISTICS Recruitment involved CDC 1988 criteria [36], where post-exertional deterioration is optional. However, according to symptom reports all but one participant did experience prolonged fatigue post-exertion.

Participants appear to have had less severe ME/CFS as mean scores on the MOS SF Physical Function Subscale [42,43] at baseline were over 60 for both groups. In contrast, a paper reporting the function of patients in Prof Peterson's clinic from around this time indicates that mean scores on this scale were in the region of 16 [52].

PREDICTORS OF OUTCOME Had patients whose health perceptions improved also experienced improvements in CMI, as identified in Study 1? As CMI was not assessed after baseline, we simply do not know.

IgG subclass levels (humoral immunity) were recorded post-treatment (see Section 3.2.2), but were not correlated with other outcome measures.

3.3. Study 3 (Australia) Rowe 1997

This trial involved the administration of IV IgG (n = 35) or placebo (n = 35) in adolescents aged 11–18 [22]. A subsequent paper [23] included further analysis of trial data.

3.3.1. Cell-Mediated Immunity at Baseline

On cutaneous delayed-type hypersensitivity (DTH) testing, a complete absence of CMI response (anergy) was found in 21% of patients tested at baseline, while a further 31% evidenced a reduced response (hypoergy). (This information relates to 63 of the 70 participants in this trial.) In contrast, normative data for healthy children indicate that only 5% are hypoergic and none are anergic [53].

3.3.2. Summary of Reported Findings

Outcomes on a range of measures were positive for both intervention and placebo control groups, but with significantly greater improvement occurring in the IgG group when assessed six months post the final infusion.

The primary outcome measure involved an overall functional score that was based on proportion of: school attendance; school work attempted; normal social activity; and of normal physical activity attempted. The young people were to compare themselves with premorbid levels and a mean of the four ratings was calculated for each participant. On a scale of 0–100, where 100 represents premorbid level of activity, the mean functional improvement from baseline was significantly greater for the IgG group (from 23.9 to 64.1, up 40.2 points) than the placebo control group (from 25.9 to 52.1, up 26.2 points). The proportion who had returned to full function was significantly greater also: 25% of IV IgG recipients, as compared to 12% in the placebo control group.

It is notable that significant differences between IgG and placebo groups emerged at the six month follow-up stage. Follow-up at three months after the final infusion found improvements of a lesser degree, which were not statistically significant. For example, the incidence of a *"clinically significant improvement"*—set at a functional improvement of at least 25% over baseline—was 52% among IgG recipients at three months and 72.2% at six months (see Table 3).

Table 3. Incidence of clinically significant improvement, Rowe 1997 [22].

Period after Final Infusion	Gammaglobulin Group (*n* = 36) [1]		Placebo Control Group (*n* = 35) [2]	
	Not Improved	**Improved**	**Not Improved**	**Improved**
3 months	17 (47.2%)	19 (52.8%)	24 (68.6%)	11 (31.4%)
6 months	10 (27.8%)	26 (72.2%)	19 (55.9%)	15 (44.1%)

[1] The therapeutic agent is referred to as 'gammaglobulin' in this paper; this was the same IgG preparation as used in Studies 1 and 4.
[2] *n* = 34 at six months as one participant was lost to follow-up after the three month assessment.

A subsequent paper [23] presented further analysis of the data, investigating the role of CMI in the form of DTH in determining outcomes. It was found that young ME/CFS patients with an appropriate CMI response had tended to experience improvement, whether treated with IgG or not, while those with impaired response fared better only if treated, as those with poor CMI treated with IgG recorded an average functioning score at six-month follow-up of 65.4, as compared to 39.3 for their counterparts in the placebo control group.

3.3.3. Discussion of Study 3

PARTICIPANT CHARACTERISTICS This remains one of very few RCTs investigating the efficacy of a pharmacological intervention specifically in young people with ME/CFS. The young patients were reported to be *"by selection at the more severe end of the spectrum"* [23]. They were required to meet 1994 CDC Criteria [39] and to be experiencing *"chronic and persisting or relapsing fatigue of a generalized nature, exacerbated by minor exercise"*. The latter condition represents a strengthening of these criteria, in which *"postexertional malaise lasting more than 24 h"* is merely optional.

PREDICTORS OF OUTCOME In addition to consideration of outcomes in relation to DTH (see Section 3.3.2 above), analysis was conducted to identify any associations between side effects and outcome. *"Severity of headache"* and *"duration of nausea"* proved to be sensitive for predicting clinical improvement. These were not specific to clinical improvement, as all but one of those who improved only minimally also experienced them, suggesting that experience of severe headache and nausea of more than three days duration following infusion may have been a necessary but not a sufficient condition of clinically effective action of IgG on the patient.

3.4. Study 4: (Australia) Vollmer-Conna et al., 1997

Several authors of Study 1, together with others, went on to conduct this further trial *"To determine whether the reported therapeutic benefit of intravenous immunoglobulin in patients with chronic fatigue syndrome is dose dependent"* [24]. Participants received either placebo or IgG infusion at one of three doses: 2 g per kg, as used in Study 1—described as 'high' (*n* = 23); 1 g per kg, as used in Study 2—described as 'medium' (*n* = 28); and 0.5 g per kg—described as 'low' (*n* = 22).

3.4.1. Cell-Mediated Immunity at Baseline

At entry, abnormal CMI in the form of reduced DTH skin responses was evident in 47 (48%) with 16 of these patients (16%) showing anergy. However, the proportion evidencing reduced DTH reponses varied greatly between the study groups, ranging from 32% in the medium-dose IgG group to 64% in the high-dose group.

3.4.2. Summary of Reported Findings

The published paper presents a stark conclusion, placed prominently in the title: *"Intravenous Immunoglobulin Is Ineffective in the Treatment of Patients with Chronic Fatigue Syndrome"* [24]. Far from showing that participants did not experience improvement, it was reported that three months after the final infusion *all* groups had improved significantly, regardless of whether infused with IgG or placebo and regardless of the IgG dose administered. For example, the increase in Karnofsky performance scores [48] pre- and post -treatment is reported as highly significant in all groups, but with no significant difference *between* the groups.

Furthermore, all groups had experienced improvement in measures of cell-mediated immunity over the same time period. Immunological findings are presented as group means for CD4 and CD8 cell counts at baseline, immediately before the final infusion and at follow-up three months later. It is reported that *"A significant, linear increase in the absolute numbers of T suppressor/cytotoxic (CD8) cells was demonstrated across the three measurement occasions (F = 17.8, p < 0.0001)."* T inducer (CD4) cells, on the other hand, had flatlined (high-dose IgG group) or reduced (all other groups) over the course of the trial, then increased during the following three months—though the mean count for the medium-dose IgG group remained below baseline level.

3.4.3. Discussion of Study 4

In view of the subsequent influence of its 'negative' conclusion, we present a more extensive commentary on this particular paper.

PARTICIPANT CHARACTERISTICS: The authors advise that participants were *"probably, more heterogeneous than that enrolled in our earlier study"*, i.e., Study 1 [13]. This assertion appears to be justified. Recruitment criteria [37,38] lacked an indicator of post-exertional deterioration. Being severely affected was not the norm: all four groups were undertaking a mean of five hours of non-sedentary activity per day at baseline, with some participants managing in excess of nine hours of non-sedentary activity.

RESULTS: Having observed that *"improvements occurred irrespective of whether patients received immunoglobulin or placebo infusions"*, the authors conclude: *"This outcome, although consistent with the results obtained by Peterson et al., does not support our own previous findings of significant clinical and immunologic improvement with high (2 g/kg) dose immunoglobulin"*. Yet, the group given 2 g/kg dose of IgG had improved significantly in respect of both clinical and immunological measures in this trial. What was different here from their previous trial (Study 1) was the occurrence of significant improvement in the placebo group also.

RATIONALE FOR IMPROVEMENT IN PLACEBO GROUP: The authors speculate that significant improvements in functioning may simply reflect the natural course of the illness. This is unconvincing: a 2005 systematic review of 14 studies found that fewer than half of patients had improved [54]. A more likely explanation may lie in the operation of the 'placebo' preparation. This contained albumin, which exerts high oncotic pressure, acting to improve brain plasma volume. (Studies 2 and 3 also involved a placebo containing albumin, while in Study 1 this is unclear.)

Regarding blinding in this study, the placebo solution was identical in appearance to the IV IgG solution (Intragram™). Both solutions were delivered to the pharmacies of participating institutions in 500-mL bottles.

PATTERN OF CHANGE: Study 1 had likewise found no significant difference between intervention and control groups on conclusion of the trial, when assessed as a whole, but had observed a dichotomous distribution of outcomes in both groups. Patients improved greatly—being referred to as 'recovered'—or scarcely at all. It is notable that the investigators did not interrogate the data from the present study for the presence of a 'response'/'non-response' pattern.

The rationale provided for not doing so is that the method used—being based on physician's assessment of substantial improvement in functional capacity and symptomatology—*"is potentially too subjective"*.

We find this unconvincing. This perspective on the primary outcome measure used in Study 1 is irrelevant: Study 4 did not employ this outcome measure. The data emerging on the outcome measures that *were* employed in Study 4 could have been interrogated to establish if a similar pattern of strong response vs. little or no response pertained.

With regard to this retrospective criticism of their own method, the issue is not whether or not a physician assessment could possibly be *"too subjective"*, but whether or not this could rightly be said of the method employed in Study 1. We are of the view that this charge would be misplaced and that the assessments of outcomes recorded in Study 1 were robust, for three reasons. Firstly, the patient had to agree with the physician's assessment that they were a 'responder'. Secondly, 'response' vs. 'non-response' designation was arrived at with reference to specific details of real-life activities. Thirdly, physician ratings were corroborated by other outcome measures.)

BETWEEN GROUP DIFFERENCES The data presented do not appear to bear out the narrative conclusion concerning the lack of any discernable difference between the groups, with the 'medium'-dose IgG group recording a median incremental change on the Karnofsky Scale [48] four times the magnitude occurring in the 'low'-dose group (10 points, as compared to 2.5 points).

PREDICTORS OF OUTCOME Study 1 had found resolution of abnormal CMI to be strongly predictive of response. There was no attempt to assess CMI against improvements in function on a patient by patient basis for degree of association in the present Study. This is remarkable as, in their discussion, the authors observe: *"If one could identify a cohort of genuine postinfective cases with CFS as well as immunological abnormalities, then it may be appropriate to reevaluate intravenous immunoglobulin therapy in that specific group only"*.

3.5. Summary of Immunoglobulin Trial Findings

STUDY 1 [13] found that *"A significant proportion of patients (43%) with well-characterized, severe, and long standing CFS responded to high dose intravenous immunoglobulin therapy"*. Likely responders could be identified in measures of cell-mediated immunity (CMI) at baseline. Change was either dramatic—being referred to as *"recovery"*—or negligible. Participants who subsequently relapsed were given further IgG therapy, *"associated with an identical remission of symptoms in improvement in function in each case"*.

Despite differing 'headlines' emerging from the following studies, none recorded outcomes that are demonstrably inconsistent with these clear findings from Study 1.

STUDY 2 [14] was small and lacked the power to identify the incidence of change observed in Study 1. The significant improvement in health perceptions, unique to the IgG group, is not discussed or referred to by the authors when reaching conclusions.

STUDY 3 [22,23] found a significantly greater improvement among adolescent ME/CFS patients treated with IV IgG in comparison with placebo. Furthermore, this improvement was of a clinically significant magnitude. It also found that young patients with normal CMI tended to improve over time, regardless of treatment, while those with subnormal CMI were significantly more likely to improve if treated with IgG.

STUDY 4 [24] involved the least well-characterised group of patients and the least severely affected. All groups, including placebo, improved significantly and all had experienced improvements in CMI. There was no attempt to ascertain if resolution of abnormal CMI was associated with individual participant improvement. Nor did the authors interrogate the data for evidence of the 'response'/'non-response' dichotomy which had emerged in Study 1, and which would have remained masked had analysis been confined to comparison of IgG and placebo group mean scores, as in Study 4.

PARTICIPANT CHARACTERISTICS It was a pre-condition of recruitment to Studies 1 and 3 that the patient's clinical picture include post-exertional deterioration. While not a recruitment criterion in Study 2, all but one participant did report experiencing *"prolonged post-exertional fatigue"*. Study 4 was the exception in this regard.

PLACEBO GROUPS In so far as some placebo group participants had improved, this was found to be associated with resolution of, or improvement in, abnormal cell-mediated immunity (investigated in Studies 1 and 3). It is further plausible that the 'placebo' infusion had intrinsic therapeutic action. Albumin, used in the placebo infusion in at least three of the four trials (Studies 2, 3 and 4), is known to exert high oncotic pressure leading to improved perfusion of vital organs, including the brain.

4. Clinical Experience

Following the publication of Study 4 [24] in November 1997 no further RCTs of immunoglobulin treatment for ME/CFS have been conducted. In contrast, reports in the peer-reviewed medical literature document clinical experience that is predominantly positive.

Published reports concerning use of IgG in clinical practice include a 2003 article from the UK 'Successful Intravenous Immunoglobulin Therapy in 3 Cases of Parvovirus B19–Associated Chronic Fatigue Syndrome' [55] and a 2013 paper documenting outcomes in respect of treatment schedules followed by patients attending the CFS Unit of the Aviano National Cancer Institute, Italy, which found that antiviral and immunological therapies (jointly reported) were the most successful, with 75% of those treated responding [56]. The authors of the 2013 paper conclude: *"our results show that a significant number of patients treated with antiviral/immunoglobulin approaches have a long positive disease free survival in comparison with other patients treated with other approaches"* and describe these treatments as *"clearly superior"*.

Conference presentations provide another source of positive reinforcement for the potential benefits of IgG. In the US, Dr James Oleske—then chair of the Chronic Fatigue Syndrome Advisory Committee (CFSAC) to Health and Human Services (HHS)—made a presentation to CFSAC in 2009 reporting on successful IgG treatment of young patients [57]. More recently, Ryan Wheelan of Simmaron Research presented to the 2020 IACFS/ME conference data from a patient who had neuropathic symptoms in the presence of an autoimmune-associated small-fibre polyneuropathy (aaSFPN) [28]. Following two IV IgG infusions, *"The patient experienced a dramatic reduction in levels of all four of the relevant autoantibodies and favorable symptom reduction."*

4.1. Brief Case Reports (NS)

One of the authors, (NS), has significant experience of using immunoglobulin in selected cases of severe ME/CFS, including three cases described elsewhere in this issue [58]. Thumbnail sketches of their IgG treatment together with the experiences of two other young patients are presented below.

These cases occurred over a sixteen-year period between 1990 and 2006. Most had clinical evidence of a viral onset. All but one received IgG treatment by intra-muscular (IM) administration, being too unwell to attend hospital.

Interest in the therapeutic use of immunoglobulin stemmed from a case where the family doctor, unbeknown to paediatrician NS, started his patient on IM immunoglobulin. NS observed that the patient made a rapid improvement over the next few months and managed to return to full time schooling after a total loss of two years; thereafter he continued to improve and was able to participate in sports.

Following this experience, NS decided on a policy of offering a trial of immunoglobulin to each of his severely affected cases, supported by the evidence in the Lloyd et al. paper [13] (Study 1, Section 3.1 above; the positive adolescent study by Rowe [22,23]—Study 3, Section 3.3—not yet having been published) and awareness of the research base on immune system defects.

Prescription of immunoglobulin was based on the clear clinical picture of ME/CFS. All would have fitted any of the main diagnostic definitions now in use. and in the course of long-term clinical monitoring and follow-up none transpired to have been misdiagnosed.

4.1.1. Case 1

This 13-year-old girl was bedridden and needed tube feeding over a period of 18 months. Testing of blood samples identified that antibody titres to Coxsackie virus were raised; this remained the case for one year. She was given monthly injections of IM immunoglobulin and made a slow but steady recovery over a two year period. In her 20s, she was fully recovered and holding down a full-time job.

4.1.2. Case 2

This 13-year-old girl had a sudden onset of very severe ME and lay in a darkened room in severe pain for over 12 months. She was then given IM immunoglobulin for a year and made a total recovery over two years.

4.1.3. Case 3

This adolescent boy had moderately severe ME for over a year. He was well enough to come to hospital for monthly IV immunoglobulin. His condition improved slowly over two years, and he eventually made a full recovery, being able to undertake strenuous athletic pursuits.

4.1.4. Case 4

This 13-year-old girl was the most severe case of this series, being virtually paralysed with shallow respirations for several months. She received IM immunoglobulin but there was no improvement over the first nine months. She then began to improve steadily, and made a total recovery over two years. Ten years later, she remained completely well.

4.1.5. Case 5

NS encountered this 19-year-old young woman through meeting her parents at an ME conference in Norway. She had been lying in a darkened room for several years and was not improving. On his suggestion, her family doctor gave her a trial of immunoglobulin, following which she improved rapidly. Within nine months, she had a functional level of 90% and was cycling and campaigning for her chosen party in elections.

4.1.6. Reflection

While this is a series of documented clinical histories that is uncontrolled, it is worth stressing that: (a) all the cases were severe or very severe; (b) all made near total recoveries; and (c) there were no treatment failures, only instances of treatment success. Treatment continued until response was achieved, in contrast to the research trials which provided shorter-term treatment. Additionally notable in a clinical context is the fact that patients in Study 1 who relapsed after the trial ceased were given further IgG treatment, which effected an identical remission in each case (as reported at Section 3.1.2 above).

5. Informed Consent: Awareness of Any Potential for Adverse Events

It is of course vital that both physician and patient are aware of potential for adverse events before commencing any treatment or therapy. Informed consent involves the patient being provided with the information required to assess any potential adverse events against the health-related impact on quality of life of their ongoing illness and to contrast this with any alternative therapeutic options.

Quality of life with this illness is low if left untreated [59–68].

Information given to patients receiving IgG on the National Health Service (NHS) in the UK advises that *"some patients may experience reactions during treatment such as fevers, shivering, skin rashes, wheezing and headaches"* [69]. According to *'Managing patients with side effects and adverse events to immunoglobulin therapy'*—a 2016 paper from the Expert Review on Clinical Pharmacology [70]—*"Most of the adverse effects associated with immunoglobulin therapy are mild, transient and self-limiting"* while, less commonly, serious side effects can occur. The latter may be inherent in the mode of operation as key issues identified include: *"The antigenicity of the IgG, large molecular weight of the IgG aggregates, and complement activation or direct release of cytokines from mononuclear cells may be underlying mechanisms of adverse reactions in immunoglobulin therapy."*

The findings of the IgG trials involving ME/CFS patients reflect these assessments, with the additional dimension that the effort required to participate could cause an exacerbation of ME/CFS.

5.1. Summary of Adverse Events Reported in IgG Trials Involving Patients with ME/CFS

Adverse events commonly took the form of a transient exacerbation of ME/CFS symptoms and for the most part were equally likely to occur in the IgG and placebo control groups, suggesting that they were attributable to the effort of participation in the research.

Some adverse experiences, notably severe headache, were more commonly reported among participants receiving IgG and as such may be treatment related. Furthermore, this may have been intrinsic to effective operation of IgG as they were found to be correlated with a positive outcome in Study 3 (discussed at Section 3.3.3 above).

A small number of unspecified *"major adverse experiences"* were reported in Study 2, with the same incidence in IgG and placebo control groups (three of 15 patients in each). It is not clear that any of these were specifically related to the administration of IgG.

Study 3 took steps to obtain feedback from the young patients regarding what might be improved in this regard, reporting: *"Seventy-five per cent thought they received positive benefit from participating in the trial but 30% thought that certain aspects could have been managed better. This included management of side effects, the option to stay overnight rather than treatment in the Day Medical Ward as many had to travel long distances, and prior information about the likely severity of symptoms after the infusions"* [23].

5.2. Discussion and Implications

As with any therapy, steps can and should be taken to minimise the potential for adverse impact of IgG treatment on a patient with ME/CFS, such as those suggested by the young patients in Study 3 (see Section 5.1 above).

Intramuscular and subcutaneous administration present alternatives to IV, avoiding the necessity of travel to hospital. These may be preferable options for the most severely affected patients.

6. Healthcare Systems and Immunoglobulin for ME/CFS

Immunoglobulin therapy remains unendorsed in healthcare guidance for ME/CFS in the USA [71,72], Australia [73], Canada [74] and the UK [75]. Negative recommendations tend to rest on two elements. Firstly, the assessed results of the four trials, historically summarised as *"mixed"* [25], *"inconclusive"* [26,76], and/or *"insufficient"* [76]. Secondly, the reported incidence of adverse events. While a full analysis is beyond the scope of this paper, both elements would appear to have been subject to a degree of misinterpretation and selective reporting, as exemplified in the justification provided for the UK's *"not recommended"* decision (described at Section 6.1.1. below). We have covered both aspects in this paper (in Sections 3 and 5, respectively) drawing on the original trial reports [13,14,22–24] and hold that a strong case for IgG therapy in selected cases of severe ME/CFS emerges.

6.1. Accessing Immunoglobulin Therapy

6.1.1. United Kingdom

Faced with increasing therapeutic use of IgG and reduced supply, guidelines limiting use on the NHS were introduced in 2008, with 'chronic fatigue syndrome' listed under *"indications for which IVIG is not recommended"* [77] and this remains the case in the current version, published in 2011 [78]. As none of the research studies on IgG for ME/CFS were among the 248 publications referenced, the evidence used to reach this decision was opaque, emerging only in reply to rapid responses challenging this assessment following publication of a summary in the British Medical Journal [79].

The reply advises: *"The principal study reviewed (a randomized, double-blind, placebo controlled trial) showed no significant benefit of intravenous immunoglobulin"* [80], referencing Study 4 [24]: this trial had found significant improvements in all groups, including placebo.

There is no acknowledgement of the clearly positive findings for IgG over placebo reported in Studies 1 [13] and 3 [22,23]. It is further asserted that findings raise the potential for adverse reactions. It is not noted that worsened symptoms had commonly been reported to occur as just commonly in the placebo control group as well as among those receiving IgG in this trial. Rather, the reference is specifically to *"potential and sometimes fatal risks associated with treatment with a blood product"*. We are not aware of any other disorder where the case for therapeutic use of IgG is inhibited on the grounds of the potential risk of blood products.

Rapid responses [81,82] highlighted a published clinical study documenting improvement in 'Parvovirus B19–Associated Chronic Fatigue Syndrome' [55]. It is notable that the authors' reply sets out conditions under which an effective case might be made for prescription of IgG to an ME/CFS patient on the NHS (see Box 2 below)

Box 2. UK NHS IgG treatment: Process of 'exceptionality'.

> *"There is a process of 'exceptionality' that offers the opportunity for treatment in cases that fall outside the broad definition of a disease state. In the case of parvovirus B19-associated CFS, the treating physician may request immunoglobulin treatment on the basis that parvovirus B19-associated CFS is different from the reference population of CFS patients. The physician would be expected to present evidence to the local immunoglobulin panel that an individual patients is likely to gain significantly more benefit from the intervention than might be expected for an average patient with that condition"* [80].

6.1.2. United States

In the US, IgG is not Food and Drug Administration (FDA) approved for ME/CFS [83]. However, data reported in a 2006 CDC review demonstrate that much of the IgG used in the US is prescribed off label [84]. A lack of RCTs does not necessarily present a significant barrier to prescribing, as the review advises: *"Reports concerning IVIG continue to grow at a tremendous pace but few high-quality randomized controlled trials have been reported"*. It is further reported that IV IgG is most commonly used in the treatment of chronic neuropathy. Given that autoimmune-associated small-fibre polyneuropathy (aaSFPN) has been identified in some ME/CFS patients [28,30], this does appear to open a door to off label prescribing of IgG for selected ME/CFS patients.

6.1.3. Australia

In Australia, where the majority of the research on use of IV Ig in ME/CFS has been conducted, access is governed by the National Blood Authority (NBA) which first produced clinical criteria for use of IgG in 2007. The current version was launched in 2018 [85]. These are broadly similar in approach to the UK criteria, with three categories of use—'established', 'emerging', and 'exceptional'—and a further category 'not funded'.

The position in respect of use for ME/CFS has remained the same throughout, with IgG 'not funded'. Regarding this decision, the NBA advises: *"Ig therapy is not supported for myalgic encephalomyelitis. Results from a single RCT in 1990, have not been reproduced, and subsequent studies have shown no evidence of benefit."* This assessment ignores the positive findings emerging from the 1997 RCT involving children and young people [22,23] (Study 3: see Section 3.3 above).

The NBA recognise that *"For a number of reasons, medical specialists may sometimes want to prescribe immunoglobulin for medical conditions that are not funded under the national blood supply arrangements"* and advise that, for such patients, Approved Health Providers may purchase imported IgG directly from the supplier at an equivalent price to that negotiated by the NBA. This requires a Jurisdictional Direct Order (see Box 3).

Box 3. Australian Jurisdictional Direct Order (JDO) arrangements.

<table><tr><td>

- Medical specialist seeks approval through a JDO for their patient, following their local processes
- Once approval is granted, the relevant Approved Health Provider (AHP) places an order for the imported IVIg or SCIg directly with the supplier, establishing a contract directly between the supplier and the AHP for the supply of the product
- Purchases are paid for in full by the AHP.

SOURCE: https://www.blood.gov.au/IgOtherAccessArrangements (accessed on 28 June 2021).

</td></tr></table>

6.2. The Present Approach: Ongoing "General Management Strategies for Chronic Illness"

Cost has also been cited as a reason not to prescribe IgG to ME/CFS patients [18]. However, in a disease causing prolonged and substantial morbidity in many patients [54,59–68], any treatment with curative potential is likely to prove cost-effective in the long run.

Additionally the alternative direction taken by healthcare systems is not devoid of expense in terms of professional salaries. For example, in Australia—where three of the four Studies of IV IgG were conducted—Dr K. Rowe (author of Study 3) reports as follows regarding the direction subsequently taken by her paediatric service: *"As immunoglobulin was a scarce resource requiring approval by a government agency, a decision was made to not allow intravenous immunoglobulin to be available for ME/CFS for young people due to some adverse effects (Study 3), as well as inconclusive trials in adults (Studies 1,2 and 4). Thus, options for treatment reverted to general management strategies for chronic illness. . . . . The service has since expanded to several pediatricians and access to a 4-week self-management program run by the Victorian Pediatric Rehabilitation Service"* [86].

7. Conclusions

From the evidence outlined in this review, we hold that immunoglobulin is a potentially curative treatment for a proportion of patients with ME/CFS and that the interpretation of some of the literature regarding the issue has been faulty. Considering the lack of any other curative treatment for this debilitating disease, we find this regrettable.

Research on IgG for ME/CFS patients has stagnated. A severe global supply shortage of IgG has not helped; however, this is set to ease [33]. While treatment with IgG is not without cost, this has to be set against the cost of providing an ongoing healthcare service to patients with a chronic illness. An illness which, for some patients, may well cease to be chronic if treated with immunotherapy.

In 1991, a British Medical Journal editorial on use of intravenous immunglobulins projected that *"Exciting developments in treatment with intravenous immunglobulin should take place in the 1990s"* [87], referring to various disorders in this context, including myalgic encephalomyelitis. In response, a letter from Dr Charles Shepherd, medical adviser at the ME Association, highlighted two pertinent areas for investigation: *"it would seem sensible to discover whether there is a specific subgroup of patients who benefit and whether this is related to any characteristic defects in immune function"* [88].

This review has demonstrated that the original research trials provide pointers in exactly that direction, indicating that a significant proportion of patients improve and that predictors of response lie in indicators of abnormal cell-mediated immunity. While there is a degree of variation between the several research trial reports (summarised at Section 3.5), nothing has emerged in the three decades since Study 1 [13] reported recovery in 43% of IgG recipients that would substantially undermine these two core conclusions.

The phrase *"a significant proportion of patients"* is crucial. Study 1 findings do not suggest that the generality of ME/CFS patients benefit. However, for those who did, the improvements experienced were considerable (see Box 1). At the same time, because the remaining participants experienced scarcely any change, comparison of *average* outcomes recorded at final end point did not reveal a statistically significant difference between the IgG and placebo control groups, masking this key finding.

Clinical observations support this interpretation. Beneficial reports from clinical practice, evident in 1990 [18], continue to this day (Section 4). The findings emerging from our analysis represent a clear demonstration of what might be gained by approaching ME/CFS patients on the basis of documented biomedical abnormalities.

It is often asserted that a disorder exists which we call 'CFS' and which is heterogeneous. In our view, it would be more accurate to say that there exist several research definitions for 'chronic fatigue syndrome' and that these definitions, to a greater or lesser degree, have the potential to encompass a heterogeneous group. It follows that for progress to be made in identifying and treating these patients, they require investigation and differential treatment as indicated.

Parallels with ME/CFS in terms of both clinical presentation and biomarkers of immune dysfunction and neuroinflammation are emerging from the study of patients experiencing post-acute sequelae of COVID-19 (PASC) [34,35,89–92]. We therefore further conclude that it is plausible that a subgroup of PASC patients may also be helped by IgG.

8. Action Points Emerging

Based on the findings of this review, we suggest action on three fronts.

8.1. Research

Further randomised controlled trials on IgG in ME/CFS should be conducted with urgency.

Given the emergence of similarities in patients experiencing post-acute sequelae of COVID-19 (PASC), further research in this area may prove fruitful to the understanding and care of these patients also. Researchers may wish to incorporate a comparison of associated groups in RCT design. For example:

- ME/CFS patients and patients experiencing PASC.
- ME/CFS patients who have gone on to suffer from COVID-19 and ME/CFS patients who have not.
- ME/CFS patients who have gone on to suffer from COVID-19, ME/CFS patients who have not and patients experiencing PASC.

8.2. Laboratory Testing

Healthcare systems should encourage the identification of ME/CFS patients with abnormal cell-mediated immunity (CMI) and/or an autoimmune-associated small-fibre polyneuropathy (aaSFPN); this may facilitate identification of likely responders to IgG.

The measures of CMI employed to assess subjects in the IgG trials are documented in Appendix B. While a full consideration of present day testing regimes is beyond the scope of this paper, further tests—such as autonomic receptor auto-antibody assessment—may be pertinent.

8.3. Treatment

Pending further research and introduction of laboratory testing, it would be reasonable to offer a therapeutic trial of IgG to selected ME/CFS patients at the more severe end of the spectrum based on clinical presentation.

Author Contributions: The review from original source papers was done by H.B.; the identification and analysis of published clinical research was done by H.B.; the series of brief case reports and related reflections were provided by N.S.; the overview and conclusions were developed by N.S. and H.B.; the manuscript was written, reviewed and edited by H.B. and N.S. All authors have read and agreed to the published version of the manuscript.

Funding: This research received no external funding.

Institutional Review Board Statement: Not applicable.

Informed Consent Statement: Consent to include case reports 1, 2, 4 and 5 was obtained; in case 3 contact having been lost, only the most general details are provided in order to ensure that identity is concealed.

Data Availability Statement: Not applicable.

Acknowledgments: We acknowledge with thanks the assistance of K. Geraghty in locating relevant literature in respect of the post-acute sequelae of COVID-19.

Conflicts of Interest: The authors have no conflicts of interest to declare.

Appendix A. Rating Methods: Assessing Symptoms and Function

Study 1 [13]

Physician recorded specific details of the degree of involvement in work or school, leisure, sporting and social activities. The improvement in function labelled as a 'response' required a substantial return to involvement in at least two of the following three areas of activity: work, leisure and social events, with each assessed in the context of that achieved prior to the onset of ME/CFS.

In addition, patients completed quality of life visual analogue scales [50], *"modified to include 10 aspects of physical and neuropsychiatric symptomatology typical of CFS (including fatigue, headaches, myalgia, and concentration impairment) and functional activity (including involvement in work and social activities)"*.

Psychiatrist carried out a 'standardized psychiatric interview' on 33 consecutive patients and rated these subjects using the Hamilton Depression Scale [40]. (No explanation is provided as to why these 33 patients were assessed by psychiatrist and the remaining 16 participants were not.) These participants also completed the Zung Scale [41], a self-report measure of depression.

Study 2 [14]

Participants completed a self-assessment form, with two components:

(1) Severity of symptoms during the previous month using a 4-point scale: none, mild, moderate, or severe (reported dichotomised, as either none/mild or moderate/severe);
(2) Medical Outcome Study Short Form [42,43], measuring: physical functioning; social functioning; health perceptions; and mental health.

Study 3 [22]

The paediatrician assessed the degree of functional participation during the previous two weeks in four domains: (a) attendance at school or work; (b) proportion of school or work attempted; (c) proportion of 'normal' physical activities attempted, including sports; and (d) proportion of 'normal' social activities attempted. Each rating was estimated as a proportion to within 5%, with 100% indicating premorbid levels of activity. It was considered *"important to assess these four domains to provide an overall score because these was considerable variation in how young people preferred to spend their time and energy."* The mean was taken of these four ratings.

This mean score was compared with an overall estimate of functional rating provided by both the subject and a parent. Where possible feedback was obtained from school or visiting teacher regarding attendance and proportion of school work attempted. In addition, subjects were required to complete a Weekly Activities Record. This was also used as a check against the mean functional estimate.

Participants completed several self-report measures: the Spielberger self-evaluation state trait anxiety inventory [44]; the Beck depression inventory [45]; and the 12 item General Health Questionnaire [46]. In addition, the patient and parents were required to complete *"appropriate report forms from the Child Behavior Check List"* [47] and subjects were required to complete a quality of life visual analogue scale (unspecified).

Study 4 [24]

The Karnofsky performance score was evaluated for each patient, reflecting ability to participate in daily activities on a 100-point scale [48].

Patient completed quality of life visual analogue scales [50], *"modified to include 10 aspects of physical and neuropsychiatric symptomatology typical of CFS (including fatigue, headaches, myalgia, and concentration impairment) and functional activity (hours of non-sedentary activity per day)"*.

The Profile of Mood States (POMS) [49] was used to assess depression, confusion, fatigue and vigor. *"A subjective 'energy' score was derived by subtracting the POMS fatigue score from the POMS vigor score for each participant."*

Appendix B. Tests of Cell-Mediated Immunity

The following tests of cell-mediated immunity (CMI) were preformed on subjects in the respective trials. Please note that not all were systematically reported in the published papers: coverage on the respective studies in Section 3 incorporates prominent reported findings on CMI at baseline and following trial participation.

Study 1 [13]—CD4 (helper T) and CD8 (suppressor/cytotoxic T) lymphocyte count; absolute count of the T-cell subsets; T lymphocyte response to phytohaemagglutinin (PHA); cutaneous delayed-type hypersensitivity (DTH).

Study 2 [14]—Peripheral blood lymphocyte subsets: CD19 (pan B); CD3 (pan T); CD4 (helper T); CD4-29 (helper/inducer T); CD4-45RA (suppressor/inducer T); CD8 (suppressor/cytotoxic T); CD56 (natural killer).

Study 3 [22,23]—T-cell subset analysis; cutaneous DTH.

Study 4 [24]—T-cell subset analysis to enumerate the CD4 (inducer) and CD8 (suppressor/cytotoxic) subpopulations; cutaneous DTH.

References

1. Lloyd, A.; Wakefield, D.; Broughton, C.; Dwyer, J. Immunological abnormalities in the chronic fatigue syndrome. *Med. J. Aust.* **1989**, *151*, 122–124. [CrossRef] [PubMed]
2. Behan, P.; Behan, W.; Bell, E. The postviral fatigue syndrome—An analysis of the findings in 50 cases. *J. Infect.* **1985**, *10*, 211–212. [CrossRef]
3. Caligiuri, M.; Murray, C.; Buchwald, D.; Levine, H.; Cheney, P.; Peterson, D.; Komaroff, A.L.; Ritz, J. Phenotypic and functional deficiency of natural killer cells in patients with chronic fatigue syndrome. *J. Immunol.* **1987**, *139*, 3306–3313. [PubMed]
4. Tosato, G.; Straus, S.; Henle, W.; Pike, S.E.; Blease, R.M. Characteristic T cell dysfunction in patients with chronic Epstein-Barr virus infection (chronic mononucleosis syndrome). *J. Immunol.* **1985**, *134*, 3082–3088.
5. Kibler, R.; Lucas, D.O.; Hicks, M.H.; Poulos, B.T.; Jones, J.F. Immune function in chronic Epstein-Barr virus infection. *J. Clin. Immunol.* **1985**, *5*, 46–54. [CrossRef]
6. Borysiewicz, L.K.; Haworth, S.J.; Cohen, J.; Mundin, J.; Rickinson, A.; Sissons, J.G.P. Epstein-Barr virus-specific immune defects in patients with persistent symptoms following infectious mononucleosis. *QJM Int. J. Med.* **1986**, *58*, 111–121. [CrossRef]
7. Murdoch, J.C. Cell mediated immunity in patients with myalgic encephalomyelitis syndrome. *N. Z. Med. J.* **1988**, *101*, 511–512.
8. Subira, M.L.; Castilla, A.; Civeira, M.-P.; Prieto, J. Deficient display of CD3 on lymphocytes of patients with chronic fatigue syndrome. *J. Infect. Dis.* **1989**, *160*, 165–166. [CrossRef]
9. Prieto, J.; Subira, M.L.; Castilla, A.; Serrano, M. Naloxone-reversible monocyte dysfunction in patients with chronic fatigue syndrome. *Scand. J. Immunol.* **1989**, *30*, 13–20. [CrossRef]
10. Read, P.; Spickett, G.; Harvey, J.; Edwards, A.J.; Larson, H.E. IgG1 subclass deficiency in patients with chronic fatigue syndrome. *Lancet* **1988**, *1*, 241–242. [CrossRef]
11. Komaroff, A.L.; Geiger, A.M.; Wormsley, S. IgG subclass deficiencies in the chronic fatigue syndrome. *Lancet* **1988**, *1*, 288–289. [CrossRef]
12. Linde, A.; Hammearstrom, L.; Smith, C. IgG subclass deficiency and chronic fatigue syndrome. *Lancet* **1988**, *1*, 885–886. [CrossRef]
13. Lloyd, A.; Hickie, I.; Wakefield, D.; Boughton, C.; Dwyer, A. A double-blind placebo controlled trial of intravenous immunoglobulin therapy in patients with chronic fatigue syndrome. *Am. J. Med.* **1990**, *89*, 561–568. [CrossRef]
14. Peterson, P.K.; Shepard, J.; Macres, M.; Schenck, C.; Crosson, J.; Rechtman, D.; Lurie, N. A controlled trial of intravenous immunoglobulin G in chronic fatigue syndrome. *Am. J. Med.* **1990**, *89*, 554–560. [CrossRef]
15. Archard, L.; Bowles, N.; Behan, P.; Bell, E.; Doyle, D. Postviral fatigue syndrome: Persistence of enterovirus RNA in muscle and elevated creatine kinase. *J. R. Soc. Med.* **1988**, *81*, 326–329. [CrossRef]

16. Yousef, G.; Mann, G.; Smith, D.; Bell, E.; Murugesan, V.; Mccartney, R.; Mowbray, J. Chronic enterovirus infection in patients with postviral fatigue syndrome. *Lancet* **1988**, *1*, 146–150. [CrossRef]
17. Wakefield, D.; Lloyd, A. Pathophysiology of myalgic encephalomyelitis. *Lancet* **1987**, *2*, 918–919. [CrossRef]
18. Strauss, S.E. Intravenous immunoglobulin treatment for the chronic fatigue syndrome. *Am. J. Med.* **1990**, *89*, 551–552. [CrossRef]
19. Dwyer, J. Intravenous therapy with gammaglobulin. *Adv. Intern. Med.* **1987**, *32*, 111–136.
20. Gelfand, E. The use of intravenous immune globulin in collagen vascular disorders: A potentially new modality of therapy. *J. Allergy Clin. Immunol.* **1989**, *84*, 613–616. [CrossRef]
21. DuBois, R.E. Gamma globulin therapy for chronic mononucleosis syndrome. *AIDS Res.* **1986**, *2* (Suppl. S1), S191–S195.
22. Rowe, K.S. Double-blind randomized controlled trial to assess the efficacy of intravenous gammaglobulin for the management of chronic fatigue syndrome in adolescents. *J. Psychiatr. Res.* **1997**, *31*, 133–147. [CrossRef]
23. Rowe, K.S. Five-year follow-up of young people with chronic fatigue syndrome following the double blind randomised controlled intravenous gammaglobulin trial. *J. Chronic Fatigue Syndr.* **1999**, *5*, 97–107. [CrossRef]
24. Vollmer-Conna, U.; Hickie, I.; Hadzi-Pavlovic, D.; Tymms, K.; Wakefield, D.; Dwyer, J.; Lloyd, A. Intravenous immunoglobulin is ineffective in the treatment of patients with chronic fatigue syndrome. *Am. J. Med.* **1997**, *103*, 38–43. [CrossRef]
25. Turnbull, N.; Shaw, E.J.; Baker, R.; Dunsdon, S.; Costin, N.; Britton, G.; Kuntze, S.; Norman, R. *Chronic Fatigue Syndrome/Myalgic Encephalomyelitis (or Encephalopathy): Diagnosis and Management of Chronic Fatigue Syndrome/Myalgic Encephalomyelitis (or Encephalopathy) in Adults and Children. National Institute for Clinical Excellence Clinical Guideline 53*; The Royal College of General Practitioners and the National Collaborating Centre for Primary Care: London, UK, 2007.
26. Department of Health. *A Report of the CFS/ME Working Group: Report to the Chief Medical Officer of an Independent Working Group; Annex 5—Management of CFS/ME—Evidence Base*; Department of Health: London, UK, 2002.
27. Hartwig, J.; Sotzny, F.; Bauer, S.; Heidecke, H.; Riemekasten, G.; Dragun, D.; Meisel, C.; Dames, C.; Grabowski, P.; Scheibenbogen, C. IgG Stimulated B2 adrenergic receptor activation is attenuated in patients with ME/CFS. *Brain Behav. Immun. Health* **2020**, *3*, 100047. [CrossRef]
28. Tucker, M.E. Small-Fiber Polyneuropathy May Underlie Dysautonomia in ME/CFS (Report Following IACFS Conference Presentation by Ryan Wheelan). *Medscape* **2020**. Available online: https://www.medscape.com/viewarticle/936745 (accessed on 14 August 2021).
29. Wakiro, S.; Ono, H.; Matsutani, T.; Nakamura, M.; Shin, I.; Amano, K.; Suzuki, R.; Tamamura, T. Skewing of the B cell receptor repertoire in myalgic encephalomyelitis/chronic fatigue syndrome. *Brain Behav. Immun.* **2021**, *95*, 245–255. [CrossRef]
30. Vallings, R.; Dalziel, S. International Association of Chronic Fatigue Syndrome/Myalgic Encephalomyelitis (IACFS/ME) Conference Report 2020. Available online: https://www.iacfsme.org/2020-conference-summaries/ (accessed on 29 June 2021).
31. US ME/CFS Clinician Coalition. Treatment Recommendations (Read More). 2021. Available online: (accessed on 7 July 2021).
32. Bateman, L.; Bested, A.C.; Bonilla, H.F.; Chheda, B.V.; Chu, L.; Curtin, J.M.; Dempsey, T.T.; Dimmock, M.E.; Dowell, T.G.; Felsenstein, D.; et al. Myalgic Encephalomyelitis / Chronic Fatigue Syndrome: Essentials of Diagnosis and Management. *Mayo Clin. Proc.* **2021**, S0025619621005139. [CrossRef]
33. UK Government. Ban Lifted to Allow UK Blood Plasma to be Used for Life-Saving Treatments Press Release 25 February 2021. Available online: https://www.gov.uk/government/news/ban-lifted-to-allow-uk-blood-plasma-to-be-used-for-life-saving-treatments (accessed on 7 July 2021).
34. García-Abellán, J.; Padilla, S.; Fernández-González, M.; García, J.A.; Agulló, V.; Andreo, M.; Ruiz, S.; Galiana, A.; Gutiérrez, F.; Masiá, M. Antibody response to SARS-CoV-2 is associated with long-term clinical outcome in patients with COVID-19: A longitudinal study. *J. Clin. Immunol.* **2021**, *41*, 1490–1501. [CrossRef] [PubMed]
35. Phetsouphanh, C.; Darley, D.; Howe, A.; Munier, C.M.L.; Patel, S.K.; Juno, J.A.; Burrell, L.M.; Kent, S.J.; Dore, G.J.; Kelleher, A.D.; et al. Immunological dysfunction persists for 8 months following initial mild-moderate SARS-CoV-2 infection. *MedRVix* **2021**. [CrossRef]
36. Holmes, G.P.; Kaplan, J.E.; Gantz, N.M.; Komaroff, A.L.; Schonberger, L.B.; Straus, S.E.; Jones, J.F.; DuBois, R.E.; Brus, I.; Tosayo, G.; et al. Chronic fatigue syndrome: A working case definition. *Ann. Intern. Med.* **1988**, *108*, 387–389. [CrossRef]
37. Lloyd, A.R.; Hickie, I.; Boughton, C.R.; Wakefield, D.; Spencer, O. Prevalence of chronic fatigue syndrome in an Australian population. *Med. J. Aust.* **1990**, *153*, 522–528. [CrossRef] [PubMed]
38. Schluederberg, A.; Straus, S.E.; Peterson, P.; Blumenthal, S.; Komaroff, A.L.; Spring, S.B.; Landay, A.; Buchwald, D. Chronic fatigue syndrome research: Definition and medical outcome assessment. *Ann. Intern. Med.* **1992**, *117*, 325. [CrossRef]
39. Fukuda, K.; Straus, S.; Hickie, I.; Sharpe, M.; Dobbins, J.; Komaroff, A.; The International CFS Study Group. The chronic fatigue syndrome: A comprehensive approach to its definition and study. *Ann. Intern. Med.* **1994**, *121*, 953–959. [CrossRef]
40. Hamilton, M. A rating scale for depression. *J. Neurol. Neurosurg. Psychiatry* **1960**, *23*, 56–62. [CrossRef]
41. Zung, W.W.K. A self-rating depression scale. *Arch. Gen. Psychiatry* **1965**, *12*, 63. [CrossRef]
42. Stewart, A.L.; Hays, R.D.; Ware, J.E. The MOS short-form general health survey: Reliability and validity in a patient population. *Med. Care* **1988**, *26*, 724–735. [CrossRef] [PubMed]
43. Stewart, A.L. Functional status and well-being of patients with chronic conditions: Results from the medical outcomes study. *JAMA* **1989**, *262*, 907. [CrossRef]
44. Spielberger, C.D. *Self-Evaluation Questionnaire State Trait Anxiety Inventory*; Consulting Psychologists Press: Paolo Alto, CA, USA, 1977.

45. Beck, A.T. An inventory for measuring depression. *Arch. Gen. Psychiatry* **1961**, *4*, 561. [CrossRef] [PubMed]
46. Goldberg, D.P.; Williams, P. *A User's Guide to the General Health Questionnaire. Reprint*; GL Assessment: London, UK, 2006; ISBN 9780700511822.
47. Achenbach, T.M.; Edelbrock, C.S. *Manual for the Child Behavior Checklist and Revised Child Behavior Profile*; Univ Vermont/Dept Psychiatry: Burlington, VT, USA, 1983; ISBN 9780961189808.
48. Karnofsky, D.; Abelmann, W.; Craver, L.; Burchenal, J. The use of nitrogen mustards in the palliative treatment of carcinoma. *Cancer* **1948**, *1*, 634–656. [CrossRef]
49. McNair, D.M.; Lorr, M.; Droppleman, L.F. *Profile of Mood States: Manual*; Educational and Industrial Testing Service: San Diego, CA, USA, 1981.
50. Gill, W.M. Monitoring the subjective well-being of chronically Ill patients over time. *Commun. Health Stud.* **2010**, *8*, 288–296. [CrossRef] [PubMed]
51. Lloyd, A.R.; Wakefield, D.; Boughton, C.; Dwyer, J. What is myalgic encephalomyelitis? *Lancet* **1988**, *331*, 1286–1287. [CrossRef]
52. Peterson, P.K.; Schenck, C.H.; Sherman, R. Chronic fatigue syndrome in Minnesota. *Minn. Med.* **1991**, *74*, 21–26.
53. Corriel, R.N.; Kniker, W.I.; McBryde, J.L.; Lesourd, B.M. Cell-mediated immunity in schoolchildren assessed by multitest skin testing: Normal values and proposed scoring system for healthy children. *Am. J. Dis. Child.* **1985**, *139*, 141. [CrossRef] [PubMed]
54. Cairns, R.; Hotopf, M. A Systematic review describing the prognosis of chronic fatigue syndrome. *Occup. Med.* **2005**, *55*, 20–31. [CrossRef] [PubMed]
55. Kerr, J.R.; Cunniffe, V.S.; Kelleher, P.; Bernstein, R.M.; Bruce, I.N. Successful intravenous immunoglobulin therapy in 3 cases of parvovirus B19—Associated chronic fatigue syndrome. *Clin. Infect. Dis.* **2003**, *36*, e100–e106. [CrossRef]
56. Tirelli, U.; Lleshi, A.; Berretta, M.; Spina, M.; Talamini, R.; Giacalone, A. Treatment of 741 Italian patients with chronic fatigue syndrome. *Eur. Rev. Med. Pharmacol. Sci.* **2013**, *17*, 2847–2852.
57. Oleske, J. Chronic Fatigue Syndrome Advisory Committee Meeting. In Proceedings of the CFSAC Meeting; Washington, DC, USA, 28 May 2009.
58. Speight, N. Severe ME in children. *Healthcare* **2020**, *8*, 211. [CrossRef]
59. Eaton-Fitch, N.; Johnston, S.C.; Zalewski, P.; Staines, D.; Marshall-Gradisnik, S. Health-related quality of life in patients with myalgic encephalomyelitis/chronic fatigue syndrome: An Australian cross-sectional study. *Qual. Life Res.* **2020**, *29*, 1521–1531. [CrossRef]
60. Kingdon, C.C.; Bowman, E.W.; Curran, H.; Nacul, L.; Lacerda, E.M. Functional status and well-being in people with myalgic encephalomyelitis/chronic fatigue syndrome compared with people with multiple sclerosis and healthy controls. *Pharmaco. Econ. Open* **2018**, *2*, 381–392. [CrossRef]
61. Roma, M.; Marden, C.L.; Flaherty, M.A.K.; Jasion, S.E.; Cranston, E.M.; Rowe, P.C. Impaired health-related quality of life in adolescent myalgic encephalomyelitis/chronic fatigue syndrome: The impact of core symptoms. *Front. Pediatr.* **2019**, *7*, 26. [CrossRef]
62. Nacul, L.C.; Lacerda, E.M.; Campion, P.; Pheby, D.; Drachler, M.D.L.; Leite, J.C.; Poland, F.; Howe, A.; Fayyaz, S.; Molokhia, M. The functional status and well being of people with myalgic encephalomyelitis/chronic fatigue syndrome and their carers. *BMC Public Health* **2011**, *11*, 402. [CrossRef]
63. Winger, A.; Kvarstein, G.; Wyller, V.B.; Ekstedt, M.; Sulheim, D.; Fagermoen, E.; Småstuen, M.C.; Helseth, S. Health related quality of life in adolescents with chronic fatigue syndrome: A cross-sectional study. *Health Qual. Life Outcomes* **2015**, *13*, 96. [CrossRef] [PubMed]
64. Kennedy, G.; Underwood, C.; Belch, J.J.F. Physical and functional impact of chronic fatigue syndrome/myalgic encephalomyelitis in childhood. *Pediatrics* **2010**, *125*, e1324–e1330. [CrossRef] [PubMed]
65. Anderson, J.S.; Ferrans, C.E. The quality of life of persons with chronic fatigue syndrome. *J. Nerv. Ment. Dis.* **1997**, *185*, 359–367. [CrossRef] [PubMed]
66. Schweitzer, R.; Kelly, B.; Foran, A.; Terry, D.; Whiting, J. Quality of life in chronic fatigue syndrome. *Soc. Sci. Med.* **1995**, *41*, 1367–1372. [CrossRef]
67. Komaroff, A.L.; Fagioli, L.R.; Doolittle, T.H.; Gandek, B.; Gleit, M.A.; Guerriero, R.T.; Kornish, R.J.; Ware, N.C.; Ware, J.E.; Bates, D.W. Health status in patients with chronic fatigue syndrome and in general population and disease comparison groups. *Am. J. Med.* **1996**, *101*, 281–290. [CrossRef]
68. Buchwald, D.; Pearlman, T.; Umali, J.; Schmaling, K.; Katon, W. Functional status in patients with chronic fatigue syndrome, other fatiguing illnesses, and healthy individuals. *Am. J. Med.* **1996**, *101*, 364–370. [CrossRef]
69. Royal Brompton & Harefield NHS Foundation Trust. Our Services: Immunoglobulin Therapy. 2011. Available online: https://www.rbht.nhs.uk/our-services/immunoglobulin-therapy (accessed on 8 August 2021).
70. Azizi, G.; Abolhassani, H.; Asgardoon, M.H.; Shaghaghi, S.; Negahdari, B.; Mohammadi, J.; Rezaei, N.; Aghamohammadi, A. Managing patients with side effects and adverse events to immunoglobulin therapy. *Expert Rev. Clin. Pharmacol.* **2016**, *9*, 91–102. [CrossRef]
71. US Centers for Disease Control. Treatment of ME/CFS. Updated 28 January 2021. Available online: https://www.cdc.gov/me-cfs/treatment/index.html (accessed on 28 June 2021).
72. US Centers for Disease Control. Healthcare Provider Toolkit. Page Reviewed 3 May 2021. Available online: https://www.cdc.gov/me-cfs/healthcare-providers/toolkit.html (accessed on 28 June 2021).

73. Beilby, J.; Torpy, D.; Gilles, D.; Burnet, R.; Casse, R.; Hundertmark, J. *Myalgic Encephalomyelitis/Chronic Fatigue Syndrome (ME/CFS) Guidelines—Management Guidelines for General Practitioners*; South Australian Department of Human Services—Primary Health Care Branch, National Health and Medical Research Council: Adelaide, Australia, 2004. Available online: https://www.nhmrc.gov.au/me-cfs#download (accessed on 28 June 2021).

74. Accelerating Change Transformation Team; Alberta Medical Association. Towards Optimized Practice: Identification and Symptom Management of Myalgic Encephalomyelitis/Chronic Fatigue Syndrome—Clinical Practice Guideline. 2016. Available online: https://actt.albertadoctors.org/CPGs/Lists/CPGDocumentList/MECFS-CPG.pdf (accessed on 28 June 2021).

75. National Institute for Health and Care Excellence. Myalgic Encephalomyelitis (or Encephalopathy)/Chronic Fatigue Syndrome: Diagnosis and Management. National Guideline 206. 2021. Available online: https://www.nice.org.uk/guidance/ng206/resources/myalgic-encephalomyelitis-or-encephalopathychronic-fatigue-syndrome-diagnosis-and-management-pdf-66143718094021 (accessed on 29 October 2021).

76. Smith, M.E.B.; Haney, E.; McDonagh, M.; Pappas, M.; Daeges, M.; Wasson, N.; Fu, R.; Nelson, H.D. Treatment of myalgic encephalomyelitis/chronic fatigue syndrome: A systematic review for a National Institutes of Health Pathways to Prevention workshop. *Ann. Intern. Med.* **2015**, *162*, 841. [CrossRef]

77. Provan, D.; Nokes, T.J.C.; Agrawal, S.; Winer, J.; Wood, P. *IVIg Guideline Development Group of the IVIg Expert Working Group. Clinical Guidelines for Immunoglobulin Use*, 2nd ed.; Department of Health: London, UK, 2008. Available online: https://assets.publishing.service.gov.uk/government/uploads/system/uploads/attachment_data/file/216672/dh_130393.pdf (accessed on 31 July 2021).

78. Wimperis, J.; Lunn, M.; Jones, A.; Herriot, R.; Wood, P.; O'Shaughnessy, D.; Qualie, M. *Clinical Guidelines for Immunoglobulin Use*, 2nd ed.; Department of Health: London, UK, 2011. Available online: https://www.gov.uk/government/publications/clinical-guidelines-for-immunoglobulin-use-second-edition-update (accessed on 31 July 2021).

79. Provan, D.; Chapel, H.M.; Sewell, W.A.C.; O'Shaughnessy, D.; on behalf of the UK Immunoglobulin Expert Working Group. Prescribing intravenous immunoglobulin: Summary of department of health guidelines. *BMJ* **2008**, *337*, a1831. [CrossRef]

80. Provan, D.; Chapel, H.M.; Sewell, W.A.C.; O'Shaughnessy, D. Department of health immunoglobulin guidelines are evidence-based and provide a framework for treatment decisions. *BMJ* **2008**. Available online: https://www.bmj.com/rapid-response/2011/11/02/department-health-immunoglobulin-guidelines-are-evidence-based-and-provide (accessed on 18 August 2021).

81. Kindlon, T. On what basis was it decided that immunoglobulin was not suitable for CFS? *BMJ* **2008**. Available online: https://www.bmj.com/rapid-response/2011/11/02/what-basis-was-it-decided-immunoglobulin-was-not-suitable-cfs (accessed on 18 August 2021).

82. Shepherd, C.B. Successful intravenous immunoglobulin treatment for parvovirus B19-associated chronic fatigue syndrome. *BMJ* **2008**. Available online: https://www.bmj.com/rapid-response/2011/11/02/successful-intravenous-immunoglobulin-treatment-parvovirus-b19-associated- (accessed on 18 August 2021).

83. Centres for Disease Control and Prevention. Information for Healthcare Providers, Clinical Care of Patients with ME/CFS. Available online: https://www.cdc.gov/me-cfs/healthcare-providers/clinical-care-patients-mecfs/ (accessed on 8 August 2021).

84. Darabi, K.; Abdel-Wahab, O.; Dzik, W.H. Current usage of intravenous immune globulin and the rationale behind it: The Massachusetts general hospital data and a Review of the literature. *Transfusion* **2006**, *46*, 741–753. [CrossRef] [PubMed]

85. National Blood Authority. Criteria for the Clinical Use of Immunoglobulin in Australia. 2018. Available online: https://www.criteria.blood.gov.au (accessed on 18 August 2021).

86. Rowe, K.S. Long term follow up of young people with chronic fatigue syndrome attending a pediatric outpatient service. *Front. Pediatr.* **2019**, *7*, 21. [CrossRef]

87. Webster, A.D. Intravenous immunoglobulins. *BMJ* **1991**, *303*, 375–376. [CrossRef] [PubMed]

88. Shepherd, C. Intravenous immunoglobulin and myalgic encephalomyelitis. *BMJ* **1991**, *303*, 716. [CrossRef] [PubMed]

89. Proal, A.D.; VanElzakker, M.B. Long COVID or post-acute sequelae of COVID-19 (PASC): An overview of biological factors that may contribute to persistent symptoms. *Front. Microbiol.* **2021**, *12*, 698169. [CrossRef]

90. Stanculescu, D.; Larsson, L.; Bergquist, J. Hypothesis: Mechanisms that prevent recovery in prolonged ICU patients also underlie myalgic encephalomyelitis/chronic fatigue syndrome (ME/CFS). *Front. Med.* **2021**, *8*, 628029. [CrossRef]

91. Komaroff, A.L.; Lipkin, W.I. Insights from myalgic encephalomyelitis/chronic fatigue syndrome may help unravel the pathogenesis of postacute COVID-19 syndrome. *Trends Mol. Med.* **2021**, *27*, 895–906. [CrossRef] [PubMed]

92. Paul, B.D.; Lemle, M.D.; Komaroff, A.L.; Snyder, S.H. Redox imbalance links COVID-19 and myalgic encephalomyelitis/chronic fatigue syndrome. *Proc. Natl. Acad. Sci. USA* **2021**, *118*, e2024358118. [CrossRef]

Review

Identifying and Managing Suicidality in Myalgic Encephalomyelitis/Chronic Fatigue Syndrome

Lily Chu [1,*], Meghan Elliott [2], Eleanor Stein [3] and Leonard A. Jason [2]

[1] Independent Consultant, Burlingame, CA 94010, USA
[2] Center for Community Research, DePaul University, Chicago, IL 60614, USA; meghan.elliott@depaul.edu (M.E.); ljason@depaul.edu (L.A.J.)
[3] Department of Psychiatry, Faculty of Medicine, University of Calgary, Calgary, AB T2T 4L8, Canada; espc@eleanorsteinmd.ca
* Correspondence: lilyxchu@gmail.com

Abstract: Adult patients affected by myalgic encephalomyelitis/chronic fatigue syndrome (ME/CFS) are at an increased risk of death by suicide. Based on the scientific literature and our clinical/research experiences, we identify risk and protective factors and provide a guide to assessing and managing suicidality in an outpatient medical setting. A clinical case is used to illustrate how information from this article can be applied. Characteristics of ME/CFS that make addressing suicidality challenging include absence of any disease-modifying treatments, severe functional limitations, and symptoms which limit therapies. Decades-long misattribution of ME/CFS to physical deconditioning or psychiatric disorders have resulted in undereducated healthcare professionals, public stigma, and unsupportive social interactions. Consequently, some patients may be reluctant to engage with mental health care. Outpatient medical professionals play a vital role in mitigating these effects. By combining evidence-based interventions aimed at all suicidal patients with those adapted to individual patients' circumstances, suffering and suicidality can be alleviated in ME/CFS. Increased access to newer virtual or asynchronous modalities of psychiatric/psychological care, especially for severely ill patients, may be a silver lining of the COVID-19 pandemic.

Keywords: severely ill; suicide screening; suicide assessment; suicide management; chronic illness; primary care; outpatient; adult

Citation: Chu, L.; Elliott, M.; Stein, E.; Jason, L.A. Identifying and Managing Suicidality in Myalgic Encephalomyelitis/Chronic Fatigue Syndrome. *Healthcare* **2021**, *9*, 629. https://doi.org/10.3390/healthcare9060629

Academic Editors: Kenneth J. Friedman, Lucinda Bateman and Kenny Leo De Meirleir

Received: 30 March 2021
Accepted: 16 May 2021
Published: 25 May 2021

Publisher's Note: MDPI stays neutral with regard to jurisdictional claims in published maps and institutional affiliations.

1. Introduction

Myalgic encephalomyelitis/chronic fatigue syndrome (ME/CFS) is a debilitating chronic illness characterized by post-exertional malaise (PEM), unrelenting fatigue, unrefreshing sleep, cognitive dysfunction, and orthostatic intolerance. This illness is estimated to affect at least 0.42% of the United States (US) adult population [1]. ME/CFS causes significant reduction in functioning, and, as with many chronic illnesses, is associated with high rates of disability and unemployment [2,3]. Up to 69% [4] are unable to work and a quarter of patients report being consistently home- or bed-bound. Unfortunately, we do not yet understand the cause(s) of ME/CFS and there are currently no effective disease-modifying treatments: management is targeted at alleviating symptoms.

Multiple studies have found that people with ME/CFS are at an increased risk of death by suicide. In the UK, people with ME/CFS had a more than a six-fold increase in suicide risk (standardized mortality ratio 6.85) compared to the general population [5]. In Spain, 12.75% of people with ME/CFS, compared to 2.3% of the general population, were at risk of suicide [6]. Another study, despite not finding an increased suicide risk, discovered an increased risk of non-fatal self-harm [7], which is a robust predictor of future suicide attempt [8]. A retrospective convenience sample implicated suicide as one of the three leading causes of death in people with ME/CFS, alongside heart failure and cancer [9]. Compared to the general population, the median age at completed suicide was also significantly lower, at 39.3 years of age compared to 48 years [9]. Surprisingly, in a study focusing on the association between suicide and physical illness of all types in the United Kingdom, 16% of the deceased in the county sampled suffered from ME/CFS [10]. Among both moderately and severely ill ME/CFS patients, 39–57.25% [6,9,10] have contemplated suicide, compared to 4% of the general US population [11] and 1–10% of primary care outpatients [12].

Despite this situation being a clear, urgent public health issue, the specific reasons behind the increased risk of suicide in ME/CFS have not been well-examined. Emerging trends in the literature reflect the impact of not only the symptomatology of ME/CFS itself, but also the stigmatization of this illness by social connections (i.e., family members, friends, employers, etc.) [13], medical professionals [14] and the general public. Thus, those with ME/CFS may be at an even higher risk of suicide and mental health comorbidity than those with other chronic physical ailments, due to the additional burden of constantly having to justify, explain, and defend their disease experience [6].

Medical professionals who do not specialize in psychology or psychiatry nor work in a mental health setting play a vital role in suicide prevention and management. It has long been known that multiple or serious medical conditions subject patients to a higher risk of suicide. In the month preceding their completed suicides, approximately half of patients [15] saw a primary care provider at least once. In contrast, 71% had not had any contact with mental health services in the preceding year [16]. Tragically, even though many of these patients expressed suicide ideation or exhibited concerning behaviors during their last medical visit, most medical professionals upon later interview admitted dismissing or downplaying patients' reports [15]. Medical professionals can not only identify people at risk, they can also prevent imminent suicides by directing high-risk patients to immediate/emergency mental health care. For patients determined to be at low- to moderate risk, they could potentially treat them within their medical clinics while collaborating with outpatient mental health providers and other specialists (e.g., medical social workers, occupational therapists, chronic pain management experts).

The purpose of this article is four-fold: (a) to perform a review of the literature on ME/CFS and suicide, (b) to identify risk and mitigating factors for suicide in ME/CFS versus other chronic physical conditions, (c) to outline a strategy for assessing suicidality and (d) to explain basic management of at-risk patients in an outpatient medical setting. Because most ME/CFS research has involved adults, this article focuses on those 18 years of age and older. When we refer to clinicians in this article, we do not mean those professionals who specialize in mental health or work in such a setting (e.g., an internist assisting with medical issues in an inpatient psychiatric unit) but instead those who manage mostly physical health conditions. Figure 1 summarizes the process to assess and manage suicidality described in this article.

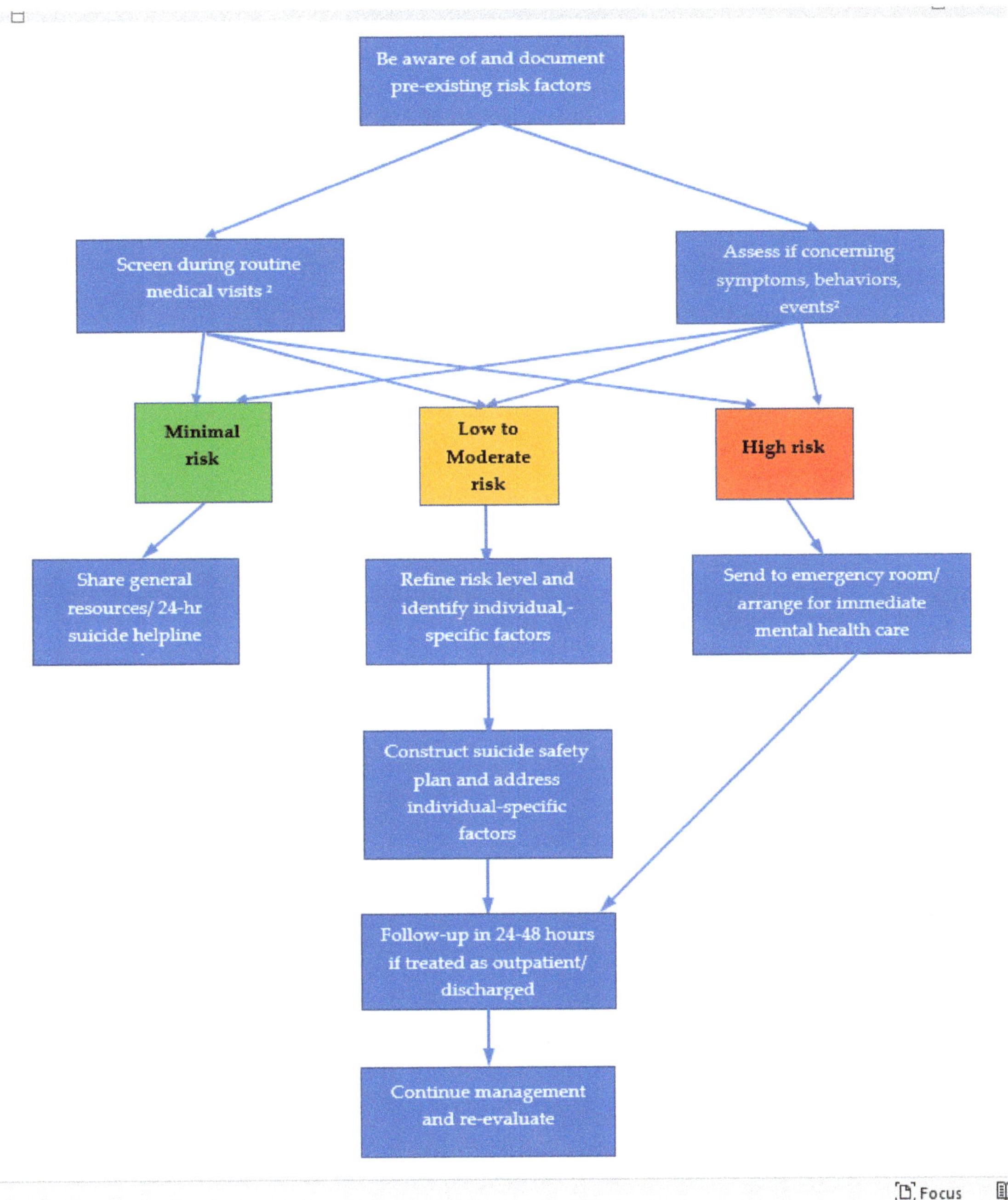

Figure 1. Overall approach to evaluation and management of suicidality in individuals. Use the Ask Suicide-Screening Questionnaire (ASQ) or Columbia—Suicide Severity Risk Scale (C-SSRS) for screening and assessment. See instruments for definitions of initial risk level and text for details.

2. Risk Factors for Suicide

To illustrate how information from this article can be applied, we have created a clinical case (Box 1), based on a composite of patients seen by one author (ES).

Box 1. Clinical case—part 1.

> Maria is a 56-year-old woman who was diagnosed with ME/CFS 10 years ago. She experiences severe fatigue, nausea and dizziness upon sitting or standing up, problems with concentration and memory, and is easily overstimulated. Despite feeling exhausted, Maria describes her sleep as broken and unrefreshing. Her symptoms limit her activities to a total of 2–3 h daily out of bed. If she does more than 2 light household tasks (e.g., washing dishes, shopping for groceries), her symptoms worsen.
>
> She has given up driving after nearly being in a serious accident due to slowed thinking and reaction time. She relies on the bus for medical appointments but the exertion of getting out of bed, dressing, waiting for the bus and being upright for the appointment leaves her exhausted and bed-bound for days. For the last 7 years, Maria has been unable to work as a manager at a telecommunications company.
>
> Maria is divorced, lives alone and has no children. She has lost many friends because she often lacks the energy to get together even by phone. There is no one she can call on for practical help with groceries or rides. She has not been out for a social occasion for a few years. She receives a disability pension but it does not cover her monthly bills. She has drained her savings to supplement her pension.

2.1. What Factors in Maria's Background Place Her at Higher Risk of Suicidality (e.g., Ideation, Attempts, Completed Suicide) Than the General Population?

Maria's situation demonstrates a variety of factors which place her at higher risk of suicidal ideation, attempt, and completion. These factors can be classified by their modifiability and specificity to ME/CFS patients (vs. factors that are common in the general population or any chronically ill patient group). See Table 1 for a list of risk factors and their classification. Factors marked with an asterisk are especially relevant to ME/CFS patients, based on one study in Spain [6] and another in the United States [17]. Demographic and historical characteristics [12,18] are often chronic and not modifiable. History of a prior suicide attempt places patients at the high risk of a future attempt, even up to 3 decades later [19]. Compared to the general population, older and female patients are at higher risk of suicidal ideation although women are less likely than men to complete suicide. In the US and Canada, rates of suicide among people identifying as Native American/Alaskan or First Nations and lesbian, gay, bisexual, trans, or queer/questioning (LGBTQ) are respectively, 1.5–3 times higher than other ethnic/minority groups and 2–6 times higher than heterosexual peers [20–22]. Maria's marital status, solo living situation, poor financial state, and lack of consistent social contact (whether via family, friends, or work) also place her at higher risk. In the United States, among non-depressed patients with ME/CFS, lack of resources, including social/financial support and occupational engagement, was the most cited reason (by 79%) contributing to suicidal ideation [17]. Among Spanish patients, lack of resources was linked to suicidal ideation, depression, and hopelessness [6]. Some of these adverse social factors can be ameliorated but will involve actions, professionals, and agencies beyond those strictly focused on healthcare.

Conversely, medically related factors may be directly modifiable by the clinician even within just their own practice. In Maria's case, persistent symptoms such as disturbed sleep, intense pain, severe limitation in function, and resultant poor quality of life can be addressed and managed further. These are factors that have been found to increase risk of suicidality across a variety of physical health conditions.

One hallmark symptom of ME/CFS is unrefreshing sleep [4], which is also a risk factor for suicide. Ahmedani et al. found that sleep disorders more than doubled the likelihood a person would die by suicide, compared to the general population [16]. In postural orthopedic tachycardia (POTS), often considered a sister disorder of ME/CFS, low sleep quality scores were significantly associated with suicidal ideation [23]. Although this is a troubling implication for suicide in ME/CFS, it also represents a salient opportunity for intervention; treating sleep dysfunction could be a practical way to reduce suicide risk and increase quality of life for people with ME/CFS. In a study of nonmalignant chronic pain—another common experience for those with ME/CFS—a significant indirect effect of chronic pain on suicide risk was found, mediated by disturbed sleep; with sleep removed from the model, the direct effect of chronic pain on suicide risk was nonsignificant [24].

Table 1. Risk factors for suicide.

Potentially Modifiable	Non-Modifiable
Chronic, serious illness [1]	Older age
Sleep disturbances/problems	Male sex
Pain	Caucasian
Other severe symptoms (e.g., cognitive dysfunction, hypersensitivity to stimuli)	Native American/Alaskan Native background
	Identifying as LGBTQ [2]
Depression	
Anxiety	History of self-harm
Substance Abuse	History of suicide attempts
Other comorbid medical conditions (e.g., fibromyalgia, orthostatic intolerance syndromes)	Recently discharged from inpatient psychiatric care
	Personality disorder
Low quality of life	Past traumatic events (e.g., adverse childhood experiences,
Limited function [1]	sexual abuse, domestic violence)
Social isolation, loneliness [1]	
Lack of supportive relationships [1]	Family history of suicide, mental health disorder
Thwarted belongingness [1]	
	Exposure to other people who have committed suicide
Unstable, challenging social circumstances (e.g., homelessness, poverty, unemployment) [1]	
Unsupportive social and healthcare provider interactions [1]	
Lack of/poor coping skills	
Personal beliefs	

[1] Risk factors specifically cited by patients with myalgic encephalomyelitis/chronic fatigue syndrome; [2] Lesbian, gay, bisexual, trans, queer/questioning.

Another highly relevant risk factor is functional limitation. Many chronic illnesses lead to reduction in daily activities; ME/CFS stipulates such a reduction as a criterion for diagnosis [4]. In a recent census-based study of people with chronic physical illnesses in Northern Ireland, the degree of functional limitation in chronic illness was the largest statistical predictor of death by suicide [25]. Those who reported that their day-to-day activities were "limited a lot" were more than three times as likely to die by suicide as those without functional limitations. In a qualitative study of people living with both multiple sclerosis (MS) and late-stage kidney disease, limitation of activities was mentioned by multiple participants as a driving factor in their suicidal ideation [26]. This finding has particularly salient implications for ME/CFS, which has been suggested in multiple studies to reduce function even more than MS [27,28]. Thus, one might expect to see an even greater risk of suicide in ME/CFS than in other chronic illnesses with less impaired functionality. Even within the ME/CFS category, there is a spectrum of physical functioning, with some individuals able to leave the house for work and recreation, while others with greater activity limitations are confined to their homes or even their beds.

The admittedly limited literature on ME/CFS and suicide seems to support functional limitation as a risk factor. Johnson et al. found that people with ME/CFS who were housebound were three times as likely to die by suicide than those who were not housebound, or, interestingly, those who were bedridden [29]. This finding suggests a floor effect, in which reduced functioning below a certain degree acts as a protective factor rather than a risk, possibly through limiting access to lethal means or due to the often-ubiquitous presence of a caregiver who, if understanding and non-stigmatizing, could provide additional social support as well as supervision. Such a floor effect could shed new light on mixed results from earlier studies of suicide in ME/CFS; for example, McManimen et al. did not find an increased risk of death by suicide for those with ME/CFS, but over half of the sample were bed-bound, perhaps explaining the lack of statistical significance in suicide risk [30]. Qualitative work in people who have ME/CFS but do not meet criteria for depression further underscores the role of functional limitation in suicide risk. Devendorf et al. qualitatively analyzed the open-ended responses of people with ME/CFS who endorsed suicidal ideation but did not meet criteria for depression. Reduced ability to participate in daily life was a common theme, with one person remarking [17]: "When crashed, I can do nothing but lie in my bed in total agony and in silence and darkness, trying not to move—sometimes for weeks on end. So, yes, that can be distressing and depressing and make it hard to concentrate or feel hopeful. But my desire for life and to participate in life has not changed. It's not that I don't want to do things; it's that I can't."

Multiple studies suggest that presence of chronic pain, common in those with ME/CFS, is a significant risk factor for suicide. Braden and Sullivan found that the presence of any chronic pain condition was associated with both suicidal ideation and likelihood of suicide attempt. This association held true for both 12-month suicide risk (OR = 1.5, 95% CI 1.1–2.0) and for lifetime suicide risk (OR = 1.3, 95% CI 1.2–1.4) [31]. Interestingly, the type of pain which was associated most strongly with lifetime ideation, plan, and attempt, was the heterogeneous "other" chronic pain category, one which might be particularly relevant to ME/CFS due to its own heterogeneous nature. Chronic pain also feeds into the activity limitation suicide risk discussed previously; the combination of pain and fatigue could limit daily life even further. Fuller-Thompson and Nimigon, in a study of depression risk in those with ME/CFS, found that those whose activities were limited by pain were approximately 1.5 times as likely (OR = 1.59, 95% CI 1.11, 2.26) to have depression as those who did not experience such pain-related limitations [32]. Although pain treatment is a potentially promising target for intervention, it can become complicated when dealing with suicide risk; many commonly used pain medications are fatal in overdose, which must be considered from a reduction of lethal means perspective when addressing chronic pain and suicide.

2.2. What Risk Factors Are Unique or More Prominent in Patients with ME/CFS Compared to Patients Affected by Other Conditions?

There are also factors which are unique to ME/CFS. The unusual symptom of post-exertional malaise (PEM) can lead to and promote other risk factors. PEM refers to the appearance of new or worsening of baseline symptoms when patients engage in ordinary physical or cognitive activities, such as sitting upright, reading a newspaper article, or walking around the house [4]. PEM can occur immediately or be delayed by hours to days and can last hours to days, decreasing a patient's function further. Thus, actions such as taking a walk outside to alleviate depression or meeting with friends to curtail isolation may not be possible. Cognitive dysfunction can manifest as a short attention span, poor memory, decreased comprehension, and word-finding difficulties, resulting in problems communicating and interacting with others. Although cognitive issues are one of the core criteria for ME/CFS, 31% to 45% of those severely affected reported this symptom as not only present but occurring at an intense level [33]. Hypersensitivity to stimuli—whether light, touch, sound, or substances (e.g., certain foods, fragrances)—has also been observed to be more common and intense in the severely ill compared to mildly and moderately affected patients [34]. In some patients, hypersensitivities play an equal or greater role than fatigue or post-exertional malaise in confining them to their homes. All these symptoms contribute to social and physical isolation and further limit function.

Moreover, since there are currently no disease-modifying treatments for ME/CFS, the root of Maria's situation cannot be addressed directly yet. Instead, treatment is concentrated on managing ME/CFS symptoms and supportive care [35]. That can lead to feelings of frustration, disappointment, and hopelessness. Hopelessness is recognized as a risk factor for suicide in chronically ill patients and was cited by 48% of non-depressed people with ME/CFS contemplating suicide [17].

Additionally, since many healthcare professionals are not knowledgeable or continue to hold misconceptions about ME/CFS, patients often feel their experiences are dismissed, downplayed, or disparaged. For decades and up until a few years ago, ME/CFS was attributed to deconditioning [36] or to an irrational fear/avoidance of activity [37]. Home-bound patients were characterized as "pervasively passive" "with a predominant belief in a somatic cause" while caregivers were blamed for "unwittingly contribut[ing] to the persistence of the condition by taking over too many activities of the patient." [38]. Thus, patients were instructed that graded exercise therapy or ignoring/de-emphasizing their own symptoms via cognitive behavioral therapy (CBT) would lead to a cure or improvement. Some researchers and groups even discouraged or warned patients about joining ME/CFS support groups because the latter opposed these treatments [39,40].

We now know those theories are erroneous and even harmful: metabolic, neurologic, and immunologic abnormalities may underlie ME/CFS [4,41–43] and between 54–74% of patients have reported that their health worsened with exercise programs [44]. Nevertheless, these ideas and treatments persist as changes in the practice of medicine frequently take years to reach frontline practitioners. Lack of understanding from healthcare providers, being labelled as "rebellious"/"noncompliant" because they disagreed with now-disproven treatments, being blamed for their own illness, and the burden of having to educate others led to suicidal feelings, depression, and hopelessness among both US and Spanish patients [6,17]. In contrast, medical conditions such as multiple sclerosis, chronic heart disease, and stroke are recognized by the great majority of health professionals as legitimate, severely disabling diseases. Patients can rely on their professionals' knowledge, experience, and sympathy. Many communities even have specialty clinics and designated support services available for these conditions.

Lack of knowledge and negative attitudes also permeate the public's view of ME/CFS. For those living with this illness, such ignorance and stigma can lead to a variety of distressing encounters, even with those considered close social contacts. McManimen et al. found that people with ME/CFS who met depression criteria or endorsed suicidal ideation were more likely than those who did not meet criteria to have experienced unsupportive social interactions—both overall and on specific distancing, minimizing, and blaming subscales—and stigma [13]. This finding suggests that the dismissive, harmful interactions experienced by those with ME/CFS might contribute to the increased suicide risk. In a study comparing people with ME/CFS and/or fibromyalgia to those with an autoimmune disorder, the overall level of unsupportive interactions did not differ, but the nature of such interactions did; those with ME/CFS were significantly more likely to report "distancing" and "minimizing" interactions [45]. People with ME/CFS were, in one study, more likely to report never having been married than those with other chronic illnesses [46], suggesting an illness-specific hindrance of social relationships. In a study of a suicide risk scale for people with POTS, a disorder related to the orthostatic intolerance symptom of ME/CFS, 79% of respondents reported "high" or "very high" loneliness on the UCLA Loneliness Scale [47], further suggesting social impoverishment in ME/CFS as a mechanism for suicide risk. The trend of lack of social support suggests that education and stigma reduction could be a powerful mechanism for suicide risk reduction at both the individual and global levels.

3. Initial Screening/Assessment of Suicide Risk: Is This Patient Currently at Risk of Suicide?

3.1. Who Should Be Assessed for Suicidality and When Should It Be Done?

Some organizations, such as the Joint Commission recommend screening all adult medical patients for suicide [48] while others such as the United States Preventive Services Task Force (USPSTF) and Canadian Centre for Addiction and Mental Health found insufficient evidence for universal action [49,50]. Universal screening may not identify more at-risk patients than selective screening and it is unclear whether earlier intervention is effective. Selective screening targets patients with one or more risk factors. As illustrated by our clinical case (Box 1), patients affected by ME/CFS would certainly fit into this category. Before screening is instituted, clinicians should prepare a reference sheet of local mental health professionals, institutions, and resources they can refer to quickly should patients screen positive and be at a high risk of suicide [51]. Some patients may be reluctant or, if extremely ill, unable to describe their circumstances: reports from family, friends and caregivers should be heeded.

Ideally, all patients with ME/CFS should be screened upon initial intake and then occasionally through the years, perhaps linked to when other preventive measures are being discussed or carried out (e.g., annual influenza vaccination, mammograms, etc.). Conducting screenings during these times can be framed as part of the process the clinician and/or clinic regularly performs for all patients. Acceptability of screening is high with between 81–95% of patients [52,53] deeming it to be an appropriate component of inpatient and outpatient medical care.

Although some patients will directly report suicidal thoughts or behaviors during an office visit, up to 81% of people who saw their physician shortly before dying did not. However, reviews of the medical records and interviews with clinicians suggest premonitory signs [12]. Certain beliefs, statements, symptoms, and actions expressed by patients should prompt more immediate screening (Table 2). Feelings of hopelessness, loneliness, disconnectedness from others, and being a burden to society [54] have been linked to increased suicide risk among the chronically ill. Statements directly or indirectly surrounding these feelings should be explored further. Mood changes encompass new onset of depression and anxiety, exacerbation of pre-existing mood disorders, and rapid/intense fluctuations. People who exhibit agitated or impulsive behaviors may be more likely to attempt suicide rather than merely confining themselves to thoughts [55]. Clinicians should also pay attention to worsening or relentless chronic pain, sleep, or other symptoms. Patients who suddenly seem more peaceful without a clear cause after a period of depression should be assessed especially carefully: their lightening of mood may be due to finally deciding to proceed with a suicidal plan.

Another trigger to query patients is when major negative events happen, singularly or in quick succession. Examples of such events are divorce, unemployment due to disability, sudden worsening of health, denial or loss of disability benefits, failure of a highly anticipated treatment, and threat of homelessness. Other times, without an inciting event, patients may simply become weary of their unrelenting symptoms and difficult circumstances. Indeed, although they may occur for reasons unrelated to suicide, abrupt cessation of treatments, withdrawal from care and avoidance of contact with health professionals have been flagged as potential warning signs [55]. If a previously engaged patient suddenly disappears, health professionals should explicitly ask why: is it due to a mood disorder and/or hopelessness or for more mundane reasons (e.g., cost of care, preference for another provider)? The period immediately after a recent suicide attempt or discharge from inpatient/outpatient psychiatric/psychological care is also acknowledged to be perilous times.

A third pattern is acute-on-chronic distress. A patient may be coping to some degree with chronic risk factors but then be blind-sided by additional events (Box 2).

In Maria's case, she appeared to be coping somewhat until development of new symptoms (widespread pain, breathlessness, rash) and disruptions to her housing situation. Through careful listening and observation of a patient, health professionals can initiate a conversation by calmly stating "The last few weeks sound really challenging. How have you been coping?" and then seeing how the patient responds. Some, such as Maria, will openly and willingly share how they feel (e.g., hopeless, weary, and anhedonic) in which case the health professional should let them talk uninterrupted.

Table 2. Concerning statements, symptoms, behaviors, and events should prompt clinicians to assess for suicidality.

Statements
Passive suicidal ideation: "I wish I could go to sleep one day and not wake up." Active suicide ideation: "I am tired of living and looking for a way out." Depression: "I feel sad/cry all the time." Feeling like a burden to family/others: "My family would be better off if I were dead." Hopelessness: "I have nothing to look forward to." "Life is meaningless." Loneliness: "There is no one I can talk to about my problems." "I don't have any friends."
Symptoms
Changes in mood, including onset/exacerbation of depression anxiety; dramatic fluctuations Worsening somatic symptoms, especially pain and insomnia Anger, irritability
Behaviors
Agitated actions: pacing, shaking, rapid/loud speech Impulsive behaviors Withdrawal from care: stopping treatments, missing appointments, avoiding contact Repetitive self-harm Drinking or abusing other substances more than usual Decreasing social contact Giving away items which are important/meaningful to patient Ceasing activities previously enjoyed
Events
Unemployment Loss of significant relationships (e.g., divorce, death of loved one) Denial of disability benefits Homelessness Anticipated treatment is not effective Recent suicide attempt Recent discharge from inpatient/outpatient psychiatric care

Box 2. Clinical case—part 2.

Recently, Maria underwent cervical spinal stabilization surgery for left arm pain, numbness, and weakness. Although she regained use of her arm, she is left with chronic pain which gradually spreads from her left arm to her whole body. Her building had a water leak and her apartment suffered water damage. Maria developed a rash, difficulty breathing and increased general unwellness. She suspects mold growth but was unable to afford to have it investigated. The building management has denied there is a problem. Maria wants to move but her limited health, financial means and social connections make it hard to do so.

Maria reports feeling hopeless at her appointment today. She says she does not want to endure additional health problems. It all feels too much to handle. She cannot think of anything she enjoys any more.

It bears repeating that asking or talking about suicide with patients does not stimulate new or encourage existing thoughts of suicide. Both a 2014 review [56] and a 2018 meta-analysis [57] addressing this issue found no significant risks from screening; suicide risk and attempts were slightly reduced instead. Many people will feel relieved that the practitioner is taking their distress seriously. Such discussions are also not futile. People who attempt or survive suicide often demonstrate ambivalence about their actions [58]. Some part of them wants to live: most dramatically, all 29 people who survived a jump from the Golden Gate Bridge recounted they regretted their decision as soon as they stepped off the bridge [59,60]. For chronically ill people, suicide may be less about ending their lives than ceasing their symptoms and the downstream consequences. For example, Anne Ortegen, afflicted with severe ME/CFS, noted she still retained her joy and curiosity about life but "unbearable [physical] suffering" with no effective treatment in sight compelled her to seek medical aid in dying [61].

Unfortunately, some physicians and authorities, erroneously attributing ME/CFS symptoms to psychiatric/psychological etiologies, have forcibly detained patients in inpatient psychiatric units. Children [62] and severely affected adults have been especially vulnerable since they are unable to advocate vigorously for themselves. For example, Sophie Mirza, whose health was declining rapidly, was labelled as suicidal and subsequently hospitalized. It took her family much time, expense, and legal maneuvering to obtain her release. When she passed away, an autopsy discovered significant inflammation of her spinal cord, which might have contributed to her condition [63]. Thus, patients may be rightfully reluctant or fearful of admitting to suicidal thoughts or actions. Clinicians can reassure patients by framing assessments within the context of chronic medical illness (e.g., "Many chronically ill patients, like my patients with heart or lung disease occasionally experience suicidal feelings. Have you felt similarly?") and informing them that suicidality can often be cooperatively managed with the patient maintaining agency in an outpatient setting. Contemporary mental health standards encourage care in the least restrictive manner and setting possible [64].

3.2. How Should Patients Be Screened or Assessed? What Issues Should Clinicians Pay Attention to or Ask about?

Since many medical professionals do not feel confident assessing suicidality [65,66], we recommend using validated, standardized instruments as much as possible. For routine screening, self-administered versions of these instruments can be used, or the questions asked by ancillary staff (e.g., medical assistants) but for more urgent situations, the healthcare professional should ask the questions themselves. Severely affected patients may need help from their caregivers with completing the questionnaires or their caregivers may have to serve as proxy respondents. Questionnaires may need to be administered over the telephone or via virtual visits. Using these tools assures that salient issues are covered using validated questions, patient responses are appropriately interpreted via an expert-vetted scale, and thorough documentation of this crucial conversation exists.

Currently, there is a lack of brief, validated instruments for screening or assessment specifically created to be used by medical professionals in busy clinic settings. Two recent candidates are the US National Institute of Mental Health's Ask Suicide-Screening Questions (ASQ) [67,68] (Figure 2) and Columbia-Suicide Severity Rating Scale (C-SSRS) [69] (Figure 3). Both instruments take only minutes to administer (5 and 6 questions respectively), address passive and active suicidal ideation (both types confer equal risk), ask simple Yes/No questions initially, and prompt patients for more details. If patients are unable or unwilling to answer any question on the ASQ, the default is to score the question as though the patient answered "Yes". Although 4 of the 5 ASQ's questions concern thoughts, the C-SSR covers thoughts, methods, intentions, and specific plans. "Yes" responses on each subsequent question are linked to escalating suicide risk. Stronger intentions and more recent, explicit plans indicate a higher risk [18,70]. Note both instruments inquire about lifetime history of suicidal actions [71]; unlike suicidal ideation in the remote past, actions anytime in the past significantly increase risk of another attempt.

Both instruments superbly separate out non-suicidal patients; however, the ability to identify at-risk patients accurately is mixed. Negative predictive values are the percentages generated when patients whose questionnaires indicate minimal risk and who truly are minimal risk (true-negatives) are divided by the total number of patients (both true- and false-negatives) determined by the questionnaire to be at minimal risk. Positive predictive values are calculated the same way except they concern patients whose questionnaires indicate some risk of suicide. For the C-SSRS, among psychiatric or non-psychiatric subjects enrolled in a variety of clinical trials, negative predictive values (NPV) for prospective suicidal behavior ranged from 97.93% to 99.63%. In contrast, the positive predictive values (PPV) are quite low, ranging from 8.97% to 16.49%, with the higher values observed in psychiatric patients [72]. For the ASQ, in medical and surgical inpatients, compared to the longer, more detailed Adult Suicidal Ideation Questionnaire, the NPV was 100% and the PPV was 32% [68].

Ask the patient:

1. In the past few weeks, have you wished you were dead? Yes No
2. In the past few weeks, have you felt that you or your family would be better off if you were dead? Yes No
3. In the past week, have you been having thoughts about killing yourself? Yes No
4. Have you ever tried to kill yourself? Yes No
   If yes, how?

   When?

   If the patient answers yes to any of the above, ask the following question:
5. Are you having thoughts of killing yourself right now? Yes No
   If yes, please describe:

Next steps:

- **If patient answers "No" to all questions 1 through 4, screening is complete** (not necessary to ask question #5). No intervention is necessary (**Note: Clinical judgment can always override a negative screen*).
- **If patient answers "Yes" to any of questions 1 through 4, or refuses to answer, they are considered a positive screen. Ask question #5 to assess acuity:**
  - **"Yes" to question #5 = acute positive screen (imminent risk identified)**
    - Patients require a STAT [1] safety/full mental health evaluation. Patient cannot leave until evaluated for safety.
    - Keep patient in sight. Remove all dangerous objects from room. Alert physician or clinician responsible for patient's care.
  - **"No" to question #5 = non-acute positive screen (potential risk identified)**
    - Patient requires a brief suicide safety assessment to determine if a full mental health evaluation is needed. Patient cannot leave until evaluated for safety.
    - Alert physician or clinician responsible for patient's care.

[1] Short for Latin, "statim," meaning immediately.

Figure 2. Ask Suicide-Screening Questions (ASQ). Reproduced with permission from Dr Lisa M. Horowitz, Arch Ped Adolesc Med, published by JAMA Network, 2012 [67].

These low PPVs are in-line with other suicide assessment instruments [73] and may not be as alarming as they first seem. Predictive values are substantially influenced by the prevalence or absence of a condition in a population. As the prevalence of a condition increases, the PPV increases whereas the NPV decreases. Suicidal attempts and completions, not ideation, in primary care outpatients are still relatively rare, with a prevalence of 1% [18] or less. Thus, PPVs will be low. The prevalence is likely higher among chronically medically ill patients (e.g., 8.9%) [74]: with that increase, the PPV will increase. Secondly, predicting future suicide attempts might not be the best measure of these instruments' effectiveness. Assessment and screening per se do not prevent or stop suicide attempts. Instead, these actions help clinicians to identify who needs further evaluation and to design individualized care plans. If the plan works as intended, suicide attempts may never be made. Conversely, if the patient is resolved to commit suicide, attempts and completed suicides will result.

Ask questions that are in bold and underlined.	Past month	
Ask Questions 1 and 2	**YES**	**NO**
1) *Have you wished you were dead or wished you could go to sleep and not wake up?*		
2) *Have you had any actual thoughts of killing yourself?*		
If YES to 2, ask questions 3, 4, 5, and 6. If NO to 2, go directly to question 6.		
3) *Have you been thinking about how you might do this?* e.g. "*I thought about taking an overdose but I never made a specific plan as to when where or how I would actually do it....and I would never go through with it.*"		
4) *Have you had these thoughts and had some intention of acting on them?* as opposed to "*I have the thoughts but I definitely will not do anything about them.*"		
5) *Have you started to work out or worked out the details of how to kill yourself? Do you intend to carry out this plan?*		
6) *Have you ever done anything, started to do anything, or prepared to do anything to end your life?* Examples: Collected pills, obtained a gun, gave away valuables, wrote a will or suicide note, took out pills but didn't swallow any, held a gun but changed your mind or it was grabbed from your hand, went to the roof but didn't jump; or actually took pills, tried to shoot yourself, cut yourself, tried to hang yourself, etc. If YES, ask: *Was this within the past 3 months?*	Lifetime Past 3 Months	

Figure 3. Columbia-Suicide Severity Rating Scale (C-SSRS): screen with triage points for primary care. Pay attention to question 2: if "Yes", ask questions 3 through 6; if "No", ask question 6. "Yes" responses in white boxes signify minimal risk; yellow boxes, low risk; orange, moderate risk; and red, high risk. Patients should be classified according to the highest risk box to which they reply "Yes". Reproduced with permission from Dr Kelly Posner Gerstenhaber. The Columbia Lighthouse Project (https://cssrs.columbia.edu/ (accessed on 17 May 2021)).

3.3. Why Should Clinicians Screen for Suicide Directly, Independent of Mood Disorders or Anxiety?

Confronting suicide directly (Box 3) and not only within the context of depression or anxiety is important. Although both the ASQ and C-SSRS's questions appear blunt, they avoid confusion or equivocation. In the past, 90% of suicides were attributed to psychiatric disorders [75]. Consequently, suicide evaluations customarily take place during examination for mood disorders. One popular method uses item 9 of the Patient Health Questionnaire-9 (PHQ-9) [76] which pertains only to the frequency of suicidal thoughts. However, up to 30% of patients endorsing absence of suicidal thoughts may currently be at risk [77] while 16–19% endorsing any presence may not be [77]. In one study, the PHQ-9 flagged 24% of subjects as suicidal versus the 6% and 1% detected, respectively, by the C-SSRS and clinical interviews [77]. Secondly, as mentioned in the section on risk factors, chronically ill patients may be suicidal for reasons unrelated to a mood disorder. The 90% figure cited previously originated from controversial "psychological autopsies" [78] (e.g., interviews of surviving family members) after completed suicides and closer examination in one study found only 17% were linked with depression [79]. More recent data from the US Centers for Disease Control and Prevention indicate that 54% of those dying by suicide in the US do not have any known mental health condition [80]. Third, steps to prevent suicide can be implemented right away and go beyond treating psychiatric disorders, as discussed later in this article. Conversely, it may take weeks to months for patients to start seeing the benefits of some depression and anxiety treatments.

Box 3. Clinical case—part 3.

> Maria's doctor administers the C-SSRS. Maria admits in the last 2 weeks, she has wished she would develop a fatal illness to put an end to what she views as a senseless existence. She occasionally thinks about taking more of the pain pills she was prescribed but explains that she does not have the courage to commit suicide. She worries that if she messes up she could be left worse off. She denies having the means to kill herself. She does not own any firearms and has not stockpiled medications. Maria has no history of suicide attempts.

4. Secondary Assessment of Suicide

4.1. Is This Patient at Low, Moderate, or High Risk of Suicide?

A clear advantage of both questionnaires is they quickly help clinicians determine if and how urgently management is needed, along with the best setting/professionals for care. If patients answer "No" to the first 4 questions of the ASQ [68] (Figure 2), the clinician does not need to ask the remaining question 5. For the C-SSRS [69] (Figure 3), a color-coding system is used with white, yellow, orange, and red boxes checked "Yes" indicating, respectively, minimal, low, moderate, and high risk. If the patient denies C-SSRS question 2, question 6 should be asked next. Absence of preparatory suicidal behaviors in their lifetime and in the last three months classifies the patient as minimal risk (i.e., only white boxes are marked). Responses should be recorded but no specific, current intervention is needed for patients falling in this category.

Conversely if patients answer "Yes" to question 5 of the ASQ or any one of questions 4, 5, and 6 (within the last 3 months) of the C-SSRS, these patients are deemed high risk. Immediate care needs to be instituted, whether that be a referral to the emergency room or a same-day appointment with an outpatient mental healthcare professional for urgent, comprehensive care. High-risk patients should not be left alone at any time and should be watched by a family or staff member whether they are at home or in the clinic. The patient's clothes, possessions, and environment should be searched and cleared of any potentially harmful/lethal implements (e.g., guns, pills, knives, ropes/rubber tubes). If transportation to the emergency room or other facility is needed, consider calling an ambulance rather than having the patient or family member drive themselves or take public transport. If the patient interview suggests they have acted, ask the patient directly if they have done anything to hurt themselves already, such as overdosing on sedatives or cutting themselves. These behaviors may require immediate medical and not just psychiatric care.

Based on Maria's answers on the C-SSRS, answering "Yes" to questions 1, 2, and 3 but "'No" to questions 4, 5, and 6 (Figure 3), she would be classified as at moderate risk (Box 4). For the ASQ (Figure 2), this category would apply to those answering "Yes" to any of the first four questions but "No" to item 5. Patients answering only "Yes" to questions 1 and/or 2 on the C-SSRS are deemed low risk. These patients should undergo further appraisal by the clinician to refine the level of risk, decide upon urgency of outpatient mental health care referral and identify modifiable risk factors. Although they may not require emergency room to services or a mental health appointment that day, there are still actions the clinician should take during the visit.

Regardless of how patients are classified during this initial stage, the clinicians should thank the patient for trusting them enough to share these intimate and often painful experiences. All patients should be provided the contact information for the national 24-h suicide prevention hotlines in their country. For the United States, this number is 800-273-8255 (https://suicidepreventionlifeline.org/ (accessed on 17 May 2021)); for Canada, 1-833-456-4566 (https://www.crisisservicescanada.ca/en/ (accessed on 17 May 2021)); and for other countries, see https://www.opencounseling.com/suicide-hotlines (accessed on 17 May 2021)).

Box 4. Clinical case—part 4.

> Based on her C-SSRS score, Maria appears to be at moderate risk of suicide.

4.2. How Can Risk Be Further Evaluated, Especially for Patients Deemed to Be at Moderate Risk?

Review factors that increase (Table 1) and reduce risk (Table 3) with the patient if they have not been documented previously or have not come up during the current visit. The developers of both the ASQ [70] and C-SSRS [81] also provide similar lists on their websites. Because the ASQ concentrates on suicidal thoughts but not current intention, plans, or actions, the secondary level of risk assessment encourages clinicians to discuss these topics with the patient.

Table 3. Protective factors for suicide.

Potentially Modifiable	Non-Modifiable
Religious background/personal beliefs	Younger age
Positive coping behaviors	Female Sex
Strong relationships	Having children
Stable social circumstances (e.g., financial status, housing)	Marriage
Supportive clinical interactions	Pregnancy

Taking a thorough social history including where the patient currently lives; who they live with; whether they are engaged in work, school, or other activities; who they can rely on for regular support, logistical and psychological; and non-medical stressors in their lives (e.g., employment, divorce, bankruptcy, homelessness) can be revealing. Because ME/CFS, by definition, substantially impairs daily function, ask particularly about any difficulties performing basic (e.g., ambulating, toileting, bathing, etc.) or instrumental (e.g., cooking, shopping, managing medications, etc.) activities of daily living [82]. Ask even if a patient appears functionally normal during an appointment. To interact with clinicians, patients often save up energy before a visit or, conversely, plan to suffer the consequences of any physical/cognitive exertion afterwards.

All patients should be examined for depression and anxiety using standardized, validated instruments such as the Patient Health Questionnaire-2 (PHQ-2), General Anxiety Disorder-7 (GAD-7), or Hospital Anxiety and Depression Scale (HADS) [83–85]. Instruments emphasizing fatigue, insomnia, decreased activity, or appetite—symptoms which may be due to ME/CFS itself—may mistakenly label patients with major depression. Furthermore, the most current definition of ME/CFS, created by the National Academy of Medicine in 2015 [4], includes the symptom of orthostatic intolerance (OI). Upon sitting or standing up for periods as short as a few minutes, patients may report dizziness, nausea, confusion, heart palpitations, and short of breath. These symptoms disappear or improve upon lying down but mimic those of anxiety. Hence, it is no surprise that patients affected by OI have been misdiagnosed with anxiety [86]. The HADS has been used in several ME/CFS studies and all three instruments focus on affective rather than somatic symptoms, thus reducing the risk of overdiagnosing psychiatric disorders.

If not expressed, assess for loneliness, thwarted belongingness, and burdensomeness. Items originating from a suicide-screening instrument by Pederson et al. [47] intended for patients specifically affected by "invisible" illnesses (i.e., those without immediate, visual clues of illness, such as weight loss or jaundiced skin) could be adapted. Clinicians can ask: "How often do you feel lonely?" "Do you feel like you are a burden to your family or caregivers?" "Do you have a sense of belonging within your family?" "Is there anyone or any group you feel connected with?"

Conversely, review protective factors (Table 3) with the patient. If not voluntarily offered (e.g., "Even though I feel bad, I would never kill myself. My children need me."), ask patients directly "What would keep you from harming yourself?" "What makes your life worth living?" Some factors such as being a parent or being married are outside of the clinician's hands while others can be introduced or reinforced, such as positive coping behaviors.

There is no threshold upon which a patient's risk of suicide can be confirmed, downgraded, or upgraded. Instead, clinicians will need to examine the quantity, nature, and intensity of risk and protective factors to decide a patient's final status (Box 5). The patient's final status and the rationale for it should be documented in the medical records. The clinician's intuition about a patient/situation should always override any answers on questionnaires or checklists.

Box 5. Clinical case—part 5.

> Maria is assessed for anxiety and depression-related disorders. She scores 18/27 on the PHQ-9, a score suggestive of moderately severe depression. Her scores on the HADS are 15 for depression and 11 for anxiety, both in the clinical range. She has never smoked, rarely drinks, and denies abuse of other substances. She denies impulsive behaviors.
> Her other risk factors are her chronic medical problems (including ME/CFS, pain, unrefreshing sleep, possible OI, and breathing problems), hopelessness, functional limitations, social isolation, poverty, and unstable housing situation. Her protective factors include her religious faith, her belief that she is not brave enough to commit suicide and a clinician she feels comfortable speaking with. Overall, her risk level of moderate remains unchanged.

5. Managing Suicidality

5.1. What Steps Would You Take Next? What Are Interventions All Suicidal Patients Should Receive?

A 2-step approach incorporating both general management of suicidality as well as management of patient-specific factors can be implemented. The approach outlined here applies mostly to patients at low to moderate risk who are suitable for outpatient treatment. For high-risk patients, their initial care may be carried out by the emergency room and/or an inpatient psychiatric unit. After discharge, similar steps can be used for these high-risk patients if they have not already been initiated by prior clinicians.

General management consists of (a) referring patients to mental health professionals and (b) collaborating with patients to create a suicide safety plan. As mentioned earlier, clinicians should generate and maintain a list of local mental health professionals, facilities, and resources so that referrals can be made as quickly and seamlessly as possible. Ideally, patients should be seen or contacted within 48–72 h. Psychiatrists can help with pharmacologic management of depression, anxiety, and some symptoms such as sleep. However, even patients without anxiety or depression can benefit from mental health care [87]. Psychological treatments using dialectical behavioral and cognitive behavioral therapy processes as well as a new therapy called Collaborative Assessment and Management of Suicidality are designed to specifically address suicidality [88]. Clinical trials and observational studies show these treatments decrease suicidal thoughts and attempts by 37.5% to 60% [88,89].

Before patients leave their office or end the visit, clinicians should collaborate with them to create a suicide safety plan [90]. Like diabetes "sick day" or asthma action plans most medical providers are already familiar with, these are written documents the patient can easily look to for guidance when suicidal thoughts and feelings surface or intensify. These plans are not just practical by themselves; they also help the patient feel more in control by reminding them of the alternative actions they compiled [90,91]. "Tunnel vision" or psychological constriction, whereby patients feel trapped and cannot see other options, is well-recognized among suicidal patients [58]. Engaging in alternative actions also allows impulsive thoughts to dissipate [92].

In 2008, Brown and Stanley produced a 6-step suicide safety plan which clinicians and patients can readily complete together [92]. Other staff (e.g., nurses, medical assistants, etc.) can also be trained to help patients fill out the form. Table 4 lists these 6 components, sample questions to elicit responses, and examples of how patients might answer. For a paper template that can be immediately printed out and used, see or the suicidesafetyplan.com website. Encourage patients to be as detailed as possible and to fill in steps using their own words. However, if a patient does not find one of the steps useful for them, they can skip it. People named in "Step 3" do not necessarily need to know of the patients' suicidality whereas those in "Step 4" and "Step 5" might already know or can be informed by the patient. Firearms are a common and lethal method of suicide in the United States, so Step 6 should always include questions about access to handguns and rifles. Since patients with ME/CFS often take sleep, pain, or other medications, ask about which drugs they have, whether they have stockpiled tablets and how they are handled. Additionally, customize the answer: if a patient brings up leaping from a bridge, make sure Step 6 addresses that method. Although not a step, ask patients to record their reasons for living on the form. Finally, ask patients how likely they are to carry out the steps and what obstacles they might encounter. If needed, revise the plan so it will be simple to actualize.

Table 4. Suicide safety plan by Brown and Stanley.

Component	Ask Patient	Example Answers	Comment
1. Warning signs	How will you know when the safety plan should be used?	"Feeling hopeless." "Thinking life is all downhill from here." "Lying in bed more than usual."	Thoughts, behaviors, moods, events that lead to suicidality.
2. Internal strategies	What activities can you do on your own if you become suicidal again, to help yourself not to act on your thoughts or urges?	Sit outside in the sun, listen to relaxing music, take a warm bath.	
3. People and settings that provide distraction	Who helps you take your mind off your problems at least for a little while? Where can you go where you will be around people in a safe environment?	Knitting group, the park near my home, online patient support group.	People named need not know about the patient's suicidal feelings. Places may allow casual interactions.
4. People whom I can contact for help	Who is supportive of you and who do you feel that you can talk with when you are under stress?	My neighbor Sarah, my church's pastor.	These are people who are aware of or could be trusted with the individual's suicidal thoughts/feelings.
5. Professionals and agencies I can call in a crisis	Who are the medical/mental health professionals that we should identify to be on your safety plan?	Springfield Emergency Room, my psychiatrist Dr Joseph Lopez, National Suicide Prevention Lifeline, 911	List contact information.
6. Making the environment safe	What items do you have around you that you might use to hurt/kill yourself? How can we make your surroundings safe for you?	Doctor/pharmacy will limit number of medications mailed to one week at a time. Place kitchen knives in locked cabinet.	Always ask about firearms. Means restriction should be matched to the methods the individual names.
7. My reasons for living [1]	What makes your life worth living? What brings joy to your life?	My children, my faith, my pets, enjoying nature.	

[1] Except for this step, all others are drawn from Brown and Stanley's work on suicide safety planning. Adapted with permission from Dr Barbara Stanley, Cognitive and Behavioral Practice, published by Elsevier, 2012 [92]. Please see Figure S1 in Supplementary Materials or suicidesafetyplan.com (accessed on 17 May 2021) for a downloadable template which can be used with patients.

Completion of the form is estimated to take between 30 to 45 min. Afterwards, several copies should be made, including one for the clinician's medical records, and several copies for the patient to be stored in convenient locations. For example, a miniature copy on their nightstand or a scanned version on their mobile phone might be easier to find than a paper form in a desk drawer. Obtain written permission to share the plan with the patient's supporters so they can reinforce the steps. For more information about how to use the template, visit suicidesafetyplan.com. Some clinicians express not knowing how to ask about means and recommending ways to restrict them [93]: the Suicide Prevention

Resource Center provides free online training via their Counseling on Access to Lethal Means program [94].

Compared to patients receiving usual care, patients introduced to suicide safety planning during an emergency room visit were half as likely to attempt suicide in the subsequent 6 months [95]. Furthermore, two thirds of this group cited the plans as instrumental in reducing their suicide risk and twice as many showed up for follow-up mental health appointments as those in the usual care group. In contrast, although they have been recommended for many years, there is no consistent evidence of effectiveness for no-suicide contracts, whereby the patient promises the clinician they will not take action [96]. Although such contracts may make clinicians feel more secure, they do not help patients in crisis and may even compel patients to conceal intense feelings, out of a misguided effort to avoid disappointing the clinician.

The 2-step approach is illustrated in Box 6.

Box 6. Clinical case—part 6.

Maria's doctor informs her that based on what she has expressed, her medical/social situation, and the questionnaire results, she is at moderate risk of suicide. She introduces Maria to the purpose of suicide safety plans and together they start completing one. Initially, Maria cannot think of calming activities nor who she can count on for support. With a little more probing, she remembers she dropped her knitting hobby after work became too busy and that her neighbor Sarah has said to call any time to chat. While they are completing the form, the doctor asks the receptionist to set up an appointment in the next 2 days with a psychiatrist, Dr Joseph Lopez, who offers virtual appointments in his practice. The medical assistant prints out materials about suicide with the national suicide prevention hotline and a local helpline on them. She gives Maria the pamphlet.

5.2. What Are Individual-Specific Suicide Risk Factors? How Should They Be Addressed?

Individual-specific factors refer to those characteristics covered in Table 1 and in the secondary suicide risk assessment evaluation. They are called "individual-specific" because not every suicidal patient is affected by them nor is their degree of influence the same for each patient. Some factors even increase risk in one patient but are protective in another. For example, a happy marriage is protective but one marred by conflict or abuse is not. A well-paying meaningful job might be protective while a low-paying, stressful position may be worse than unemployment. Thus, based on what the patient reports, the clinician will need to make a judgment about how to classify a factor. Table 5 shows one way to categorize patient-specific factors, examples of factors, and examples of interventions to address them.

Two factors that require immediate treatment regardless of the presence of others are anxiety and depression. As mentioned above, while depression does not explain all suicides, it may be linked to 17% to 54% of them [79,80]. Between 21–88% and 17–47% [97,98] of patients with ME/CFS may be afflicted by anxiety or mood disorders, respectively. HADS scores above 11 indicate presence of anxiety or depression: Maria's results indicate she is affected by both. Pharmacologic treatment of depression and anxiety for people with ME/CFS is no different from those without it [97,99]. As with treatment of any patient with a chronic medical condition, clinicians should avoid medications that could interact with existing medication and exacerbate ME/CFS symptoms (e.g., cognition) while favoring medications that may serve dual purposes (e.g., citalopram can be used for both anxiety and depression). Some patients may react strongly to medications, especially the severely ill: thus, starting at a lower dosage than usual and titrating up slowly is wise.

Although psychological treatments such as CBT are not recommended for ME/CFS, they are moderately effective for mood disorders and anxiety disorders [100,101]. Maria should be referred to a counselor or therapist who can support her over the short to medium term. If possible, select mental health professionals who regularly treat patients with chronic, disabling medical illnesses and who are familiar with or willing to learn about ME/CFS. Patients may need to be convinced that psychological treatment is not being prescribed for ME/CFS itself. Tell them you can refer them to another mental health professional if the initial practitioner is not compatible with them.

Table 5. Interventions addressing individual-specific risk factors for suicide.

Category	Examples of Specific Factor	Examples of Interventions	Comments
ME/CFS [1] symptoms	Sleep	Cognitive behavioral therapy—insomnia Blue light filters Exposure to natural light [2] Amitriptyline [3] Trazodone [3]	Evaluate for pain and sleep conditions with specific treatments (e.g., obstructive sleep apnea, migraine).
	Pain	Re-positioning Massage Heat/ice Gabapentin [3] Tricyclic antidepressant [3]	
Comorbid psychiatric conditions	Major depressive disorder	Referral to mental health professional CBT [4] Citalopram [3] Venlafaxine [3]	
Comorbid medical conditions	Multiple chemical sensitivity	Avoid/reduce exposure to concerning stimuli	Exercise may not be suitable for many patients. If used, start at a low level and continue/increase only if patient tolerates.
	Postural orthostatic tachycardia syndrome (POTS)	Isotonic fluids, support hose, awareness/prevention of exacerbating factors, recumbent exercises, fluoxetine [3]	
Isolation/loneliness/ social support	Healthcare professionals	Validation of patient experience Reflective listening Caring contacts	Caring contacts are brief, intermittent e-mails, cards, phone calls to patients by staff between visits.
	Family/caregiver	Educate about ME/CFS Educate about caregiver stress	
	Community support	In-person activity/support groups Electronic forums specific for ME/CFS Virtual support groups	Caregivers need respite/ support to provide support.
Functional Limitations	Ambulation	Refer to physical therapy Bedside commode Wheelchair	
	Bathing	Refer to occupational therapy Hand-held shower head Shower chair	
Other Support	Poverty	Food banks, vouchers Apply for disability financial support	Clinic/facility-based medical social workers can help patients find and apply for programs.
	Homelessness	Home-sharing/roommate arrangements Government-supported housing vouchers	

[1] Myalgic encephalomyelitis/chronic fatigue syndrome. [2] For some patients, especially the severely ill, bright light may worsen their ME/CFS. For others, light sensitivity is not a problem or is tolerable with sunglasses. [3] Start all medications at lower dosages and titrate up slowly. Pain, sleep, and sedative medications may need to be given in smaller quantities (e.g., a week's supply) initially due to risk of suicide. [4] Cognitive behavioral therapy.

Ideally, mental health professionals should be notified that common solutions for other patients such as physical exercise, increased socialization, intense "homework" (i.e., complicated, long workbooks), and relaxing music might not be possible or will need to be adapted to patients with ME/CFS due to symptoms such as PEM, cognitive dysfunction, and hypersensitivity to sound. Due to their decreased mobility, moderately and severely ill patients would benefit greatly from therapy delivered remotely via telephone and video-conferencing: fortunately, these modalities appear to be comparably effective [102] to sessions administered in person. Shorter but more frequent visits, asynchronous interaction, and/or written vs. oral may be beneficial. For an in-depth discussion of assessment and treatment of psychiatric issues in ME/CFS, see author ES's articles [97,99].

As with any patient affected by multiple, chronic, and/or complex conditions, care will often take place over time and multiple visits. The clinician will need to prioritize which factors to address and which interventions to start right away versus which ones can wait. First, ask patients "What can we do or change that would make your life worth living?" The answers given might be surprising and engages the patient in planning care. Other criteria might be the urgency of a factor (e.g., impending homelessness), how quickly the intervention may start working (e.g., pain medication), and how common the factor has been found to influence suicide risk among people with ME/CFS (e.g., lacking healthcare providers knowledgeable about ME/CFS).

There are a few caveats when introducing remedies similar to those in Table 5. Patients affected by ME/CFS may be exquisitely sensitive to the active or inactive components (e.g., coloring agents, preservatives) in a medication. Thus, start low and titrate up any medication slowly. Physical activity performed to alleviate co-morbidities (e.g., fibromyalgia, postural orthostatic intolerance syndrome) or during assessments (e.g., physical therapy) should be adapted so patients avoid triggering PEM. For severely ill patients, physical activity may not be possible without making the patient sicker. Although pain (e.g., joint/muscle aches, sore throats, headaches) is common in ME/CFS [103] and sleep disturbances are part of diagnostic criteria for ME/CFS [4,104], clinicians should evaluate the cause of symptoms before attributing them solely or entirely to ME/CFS, especially if the pain is new or worsening. For example, fibromyalgia, migraine headaches, and obstructive sleep apnea are common comorbid conditions yet each condition has specific treatments. There have been rare cases where late-stage cancer was discovered to be the source of pain. If possible, attempt to refer patients to other professionals who are knowledgeable or open to learning about ME/CFS.

The significance of validation and understanding conveyed by even one supportive clinician cannot be emphasized enough. Most patients affected by ME/CFS have endured years of indifferent or degrading medical professionals [6]. Even if a clinician is not adept at caring for ME/CFS patients, any good-faith efforts to learn about ME/CFS and communicate sympathy will be appreciated. During a short appointment, clinicians can earn the trust of patients by carefully listening to them, reflecting back to them their understanding of what was said, and honestly admitting what they know or do not know. There are many steps (Box 7) clinicians can take to improve their relationship with patients [105]

Box 7. Clinical case—part 7.

The doctor considers starting Maria on an antidepressant but decides to let the psychiatrist start a medication. When asked which symptoms are the most problematic, Maria nominates pain and sleep. When asked what else would improve her quality of life immediately, Maria desires help with daily tasks like bathing and the ability to be a little more social.

A week's supply of a low dose of amitriptyline, which addresses both sleep and chronic neuropathic pain, is called in by the medical assistant to a pharmacy that delivers. Maria is told the medication may take up to 4 weeks to start working. The doctor also tells Maria she will ask a physical therapist, occupational therapist, and medical social worker to visit her at home.

On her way out of the office, Maria is stopped by the receptionist and told she will be hearing from the clinic in 2 days to check on how she is doing.

5.3. How Should Suicidal Patients Be Followed-Up?

As with any acute medical condition, clinicians are responsible for assuring that care plans are executed in an appropriate and timely manner. Aside from arranging for subsequent appointments within their own clinic and reviewing whether the interventions they personally initiated are working, coordinating care with ancillary and specialist providers/facilities and communicating with the patient and their supporters are also vital (Box 8). If the patient is sent to the emergency room or for immediate outpatient mental healthcare and nothing is reported back, the clinician's office should verify the patient arrived, was seen, and was admitted for inpatient care or discharged with subsequent psychiatric aftercare. If solely outpatient treatment is indicated, the clinician or their staff should contact the patient 24–48 h after the initial visit. Ask patients how they are feeling, review and adjust the safety plan as needed, scrutinize whether access to lethal means has been blocked [106], and check that they either have already been seen by a mental health professional or will be keeping the appointment made. Currently, up to 50% of suicidal patients do not show up for their psychiatric/psychological appointments [90]. These "caring contacts"—as fleeting and minimal as they might seem—have been shown to reduce suicidal ideation and behavior in some studies [90,107].

Box 8. Clinical case—part 8.

Two days later, Maria receives a telephone call from the medical assistant. The assistant asks her about how she is feeling and whether the psychiatrist or other clinicians have contacted her. Maria states that last night was the first night in 6 months she was able to obtain more than 4 h of continuous sleep. Her mouth feels a bit dry this morning, a mild side effect of the amitriptyline that the assistant tells her can be alleviated by chewing gum or drinking fluids. Maria forced herself to sit outside in the sun for 15 min and felt more cheerful afterwards. Dr Lopez's office has scheduled her for a telephone call tomorrow. She has to return the social worker's and occupational therapist's voice messages. The assistant compliments Maria for spending time outside (part of her self-generated plan to address suicidal feelings) and for taking action.

The call is then transferred to the receptionist. Based on the severity of her ME/CFS, Maria is offered a virtual rather than in-person visit with the doctor in one week.

In Maria's situation, the clinician should make a formal follow-up appointment (Box 9) with her for medication management (e.g., Was she able to obtain the amitriptyline? Has she tried them? What have the effects been?) and at the next visit, acknowledge any efforts she has put forth (e.g., "I know how hard it must be to reach out to strangers for help. You should be proud of yourself."). Speaking briefly with or reviewing notes from the occupational therapist and social worker can update the clinician regarding the patient's functional status, support network, and housing situation. Asking open-ended questions, letting the patient lead the conversation sometimes, and showing a personal interest in the patient can also support the patient on their path to better mental health. If a patient is discharged from inpatient psychiatric care, an outpatient appointment should be made within 7 days at most [108]. During the first week and first month after hospitalization, the risk of suicide is 200 and 300 times higher than the general population, respectively [109]. Even after 5 to 10 years, the rate of suicide was 30 times than of the general population. Thus, suicidality should be considered a chronic condition that the clinician inquires about occasionally even when the patient is stable, much as they would with hypertension or diabetes. Clinicians should also be aware of when patients stop or are discharged from outpatient psychiatric/psychological care.

Box 9. Clinical case—part 9.

> The next week, Maria sees her doctor virtually, using online video-conferencing software. Her sleep continues to improve, especially as pain no longer wakes her up, and she has a little bit more energy to take care of herself and her household chores. She reports that during Dr Lopez's phone visit, they discussed positive coping behaviors such as focusing on what she can do (vs. what she cannot do), setting small goals and working towards them. Her "homework" includes planning at least one social interaction (which is tolerable for her level of ME/CFS activity) every week and noting something she is grateful for every night in a journal. She will be talking to Dr Lopez again next week.
>
> The doctor reviews the occupational therapist's notes. Maria qualifies for bath and shower bars and a shower stool to decrease dizziness and exhaustion while bathing. A wheelchair is recommended to facilitate travel outside the home. The OT also teaches Maria to balance her activities with rest and to save her most challenging activities for the times of the day when she is likely to have more energy. This has allowed Maria to start completing an application for housing support mailed to her by the social worker.
>
> Maria states she is starting to feel more optimistic about her future: if her symptoms continue to abate a little more and she can get around in a wheelchair, she might be able to attend a knitting group that her neighbor Sarah hosts twice a month.
>
> The doctor celebrates her progress with her causing Maria to feel confident enough to ask the doctor whether orthostatic intolerance—which she has been able to read more about—might account for some of her nausea and dizziness. The doctor admits she does not know much about OI but will try to learn more about it. A follow-up appointment is scheduled in 2 weeks.

6. Barriers, Gaps, and Opportunities

We recognize that the process of assessment and treatment detailed in this article may be challenging to implement. Research, clinical, and societal barriers exist.

6.1. Research Barriers

Much more investigation into the relationships among suicide, chronic physical illness, and ME/CFS are needed. Oftentimes, concepts, assessments, and interventions had to be extrapolated from one field to another because studies were absent or lacking. For example, risk factors for suicidal ideation may not be the same as those for suicide attempts or completion. Motivation for suicide in ME/CFS or in chronic medical illnesses might be different from that of psychiatric patients or the general population. Epidemiological studies of suicide in ME/CFS have been retrospective in nature, either medical record reviews or psychological autopsies. That has meant the true incidence and prevalence of suicide are not known and drivers of suicide could be better elucidated. Most of what we know about ME/CFS itself is based on adult patients who are given an ME/CFS diagnosis and possess the health, financial and social status to access the few specialists scattered globally. Although the C-SSRS and the ASQ are helpful, their use by non-mental health professionals in adult patients seen in community-based, outpatient medical settings has not been well-established. Self-report assessments also have weaknesses: up to 25% of patients who attempted suicide actively denied suicidality during appointments the week before [110] due to shame, embarrassment and concerns about losing autonomy if they were hospitalized. Refinement of risk also relies on clinician interpretation of patients' words. Consequently, some scientists are eagerly pursuing biomarkers which can predict risk [111]. There is adequate research in some areas of treatment, such as the efficacy of antidepressants or suicide-specific dialectical behavioral therapy, but less in others, such as the impact of occupational therapy on suicide risk when functional restrictions are present. Reviewing research for this paper reveals a plethora of questions and issues that have yet to be answered.

6.2. Clinical Care Barriers

Educational, attitudinal, logistical, financial, and legal barriers impede optimal care. Healthcare providers readily admit they are not confident about diagnosing and managing both ME/CFS [4] and suicidality [65,66]. Hopefully, the process outlined (Figure 1) and detailed in this article advances practitioners' knowledge about suicidality and supplies them with a straightforward care plan. Widespread education of healthcare providers and the public about ME/CFS is a critical step in reducing stigma and the unsupportive social interactions driving suicidality in ME/CFS. The 2015 US National Academy of Medicine Report for Clinicians [4], 2014 International Association for Chronic Fatigue Syndrome/Myalgic Encephalomyelitis (IACFS/ME) Primer [33], US Centers for Disease Control and Prevention (CDC) ME/CFS website [112], and US ME/CFS Clinician Coalition short summary [113] provide guidance on the clinical care of ME/CFS.

As presented in this article, optimal outcomes necessitate multiple professionals working together with the patient. Coordination and communication among these individuals are vital yet often neglected. Multiple appointments for care separated by time and physical location are especially challenging for severely ill populations. Some organizations are examining whether mental health professionals embedded in medical practices, where they can see patients immediately, or specially trained personnel (e.g., a nurse or social worker) assigned to outpatient clinics can help [12,18]. Clinicians have also been concerned about whether protocols addressing suicide will take up too much time or resources. For both the ASQ and C-SSRS, in non-psychiatric settings, over 95% of screenings are rated as "no" or "minimal" risk (meaning no further action is needed), 1.9%, as moderate risk, and 0.2–0.5% as high risk [68,114]. The latter two percentages are likely higher among patients afflicted by ME/CFS yet may not be as overwhelming as expected. From an institutional point of view, standardized protocols have resulted in more cost-effective care as emergency rooms and mental health consultants channel their immediate energy towards patients at the highest risk rather than dispersing it among all patients at risk of suicide.

Health insurance reimbursement and coverage for mental health care continue to be obstacles. Despite mental health parity laws passed in the United States in 2008 and 2013 advocating for equal treatment of mental and physical health conditions [115], enforcement of regulations has not been consistent nor uniform across the country. For example, reimbursement for primary medical care is 30–50% higher than that for behavioral health care and prior authorizations obstruct timely access to care [116]. Circumstances are also challenging in Canada: the publicly funded health care system provides limited access to mental health professionals, such as psychologists and counselors [117]. Waiting lists and inability to access someone with illness-specific expertise are the norm. During 2020, the Canadian Alliance on Mental Illness and Mental Health introduced a new Mental Health Parity Act [118].

Some clinicians worry they will be held responsible for a patient's completed suicide and thus avoid asking about suicide entirely. In the United States, the concepts governing liability in suicide are the same as to those affecting other medical conditions: existence of a duty, negligent breach of that duty, proven damage to the patient, and a proximate link between breach and damage [119]. Unfavorable outcomes by themselves do not necessarily lead to finding the clinician at fault. Instead, judgments are based on whether the professional acted reasonably according to community-accepted standards of care. The steps and questionnaires discussed in this article are based on the scientific literature and align with the 2018 guidelines from the US National Action Alliance for Suicide Prevention [90]. Constructing suicide safety plans may decrease legal exposure [120]. Careful, timely documentation of what was done and the rationale for decisions [15] as well as communicating the care plan to the patient's family or supporters (with the patient's written permission) can further protect against liability.

6.3. Societal Barriers

Ultimately, to decrease suffering and suicide in ME/CFS, steps must be taken on a larger, systemic level, beyond those of the research and clinical care realms. Education about ME/CFS must be extended to lawmakers and disability benefit providers, who have the authority to address the lack of resources so commonly experienced by those with ME/CFS and directly cited as a major factor in suicidal ideation [17]. Recently, some mental health professionals have pushed for programs responding to the external roots of suicide, including poverty, social connectedness, unemployment, firearm availability, and homelessness [121–123]. Expansion of such programs will also benefit patients with ME/CFS although some existing programs (e.g., Meals on Wheels) are unfamiliar with ME/CFS and thus, patients face skepticism when applying for them. Education of the public is also needed to reduce stigma and make the unsupportive social interactions driving suicidality in ME/CFS increasingly infrequent.

6.4. Emerging Opportunities

A potentially positive effect of the COVID-19 pandemic and lockdown is the expanded use of virtual care: teletherapy, telemedicine, etc. Home-based care, either in-person or virtual, is already being used in some homebound geriatric patients [124]; similar adaptations to those with ME/CFS who are home- or bed-bound could make accessible mental health services that might otherwise trigger too much post-exertional malaise to attend. In a randomized controlled trial of home-based care for people with multiple sclerosis, people who received care in the home showed significant improvements in multiple domains of quality of life, including the mental health-related role-emotional and social functioning domains [125]. Critically, this home-based care model integrated physical and mental healthcare, something important in ME/CFS given that physical symptoms such as sleep disturbance and pain are major risk factors for suicide, as previously discussed.

Even asynchronous online therapy has been shown in some studies to reduce suicidal ideation in primary care populations [126,127], and internet-based programs for comorbid depression and chronic illness show some success in reducing depression rates in meta-analysis [128]. Interventions delivered by telephone have also shown some success; an intervention for emergency department patients consisting of a safety plan, provision of crisis resources, and a series of telephone follow-ups reduced suicide attempts in the following year by 30% [129]. There are thus multiple feasible methods of mental healthcare delivery other than traditional in-person office visits, which could improve access to such care in the future. Adapting mental and physical healthcare to the energy limitations of people with ME/CFS represents a logical next step in treating this illness and is not unprecedented elsewhere in suicide prevention and chronic illness literature.

7. Conclusions

Like other chronic, debilitating illnesses, ME/CFS places individuals at an increased risk of death by suicide. Several characteristics prominent in ME/CFS exacerbate this risk and make diagnosis and management of suicidality demanding. These include absence of any disease-modifying treatments, severe functional limitations confining sizable numbers of patients at home, and symptoms (e.g., PEM, medication sensitivities, cognitive dysfunction) limiting certain therapies. Decades-long misattribution of ME/CFS to physical deconditioning or irrational, hypochondriacal beliefs combined with conflation of ME/CFS with depression or anxiety have also resulted in an uneducated healthcare workforce at best and a skeptical, dismissive one at worst. Severity of impairment is often not acknowledged. Consequently, some patients are reluctant to engage in psychiatric/psychological care despite sometimes desperately needing it. Lack of proper recognition by medical professionals and authorities in turn has meant an absence and scarcity of resources targeted or available to patients, whether medical/psychiatric/psychological care, social support from family members or friends, or disability benefits.

Outpatient medical professionals play a vital role in ameliorating this cascade of effects. We have provided a framework for identifying and managing adult suicidal patients afflicted by ME/CFS through adapting current recommendations to this neglected population. Through both applying evidence-based interventions aimed at all suicidal patients and tailoring interventions specific to an individual patient's circumstances (Box 10), we believe that suffering and suicidality can be alleviated.

Box 10. Clinical case—part 10.

Three months later, Maria returns for a follow-up visit. Although her neck and arm pain persist, a higher dose of amitriptyline has dulled it considerably and she is able to sleep through the night now. Dr Lopez has her taking a stable dose of citalopram; she continues to see him. After being diagnosed with OI by her doctor, her doctor teaches her to mix up a homemade oral rehydration solution. Drinking this regularly helps control her dizziness and she is now able to sit up for 2 h at a time. With the help of the wheelchair, she is able now to attend Sarah's knitting group regularly. After her housing voucher application is approved, she is able to move to a new place, decreasing her respiratory symptoms. The extra financial assistance also allows her to save money each month. Although Maria's ME/CFS remains, symptom relief, treatment of depression, mild functional improvement, social connection, and a change in housing result in a decrease in suicidal ideation. Eventually, Maria joins a Facebook support group for patients with ME/CFS. As time passes, she is able to offer support and hope to new members. This gives her a renewed sense of purpose.

Supplementary Materials: The following are available online at https://www.mdpi.com/article/10.3390/healthcare9060629/s1, Figure S1: Suicide Safety Plan Template. Reproduced with permission from Dr Barbara Stanley, Cognitive and Behavioral Practice, published by Elsevier, 2012 [92].

Author Contributions: Conceptualization, L.C., M.E., E.S. and L.A.J.; formal analysis, L.C., M.E., E.S. and L.A.J.; investigation, L.C., M.E., E.S. and L.A.J.; resources, L.C., M.E., E.S. and L.A.J.; writing—original draft preparation, L.C., M.E., E.S. and L.A.J.; writing—review and editing, L.C., M.E., E.S. and L.A.J. All authors have read and agreed to the published version of the manuscript.

Funding: This research received no external funding.

Institutional Review Board Statement: Not applicable.

Informed Consent Statement: Not applicable.

Conflicts of Interest: The authors declare no conflict of interest.

References

1. Jason, L.A.; Richman, J.A.; Rademaker, A.W.; Jordan, K.M.; Plioplys, A.V.; Taylor, R.R.; McCready, W.; Huang, C.-F.; Plioplys, S. A Community-Based Study of Chronic Fatigue Syndrome. *Arch. Intern. Med.* **1999**, *159*, 2129–2137. [CrossRef]
2. Dimmock, M.E.; Mirin, A.A.; Jason, L.A. Estimating the disease burden of ME/CFS in the United States and its relation to research funding. *J. Med.* **2016**, 1. [CrossRef]
3. Jason, L.; Mirin, A. Updating the National Academy of Medicine ME/CFS prevalence and economic impact figures to account for population growth and inflation. *Fatigue Biomed. Health Behav.* **2021**, *9*, 9–13. [CrossRef]
4. Institute of Medicine. *Beyond Myalgic Encephalomyelitis/Chronic Fatigue Syndrome: Redefining An Illness*; Institute of Medicine (U.S.), Ed.; The National Academies Press: Washington, DC, USA, 2015; ISBN 9780309316897.
5. Roberts, E.; Wessely, S.; Chalder, T.; Chang, C.-K.; Hotopf, M. Mortality of people with chronic fatigue syndrome: A retrospective cohort study in England and Wales from the South London and Maudsley NHS Foundation Trust Biomedical Research Centre (SLaM BRC) Clinical Record Interactive Search (CRIS) Register. *Lancet* **2016**, *387*, 1638–1643. [CrossRef]
6. Ortiz, J.J.J. *Depresión y Desesperanza en Personas Enfermas de Encefalomielitis Miálgica/Síndrome de Fatiga Crónica: Factores de Riesgo y de Protección*; Facultad de Educación y Trabajo Social, Universidad de Valladolid: Valladolid, Spain, 2019.
7. Carr, M.J.; Ashcroft, D.M.; White, P.D.; Kapur, N.; Webb, R.T. Prevalence of comorbid mental and physical illnesses and risks for self-harm and premature death among primary care patients diagnosed with fatigue syndromes. *Psychol. Med.* **2019**, *50*, 1156–1163. [CrossRef]
8. Ribeiro, J.D.; Franklin, J.C.; Fox, K.R.; Bentley, K.H.; Kleiman, E.M.; Chang, B.P.; Nock, M.K. Self-injurious thoughts and behaviors as risk factors for future suicide ideation, attempts, and death: A meta-analysis of longitudinal studies. *Psychol. Med.* **2016**, *46*, 225–236. [CrossRef]
9. Jason, L.A.; Corradi, K.; Gress, S.; Williams, S.; Torres-Harding, S. Causes of Death among Patients with Chronic Fatigue Syndrome. *Health Care Women Int.* **2006**, *27*, 615–626. [CrossRef]
10. Bazalgette, L.; Bradley, W.; Ousbey, J. The Truth about Suicide. Available online: https://demosuk.wpengine.com/files/Suicide_-_web.pdf?1314370102 (accessed on 17 May 2021).
11. Substance Abuse and Mental Health Services Administration. *Results from the 2013 National Survey on Drug Use and Health: Mental Health Findings*; NSDUH Series H-49, HHS Publication No. (SMA) 14-4887; Substance Abuse and Mental Health Services: Rockville, MD, USA, 2014. Available online: http://www.samhsa.gov/data/sites/default/files/NSDUHmhfr2013/NSDUHmhfr2013.pdf (accessed on 17 May 2021).
12. Schulberg, H.C.; Bruce, M.L.; Lee, P.W.; Williams, J.W.; Dietrich, A.J. Preventing suicide in primary care patients: The primary care physician's role. *Gen. Hosp. Psychiatry* **2004**, *26*, 337–345. [CrossRef] [PubMed]

13. McManimen, S.L.; McClellan, D.; Stoothoff, J.; Jason, L.A. Effects of unsupportive social interactions, stigma, and symptoms on patients with myalgic encephalomyelitis and chronic fatigue syndrome. *J. Community Psychol.* **2018**, *46*, 959–971. [CrossRef] [PubMed]
14. Geraghty, K.J.; Blease, C. Myalgic encephalomyelitis/chronic fatigue syndrome and the biopsychosocial model: A review of patient harm and distress in the medical encounter. *Disabil. Rehabil.* **2018**, *41*, 3092–3102. [CrossRef]
15. Weber, A.N.; Michail, M.; Thompson, A.; Fiedorowicz, J.G. Psychiatric Emergencies. *Med. Clin. N. Am.* **2017**, *101*, 553–571. [CrossRef] [PubMed]
16. Ahmedani, B.K.; Peterson, E.L.; Hu, Y.; Rossom, R.C.; Lynch, F.; Lu, C.Y.; Waitzfelder, B.E.; Owen-Smith, A.A.; Hubley, S.; Prabhakar, D.; et al. Major Physical Health Conditions and Risk of Suicide. *Am. J. Prev. Med.* **2017**, *53*, 308–315. [CrossRef]
17. Devendorf, A.R.; McManimen, S.L.; Jason, L.A. Suicidal ideation in non-depressed individuals: The effects of a chronic, misunderstood illness. *J. Health Psychol.* **2018**, *25*, 2106–2117. [CrossRef] [PubMed]
18. McDowell, A.K.; Lineberry, T.W.; Bostwick, J.M. Practical Suicide-Risk Management for the Busy Primary Care Physician. *Mayo Clin. Proc.* **2011**, *86*, 792–800. [CrossRef]
19. Probert-Lindström, S.; Berge, J.; Westrin, Å.; Öjehagen, A.; Pavulans, K.S. Long-term risk factors for suicide in suicide attempters examined at a medical emergency in patient unit: Results from a 32-year follow-up study. *BMJ Open* **2020**, *10*, e038794. [CrossRef] [PubMed]
20. Government of Canada. S.C. Suicide among First Nations People, Métis and Inuit (2011–2016): Findings from the 2011 Canadian Census Health and Environment Cohort (CanCHEC). Available online: https://www150.statcan.gc.ca/n1/pub/99-011-x/99-011-x2019001-eng.htm (accessed on 17 May 2021).
21. Grunbaum, J.A.; Kann, L.; Kinchen, S.; Ross, J.; Hawkins, J.; Lowry, R. Youth Risk Behavior Surveillance–United States, 2003. *Psycextra Dataset* **2004**, *63*, 1–168. [CrossRef] [PubMed]
22. Movement Advancement Project Talking About Suicide & LGBT Populations. Available online: https://www.lgbtmap.org/talking-about-suicide-and-lgbt-populations (accessed on 10 March 2021).
23. Pederson, C.L.; Brook, J.B. Sleep disturbance linked to suicidal ideation in postural orthostatic tachycardia syndrome. *Nat. Sci. Sleep* **2017**, *9*, 109–115. [CrossRef] [PubMed]
24. Owen-Smith, A.A.; Ahmedani, B.K.; Peterson, E.; Simon, G.E.; Rossom, R.C.; Lynch, F.L.; Lu, C.Y.; Waitzfelder, B.E.; Beck, A.; DeBar, L.L.; et al. The mediating effect of sleep disturbance on the relationship between nonmalignant chronic pain and suicide death. *Pain Pract.* **2019**, *19*, 382–389. [CrossRef]
25. Onyeka, I.N.; Maguire, A.; Ross, E.; O'Reilly, D. Does physical ill-health increase the risk of suicide? A census-based follow-up study of over 1 million people. *Epidemiol. Psychiatr. Sci.* **2020**, *29*, 140. [CrossRef]
26. Karasouli, E.; Latchford, G.; Owens, D. The impact of chronic illness in suicidality: A qualitative exploration. *Health Psychol. Behav. Med.* **2013**, *2*, 899–908. [CrossRef]
27. Jason, L.A.; Ohanian, D.; Brown, A.; Sunnquist, M.; McManimen, S.; Klebek, L.; Fox, P.; Sorenson, M. Differentiating multiple sclerosis from myalgic encephalomyelitis and chronic fatigue syndrome. *Insights Biomed.* **2017**, *2*, 11. [CrossRef] [PubMed]
28. Kingdon, C.; Bowman, E.W.; Curran, H.; Nacul, L.; Lacerda, E.M. Functional status and well-being in people with myalgic encephalomyelitis/chronic fatigue syndrome compared with people with multiple sclerosis and healthy controls. *Pharm. Open* **2018**, *2*, 381–392. [CrossRef]
29. Johnson, M.L.; Cotler, J.; Terman, J.M.; Jason, L.A. Risk factors for suicide in chronic fatigue syndrome. *Death Stud.* **2020**, 1–7. [CrossRef] [PubMed]
30. McManimen, S.L.; Devendorf, A.R.; Brown, A.A.; Moore, B.C.; Moore, J.H.; Jason, L.A. Mortality in patients with myalgic encephalomyelitis and chronic fatigue syndrome. *Fatigue Biomed. Health Behav.* **2016**, *4*, 195–207. [CrossRef] [PubMed]
31. Braden, J.B.; Sullivan, M.D. Suicidal Thoughts and Behavior among Adults with Self-Reported Pain Conditions in the National Comorbidity Survey Replication. *J. Pain* **2008**, *9*, 1106–1115. [CrossRef] [PubMed]
32. Fuller-Thomson, E.; Nimigon, J. Factors associated with depression among individuals with chronic fatigue syndrome: Findings from a nationally representative survey. *Fam. Pract.* **2008**, *25*, 414–422. [CrossRef]
33. Kingdon, C.; Giotas, D.; Nacul, L.; Lacerda, E. Health care responsibility and compassion-visiting the housebound patient severely affected by ME/CFS. *Healthcare* **2020**, *8*, 197. [CrossRef]
34. Strassheim, V.; Lambson, R.; Hackett, K.L.; Newton, J.L. What is known about severe and very severe chronic fatigue syndrome? A scoping review. *Fatigue Biomed. Health Behav.* **2017**, *5*, 167–183. [CrossRef]
35. Chronic Fatigue Syndrome/Myalgic Encephalomyelitis. Primer for Clinical Practitioners. 2014 Edition. International Association for Chronic Fatigue Syndrome/Myalgic Encephalomyelitis. Available online: https://growthzonesitesprod.azureedge.net/wp-content/uploads/sites/1869/2020/10/Primer_Post_2014_conference.pdf (accessed on 17 May 2021).
36. Bavington, J.; Darbishire, L.; White, P. Graded Exercise Therapy for CFS/ME; Version 7 (MREC Version 2; PACE Trial Management Group. 2004. Available online: https://me-pedia.org/images/8/89/PACE-get-therapist-manual.pdf (accessed on 17 May 2021).
37. Geraghty, K.; Jason, L.; Sunnquist, M.; Tuller, D.; Blease, C.; Adeniji, C. The 'cognitive behavioural model' of chronic fatigue syndrome: Critique of a flawed model. *Health Psychol. Open* **2019**, *6*, 6. [CrossRef]
38. Wiborg, J.F.; Van Der Werf, S.; Prins, J.B.; Bleijenberg, G. Being homebound with chronic fatigue syndrome: A multidimensional comparison with outpatients. *Psychiatry Res.* **2010**, *177*, 246–249. [CrossRef]

39. Bentall, R.P.; Powell, P.; Nye, F.J.; Edwards, R.H.T. Predictors of response to treatment for chronic fatigue syndrome. *Br. J. Psychiatry* **2002**, *181*, 248–252. [CrossRef]

40. CDC. Improving Health and Quality of Life—Chronic Fatigue Syndrome (CFS). Available online: https://web.archive.org/web/20150905080404/http://www.cdc.gov/cfs/management/quality-of-life.html (accessed on 10 March 2021).

41. Maksoud, R.; Du Preez, S.; Eaton-Fitch, N.; Thapaliya, K.; Barnden, L.; Cabanas, H.; Staines, D.; Marshall-Gradisnik, S. A systematic review of neurological impairments in myalgic encephalomyelitis/ chronic fatigue syndrome using neuroimaging techniques. *PLoS ONE* **2020**, *15*, e0232475. [CrossRef] [PubMed]

42. Eaton-Fitch, N.; Du Preez, S.; Cabanas, H.; Staines, D.; Marshall-Gradisnik, S. A systematic review of natural killer cells profile and cytotoxic function in myalgic encephalomyelitis/chronic fatigue syndrome. *Syst. Rev.* **2019**, *8*, 1–13. [CrossRef] [PubMed]

43. Rutherford, G.; Manning, P.; Newton, J.L. Understanding Muscle Dysfunction in Chronic Fatigue Syndrome. *J. Aging Res.* **2016**, *2016*, 1–13. [CrossRef] [PubMed]

44. Geraghty, K.; Hann, M.; Kurtev, S. Myalgic encephalomyelitis/chronic fatigue syndrome patients' reports of symptom changes following cognitive behavioural therapy, graded exercise therapy and pacing treatments: Analysis of a primary survey compared with secondary surveys. *J. Health Psychol.* **2017**, *24*, 1318–1333. [CrossRef]

45. McInnis, O.A.; Matheson, K.; Anisman, H. Living with the unexplained: Coping, distress, and depression among women with chronic fatigue syndrome and/or fibromyalgia compared to an autoimmune disorder. *Anxiety Stress. Coping* **2014**, *27*, 601–618. [CrossRef]

46. Hvidberg, M.F.; Brinth, L.S.; Olesen, A.V.; Petersen, K.D.; Ehlers, L. The Health-Related Quality of Life for Patients with Myalgic Encephalomyelitis / Chronic Fatigue Syndrome (ME/CFS). *PLoS ONE* **2015**, *10*, e0132421. [CrossRef]

47. Pederson, C.L.; Gorman-Ezell, K.; Mayer, G.H.; Brookings, J.B. Development and Preliminary Validation of a Tool for Screening Suicide Risk in Chronically Ill Women. *Meas. Eval. Couns. Dev.* **2021**, *54*, 130–140. [CrossRef]

48. The Joint Commission. Detecting and treating suicide ideation in all settings. *Sentin. Event Alert* **2016**, *56*, 1–7.

49. United States Preventive Services Taskforce. Recommendation: Suicide Risk in Adolescents, Adults and Older Adults: Screening. Available online: https://www.uspreventiveservicestaskforce.org/uspstf/recommendation/suicide-risk-in-adolescents-adults-and-older-adults-screening (accessed on 19 March 2021).

50. Suicide Prevention: A Review and Policy Recommendations; CAMH. 2020. Available online: https://www.camh.ca/-/media/files/pdfs---public-policy-submissions/suicide-prevention-review-and-policy-recommendations-pdf.pdf?la=en&hash=A43E96FBFEFDEC87F40EB0E203D2A605323407E0 (accessed on 17 May 2021).

51. Yawn, B.; Dietrich, A.; Wollan, P.; Bertram, S.; Kurland, M.; Pace, W.; Graham, D.; Huff, J. Immediate action protocol: A tool to help your practice assess suicidal patients. *Fam. Pr. Manag.* **2009**, *16*, 17.

52. Snyder, D.J.; Ballard, E.; Stanley, I.H.; Ludi, E.; Kohn-Godbout, J.; Pao, M.; Horowitz, L.M. Patient Opinions About Screening for Suicide Risk in the Adult Medical Inpatient Unit. *J. Behav. Health Serv. Res.* **2017**, *44*, 364–372. [CrossRef] [PubMed]

53. LeCloux, M.A.; Weimer, M.; Culp, S.L.; Bjorkgren, K.; Service, S.; Campo, J.V. The Feasibility and Impact of a Suicide Risk Screening Program in Rural Adult Primary Care: A Pilot Test of the Ask Suicide-Screening Questions Toolkit. *J. Psychosom. Res.* **2020**, *61*, 698–706. [CrossRef] [PubMed]

54. Pederson, C.L. The importance of screening for suicide risk in chronic invisible illness. *J. Health Sci. Educ.* **2018**, *2*, 141. [CrossRef]

55. Busch, K.A.; Fawcett, J.; Jacobs, D.G. Clinical Correlates of Inpatient Suicide. *J. Clin. Psychiatry* **2003**, *64*, 14–19. [CrossRef]

56. Dazzi, T.; Gribble, R.; Wessely, S.; Fear, N.T. Does asking about suicide and related behaviours induce suicidal ideation? What is the evidence? *Psychol. Med.* **2014**, *44*, 3361–3363. [CrossRef]

57. Blades, C.A.; Stritzke, W.G.; Page, A.C.; Brown, J.D. The benefits and risks of asking research participants about suicide: A meta-analysis of the impact of exposure to suicide-related content. *Clin. Psychol. Rev.* **2018**, *64*, 1–12. [CrossRef]

58. Common Characteristics of Suicide. Suicide & Crisis Center of North Texas. Available online: https://www.sccenter.org/facts-and-resources/common-characteristics-of-suicide/ (accessed on 23 March 2021).

59. Firestone, L. Busting the Myths about Suicide. Available online: https://www.psychalive.org/busting-the-myths-about-suicide/ (accessed on 17 May 2021).

60. KGO Second Chances: I Survived Jumping Off the Golden Gate Bridge. Available online: https://abc7news.com/2010562/ (accessed on 26 March 2021).

61. Ortegren, A. Farewell—A Last Post from Anne Örtegren. Available online: https://www.healthrising.org/blog/2018/01/10/farewelll-last-post-anne-ortegren/ (accessed on 17 May 2021).

62. Colby, J. False Allegations of Child Abuse in Cases of Childhood Myalgic Encephalomyelitis (ME). Available online: https://www.argumentcritique.com/uploads/1/0/3/1/10317653/colby_j.pdf (accessed on 17 May 2021).

63. Neuropathology Report: Sophia Mirza. Available online: http://www.sophiaandme.org.uk/neuropathologicalreport.html (accessed on 19 March 2021).

64. Ahmedani, B.K.; Vannoy, S. National Pathways for Suicide Prevention and Health Services Research. *Am. J. Prev. Med.* **2014**, *47*, S222–S228. [CrossRef]

65. Elzinga, E.; De Kruif, A.J.T.C.M.; De Beurs, D.P.; Beekman, A.T.F.; Franx, G.; Gilissen, R. Engaging primary care professionals in suicide prevention: A qualitative study. *PLoS ONE* **2020**, *15*, e0242540. [CrossRef]

66. Leavey, G.; Mallon, S.; Rondon-Sulbaran, J.; Galway, K.; Rosato, M.; Hughes, L. The failure of suicide prevention in primary care: Family and GP perspectives—A qualitative study. *BMC Psychiatry* **2017**, *17*, 369. [CrossRef]

67. Horowitz, L.M.; Bridge, J.A.; Teach, S.J.; Ballard, E.; Klima, J.; Rosenstein, D.L.; Wharff, E.A.; Ginnis, K.; Cannon, E.; Joshi, P.; et al. Ask Suicide-Screening Questions (ASQ). *Arch. Pediatr. Adolesc. Med.* **2012**, *166*, 1170–1176. [CrossRef]

68. Horowitz, L.M.; Snyder, D.J.; Boudreaux, E.D.; He, J.-P.; Harrington, C.J.; Cai, J.; Claassen, C.A.; Salhany, J.E.; Dao, T.; Chaves, J.F.; et al. Validation of the Ask Suicide-Screening Questions for Adult Medical Inpatients: A Brief Tool for All Ages. *J. Psychosom. Res.* **2020**, *61*, 713–722. [CrossRef] [PubMed]

69. Posner, K.; Brent, D.; Lucas, C.; Gould, M.; Stanley, B.; Brown, G.; Mann, J. *Columbia-Suicide Severity Rating Scale (C.-SSRS)*; Columbia University Medical Center: New York, NY, USA, 2008; p. 10.

70. NIMH. Adult Outpatient Brief Suicide Safety Assessment Worksheet. Available online: https://www.nimh.nih.gov/research/research-conducted-at-nimh/asq-toolkit-materials/adult-outpatient/adult-outpatient-brief-suicide-safety-assessment-worksheet.shtml (accessed on 24 March 2021).

71. C-SSRS Training—English (USA) (Most Recent Version). Available online: https://www.youtube.com/watch?v=epTDFFv3uwc&list=PLZ6DpvOfzN1kV1F_lDw9-26JifBSDlIbF&index=2 (accessed on 23 March 2021).

72. Greist, J.H.; Mundt, J.C.; Gwaltney, C.J.; Jefferson, J.W.; Posner, K. Predictive Value of Baseline Electronic Columbia–Suicide Severity Rating Scale (eC–SSRS) Assessments for Identifying Risk of Prospective Reports of Suicidal Behavior During Research Participation. *Innov. Clin. Neurosci.* **2014**, *11*, 23–31. [PubMed]

73. Lukaschek, K.; Frank, M.; Halfter, K.; Schneider, A.; Gensichen, J. *A Systematic Review of Brief Screeners for Suicidal Behaviour in Primary Care*, 2019; in review.

74. Druss, B.; Pincus, H. Suicidal Ideation and Suicide Attempts in General Medical Illnesses. *Arch. Intern. Med.* **2000**, *160*, 1522–1526. [CrossRef] [PubMed]

75. Cavanagh, J.T.O.; Carson, A.J.; Sharpe, M.; Lawrie, S.M. Psychological autopsy studies of suicide: A systematic review. *Psychol. Med.* **2003**, *33*, 395–405. [CrossRef]

76. Kroenke, K.; Spitzer, R.L.; Williams, J.B.W. The PHQ-9. *J. Gen. Intern. Med.* **2001**, *16*, 606–613. [CrossRef]

77. Viguera, A.C.; Milano, N.; Laurel, R.; Thompson, N.R.; Griffith, S.D.; Baldessarini, R.J.; Katzan, I.L. Comparison of Electronic Screening for Suicidal Risk With the Patient Health Questionnaire Item 9 and the Columbia Suicide Severity Rating Scale in an Outpatient Psychiatric Clinic. *J. Psychosom. Res.* **2015**, *56*, 460–469. [CrossRef]

78. Hjelmeland, H.; Dieserud, G.; Dyregrov, K.; Knizek, B.L.; Leenaars, A.A. Psychological Autopsy Studies as Diagnostic Tools: Are they Methodologically Flawed? *Death Stud.* **2012**, *36*, 605–626. [CrossRef]

79. Shahtahmasebi, S. Homicides and Suicides by Mentally Ill People. *Sci. World J.* **2003**, *3*, 684–693. [CrossRef]

80. CDC More than a Mental Health Problem. Available online: https://www.cdc.gov/vitalsigns/suicide/index.html (accessed on 24 March 2021).

81. Risk Assessment Page. Available online: https://cssrs.columbia.edu/documents/risk-assessment-page/ (accessed on 24 March 2021).

82. Edemekong, P.F.; Bomgaars, D.L.; Sukumaran, S.; Levy, S.B. *Activities of Daily Living*; StatPearls Publishing: Treasure Island, FL, USA, 2021.

83. Kroenke, K.; Spitzer, R.L.; Williams, J.B.W. The patient health questionnaire-2. *Med. Care* **2003**, *41*, 1284–1292. [CrossRef]

84. Spitzer, R.L.; Kroenke, K.; Williams, J.B.W.; Löwe, B. A Brief measure for assessing generalized anxiety disorder. *Arch. Intern. Med.* **2006**, *166*, 1092–1097. [CrossRef]

85. Zigmond, A.S.; Snaith, R.P. The hospital anxiety and depression scale. *Acta Psychiatr. Scand.* **1983**, *67*, 361–370. [CrossRef] [PubMed]

86. Schmidt, L.L.; Karabin, B.L.; Malone, A.C. Postural Orthostatic Tachycardia Syndrome (POTS): Assess, diagnose, and evaluate for POTS treatment (ADEPT). *Integr. Med. Int.* **2017**, *4*, 142–153. [CrossRef]

87. Brown, G.K.; Jager-Hyman, S. Evidence-Based Psychotherapies for Suicide Prevention. *Am. J. Prev. Med.* **2014**, *47*, S186–S194. [CrossRef]

88. Collaborative Assessment & Management of Suicidality (CAMS). About. Available online: https://cams-care.com/about-cams/ (accessed on 24 March 2021).

89. Méndez-Bustos, P.; Calati, R.; Rubio-Ramírez, F.; Olié, E.; Courtet, P.; Lopez-Castroman, J. Effectiveness of psychotherapy on suicidal risk: A systematic review of observational studies. *Front. Psychol.* **2019**, *10*. [CrossRef] [PubMed]

90. National Action Alliance for Suicide Prevention, Transforming Health Systems Initiative Work Group. *Recommended Standard Care for People with Suicide Risk: Making Health Care Suicide Safe.* 2018. Available online: https://theactionalliance.org/sites/default/files/action_alliance_recommended_standard_care_final.pdf (accessed on 17 May 2021).

91. Pavulans, K.S.; Bolmsjö, I.; Edberg, A.-K.; Öjehagen, A. Being in want of control: Experiences of being on the road to, and making, a suicide attempt. *Int. J. Qual. Stud. Health Well-Being* **2012**, *7*, 16288. [CrossRef] [PubMed]

92. Stanley, B.; Brown, G.K. Safety Planning Intervention: A Brief Intervention to Mitigate Suicide Risk. *Cogn. Behav. Pr.* **2012**, *19*, 256–264. [CrossRef]

93. Moscardini, E.H.; Hill, R.M.; Dodd, C.G.; Do, C.; Kaplow, J.B.; Tucker, R.P. Suicide Safety Planning: Clinician Training, Comfort, and Safety Plan Utilization. *Int. J. Environ. Res. Public Health* **2020**, *17*, 6444. [CrossRef]

94. CALM: Counseling on Access to Lethal Means | Suicide Prevention Resource Center. Available online: https://www.sprc.org/resources-programs/calm-counseling-access-lethal-means (accessed on 24 March 2021).

95. Stanley, B.; Brown, G.K.; Brenner, L.A.; Galfalvy, H.; Currier, G.W.; Knox, K.L.; Chaudhury, S.R.; Bush, A.L.; Green, K.L. Comparison of the Safety Planning Intervention With Follow-up vs Usual Care of Suicidal Patients Treated in the Emergency Department. *JAMA Psychiatry* **2018**, *75*, 894–900. [CrossRef]
96. Rudd, M.D.; Mandrusiak, M.; Joiner, T.E., Jr. The case against no-suicide contracts: The commitment to treatment statement as a practice alternative. *J. Clin. Psychol.* **2005**, *62*, 243–251. [CrossRef]
97. Stein, E. Psychiatric conditions comorbid with myalgic encephalomyelitis and/or fibromyalgia. *Psychiatr. Times* **2013**, *30*, 14.
98. Chu, L.; Valencia, I.J.; Garvert, D.W.; Montoya, J.G. Onset patterns and course of myalgic encephalomyelitis/chronic fatigue syndrome. *Front. Pediatr.* **2019**, *7*, 7. [CrossRef] [PubMed]
99. Stein, E. Assessment and treatment of patients with ME/CFS: Clinical guidelines for psychiatrists. *Hentet* **2005**, *11*, 13.
100. Cuijpers, P.; Quero, S.; Dowrick, C.; Arroll, B. Psychological treatment of depression in primary care: Recent developments. *Curr. Psychiatry Rep.* **2019**, *21*, 129. [CrossRef] [PubMed]
101. Van Dis, E.A.M.; Van Veen, S.C.; Hagenaars, M.A.; Batelaan, N.M.; Bockting, C.L.H.; Heuvel, R.M.V.D.; Cuijpers, P.; Engelhard, I.M. Long-term Outcomes of Cognitive Behavioral Therapy for Anxiety-Related Disorders. *JAMA Psychiatry* **2020**, *77*, 265–273. [CrossRef] [PubMed]
102. Mohr, D.C.; Ho, J.; Duffecy, J.; Reifler, D.; Sokol, L.; Burns, M.N.; Jin, L.; Siddique, J. Effect of Telephone-Administered vs Face-to-face Cognitive Behavioral Therapy on Adherence to Therapy and Depression Outcomes Among Primary Care Patients. *JAMA* **2012**, *307*, 2278–2285. [CrossRef]
103. Marshall, R.; Paul, L.; McFadyen, A.K.; Rafferty, D.; Wood, L. Pain Characteristics of People with Chronic Fatigue Syndrome. *J. Musculoskelet. Pain* **2010**, *18*, 127–137. [CrossRef]
104. Carruthers, B.M.; Jain, A.K.; De Meirleir, K.L.; Peterson, D.L.; Klimas, N.G.; Lerner, A.M.; Bested, A.C.; Flor-Henry, P.; Joshi, P.; Powles, A.C.P.; et al. Myalgic encephalomyelitis/chronic fatigue syndrome. *J. Chronic Fatigue Syndr.* **2003**, *11*, 7–115. [CrossRef]
105. Stein, E.; Stormorken, E.; Karlsson, B. How to Improve Therapeutic Encounters between Patients with Myalgic Encephalomyelitis/Chronic Fatigue Syndrome and Health Care Practitioners. Available online: https://s3.amazonaws.com/kajabi-storefronts-production/sites/90617/themes/1534791/downloads/4t2GsiASLGVtbwaM1qDc_Improving-Therapeutic-Relationships.pdf (accessed on 17 May 2021).
106. Brodsky, B.S.; Spruch-Feiner, A.; Stanley, B. The Zero Suicide model: Applying evidence-based suicide prevention practices to clinical care. *Front. Psychiatry* **2018**, *9*, 33. [CrossRef]
107. Luxton, D.D.; June, J.D.; Comtois, K.A. Can Postdischarge Follow-Up Contacts Prevent Suicide and Suicidal Behavior? *Crisis* **2013**, *34*, 32–41. [CrossRef]
108. National Action Alliance for Suicide Prevention. *Best Practices in Care Transitions for Individuals with Suicide Risk: Inpatient Care to Outpatient Care*; Education Development Center, Inc.: Washington, DC, USA, 2019.
109. Chung, D.; Hadzi-Pavlovic, D.; Wang, M.; Swaraj, S.; Olfson, M.; Large, M. Meta-analysis of suicide rates in the first week and the first month after psychiatric hospitalisation. *BMJ Open* **2019**, *9*, e023883. [CrossRef]
110. Richards, J.E.; Whiteside, U.; Ludman, E.J.; Pabiniak, C.; Kirlin, B.; Hidalgo, R.; Simon, G. Understanding why patients may not report suicidal ideation at a health care visit prior to a suicide attempt: A qualitative study. *Psychiatr. Serv.* **2019**, *70*, 40–45. [CrossRef]
111. Sudol, K.; Mann, J.J. Biomarkers of suicide attempt behavior: Towards a biological model of risk. *Curr. Psychiatry Rep.* **2017**, *19*, 31. [CrossRef] [PubMed]
112. CDC. Myalgic Encephalomyelitis/Chronic Fatigue Syndrome (ME/CFS). Available online: https://www.cdc.gov/me-cfs/index.html (accessed on 25 March 2021).
113. US ME/CFS Clinician Coalition Diagnosing and Treating Myalgic Encephalomyelitis/Chronic Fatigue Syndrome (ME/CFS). 2020. Available online: https://batemanhornecenter.org/wp-content/uploads/filebase/Diagnosing-and-Treating-MECFS-Handout-V2.pdf (accessed on 5 March 2021).
114. Roaten, K.; Johnson, C.; Genzel, R.; Khan, F.; North, C.S. Development and implementation of a universal suicide risk screening program in a safety-net hospital system. *Jt. Comm. J. Qual. Patient Saf.* **2018**, *44*, 4–11. [CrossRef]
115. National Alliance on Mental Illness. Understanding Health Insurance, What Is Mental Health Parity? Available online: https://www.nami.org/Your-Journey/Individuals-with-Mental-Illness/Understanding-Health-Insurance/What-is-Mental-Health-Parity. (accessed on 25 March 2021).
116. Mental Health Parity in the US: Have We Made Any Real Progress? Available online: https://www.psychiatrictimes.com/view/mental-health-parity-in-the-us-have-we-made-any-real-progress (accessed on 25 March 2021).
117. Canadian Mental Health Association. *Mental Health in the Balance: Ending the Health Care Disparity in Canada*; Canadian Mental Health Association: Toronto, ON, Canada, 2018.
118. Mental Health Action Plan: Better Access and System Performance for Mental Health Services in Canada. Available online: https://cmha.ca/news/mental-health-action-plan-better-access-and-system-performance-for-mental-health-services-in-canada (accessed on 25 March 2021).
119. The Right Way to Avoid Malpractice Lawsuits. Available online: https://www.psychiatrictimes.com/view/the-right-way-to-avoid-malpractice-lawsuits (accessed on 25 March 2021).
120. Zonana, J.; Simberlund, J.; Christos, P.; Information, R. The impact of safety plans in an outpatient clinic. *Crisis* **2018**, *39*, 304–309. [CrossRef]

121. Murphy, C. Gun Laws Are the Key to Addressing America's Suicide Crisis. Available online: https://www.theatlantic.com/ideas/archive/2020/09/gun-control-key-addressing-americas-suicide-crisis/615889/ (accessed on 25 March 2021).
122. Barnhorst, A. Opinion. In *The Empty Promise of Suicide Prevention*; The New York Times: New York, NY, USA, 2019.
123. Cherkis, J. Opinion. In *What Happens to Your Mental Health When You Can't Pay Your Rent?* The New York Times: New York, NY, USA, 2021.
124. Reckrey, J.M.; DeCherrie, L.V.; Dugue, M.; Rosen, A.; Soriano, T.A.; Ornstein, K.A. Meeting the Mental Health Needs of the Homebound: A Psychiatric Consult Service Within a Home-Based Primary Care Program. *Care Manag. J.* **2015**, *16*, 122–128. [CrossRef] [PubMed]
125. Pozzilli, C.; Brunetti, M.; Amicosante, A.M.V.; Gasperini, C.; Ristori, G.; Palmisano, L.; Battaglia, M. Home based management in multiple sclerosis: Results of a randomised controlled trial. *J. Neurol. Neurosurg. Psychiatry* **2002**, *73*, 250–255. [CrossRef] [PubMed]
126. Watts, S.; Newby, J.M.; Mewton, L.; Andrews, G. A clinical audit of changes in suicide ideas with internet treatment for depression. *BMJ Open* **2012**, *2*, e001558. [CrossRef]
127. Williams, A.D.; Andrews, G. The Effectiveness of Internet Cognitive Behavioural Therapy (iCBT) for Depression in Primary Care: A Quality Assurance Study. *PLoS ONE* **2013**, *8*, e57447. [CrossRef]
128. Charova, E.; Dorstyn, D.; Tully, P.; Mittag, O. Web-based interventions for comorbid depression and chronic illness: A systematic review. *J. Telemed. Telecare* **2015**, *21*, 189–201. [CrossRef]
129. Miller, I.W.; Camargo, C.A.; Arias, S.A.; Sullivan, A.F.; Allen, M.H.; Goldstein, A.B.; Manton, A.P.; Espinola, J.A.; Jones, R.; Hasegawa, K.; et al. Suicide Prevention in an Emergency Department Population. *JAMA Psychiatry* **2017**, *74*, 563–570. [CrossRef] [PubMed]

 *healthcare*

Case Report

Life-Threatening Malnutrition in Very Severe ME/CFS

Helen Baxter [1,*], Nigel Speight [2] and William Weir [3]

1   25% ME Group, Troon KA10 6HT, UK
2   Medical Advisor to 25% ME Group, Durham DH1 1QN, UK; speight@doctors.org.uk
3   NHS Consultant Physician, London W1G 9PF, UK; wrcweir@hotmail.com
*   Correspondence: hbaxter@25megroup.org

Abstract: Very severe Myalgic Encephalomyelitis (ME), (also known as Chronic Fatigue Syndrome) can lead to problems with nutrition and hydration. The reasons can be an inability to swallow, severe gastrointestinal problems tolerating food or the patient being too debilitated to eat and drink. Some patients with very severe ME will require tube feeding, either enterally or parenterally. There can often be a significant delay in implementing this, due to professional opinion, allowing the patient to become severely malnourished. Healthcare professionals may fail to recognize that the problems are a direct consequence of very severe ME, preferring to postulate psychological theories rather than addressing the primary clinical need. We present five case reports in which delay in instigating tube feeding led to severe malnutrition of a life-threatening degree. This case study aims to alert healthcare professionals to these realities.

Keywords: Myalgic Encephalomyelitis (ME); Chronic Fatigue Syndrome (CFS); enteral feeding; Nasogastric Tube (NGT); Nasojejunal Tube (NJT); Percutaneous Endoscopic Gastrostomy (PEG); Total Parenteral Nutrition (TPN); Home Enteral Nutrition Service (HENS); Mast Cell Activation Disorder (MCAD)

Citation: Baxter, H.; Speight, N.; Weir, W. Life-Threatening Malnutrition in Very Severe ME/CFS. *Healthcare* **2021**, *9*, 459. https://doi.org/10.3390/healthcare9040459

Academic Editors: Kenneth J. Friedman, Lucinda Bateman and Kenny Leo De Meirleir

Received: 10 March 2021
Accepted: 7 April 2021
Published: 14 April 2021

Publisher's Note: MDPI stays neutral with regard to jurisdictional claims in published maps and institutional affiliations.

1. Introduction

Some of the most severely affected ME patients experience serious difficulties in maintaining adequate nutrition and hydration and will require feeding enterally. There are a variety of mechanisms whereby nutritional difficulties arise and they are not uncommon in the most severe cases of ME/CFS. Perhaps the commonest is that the patient is just so debilitated that the sheer effort of eating and drinking is too much for them [1]. Another potential mechanism is that there are genuine difficulties with swallowing [2]; the causes for this are currently unclear. Finally, there may be problems lower down the alimentary tract such as gastroparesis, or features of malabsorption, with the additional possibility of Mast Cell Activation Disorder (MCAD). In these latter instances, enteral tube feeding may fail and recourse to Total Parenteral Nutrition (TPN) may be necessary. A lack of awareness by healthcare professionals of this problem may constitute a barrier to the patient receiving timely artificial nutrition (AN).

There is limited literature on this subject. The 2007 NICE Guidelines [3] make passing reference to the issue, as do the Paediatric Guidelines from The Royal College of Paediatricians and Child Health (RCPCH) in 2004 [4]. It is also mentioned in a recent paper on severe ME in young people [1]. Both ME/CFS [5] and nutrition [6] are poorly covered in the curriculum at medical schools.

The potential adverse consequences for the patient of their problem not being promptly recognized and responded to are considerable. There is often a significant delay in implementing tube feeding in patients experiencing difficulties obtaining nutrition and hydration. Tube feeding is often not instigated until the malnutrition becomes life threatening. Healthcare professionals seem to fail to recognize that the inability to eat and drink is a direct consequence of the severity of the ME, instead preferring to postulate psychological theories. Apart from the medical issues, it can be emotionally very upsetting for the patients to have their problems not recognized, or misdiagnosed.

As staff and volunteers with the UK's 25% ME Group—the charity supporting people living with severe and very severe ME—we have become aware of these factors and have become very concerned by the clinical response both in the community and in acute settings.

This case report documents current NHS responses to patients in this situation, with the aim of raising awareness amongst healthcare professionals.

2. Methods

With the awareness of the difficulties people with very severe ME can have obtaining nutrition and hydration and the clinical response, the 25% ME Group devised a questionnaire for members who had experience of being enterally or parenterally fed. An invitation to participate was placed in the 25% ME Group charity's newsletter, 'The Quarterly', in summer 2019, This was available either on paper, by post or via email. The questionnaire contained a range of questions such as age, reason for AN, type of AN, duration and an open-ended section for noting 'any other relevant information' (see Supplementary Materials).

Reasonable adjustments were put in place to maximize participation whilst attempting to minimize the likelihood of post exertional malaise. These included flexibility in the method of communication with patients and family/paid home worker staff and no fixed end date to return the questionnaire. Two of the patients' ME was so severe that they were unable to read, write or type and in one case speak. A member of staff from 25% ME Group made direct contact with them and communicated via telephone, text or with their representative to gather information. As the member of staff has experience of talking to people with severe ME, she listened for signs of tiredness in the patient's voice and terminated telephone calls at the first sign of fatigue.

Questionnaire responses provided a foundation; additional information was obtained using the patients' preferred method of communication. From the information given, a series of anonymized case reports were developed. The case reports represent an opportunity sample.

3. Results

3.1. Case 1

This patient was diagnosed with severe ME as a child and has never fully recovered. She has been very severely affected for the past nine years. She has received care in a nursing home setting, hospital and now in her forties has around the clock care in the community.

Her difficulty swallowing started in 2015 whilst a resident in a nursing home. A speech and language therapist (SALT) diagnosed dysphagia, cause unspecified. A community dietitian visited and prescribed oral nutritional supplements (ONS). The patient was unable to tolerate these and became increasingly malnourished.

Although she was already significantly underweight, the primary health care team failed to recognize the severity of the situation. Over a seven week period she was almost completely unable to ingest any nutrition or hydration. As the situation deteriorated the patient was told that she would be sectioned under the Mental Health Act if she did not go to hospital. She reluctantly agreed to a voluntary admission.

Over a two week period in hospital she was intermittently given intravenous fluids, but no nutrition. She was screened using the malnutrition universal screening tool (MUST); her MUST score was 4 (2 or above denotes high risk of malnutrition [7]).

Her condition deteriorated further to the extent that she had to be admitted to a High Dependency Unit (HDU) because no Intensive Care Unit (ICU) bed was available, after which a Nasogastric Tube (NGT) was inserted. By this time, she was found to be suffering from a severe electrolyte imbalance, which further delayed the establishment of a feeding regime due to the risk of re-feeding syndrome.

Two months later with an established NGT feeding regime in place her MUST was still only 2.

The patient gained the strong impression that her doctors regarded her problems as psychological in origin and that she was being treated as if she was suffering from an eating disorder.

An NGT was sited prior to discharge. The patient was told another NGT would not be sited.

The tube remained in situ far longer than recommended until it became unusable. It was re-sited as an emergency in hospital. Subsequently the tube became blocked on several occasions and each time the patient had to attend the hospital as an emergency to have it replaced. This was especially concerning as there was no guarantee the tube would be re-sited, despite a clinical need.

In 2019 this situation improved after a Percutaneous Endoscopic Gastrostomy (PEG) was sited, and finally the patient had an improvement in her quality of life with the allocation of a Home Enteral Nutrition Service (HENS) dietician who made changes to the feed to ameliorate pain whilst being fed.

The patient still feels that there has been a failure to acknowledge her dysphagia. Today she has a normal BMI and continues to receive her preferred choice of care in the community.

3.2. Case 2

This female patient is in her 50s. She was diagnosed with ME following viral encephalitis in her twenties and has had very severe ME from the outset, being bed bound and needing around the clock care. She has required NGT feeding for over twenty five years. Following a diagnosis of intestinal failure two years ago she now requires parenteral feeding.

Within the first eighteen months of diagnosis, over a five month period, the patient started to experience swallowing difficulties complicated by facial paralysis leading to an inability to maintain adequate nutrition and hydration, and to significant weight loss. There was little intervention by primary care, but after five months of malnutrition the GP prescribed ONS. The nutritional difficulties and associated weight loss necessitated an emergency hospital admission.

NGT feeding was instigated following the intervention of a psychiatrist who stated that her nutritional difficulties were not psychological in origin. A SALT assessment confirmed the swallowing reflex was absent. Despite the diagnosis of dysphagia, a second psychiatrist told her that her inability to swallow was psychological. Three months later, with little weight gain, her discharge home was planned without an NGT. With family pressure she was finally sent home with an NGT in place.

For over twenty five years NGT feeding continued with the support of a HENS dietician. The GP arranged for an anesthetist to re-site the tube at home until this service was withdrawn. A private arrangement was made by the family to avoid travel to hospital, which would have been detrimental to the patient's condition.

Recent abdominal surgery, complicated by adhesions, led to a bowel obstruction and another emergency admission to hospital due to extreme vomiting. Total Parenteral Nutrition (TPN) was commenced.

A peripherally inserted central catheter (PICC line) infection necessitated its removal and the patient was transferred to a specialist unit in another hospital. Despite the success of TPN, a trial of NGT feeding was enforced causing further weight loss with vomiting and abdominal pain. A barium meal showed gastroparesis and intestinal failure. An Nasojejunal Tube (NJT) was then sited but the patient experienced the same effects as with the NGT. Nutrition via TPN was then started, but despite this, further weight loss occurred. When this was brought to the attention of the medical team, they accused the patient of interfering with the pump despite this being physically impossible. By this point the patient's BMI was 11.4. The family monitored the feed administered, and showed it was less than prescribed. A second opinion was requested by the family and Mast Cell Activation Disorder (MCAD) diagnosed. Medication was prescribed to treat this which stabilized and then increased the patient's weight. On discharge from hospital her BMI was 13.8.

Currently, TPN is still being administered via a PICC line. Her BMI is 18.6. Unfortunately, there is currently no provision for community based TPN within the NHS, so administration and supply of the feed has had to be outsourced to a private company.

3.3. Case 3

This patient is a young adult with severe ME who had been cared for at home for over three years before becoming nutritionally deficient.

At home the patient was cared for in bed in a darkened room. Her family became concerned about her poor oral intake and the associated weight loss. Her mother tried repeatedly to get professional help in the community but it was not forthcoming; the patient required an emergency admission to hospital.

Once in hospital only IV fluids were given, with a delay of nine days before nutritional feed was given via a NGT. A SALT assessment failed to find a reason for the swallowing problems. An eating disorder was excluded. With the apparent lack of an organic reason for the situation, psychological problems and lack of motivation were suggested as a cause by hospital staff. Reluctance to allow long-term feeding via NGT was expressed, the reason cited being that she might become "dependent" on this form of treatment.

Accordingly, attempts were made to wean her off the NGT, in the form of longer waits between feeds. This was observed and challenged by the patient's mother.

Two months into her admission the need for long term nutritional support in the community was accepted by the nutrition support team (NST). An NJT was inserted on the grounds that there was no domiciliary service for re-siting NGTs. She was then discharged and is maintaining a normal BMI while being cared for at home.

3.4. Case 4

This patient was diagnosed with ME as a child. Several years later the ME became very severe. She became virtually unable to eat and had severe nausea. An ME specialist recommended tube-feeding. The patient and her parents agreed to it. There was a lengthy delay because the local pediatricians refused it. Her inability to eat and the delay caused significant weight loss and the situation became life threatening. The patient was admitted to hospital where NGT feeding commenced. The local pediatricians claimed that she had anorexia nervosa and persuasive refusal syndrome and threatened admission to a psychiatric unit. A psychiatrist found that she did not have any psychiatric disorders. She was NGT fed for nine months before a PEG was sited.

3.5. Case 5

ME was diagnosed in childhood in this patient; she is now in her forties. She has had two episodes of enteral feeding, firstly in her thirties and more recently for a longer period of four years. She has needed emergency admissions to hospital on both occasions in order to access enteral feeding. As with the other cases there was no support in the community for the provision of NGT feeding.

At the first emergency admission, her BMI was 13.7 She was fed by NGT for three weeks, following which the NGT was removed prior to discharge. The patient felt she had been labelled as having an eating disorder. On discharge she was allocated an eating disorders dietician who alluded to the likelihood that she would be sectioned under the Mental Health Act if she lost more weight.

Following a move to a new area, a different consultant looked at her case and made a diagnosis of gastroparesis.

A worsening of her ME, a fall in her BMI to 14.8 and being barely able to consume any ONS led to another emergency admission.

On this occasion she was provided with a long-term plan. She was discharged with an NGT in situ and with the involvement of a HENS dietician. While in hospital she had been taught how to reinsert an NGT herself which she was then able to do at home, avoiding the need for further hospital admissions.

Working in conjunction with the HENS dietitian, the patient has progressed to modifying her diet to food which was more easily digested. She has been able to remove the NGT but still requires ONS to meet her calorific requirements. Her BMI is currently 20.5.

3.6. Summary

The experiences of all five participants share some strikingly similar features.

All had been allowed to become and remain severely malnourished and dehydrated. The experience was frightening and emotionally upsetting for the patients. The extent of the malnutrition could have further consequences, both short- and long-term. The re-introduction of nutrition poses an immediate risk of re-feeding syndrome [8] and there are possible longer-term consequences for the patients' health status including poor wound healing, neurological damage, and osteoporosis [9].

Features emerging from these five cases include:

All were allowed to become severely malnourished and dehydrated to a life-threatening extent.

The inability to swallow in all cases was believed to be psychological in origin and psychiatrists became involved in all cases.

All five were considered to be suffering from anorexia nervosa; had this been the case it would have warranted tube feeding.

Two were advised that their enteral nutrition would be stopped, despite a clinical need.

Two were threatened that they would be sectioned under the Mental Health Act, if they did not eat and drink or if they lost weight again.

Scope and Limitations of Study

This case report documents the experiences of five people. While the number of cases reported here is small, our experience of supporting other severely affected ME patients has shown similar failures in a much larger number of cases.

4. Discussion

This series of cases demonstrate a common set of problems. The clinicians involved seemed unaware that severe ME can lead to serious problems maintaining adequate nutrition and hydration. Perhaps this is understandable, as many clinicians will only meet one or two cases of severe ME in their careers, and the subject is poorly taught at both undergraduate and postgraduate levels [5].

The doctors failed to recognize the severity of the malnutrition or to provide appropriate nutritional support in a timely manner [10]. Each case developed life-threatening problems as a result and were only saved by the late introduction of some form of nutritional support. Clinical inertia was evident throughout. In respect of the repeated finding that patients were wrongly regarded as having an eating disorder as a cause for their nutritional problems, it is lacking in logic for the doctors concerned not to have treated this on its own merit. Tube feeding, with or without a court order, is frequently resorted to in cases of an eating disorder. Either the doctors were not serious in making this diagnosis or they were somehow generally prejudiced against the patients on account of their being cases of ME/CFS. In each case, the doctors resorted to making inappropriate psychological diagnoses without positive evidence of psychopathology.

NICE Clinical Guideline 53 on ME/CFS, lists the nutritional problems which may be experienced by a person with severe ME/CFS and also notes the possible need for tube feeding in patients with severe ME:

"this may include the use of tube feeding, if appropriate [3]."

NICE suggest using Clinical Guideline 32 'Nutrition support for adults: oral nutrition support, enteral feeding and parenteral nutrition' [11] for any patients who are nutritionally compromised. This document is referenced in CG53 [3]. Each of the patients met the criteria for enteral feeding set out in CG32. Given the nutritional status of the patient, the clinicians should have followed the NICE guidance.

Case 2 highlights an important issue. If a patient is failing to respond to enteral feeding, the possibility of MCAD needs to be considered. This is a recognized complication of severe ME and effective treatment exists in the form of oral cromoglycate and antihistamines. It has probably contributed to several deaths of severe ME sufferers.

In every case, the most positive improvement in their management came about as the result of the allocation of a named HENS dietician whose advanced training in enteral nutrition enabled them to make changes to the patient's diet. In one case it enabled the patient to get to a healthy weight using enteral nutrition whilst making changes to the oral nutrition such that enteral feeding is now no longer required. In another case, dietary changes ameliorated suffering. All patients felt supported by their HENS dietician. For patients with very severe ME connecting with a knowledgeable healthcare professional who does domiciliary visits is very important. Such a policy would reduce the need for hospital admissions which would be to the benefit of all. All patients with very severe ME should be allocated a HENS dietician as soon as nutritional difficulties become apparent.

An early warning system needs to be put in place for patients with severe ME so that when they or their representatives become aware of the development of problems with oral intake prompt action is taken and tube feeding started thereby avoiding undernutrition in patients with very severe ME. Early intervention in the form of tube feeding has been shown to be beneficial in patients with severe ME [1].

Patients with very severe ME are bedridden and require around the clock care. They are best cared for at home where the environment can be adapted to best meet their needs. These patients will have extreme sensitivity to noise and light, such that they need to be cared for in a darkened room. People with very severe ME invariably report travel to hospital and the hospital environment significantly exacerbates their condition. If an admission to hospital is necessary, and this should only be done for emergency treatment, they will require admission directly into a side room and to be cared for by a small number of staff who understand ME as an organic illness.

For the patient with very severe ME, it appears to be common practice for Clinical Commissioning Groups (CCGs) to adopt a 're-site in hospital' policy despite a large study showing that with protocols in place trained nurses in the community can identify the position of NGT's correctly without the need for hospital attendance [12].

Nonetheless it is stated:

'Local protocols should address the clinical criteria that permit enteral feeding [13].'

None of the participants were offered NGT re-sites at home, instead they went to significant lengths to avoid trips to hospital if at all possible; re-siting their own NGTs or opting to have NJTs or PEGs. A constructive change would an implementation of national guidelines allowing NGT re-sites to be carried out in the community by appropriately trained professionals. A community-based service could bring potential savings to the NHS and certainly benefit patients with very severe ME. The treatment of serious undernutrition issues in ME needs to be included in national and local guidelines for use by health care professionals.

5. Conclusions

To remedy the problems identified in this survey, the most important first step remains to improve medical education for healthcare professionals regarding the fact that severe ME can cause nutritional problems, and that these may require early intervention with tube feeding. Progress has been made in that a Continuing Professional Development (CPD) Module on ME has been developed [14] and launched in May 2020. The uptake was very poor with fewer than two thousand clinicians taking the module to date. Medical education around ME needs to be made part of the core curriculum for undergraduate students and should also be included in postgraduate education. It is necessary for the clinician to recognize ME/CFS as an organic illness. It can only be hoped that the new NICE Guidelines aid clinicians' understanding and provide guidance on dealing with nutritional problems such as those described in this series.

Supplementary Materials: The following are available online at https://www.mdpi.com/article/10.3390/healthcare9040459/s1. Enteral and parenteral feeding questionnaire.

Author Contributions: The conceptualisation, methodology and investigation was done by H.B. The manuscript was written, reviewed and edited by H.B., N.S. and W.W. All authors have read and agreed to the published version of the manuscript.

Funding: This research received no external funding.

Informed Consent Statement: Consent was given by all participants.

Data Availability Statement: The data presented in this study are available on request from the corresponding author.

Acknowledgments: The authors would like to thank the participants and their families.

Conflicts of Interest: The authors declare no conflict of interest.

References

1. Speight, N. Severe ME in Children. *Healthcare* **2020**, *8*, 211. [CrossRef] [PubMed]
2. CFS/ME Working Group. Chapter 4 Management of CFS/ME. In *Report to the Chief Medical Officer of an Independent Working Group*; CFS/ME Working Group: London, UK, 2002; Volume 4.2.1.2, p. 37.
3. NICE Guideline 53 Chronic Fatigue Syndrome/Myalgic Encephalomyelitis (Or Encephalopathy): Chapter 7: People with Severe CFS/ME; Section 7.6.3 Dietary Interventions & Supplements. Available online: www.nice.org.uk/guidance/cg53 (accessed on 13 December 2020).
4. Royal College of Paediatrics and Child Health. *Evidence Based Guideline for the Management of CFS/ME (Chronic Fatigue Syndrome/ Myalgic Encephalopathy) in Children and Young People. p46*; Royal College of Paediatrics and Child Health: London, UK, 2004.
5. Myalgic Encephalomyelitis (or Encephalopathy)/Chronic Fatigue Syndrome: Diagnosis and Management Appendix 3: Expert Testimonies. NICE Guideline Appendix November 2020 Supporting Documentation Expert Testimonies p18. Available online: www.nice.org.uk/guidance/indevelopment/gid-ng10091/documents (accessed on 23 December 2020).
6. Pryke, R.; Lopez, L. Managing malnutrition in the community: We will all gain from finding and feeding the frail. *Br. J. Gen. Pract.* **2013**, *63*, 233–234. [CrossRef] [PubMed]
7. BAPEN. Malnutrition Universal Screening Tool. Available online: www.bapen.org.uk/must/must_full.pdf (accessed on 15 January 2021).
8. Fuentebella, J.; Kerner, J.A. Refeeding syndrome. *Pediatr. Clin. N. Am.* **2009**, *56*, 1201–1210. [CrossRef] [PubMed]
9. Stratton, R.J.; Green, C.J.; Elia, M. *Disease Related Malnutrition: An Evidence Based Approach to Treatment*; CABI Publishing: Wallingford, UK, 2003; ISBN 9780851996486.
10. Stoud, M.; Duncan, H.; Nightingale, J. Guidelines for enteral feeding in adult hospital patients. *Gut* **2003**, *52* (Suppl. 7), vii1–vii12. Available online: https://gut.bmj.com/content/gutjnl/52/suppl_7/vii1 (accessed on 15 January 2021).
11. NICE Clinical Guideline 32: Nutrition Support for Adults: Oral Nutrition Support, Enteral Feeding and Parenteral Nutrition. Available online: https://www.nice.org.uk/guidance/cg32 (accessed on 15 November 2020).
12. Jones, B.J.M.; Jones, A.; Skelton, J. Community replacement of nasogastric feeding tubes is safe and prevents need for hospital referrals: A repeat audit. *Clin. Nutr. ESPEN* **2020**, *35*, 230. [CrossRef]
13. Jones, B.J.M. *A Position Paper on Nasogastric Tube Safety 'Time to Put Patient Safety First'*. 2020, p. 25. Available online: https://www.bapen.org.uk/pdfs/ngsig/a-position-paper-on-nasogastric-tube-safety.pdf (accessed on 20 January 2021).
14. Available online: https://www.studyprn.com/p/chronic-fatigue-syndrome (accessed on 20 May 2020).

 *healthcare*

Perspective

Caring for the Patient with Severe or Very Severe Myalgic Encephalomyelitis/Chronic Fatigue Syndrome

Jose G. Montoya [1], Theresa G. Dowell [2], Amy E. Mooney [3], Mary E. Dimmock [4,*] and Lily Chu [5]

[1] Dr. Jack S. Remington Laboratory for Specialty Diagnostics, Palo Alto Medical Foundation, Palo Alto, CA 94301, USA
[2] Four Peaks Healthcare Associates, Flagstaff, AZ 86004, USA
[3] Independent Consultant, Riverside, IL 60546, USA
[4] Independent Consultant, Waterford, CT 06385, USA
[5] Independent Consultant, Burlingame, CA 94010, USA
*   Correspondence: maryedimmock@yahoo.com

Citation: Montoya, J.G.; Dowell, T.G.; Mooney, A.E.; Dimmock, M.E.; Chu, L. Caring for the Patient with Severe or Very Severe Myalgic Encephalomyelitis/Chronic Fatigue Syndrome. *Healthcare* **2021**, *9*, 1331. https://doi.org/10.3390/healthcare9101331

Academic Editors: Kenneth J. Friedman, Lucinda Bateman and Kenny Leo De Meirleir

Received: 30 August 2021
Accepted: 29 September 2021
Published: 6 October 2021

Publisher's Note: MDPI stays neutral with regard to jurisdictional claims in published maps and institutional affiliations.

Abstract: Myalgic encephalomyelitis/chronic fatigue syndrome (ME/CFS) can cause a wide range of severity and functional impairment, leaving some patients able to work while others are homebound or bedbound. The most severely ill patients may need total care. Yet, patients with severe or very severe ME/CFS struggle to receive appropriate medical care because they cannot travel to doctors' offices and their doctors lack accurate information about the nature of this disease and how to diagnose and manage it. Recently published clinical guidance provides updated information about ME/CFS but advice on caring for the severely ill is limited. This article is intended to fill that gap. Based on published clinical guidance and clinical experience, we describe the clinical presentation of severe ME/CFS and provide patient-centered recommendations on diagnosis, assessment and approaches to treatment and management. We also provide suggestions to support the busy provider in caring for these patients by leveraging partnerships with the patient, their caregivers, and other providers and by using technology such as telemedicine. Combined with compassion, humility, and respect for the patient's experience, such approaches can enable the primary care provider and other healthcare professionals to provide the care these patients require and deserve.

Keywords: myalgic encephalomyelitis/chronic fatigue syndrome (ME/CFS); chronic fatigue syndrome (CFS); myalgic encephalomyelitis (ME); severely affected (housebound) patients; very severely affected (bedbound) patients; chronic disease; primary care; home health care

1. Introduction

Myalgic encephalomyelitis/chronic fatigue syndrome (ME/CFS) is a chronic, multi-system, debilitating disease that affects a million or more Americans of all ages, ethnicities, nationalities, genders, and socioeconomic backgrounds [1]. ME/CFS causes profound fatigue, unrefreshing sleep, cognitive impairment, orthostatic intolerance, pain, sensory sensitivities, gastrointestinal issues, and other bodily symptoms leading to substantial impairment in function. The hallmark symptom is post-exertional malaise (PEM), an exacerbation of symptoms and a further relapse in functioning, following even small physical, cognitive, orthostatic, emotional, or sensory challenges that were previously tolerated. Studies have demonstrated neurological, autonomic, immunological, and energy metabolism dysfunction in ME/CFS [1–3].

Prevalence estimates for ME/CFS vary considerably because of factors such as the case definition used, how cases were assessed, and whether the study was community based or not. In their 2020 systematic review and meta-analysis, Lim et al. found an average meta-analysis prevalence of 0.65% in adults with a range of 0.38% to 1.45% [4]. The US prevalence estimate of 1–2.5 M [1] is less than 1% of its population.

Based on studies in the US, UK, and Norway, an estimated 25% of people with ME/CFS are mild and able to work [5] while an estimated 25% are homebound (severe) or bedbound (very severe) [6]. Because they are unable to leave their homes, those who are homebound or bedbound are rarely included in research studies or seen by primary care and other healthcare providers unless there is a crisis, such as a very severe relapse [7] or life-threatening malnutrition [8]. All ME/CFS patients struggle to obtain a correct diagnosis and access the clinical care and support they need because of misunderstanding, a lack of a biomarker, and a lack of proper clinical guidance [1]. As a result, many people with ME/CFS are often not diagnosed or are misdiagnosed. This is especially challenging for severe and very severe patients because they are typically not seen in doctors' offices. For their part, clinicians may not have seen this level of debility and may not recognize or believe in the disease [8–10].

Updated ME/CFS clinical guidance has been published but these have primarily focused on the less severely ill or on children [11–15]. This article combines the published information available on severe and very severe ME in these and other sources [8,10,16,17] with the authors' experiences with this subgroup of patients and their caretakers. The emphasis on multi-disciplinary care is reflective of the authoring team, which includes two physicians, a health services researcher, an advanced practice nurse, an occupational therapist, and a physical therapist. Our expertise encompasses general medicine, geriatric medicine, public health, and infectious diseases. We have participated in the care of patients severely affected by ME/CFS and other chronic medical conditions in different settings, including in their homes. Four of us also have personal experience as patients or caregivers.

This article is intended to fill the gap in clinical knowledge and guidance for severe and very severe ME/CFS in adults and reinforce the importance of compassion, humility, and respect in all clinical interactions. Using already existing interventions and resources, primary care providers and other healthcare professionals can meet patients where they are in their sickness and potentially significantly improve their health, quality of life, and function.

2. Spectrum of ME/CFS Severity

The severity of specific symptoms and the level of functional impairment seen in ME/CFS can vary widely from person to person and over time. A given patient may experience a combination of symptoms of differing levels of severity-for instance, very severe cognitive and physical impairment coupled with somewhat less severe orthostatic intolerance. A mildly affected patient may be able to work or attend school with accommodations while the most severely affected patient may be bedbound and need total care. This spectrum of severity is described in Table 1:

Table 1. Spectrum of Severity in ME/CFS [12,18,19].

Level of Severity	Description of Level of Functioning and Disease Severity
Mild	Mobile and able to self-care. May be working or attending school, but often with accommodations and by reducing other domestic and social activities.
Moderate	Reduced mobility and restricted activities of daily living. Requires frequent rest periods and typically not working or attending school.
Severe	Mostly homebound. Limited activities of daily living (e.g., self-care, showering, dressing). Severe cognitive difficulties. May be wheelchair dependent.
Very Severe	Bedbound. Unable to carry out most activities of daily living for themselves. Often extreme sensory sensitivity to light, sound, touch, etc. May need total care.

These are general categories intended to convey the wide spectrum of disease severity and functional impairment seen in ME/CFS. The assessment of a given patient should be based on their particular level of disease severity and functional impairment.

Recent research studies provide evidence that this classification system is valid and useful [20,21]. Compared to the non-homebound ME/CFS population, homebound patients report more severe and frequent symptoms and greater functional impairment [6]. For instance, compared to less affected patients, the severely ill ambulate fewer steps daily and demonstrate a lower exercise capacity as measured by cardiopulmonary exercise test parameters such as percent peak oxygen consumption [21,22]. The prevalence of psychiatric diagnoses is similar to that observed in other chronic diseases and psychological well-being/functioning may remain relatively intact [23–25].

As with ME/CFS in general, recovery is not common [26] and patients with severe and very severe ME/CFS can remain ill for years or decades. Even so, compassionate, high-quality clinical care can help improve the quality of life, decrease the overall symptom burden, and prevent a worsening of the disease.

3. Clinical Features Prominent in Severe and Very Severe ME/CFS

The clinical presentation of severe or very severe ME/CFS includes the features seen in those with milder disease, but some features are more prevalent, and all are much more extreme. This includes [8,10,13–17]:

- Profound weakness. May be unable to move or turn over in bed, eat, get to the toilet, etc.
- Reduced or lack of ability to speak or swallow.
- Severe and often almost constant, widespread pain, severe headaches, and hyperesthesia.
- Extreme intolerance to small amounts of physical, mental, emotional, or orthostatic stressors such as sitting, bathing, toileting, eating, speaking. These can trigger post-exertional malaise and increased weakness.
- Hypersensitivity, sometimes extreme, to light, sound, touch, chemicals, or odors. Exposure can increase pain and other symptoms.
- Severe cognitive impairment that may impede the patient's ability to communicate and understand written materials.
- Severe gastrointestinal disturbances (e.g., nausea, abdominal pain), early satiety, and food intolerances which can impair adequate nutrition.
- Orthostatic intolerance severe enough to prevent upright posture.
- Sleep dysfunction such as unrefreshing sleep, shifted sleep cycles, and fractured sleep.
- Increased prevalence of comorbidities common to ME/CFS (e.g., mast cell activation syndrome, postural orthostatic tachycardia syndrome) and/or complications of being homebound or bedbound (e.g., osteoporosis, constipation, pressure ulcers, aspiration pneumonia, depression, and deconditioning). These can increase disease burden and complicate management.

Compounding the physical debility, patients with severe or very severe ME/CFS are often isolated, sometimes from their own families, and must deal with the complete loss of their former lives and all that defined them [27]. For those who became ill as children or young adults, that is particularly cruel.

4. Providing Compassionate Care for Severe and Very Severe ME/CFS

Primary care providers and other healthcare providers may not have seen patients with this level of severity before. The extreme levels of energy limitations, cognitive impairment, pain, and sensory/substance hypersensitivities may be surprising. At the same time, these patients and their caregivers may have been neglected or treated poorly by previous medical providers [1]. As a result, they have had to become their own experts [10]. Recognize and express sympathy for the challenging experiences patients may have faced previously. A patient-centered, collaborative approach to care that is grounded in compassion and respect for the patient in all interactions will be of benefit to everyone [10]. The following approaches can help:

- Plan for the need to see the patient in their home [10,16]. Currently, most severely affected patients live at home. In our experience, home-based care can be more individualized and is preferred by this subgroup and their families.
- Respect the nature and severity of the patient's disease in all clinical interactions [10]. Ask patients and caregivers beforehand about any factors (e.g., fragrances, fast movements, brightly colored clothes, loud noises, bright lights, and touch) that exacerbate the patient's specific sensory sensitivities. Minimize these factors as much as possible. Interact with patients at a pace, time of day, and length of time the patient can manage. Even home visits may tax the patient so leverage the caretaker where possible to conserve the patient's limited energy. Creative approaches may be required if the patient's ability to speak is limited.
- Accept the validity of the patient's report of symptoms. Gain the trust of the patient, caregiver, and family. Listen to what they report with understanding and compassion.
- Be honest about the limits of medical knowledge but reassure the patient that you will do what you can to help them.
- Partner closely with the caregiver, if one is involved, and if needed, other healthcare professionals to provide the resources, services, education, and practical help needed by the patient and caregiver. A specialty consultation may help diagnose and manage those aspects of ME/CFS with which you are unfamiliar. Engage a targeted set of other professionals as necessary and as tolerated by the patient. These could include physical therapists, occupational therapists, nurses, home health aides, social workers, and mental health experts. Home visits by optometrists/ophthalmologists and dentists may be required. Ensure these other professionals are knowledgeable about ME/CFS.
- While providing access to essential healthcare providers, care must be taken not to overwhelm the patient with too many providers or too many visits. Where feasible, leverage the caregiver to save the patient's energy. For example, capitalize on the caregiver's intimate knowledge of the patient's needs, preferences, and status. Teach them to provide certain services to minimize the need for additional healthcare providers. Reserve patient visits for those times where patient input is required or there is a need to examine the patient in-person.
- Be alert to caregiver stress. Community resources, local support groups, and respite services for those caring for people with ME/CFS or other chronic diseases may be helpful.
- Some severely ill ME/CFS patients may not have caregivers. Be alert to their non-medical needs, such as their ability to obtain and prepare food.

5. Diagnosis and Assessment

- In addition to the diagnostic approaches used for all ME/CFS patients [1,12,13,28,29], the following assessments are particularly important for the person with severe or very severe ME/CFS [8,10,13,14,16,17,30]: Evaluate the patient's basic and instrumental activities of daily living (ADLs and IADLs) (Table 2). Documenting ADLs has the added benefit of supporting applications for disability.
- Assess the patient's individual energy limits (their "energy envelope") [31] and the energy they expend on ADLs and IADLs.
- Investigate medical issues that may be impacting the patient's symptom burden or level of functioning. These could include over-exertion resulting in PEM, untreated orthostatic intolerance, pain, sleep difficulties, gastrointestinal issues, unrecognized sensory hypersensitivities, recurrent infections, comorbidities, or complications from being homebound or bedbound. Each symptom should be assessed individually to determine whether it is the result of another specific diagnosis that needs to also be treated [8,12,13,16].
- Assess the patient's psychological status using methods appropriate for chronic disease. Pay attention to affective symptoms (e.g., sadness, worry) and be careful about

attributing somatic symptoms (e.g., fatigue, insomnia, gastrointestinal disturbances) to psychological/psychiatric conditions.

- Assess non-medical issues that contribute to the patient's level of morbidity. Examples include lack of social services, caretaking, transportation, finances, food, and/or supportive devices.

Because ME/CFS is often unrecognized clinically [1], people with severe or very severe ME/CFS have sometimes been stigmatized or misdiagnosed with a mental illness such as anorexia nervosa. Their caregivers have sometimes been accused of neglect or abuse [8,32]. As with other chronic diseases, ME/CFS patients can experience secondary depression and anxiety. They can also be at an increased risk of suicide from the severe functional limitations and severity of symptoms, particularly in the face of medical disbelief and lack of support [33]. However, ME/CFS is not a mental illness [1]. A careful differential diagnosis is required to ensure an accurate diagnosis [13,34]. Concerns for neglect or abuse must be evaluated with full comprehension of the nature of ME/CFS and the level and types of debility that can result. For instance, weight loss or decreased consumption of food and fluids may not be due to intentional self-harm or anorexia nervosa but rather due to undiagnosed gastrointestinal issues that impede nutrition [8].

6. Recommendations for Treatment and Management of Severe and Very Severe ME/CFS

Historically, the debility of ME/CFS was incorrectly assumed to be the result of deconditioning that could be treated with graded exercise therapy. However, studies have demonstrated that ME/CFS is not deconditioning [35] and that overexertion can cause harm to patients [36]. This is especially true for people with very severe ME/CFS, for whom even basic ADLs may exceed their extreme energy limits. Thus, recommendations for treatment and management of severe or very severe ME/CFS must be individually tailored to each patient [13–15]. These recommendations should be implemented if ME/CFS is suspected, even if the patient has not reached the six-month requirement typical of ME/CFS criteria.

6.1. Recommendations for Minimizing Post-Exertional Malaise and Sensory Sensitivities

The following approaches can be used to help manage post-exertional malaise and sensory sensitivities:

- Ensure the patient and caregiver understand post-exertional malaise. Educate them about energy conservation strategies, such as pacing, to minimize the physical, mental, orthostatic, and emotional stressors that could trigger post-exertional malaise with its consequent worsening of symptoms and functioning [11–13,37].
- Minimize those stimuli to which the patient is sensitive, such as light, noise, touch, movement, chemicals, and odors, Exposure to these could increase pain and other symptoms (Table 2). The most severe patients may not be able to tolerate any touch, light or noise.
- Accommodate the patient's restricted energy (Table 2). In the most severe patient, specialized beds, wheelchairs, bedpans, feeding tubes, and catheters may be needed to conserve their extremely limited energy [13,14,16,17].

6.2. Recommendations for Treatment and Management Approaches

The following pharmacological and non-pharmacological treatment and management practices can be used to conserve energy, to treat symptoms and comorbidities, and to minimize medical complications [11–13,16]. Where possible, leverage the caregiver to minimize the number of providers directly engaging the patient.

- Drugs should be used conservatively and parsimoniously [12,13,16]. When drugs are used, start with very low doses and titrate up slowly as tolerated. For instance, naltrexone is commonly used at 50 mg for opioid overdose but for pain in ME/CFS, the dose starts at 0.1 mg daily and titrates up to 4.5 mg daily. Decrease the risk of side effects and drug–drug interactions by favoring medications which may treat more than one symptom or condition, e.g., both pain and sleep.

- If other pain medications have not been effective or cause significant side effects, it may be necessary to consider opioid medications. Consider starting a medication to counter constipation at the time opioids are prescribed.
- Oral feeding and hydration are preferred and should be tried first. However, tube feeding may be required to ensure nutrition and to conserve the patient's energy [8]. Intravenous saline may be needed for hydration. If necessary, intravenous feeding may be required as a last resort.
- Physical therapists may help with energy conservation approaches, pain management, joint protection to prevent joint contracture, body positioning, and gentle range of motion, stretching, and strength exercises to help address the effects of being inactive and bedbound (Table 2). The approaches used must be done in such a way that they do not trigger PEM or sensory sensitivities (e.g., to touch) [11–13,37]. Caution is advised as even passive straight leg lifts performed by a therapist have been shown to trigger PEM [38]. Caution on stretching is advised for patients with comorbid hypermobile Ehlers-Danlos syndrome.
- Occupational therapists can utilize modification and adaptation strategies for ADLs to conserve energy (Table 2) and to provide patient and caregiver education on techniques for pacing and nonpharmacological approaches to manage symptoms [12,13,37].
- Speech language therapists can help evaluate and treat problems with eating/swallowing as well as problems with communication, whether they stem from anatomical or functional abnormalities in the oral/gastrointestinal tract or in the brain.
- Mental health providers may be able to help patients better cope with the debility of the disease [34].
- Educate the patient, family, and caregiver about helpful behavioral measures. For example, space out caregiving tasks to avoid overstimulation of the patient, adjust/turn the patient occasionally to decrease pressure ulcers, and lower expectations such as the need for a daily bath.

6.3. Recommendations for Follow-Up Visits, Advance Care Directives, and Hospitalization

The health status of a severe or very severe ME/CFS patient can change over time, sometimes rapidly and potentially requiring hospitalization. The primary care provider should schedule regular visits, be prepared to provide guidance to hospital staff, and encourage patients to maintain advance directives and contingency plans as follows:

- Schedule follow-up visits on a regular basis. Monitor for emerging comorbidities and complications and whether changes in management practices could help. Do not assume any new issues are caused by ME/CFS or are intractable.
- In the event of a hospital admission, advise staff of the need to provide a low sensory environment and limit the tests and encounters with hospital staff to the extent possible [7,14]. Advise surgeons of necessary precautions for those patients undergoing surgery [39]. Hospitals and clinics may also need information on how to differentiate between ME/CFS and mental illness [13,34].
- Encourage the patient and family to establish a living will, appoint a healthcare proxy, and consider a power of attorney to manage finances if needed. Additionally, encourage them to establish a contingency plan and maintain a summary of their health issues and medications in the event that hospitalization is necessary, or an emergency issue arises. Examples of emergency issues include a fire, loss of a caregiver (e.g., through death or illness), or a very severe relapse in which the patient can no longer communicate their needs.

Table 2. Practical Recommendations for Energy Conservation and Management: Summary of ADLs to be evaluated in the bedbound patient and examples of corresponding activity modifications and adaptations that can be employed [1].

ADL Domain/Tasks:	Recommendations, Including Modifications/Adaptations [2]
Grooming/ Washing	<ul><li>Provide shower chair and grab bars. A transfer board can be used to transfer patients from the chair to the tub. Eliminate bathroom mats and rugs that pose a fall risk.</li><li>Use a tub with a pillow/neck support. Elevate feet and begin with lukewarm water temperatures.</li><li>Perform sponge bath bedside or in bed to conserve energy.</li><li>Wash body parts at separate times (e.g., face one day, hair another).</li><li>Use soaps with low fragrance and that are hypoallergenic.</li><li>Use dry shampoo. Consider short hair.</li><li>Examine skin integrity and look for any lesions while bathing.</li><li>Rest immediately after washing and before dressing if needed. Wrap in blankets, dry towel, or robe and return to bed.</li><li>Consider bathing every few days instead of daily.</li><li>Consider remodeling bathrooms to increase accessibility.</li></ul>
Grooming/ Tooth Brushing	<ul><li>Conserve energy by performing activity in bed if needed.</li><li>Use mild flavor paste or just water.</li><li>Use a soft-bristle brush. If an electric toothbrush is used, choose one with control for vibration and intensity.</li></ul>
Grooming/ Dressing	<ul><li>Perform activity in bed, if needed to conserve energy.</li><li>Use fragrance/chemical free laundry detergents.</li><li>Wear loose fitting clothing made of soft, lightweight, breathable materials. Wear solid colors (no patterns) as these may be less stimulating.</li><li>Consider adaptive clothing—e.g., slip on, no closures or buttons as these are easier to don (put on)/doff (take off).</li><li>Don garment on the affected side (e.g., weakest, sorest) first, doff garment on the affected side last.</li><li>Dress in stages. May not be able to complete all at once.</li><li>Assess the cause of any sensitivity to clothes—e.g., small fiber neuropathy, contact dermatitis, etc.</li><li>Change clothes for comfort/cleanliness, not necessarily daily.</li></ul>
Toileting	<ul><li>Use a raised toilet seat and install handrails near the toilet. If needed, a bedside commode can conserve steps for meaningful activity.</li><li>Use adult diapers, bedpan or catheter when unable to transfer or maintain upright posture. If a catheter is needed, try condom catheters and/or intermittent catheterization first before using long-term in-dwelling catheters.</li><li>Ask about and plan toileting on a scheduled basis. This can help decrease urgent visits and bladder/bowel accidents.</li></ul>
Feeding and Drinking	<ul><li>Assess whether a patient has food insecurity due to financial, transportation, preparation, or other problems and address as needed. If preparation is the issue, home delivery of meals and/or a supply of frozen or canned foods requiring minimal preparation can be critical, particularly when patients experience bad days. Prepare large quantities of food when able and store for future use.</li><li>Provide foods that are nutritionally dense and do not need any/much preparation, such as shakes, bars, soft or liquid foods. Referral to a nutritionist may be needed.</li><li>Provide a variety of snacks that can be easily accessed by the patient.</li><li>Eat or drink in bed, if needed, to conserve energy. Less severely ill patients may prefer to have a meal(s) with their family for social interaction.</li><li>Assist with feeding and managing the meal setup if needed.</li><li>Use lightweight bowls, plates, and utensils (e.g., plasticware, bamboo or other lightweight materials).</li><li>Use a small, lightweight cup. Use a short straw for less effort to suck. Use a non-spill water bottle or a hydration pack or bag (cut the length of the straw).</li><li>May require tube feeding for nutrition and hydration or intravenous saline for hydration if oral nutrition and hydration is not adequate.</li></ul>

Table 2. *Cont.*

ADL Domain/Tasks:	Recommendations, Including Modifications/Adaptations [2]
Positioning and Range of Motion	To protect the patient from pressure sores, joint contractures, skin and joint irritation, and poor alignment: • Utilize wedges, bolsters, pillows for support and positioning or consider a specialized/adjustable bed to provide needed support. • Switch the head/foot of bed (if needed and possible) to decrease repetitive movements and reaches. • Utilize a reclining chair with footrest. Maintain proper neck and lumbar support for proper alignment (e.g., zero gravity chair, lounge chair). • Educate caregivers about the need for regular, scheduled re-positioning as tolerated. • Utilize passive or active range of motion to help avoid contractures and maintain some flexibility. This must be done in a way that it does not trigger PEM.

Environment/ Room Setup	To protect the patient from undue physical, cognitive, or emotional exertion:
	• Provide a low sensory environment:
	    ○ Hang black-out shades and/or plain curtains (no patterns);
	    ○ Control room temperature and humidity;
	    ○ Limit sounds from inside and outside the home to the extent possible;
	    ○ Do not use products, such as cleaning supplies or perfumes, that have a strong smell.
	• Provide assistive technology such as call buttons; remotes for light, fan and tv control; smart light bulbs (dim/color changing) with remotes; and wireless remote-control electrical outlet switches for fan/lights.
	• Utilize a bedside table with adjustable height, tilt, and swivel top.
	• For ease of reach, use "hook and loop" or similar technology to attach items to the wall and headboard and to position baskets with supplies/snacks/tools within reach.
	• Use magnetic boards, bulletin boards or boards with symbols that people can point to as a communication aid.
	• Assess balance issues, fall risks and hazards (stairs, rugs, home entry, etc.). Remove obstacles to keep pathways open and recommend other mitigation strategies as needed.
	• Provide blankets, fans, and other warming and cooling devices if patients experience poor temperature regulation.
	• If the patient needs to prepare their own meals, organize the kitchen for safety and energy conservation, e.g., provide a stool, position most commonly used dishes and utensils for easy access, etc.

Mobility and Transfers	• Provide transfer and mobility devices (e.g., Hoyer lift, slide boards, other assistive devices, wheelchairs, canes, walkers) as required. • Use planned, controlled, and slow position changes, especially for people affected by orthostatic intolerance or hypersensitivity to touch. • Consider installing a stairlift and/or moving the patient to a more accessible room. • Use a wheelchair for transitions between rooms if required and possible. • Teach caregivers how to move patients safely. • Ask private (e.g., taxi, ride-share) and public (e.g., paratransit, ambulance, fire department) transport services about transport options.
Support and Socialization	• Ensure the patient has adequate caregiver support. Help facilitate access to needed community resources. • Consider the patient's desire and need for socialization when recommending energy management approaches.
Medical Management and Emergency Preparedness	• Recommend the patient or caregiver create and maintain a summary of their health issues (e.g., symptoms, sensitivities/allergies, cautions for medical services, etc. Tas), current medications (including over-the-counter drugs, supplements, vitamins, etc.), and physician contact information. • Recommend advanced directives and a health care proxy for when the patient is unable to convey their intent. • Assess emergency preparedness including emergency alert, fire extinguishers, safe exit route. Recommend the patient or caregiver maintain a pack with essential medicine, clothes, and supplies. • Recommend emergency alert technology (iWatch, Life Alert, Alexa, etc.) and a cell phone with programmed numbers. • Notify emergency services (fire department, police) of resident's mobility concerns and identify the location as high priority for utility services.

[1] These recommendations are geared primarily to the bedbound ME/CFS patient but can be tailored as appropriate for patients who are homebound but not bedbound. [2] Many of the recommendations address energy conservation and safety issues as these are of particular concern for the severe and very severe ME/CFS patient. For patients who do not have caregivers, the provider will also need to evaluate IADLs such as shopping, cooking, managing medications, and doing laundry and housework to assess the level of support needed [40].

7. Practical Considerations for Busy Providers

Caring for such ill patients may be challenging for busy primary care and other healthcare providers. The following approaches can help manage the demands on your time and ensure reimbursement:

- Document ADLs and IADLs to demonstrate the need for home care [40].
- Leverage a combination of home visits, telemedicine, written communications, partnerships with home health care services, partnership with the caregiver if one exists, and emerging remote monitoring technologies to best manage both the needs of the patient and the demands on your time. Delegate tasks which do not need your specific input (e.g., completing repetitive forms, gathering basic information) to clinic staff such as receptionists and medical assistants.
- Be aware of any regulatory or insurance requirements for providing home visits.
- Maximize reimbursement by diagnosing any comorbidities such as postural orthostatic tachycardia syndrome (POTS), Ehlers-Danlos syndrome (EDS), or mast cell activation syndrome (MCAS).

8. Pathology in Severe and Very Severe ME/CFS

Research in less severely ill patients has demonstrated dysfunction in neurological, immunological, autonomic, and energy metabolism systems [1,2]. However, few published studies have focused specifically on severe or very severe ME/CFS patients because they are often unable to travel outside their home to participate in research studies. A 2017 review found that only 21 articles had been published on severe or very severe ME/CFS over a span of three decades [41]. Differences in case definition, disease characterization, outcome measures, and small sample sizes made it difficult to draw strong conclusions.

Thus, what is known about severe and very severe ME/CFS has been based largely on clinical experience [8,10,12–14,16,17,30] and extrapolated from research utilizing the less severely ill [1–3]. However, recent studies in more severe ME/CFS have demonstrated some differences. For instance, compared to mild or moderately ill patients, severely ill patients show lower levels of glycolysis [42] and increased abnormalities in immune markers [43,44]. They also show a reduced number of steps walked per day and lower exercise capacity as measured by percent peak predicted oxygen consumption [22] and a reduction in cerebral blood flow in severe ME/CFS when sitting up at a slight incline [45] or following a 20-degree head-up tilt from a supine position [46].

Future research needs to enable the participation of severe and very severe ME/CFS patients. Methods and tools should be developed to accurately capture the full spectrum of severity. Studies must also assess the pathophysiology, natural history, risk factors and prognosis of severe and very severe ME/CFS. Finally, clinical research should focus on aspects of care unique to or prominent in severe or very severe ME/CFS and how to best address these patients' needs.

9. Conclusions

ME/CFS can cause a wide range of severity and functional impairment with the most severely ill homebound and bedbound, sometimes in need of total care. Yet, as sick as they are, these patients are often not seen by medical providers because they cannot travel to doctors' offices. Some patients no longer try because they have previously faced disbelief or received treatment recommendations that made them worse. For their part, primary care providers may not have seen this level of debility before. They often lack accurate information on the nature of the disease and how to care for patients with severe or very severe ME/CFS, a problem compounded by the lack of research on these patients.

Caring for such vulnerable patients requires a patient-centered, collaborative approach in all clinical interactions, one that is grounded in compassion, humility, and respect for the nature and severity of the patient's disease. Use of carefully selected pharmacological and non-pharmacological treatments and management approaches can help protect against worsening of the patient's health while decreasing symptom burden and improving the patient's quality of life. Partnerships with the patient, the caregiver and a targeted network of providers along with use of enablers such as telemedicine and remote monitoring are key to providing the needed care without overwhelming either the patient or the busy provider. Using these approaches, the primary care provider can make a significant difference in the lives of these underserved patients.

Author Contributions: Conceptualization: J.G.M., T.G.D., A.E.M., M.E.D., L.C. Formal analysis: J.G.M., T.G.D., A.E.M., M.E.D., L.C. investigation: J.G.M., T.G.D., A.E.M., M.E.D., L.C. Writing/original draft preparation: J.G.M., T.G.D., A.E.M., M.E.D., L.C. writing/review and editing: J.G.M., T.G.D., A.E.M., M.E.D., L.C. All authors have read and agreed to the published version of the manuscript.

Funding: No external funding was received for this project.

Institutional Review Board Statement: Not applicable.

Informed Consent Statement: Not applicable.

Data Availability Statement: Not applicable.

Acknowledgments: The authors wish to acknowledge the severely ill ME/CFS patients and their caregivers whose stories and experiences informed this paper. The authors also wish to thank the Open Medicine Foundation for supporting publication costs.

Conflicts of Interest: The authors declare no conflict of interest.

References

1. Institute of Medicine. *Beyond Myalgic Encephalomyelitis/Chronic Fatigue Syndrome: Redefining An Illness*; Institute of Medicine (U.S.), Ed.; The National Academies Press: Washington, DC, USA, 2015; ISBN 9780309316897.
2. Komaroff, A.L. Advances in understanding the pathophysiology of chronic fatigue syndrome. *JAMA* **2019**, *322*, 499–500. [CrossRef]
3. Komaroff, A.L.; Lipkin, W.I. Insights from myalgic encephalomyelitis/chronic fatigue syndrome may help unravel the pathogenesis of postacute COVID-19 syndrome. *Trends Mol. Med.* **2021**, *27*, 895–906. [CrossRef] [PubMed]
4. Lim, E.-J.; Ahn, Y.-C.; Jang, E.-S.; Lee, S.-W.; Lee, S.-H.; Son, C.-G. Systematic Review and Meta-Analysis of the Prevalence of Chronic Fatigue Syndrome/Myalgic Encephalomyelitis (CFS/ME). *J. Transl. Med.* **2020**, *18*, 100. [CrossRef]
5. Unger, E.R.; Lin, J.-M.S.; Tian, H.; Natelson, B.H.; Lange, G.; Vu, D.; Blate, M.; Klimas, N.G.; Balbin, E.G.; Bateman, L.; et al. Multi-site clinical assessment of myalgic encephalomyelitis/chronic fatigue syndrome (MCAM): Design and implementation of a prospective/retrospective rolling cohort study. *Am. J. Epidemiol.* **2017**, *185*, 617–626. [CrossRef]
6. Pendergrast, T.; Brown, A.; Sunnquist, M.; Jantke, R.; Newton, J.L.; Strand, E.B.; Jason, L.A. Housebound versus nonhousebound patients with myalgic encephalomyelitis and chronic fatigue syndrome. *Chronic Illn.* **2016**, *12*, 292–307. [CrossRef] [PubMed]
7. Timbol, C.R.; Baraniuk, J.N. Chronic fatigue syndrome in the emergency department. *Open Access Emerg. Med.* **2019**, *11*, 15–28. [CrossRef]
8. Baxter, H.; Speight, N.; Weir, W. Life-threatening malnutrition in very severe ME/CFS. *Healthcare* **2021**, *9*, 459. [CrossRef]
9. Williams, L.R.; Isaacson-Barash, C. Three cases of severe ME/CFS in adults. *Healthcare* **2021**, *9*, 215. [CrossRef] [PubMed]
10. Kingdon, C.; Giotas, D.; Nacul, L.; Lacerda, E. Health care responsibility and compassion-visiting the housebound patient severely affected by ME/CFS. *Healthcare* **2020**, *8*, 197. [CrossRef]
11. Chronic Fatigue Syndrome/Myalgic Encephalomyelitis. Primer for Clinical Practitioners. 2014 Edition. International Association for Chronic Fatigue Syndrome/Myalgic Encephalomyelitis. Available online: https://growthzonesitesprod.azureedge.net/wp-content/uploads/sites/1869/2020/10/Primer_Post_2014_conference.pdf (accessed on 21 June 2021).
12. Bateman, L.; Bested, A.C.; Bonilla, H.F.; Chheda, B.V.; Chu, L.; Curtin, J.M.; Dempsey, T.T.; Dimmock, M.E.; Dowell, T.G.; Felsenstein, D.; et al. Myalgic encephalomyelitis/chronic fatigue syndrome: Essentials of diagnosis and management. *Mayo Clin. Proc.* **2021**. [CrossRef] [PubMed]
13. Rowe, P.C.; Underhill, R.A.; Friedman, K.J.; Gurwitt, A.; Medow, M.S.; Schwartz, M.S.; Speight, N.; Stewart, J.M.; Vallings, R.; Rowe, K.S. Myalgic encephalomyelitis/chronic fatigue syndrome diagnosis and management in young people: A primer. *Front. Pediatr.* **2017**, *5*, 121. [CrossRef]
14. UK National Institute for Health and Care Excellence. Guideline: Myalgic Encephalomyelitis (or Encephalopathy)/Chronic Fatigue Syndrome: Diagnosis and Management. Draft for Consultation. November 2020. Available online: https://www.nice.org.uk/guidance/indevelopment/gid-ng10091 (accessed on 26 June 2021).
15. Centers for Disease Control and Prevention. Myalgic Encephalomyelitis/Chronic Fatigue Syndrome. Information for Healthcare Providers: Severely Affected Patients. Available online: https://www.cdc.gov/me-cfs/healthcare-providers/clinical-care-patients-mecfs/severely-affected-patients.html (accessed on 26 June 2021).

16. Speight, N. Severe ME in children. *Healthcare* **2020**, *8*, 211. [CrossRef] [PubMed]
17. Dialogues of a Neglected Illness. Severe and Very Severe ME/CFS. 2020. Available online: https://www.dialogues-mecfs.co.uk/films/severeme/ (accessed on 26 June 2021).
18. Cox, D.L.; Findley, L.J. The management of chronic fatigue syndrome in an inpatient setting: Presentation of an approach and perceived outcome. *Br. J. Occup. Ther.* **1998**, *61*, 405–409. [CrossRef]
19. Carruthers, B.M.; van de Sande, M.I.; De Meirleir, K.L.; Klimas, N.G.; Broderick, G.; Mitchell, T.; Staines, D.; Powles, A.C.P.; Speight, N.; Vallings, R.; et al. Myalgic encephalomyelitis: International consensus criteria. *J. Intern. Med.* **2011**, *270*, 327–338. [CrossRef]
20. Conroy, K.; Bhatia, S.; Islam, M.; Jason, L.A. Homebound versus bedridden status among those with myalgic encephalomyelitis/chronic fatigue syndrome. *Healthcare* **2021**, *9*, 106. [CrossRef]
21. van Campen, C.M.C.; Rowe, P.C.; Visser, F.C. Two-day cardiopulmonary exercise testing in females with a severe grade of myalgic encephalomyelitis/chronic fatigue syndrome: Comparison with patients with mild and moderate disease. *Healthcare* **2020**, *8*, 192. [CrossRef] [PubMed]
22. van Campen, C.M.C.; Rowe, P.C.; Visser, F.C. Validation of the severity of myalgic encephalomyelitis/chronic fatigue syndrome by other measures than history: Activity bracelet, cardiopulmonary exercise testing and a validated activity questionnaire: SF-36. *Healthcare* **2020**, *8*, 273. [CrossRef]
23. Kingdon, C.C.; Bowman, E.W.; Curran, H.; Nacul, L.; Lacerda, E.M. Functional status and well-being in people with myalgic encephalomyelitis/chronic fatigue syndrome compared with people with multiple sclerosis and healthy controls. *PharmacoEcon. Open* **2018**, *2*, 381–392. [CrossRef] [PubMed]
24. Komaroff, A.L.; Fagioli, L.R.; Doolittle, T.H.; Gandek, B.; Gleit, M.A.; Guerriero, R.T.; Kornish, R.J.; Ware, N.C.; Ware, J.E.; Bates, D.W. Health status in patients with chronic fatigue syndrome and in general population and disease comparison groups. *Am. J. Med.* **1996**, *101*, 281–290. [CrossRef]
25. Natelson, B.H.; Lin, J.-M.S.; Lange, G.; Khan, S.; Stegner, A.; Unger, E.R. The effect of comorbid medical and psychiatric diagnoses on chronic fatigue syndrome. *Ann. Med.* **2019**, *51*, 371–378. [CrossRef] [PubMed]
26. Cairns, R.; Hotopf, M. A Systematic review describing the prognosis of chronic fatigue syndrome. *Occup. Med. (Lond.)* **2005**, *55*, 20–31. [CrossRef]
27. Dafoe, W. Extremely severe ME/CFS-A personal account. *Healthcare* **2021**, *9*, 504. [CrossRef]
28. Centers for Disease Control and Prevention. Myalgic Encephalomyelitis/Chronic Fatigue Syndrome. Available online: https://www.cdc.gov/me-cfs/index.html (accessed on 26 June 2021).
29. US ME/CFS Clinician Coalition Website. Available online: https://mecfscliniciancoalition.org/ (accessed on 26 June 2021).
30. Collingridge, Emily. Severe ME/CFS. A Guide to Living. 2010. Available online: http://www.severeme.info/ (accessed on 26 June 2021).
31. Jason, L.A.; Brown, M.; Brown, A.; Evans, M.; Flores, S.; Grant-Holler, E.; Sunnquist, M. Energy conservation/envelope theory interventions. *Fatigue* **2013**, *1*, 27–42. [CrossRef]
32. Colby, J. Special problems of children with myalgic encephalomyelitis/chronic fatigue syndrome and the enteroviral link. *J. Clin. Pathol.* **2006**, *60*, 125–128. [CrossRef] [PubMed]
33. Chu, L.; Elliott, M.; Stein, E.; Jason, L.A. Identifying and managing suicidality in myalgic encephalomyelitis/chronic fatigue syndrome. *Healthcare* **2021**, *9*, 629. [CrossRef] [PubMed]
34. Stein, E. Chronic Fatigue Syndrome. Assessment and Treatment of Patients with ME/CFS: Clinical Guidelines for Psychiatrists. 2005. Available online: https://s3.amazonaws.com/kajabi-storefronts-production/sites/90617/themes/1513565/downloads/TSGDbZnFSWOjdt3msZgv_Guidelines-Paper-English.pdf (accessed on 26 June 2021).
35. Keller, B.A.; Pryor, J.; Giloteaux, L. Inability of myalgic encephalomyelitis/chronic fatigue syndrome patients to reproduce VO2peak indicates functional impairment. *J. Transl. Med.* **2014**, *12*, 104. [CrossRef] [PubMed]
36. Geraghty, K.; Hann, M.; Kurtev, S. Myalgic encephalomyelitis/chronic fatigue syndrome patients' reports of symptom changes following cognitive behavioural therapy, graded exercise therapy and pacing treatments: Analysis of a primary survey compared with secondary surveys. *J. Health Psychol.* **2019**, *24*, 1318–1333. [CrossRef]
37. Campbell, B.; Lapp, C. Treating Chronic Fatigue Syndrome and Fibromyalgia. Guidance on Pacing. Available online: http://www.treatcfsfm.org/menu-Pacing-7.html (accessed on 26 June 2021).
38. Rowe, P.C.; Fontaine, K.R.; Lauver, M.; Jasion, S.E.; Marden, C.L.; Moni, M.; Thompson, C.B.; Violand, R.L. Neuromuscular strain increases symptom intensity in chronic fatigue syndrome. *PLoS ONE* **2016**, *11*, e0159386. [CrossRef]
39. Lapp, C. Recommendations for Persons with Myalgic Encephalomyelitis/Chronic Fatigue Syndrome (ME/CFS) Who Are Anticipating Surgery or Anesthesia. Available online: https://docs.google.com/document/d/1MSiO2EInryV3uCrLpUWzGW4HmlLIghsJnDjGQAo3Qaw/edit (accessed on 26 June 2021).
40. Occupational therapy practice framework: Domain and process (3rd ed.). *Am. J. Occup. Ther.* **2017**, *68* (Suppl. 1), S1. Available online: https://www.aota.org/~{}/media/Corporate/Files/Advocacy/Federal/coding/OT-Practice-Framework-Table-1-Occupations.pdf (accessed on 26 June 2021).
41. Strassheim, V.; Lambson, R.; Hackett, K.L.; Newton, J.L. What Is known about severe and very severe chronic fatigue syndrome? A scoping review. *Fatigue* **2017**, *5*, 167–183. [CrossRef]

42. Tomas, C.; Elson, J.L.; Strassheim, V.; Newton, J.L.; Walker, M. The effect of myalgic encephalomyelitis/chronic fatigue syndrome (ME/CFS) severity on cellular bioenergetic function. *PLoS ONE* **2020**, *15*, e0231136. [CrossRef] [PubMed]
43. Montoya, J.G.; Holmes, T.H.; Anderson, J.N.; Maecker, H.T.; Rosenberg-Hasson, Y.; Valencia, I.J.; Chu, L.; Younger, J.W.; Tato, C.M.; Davis, M.M. Cytokine signature associated with disease severity in chronic fatigue syndrome patients. *Proc. Natl. Acad. Sci. USA* **2017**, *1114*, E7150–E7158. [CrossRef] [PubMed]
44. Hardcastle, S.L.; Brenu, E.W.; Johnston, S.; Nguyen, T.; Huth, T.; Ramos, S.; Staines, D.; Marshall-Gradisnik, S. Longitudinal analysis of immune abnormalities in varying severities of chronic fatigue syndrome/myalgic encephalomyelitis patients. *J. Transl. Med.* **2015**, *13*, 299. [CrossRef] [PubMed]
45. Van Campen, C.M.C.; van Rowe, P.C.; Visser, F.C. Reductions in cerebral blood flow can be provoked by sitting in severe myalgic encephalomyelitis/chronic fatigue syndrome patients. *Healthcare* **2020**, *8*, 394. [CrossRef]
46. Van Campen, C.M.C.; Rowe, P.C.; Visser, F.C. Cerebral blood flow Is reduced in severe myalgic encephalomyelitis/chronic fatigue syndrome patients during mild orthostatic stress testing: An exploratory study at 20 degrees of head-up tilt testing. *Healthcare* **2020**, *8*, 169. [CrossRef]

 *healthcare*

Perspective

Health Care Responsibility and Compassion-Visiting the Housebound Patient Severely Affected by ME/CFS

Caroline Kingdon *, Dionysius Giotas, Luis Nacul and Eliana Lacerda

Department of Clinical Research, London School of Hygiene & Tropical Medicine, Faculty of Infectious and Tropical Diseases, London WC1E 7HT, UK; dio.giotas@lshtm.ac.uk (D.G.); luis.nacul@lshtm.ac.uk (L.N.); Eliana.Lacerda@lshtm.ac.uk (E.L.)
* Correspondence: caroline.kingdon@lshtm.ac.uk

Received: 9 June 2020; Accepted: 2 July 2020; Published: 4 July 2020

Abstract: Many people with severe Myalgic Encephalopathy/Chronic Fatigue Syndrome (ME/CFS) commonly receive no care from healthcare professionals, while some have become distanced from all statutory medical services. Paradoxically, it is often the most seriously ill and needy who are the most neglected by those responsible for their healthcare. Reasons for this include tensions around the complexity of making an accurate diagnosis in the absence of a biomarker, the bitter debate about the effectiveness of the few available treatments, and the very real stigma associated with the diagnosis. Illness severity often precludes attendance at healthcare facilities, and if an individual is well enough to be able to attend an appointment, the presentation will not be typical; by definition, patients who are severely affected are home-bound and often confined to bed. We argue that a holistic model, such as "Compassion in Practice", can help with planning appointments and caring for people severely affected by ME/CFS. We show how this can be used to frame meaningful interactions between the healthcare practitioners (HCPs) and the homebound patient.

Keywords: ME/CFS; severe ME/CFS; validation; engagement; health encounters; housebound; bedbound

---

1. Introduction

1.1. ME/CFS and the Needs of the Severely Affected Individual

ME/CFS is a complex and multifactorial disease; the World Health Organisation has described ME/CFS as a neurological disorder in the International Classification of Functioning, Disability and Health since 1969 [1]. It is also highly stigmatised [2], and there is currently neither a validated biomarker nor an effective treatment. Disease manifestation can vary between affected individuals, and some routine clinical examinations and blood tests may not evidence abnormalities [3]. However, a careful clinical examination, which should be part of a clinical assessment, may show subtle abnormalities [4]. People with severe ME/CFS, estimated as 25% of those affected [5], have been defined as being almost exclusively housebound and unable to attend healthcare consultations, and often bedbound all or some of the time [4]. Multiple studies worldwide [6] have uncovered abnormalities in the central and autonomic nervous system, as well as changes to the metabolic and immunological systems. Severe or striking fatigue, which is often disabling, must have been present for at least six consecutive months (in adults) in order to meet currently accepted diagnostic guidelines [7–9]. While ME/CFS can affect multiple systems, many people with mild/moderate ME/CFS continue to function "deceptively normally" by carefully planning their activities and periods of rest. This may involve, for example, managing to hold down a job but not taking part in any social activities in the evenings and at weekends so that they can rest, significantly impacting friendships and quality of life.

However, other patients are severely functionally impaired, and physical or mental exertion beyond a personal threshold can exacerbate symptoms; this is commonly known as post-exertional malaise (PEM). PEM can be debilitating and increases the individual's dependency on others for hours, days, weeks or even months. The vocabulary to describe the experience of people with severe ME/CFS is inadequate, and words such as malaise or fatigue diminish and deny the grim reality of the disease.

ME/CFS has a prevalence of between 0.2 and 0.4%; recent estimates of those affected in the UK give a figure of between 150,000 and 250,000 people [10]. Two to four times as many women as men are affected, which is similar to multiple sclerosis (MS) and some other autoimmune diseases, where major fatigue is also a symptom. A GP practice of 10,000 patients could have up to 40 people with ME/CFS (PWME), with 25% severely affected [5]. PWME have been found to be more functionally impaired than people with cancer or other chronic diseases [11]. ME/CFS is potentially life-shortening, although registered deaths from ME/CFS are rare; between 2001 and 2016, 88 deaths in England and Wales were partly or fully attributable to ME/CFS [12]. This is a disease of low prestige, and PWME often experience damaging negative encounters with GPs, nurses, and other healthcare practitioners (HCPs) [13]; disrespectful treatment and trivialising of legitimate symptoms may lead to loss of agency and alienation from all medical services. PWME are often made to feel, in the words of one parent carer, *"that they are second-class citizens, malingerers or at least somehow responsible for their own misfortune"* [14]. In the face of such discrimination, we strongly believe that HCPs are duty-bound to ensure that the person with severe ME/CFS receives the person-centred care she/he needs, despite the fact that, as is understood currently, little can be done to change the pattern of the illness.

As part of the CureME research team at the London School of Hygiene and Tropical Medicine, we have visited a cohort of nearly 100 patients housebound by ME/CFS on multiple occasions, in order to enrol them in the UK ME/CFS Biobank (UKMEB) [15]. After considering our experiences and insights from these unique encounters in a methodical way, we argue that a holistic model, such as "Compassion in Practice" [16], can help with planning and caring for people severely affected by ME/CFS. We show how this can be used to frame meaningful interactions between the HCPs and the homebound patient.

To substantiate this discussion, we present in Table 1 an analysis of the frequency and severity of symptoms reported by data available from an early subsample of our cohort with severe presentation of ME/CFS ($n = 57$), whose gender breakdown reflects the accepted 3:1, F:M ratio [10]. These are accompanied by revealing quotes from those visited by the CureME team:

Table 1. Common symptoms affecting people with severe Myalgic Encephalopathy/Chronic Fatigue Syndrome (ME/CFS) seen by CureME.

Symptom		PWSME * with Symptom ($n = 57$; 44 F, 13M)	Symptom Reported as Severe (%)
a	Unrefreshing sleep	100	70
	"When I waken, often in the very late morning, I feel no more rested than I did at the end of the day before; everything is an effort no matter how long the time spent asleep." (F aged 20–29)		
b	Disabling fatigue	100	n/a
	Sleep problems	83	65
	"I can only describe a good day as trying to walk through water; on a bad day, the water becomes treacle." (M aged 40–49)		
c	Exercise intolerance (PEM)	100	90
	"If my daughter cleans her teeth herself, that is it for a week. She can't do anything more." (Parent describing F aged 40–49)		
d	Pain after exertion/activity	84	70
	Muscle pain	96	50
	"I would love to know what one, single day without pain was like; I have quite forgotten. This pain is relentless." (F aged 18–30)		

Table 1. *Cont.*

Symptom		PWSME * with Symptom (*n* = 57; 44 F, 13M)	Symptom Reported as Severe (%)
e	Intolerance to standing	81	50
"Standing is a nightmare–my heart pounds, my legs shake within minutes, I become dizzy and feel that I will faint." (M aged 50–59)			
f	Concentration problems	96	43
"Reading is like trying to listen to the radio when there is so much interference that you can only hear the odd word … " (F aged 40–49)			
g	Difficulty in finding words	96	32
	Difficulty in making decisions	75	35
"When someone visits, I have to choose. I can either eat a biscuit and have a cup of tea or listen and respond. I cannot do both." (F aged 30–39)			
h	Brain fog/cognitive dysfunction	95	45
	Slow thinking	88	32
"Sometimes I grasp at words, feeling like a toddler learning to speak all over again. I was studying for my PhD when I got ill." (F aged 30–39)			
i	Short-term memory problems	95	33
"I am so sorry. I know you just asked me to do something, but I can't remember what." (F aged 40–49)			
j	Difficulty in understanding	86	31
	Difficulty in retaining information	89	35
"I have to write that down straight away or I forget what you said. And I forget I need to write it down!" (M aged 30–39)			
k	Unusual sensitivity to light and/or noise	91	45
"Would you mind just shining a light on my arm to take blood? Away from my eyes? I just can't tolerate any light." (M aged 20–29)			
"The whole family suffers because I can't bear music playing or the chatter of voices–and yet I long to enjoy those things again." (F aged 20–29)			
l	New sensitivities to food, medication, chemicals or odours	88	43
"Please don't wear freshly washed clothes as the smell of detergent or softener can worsen her symptoms." (Parent of F aged 50–59)			
m	Intolerance to heat and cold	81	57
"I sometimes think this must be what the menopause is like for women. I vacillate between hot sweats and feeling icy cold within minutes–for no apparent reason" (M aged 40–49)			
n	Allergies/hypersensitivities	80	63
"I have had to rearrange the appointment as she is still unable to leave the house due to the severe reaction to clothes and anything that touches her skin–you can understand that is a big, big problem." (Parent of F aged 18–30)			
o	Gastrointestinal symptoms	79	40
	Sickness and nausea	79	16
"I became unable to tolerate different types of food–even the smell of food being cooked. Even water made me feel sick" (F aged 30–39)			

* PWSME—people with severe Myalgic Encephalomyelitis/Chronic Fatigue Syndrome. The letter to the left of a particular symptom is used to reference the symptom in the text following the table.

1.2. Responsibility and Compassion

Responsibility and compassion should guide all interactions between HCPs and patients and are critical to the therapeutic relationship; they are particularly important when HCPs interact with people with severe ME/CFS. The following approaches exemplify the ways that attributes, including empathy, respect and dignity, can be evident in many practical tasks when medical encounters are carefully planned and executed.

1.2.1. Arranging a Visit

Most appointments, either face-to-face or virtual, are ideally scheduled for after midday as many people with severe ME/CFS need to rest until then due to erratic sleep patterns (a). Connecting with those most severely affected may require some persistence, and the practitioner should always ensure that his/her tone of voice (k) is appropriate for someone seriously ill, and that questions are simple, requiring short answers. People with ME/CFS often make an enormous effort to prepare for the HCP visit, and the post visit malaise and exhaustion (c, d) is never witnessed by the visiting HCP but only by those who remain with the patient. The PEM caused by such efforts may continue for hours, days or even weeks.

Looking after people with severe ME/CFS at home may require ongoing commitment from the HCP over many years, as improvement in the condition is uncommon; the HCP may need to schedule short visits at regular intervals, aiming to address one or two problems at each visit commensurate with the health of the individual.

1.2.2. Preparing for a Visit

When planning the visit to the person with severe ME/CFS, the HCP should avoid smoking or the use of perfumes or fragranced lotions, as sensitivities to chemicals or odours (l) are common (4, 5); patients may even request that clothes are not freshly laundered because of the effect of fragranced detergents. HCPs with symptoms of viral infections should avoid visits, as there is evidence that the altered immunological status in ME/CFS makes people more susceptible to infection [17]. On entering the home of any person with severe ME/CFS, the HCP should offer to remove shoes and be seen to undertake thorough hand hygiene.

Visiting someone with severe ME/CFS necessitates use of the HCP's experience and expertise to priorities the patient's agenda according to need, to make the visit profitable for both parties. In the current absence of effective treatments for ME/CFS, the person with severe ME/CFS is often very aware of which strategies help or hinder their ability to function. ME/CFS charities (see Table 2) are a rich resource, and the well-informed HCP should know how to access websites, printing information in advance to hand out if appropriate and helpful.

Table 2. Examples of UK charities providing guidance for healthcare practitioners and/or patient resources for ME/CFS.

ME/CFS Charity	Webpage
ME Association	https://www.meassociation.org.uk/
Action for M.E.	https://www.actionforme.org.uk/
The ME Trust	https://www.metrust.org.uk/
25% M.E. Group-Supporting Those with Severe M.E.	https://25megroup.org

1.2.3. The Home Visit

The visiting practitioner should assess the disease severity and what that means for the patient using observational skills, without judgment or agenda, to better understand the health and social needs of the individual. The observant HCP will seek clues about who the individual was before the onset of disease; patients have often had to redefine their sense of self out of necessity [18]. A common consequence of the disease is physical, emotional and intellectual isolation; family and friends are often estranged, partnerships and marriages flounder, and studies are curtailed and jobs are lost, with ensuing loss of income [19].

HCPs without appropriate expertise may be reluctant to engage with PWME leading to inequity in treatment. PWME are often better informed about the disease than the visiting practitioner. Lack of support from established medical services force many to draw on e-resources, including support groups and online forums. In the absence of validation and support from the medical community, some PWME self-manage the illness, gathering information to legitimise the illness [20] and in an attempt to find sources of help. Taking a proactive stance with their own health differentiates PWME from people with clinical depression, and HCPs need to understand that it is the disease that inhibits activity (b, c) rather than lack of motivation. This can challenge the practitioner used to steering the therapeutic relationship who may need to acknowledge their limitations. People with severe ME/CFS are as susceptible to other diseases as the general population, and may be at higher risk of some diseases [21]. It is therefore important that PWSME are thoroughly medically reviewed on a regular basis, as well as when there is deterioration in their health status, or when presenting signs or symptoms that could be due to new comorbidities, which could be treated.

"Brain fog" (h) or cognitive dysfunction is common in PWME; time and patience are required to help the patient to make sense of his/her illness. It is often difficult and time-consuming to take a coherent medical history, not only because information processing and concentration skills (f, g, h, i, j) may be diminished by the disease but also because this is a disease with no clear natural history, in which symptoms can vary both from person to person and from day to day, over the course of the disease. Listening and hearing imply trust, which is essential to any therapeutic relationship. The individual with severe ME/CFS may be familiar with disbelief and skepticism, and therapeutic progress often involves belief in the patient's symptoms and feelings.

Some people with severe ME/CFS may lie in a darkened room with earplugs in and may exhibit increased sensitivity to light, sound and touch (k, m, n); struggle to share information (f, h); or be so wary of PEM that they are fearful of articulating the things that they want or need to share (c). Others may seem animated and engaged, making it difficult for the practitioner to recognize the extent of their illness. Practitioners should use tone of voice (k) and pressure of touch appropriate to the individual's sensitivities.

Apart from frailty, physical examination generally reveals little, and blood work, important to exclude other diagnoses, is often normal. Some severely affected individuals may appear undernourished; this often follows food intolerance and allergies (o) and is rarely a manifestation of anorexia nervosa [22]. People with severe ME/CFS may only be able to stand (e) for a few moments, if at all, and signs of weakness may be obvious. Handgrip measures are markedly reduced in the severely affected and rapidly decline when repeated [23].

The HCP should encourage balance between activity and rest, including during the encounter, to minimize any subsequent PEM and pain (c, d). There are times when the patient may risk PEM to accomplish something, and the HCP can guide the planning of such activities to minimize the risk of PEM. Such activities may, for the most severely affected, simply include producing a urine sample or raising an arm to have a BP cuff put on.

People with severe ME/CFS are disabled by their symptoms (b), and this may be an opportunity for the HCP to offer information about and access to disability aids and allowances. In the UK, benefits include Attendance Allowance and Personal Independence Payments. The ME/CFS charities are a rich resource for further information (Table 2).

Living with severe ME/CFS requires enormous courage. The inevitable reduction or loss of a societal role means that people with ME/CFS are unable fully to engage with family and friends. Not only do people with severe ME/CFS have to come to terms with a long-term, disabling illness, but they may also have to cope with doubts about its authenticity. Reactive depression and anxiety are common for people with severe ME/CFS as in any long-term, disabling illness, and suicide is a possible outcome in this patient group [24], so the HCP should always be sensitive to signs of deteriorating mental health and treat it appropriately.

The therapies and alternative treatments that people with severe ME/CFS sometimes embrace can seem unscientific to the HCP, but are sometimes turned to out of desperation; unless they could be detrimental to the individual's health or involve unreasonable expense, the HCP should endeavor to support the patient in his/her chosen path. When patients face financial difficulties, the HCP must continue to support the severely affected individual to help them to manage that possible additional burden.

2. Conclusions

We believe that compassion is central to the care of people with ME/CFS. Despite the current absence of curative treatments for people with severe ME/CFS, the HCP has a responsibility to provide care through a relationship based on empathy, respect and dignity. By supporting the individual with compassion and competence and acknowledging and learning from the patient's experience, the encounter with the housebound patient can be both effective and worthwhile. The first step in the therapeutic relationship is to believe and trust the individual: to articulate that you, the practitioner, hear what your patient is saying and recognise that their experience is legitimate.

This severe, complex multisystem disease has long been misjudged by the healthcare profession. Educating practitioners about the needs of those most severely affected by ME/CFS will help drive the step change in understanding and belief, compassion and empathy required to care for all patients with ME/CFS.

Author Contributions: C.K. wrote the manuscript after examining data and sharing experiences with D.G., E.L. contributed to the statistical analysis; L.N. reviewed and edited the paper. All authors have read and agreed to the published version of the manuscript.

Funding: Research reported in this paper was mainly funded by the National Institute of Allergy and Infectious Diseases (NIAID) of the National Institutes of Health (NIH) under Award Number: R01AI103629.

Acknowledgments: We would like to thank all the participants with severe ME/CFS for affording us the privilege of visiting their homes and for sharing their experiences. We would also like to thank the charities The ME Association, Action for ME, and ME Research UK, as well as a private donor, who founded the UKMEB start-up, and the ME Association for continued funding.

Conflicts of Interest: The authors declare no conflict of interest.

References

1. WHO. *International Classification of Functioning, Disability and Health (ICF)*; WHO: Geneva, Switzerland, 2018.
2. McManimen, S.L.; McClellan, D.; Stoothoff, J.; Jason, L.A. Effects of unsupportive social interactions, stigma, and symptoms on patients with myalgic encephalomyelitis and chronic fatigue syndrome. *J. Community Psychol.* **2018**, *46*, 959–971. [CrossRef] [PubMed]
3. CDC. Evaluation of ME/CFS. Available online: https://www.cdc.gov/me-cfs/healthcare-providers/diagnosis/evaluation.html. (accessed on 8 April 2020).
4. Shepherd, C.; Chaudhuri, A. *ME/CFS/PVFS: An Exploration of the Key Clinical Issues*, 11th ed.; The ME Association: Gawcott, UK, 2019.
5. CFS/ME Working Group. *Report to the Chief Medical Officer of an Independent Working Group*; CFS/ME Working Group: London, UK, 2002.
6. Komaroff, A.L. Advances in Understanding the Pathophysiology of Chronic Fatigue Syndrome. *JAMA* **2019**, *322*, 499–500. [CrossRef] [PubMed]
7. Centers for Disease Control and Prevention. CFS Case Definition. 1994. Available online: http://www.cdc.gov/cfs/case-definition/ (accessed on 3 July 2020).
8. Carruthers, B.M. Definitions and aetiology of myalgic encephalomyelitis: How the Canadian consensus clinical definition of myalgic encephalomyelitis works. *J. Clin. Pathol.* **2006**, *60*, 117–119. [CrossRef] [PubMed]
9. Institute of Medicine (IOM). *Beyond Myalgic Encephalomyelitis/Chronic Fatigue Syndrome: Redefining an Illness*; The National Academies Press: Washington, DC, USA, 2015.
10. Nacul, L.C.; Lacerda, E.M.; Pheby, D.; Campion, P.; Molokhia, M.; Fayyaz, S.; Leite, J.C.; Poland, F.; Howe, A.; Drachler, M.L. Prevalence of myalgic encephalomyelitis/chronic fatigue syndrome (ME/CFS) in three regions of England: A repeated cross-sectional study in primary care. *BMC Med.* **2011**, *9*, 91. [CrossRef] [PubMed]

11. Nacul, L.C.; Lacerda, E.M.; Campion, P.; Pheby, D.; Drachler, M.D.L.; Leite, J.C.; Poland, F.; Howe, A.; Fayyaz, S.; Molokhia, M. The functional status and well being of people with myalgic encephalomyelitis/chronic fatigue syndrome and their carers. *BMC Public Health* **2011**, *11*, 402. [CrossRef] [PubMed]

12. Mentions of Postviral Fatigue Syndrome (Benign Myalgic Encephalomyelitis), Deaths Registered in England and Wales, 2001 to 2016—Office for National Statistics. 2018. Available online: https://www.ons.gov.uk/peopl epopulationandcommunity/birthsdeathsandmarriages/deaths/adhocs/008461mentionsofpostviralfatigu esyndromebenignmyalgicencephalomyelitisdeathsregisteredinenglandandwales2001to2016 (accessed on 31 July 2019).

13. Lian, O.S.; Robson, C. "It's incredible how much I've had to fight." Negotiating medical uncertainty in clinical encounters. *Int. J. Qual. Stud. Health Well-Being* **2017**, *12*, 1392219. [CrossRef]

14. Mihelicova, M.; Siegel, Z.; Evans, M.; Brown, A.; Jason, L. Caring for people with severe myalgic encephalomyelitis: An interpretative phenomenological analysis of parents' experiences. *J. Health Psychol.* **2016**, *21*, 2824–2837. [CrossRef] [PubMed]

15. Lacerda, E.M.; Mudie, K.; Kingdon, C.C.; Butterworth, J.D.; O'Boyle, S.; Nacul, L. The UK ME/CFS Biobank: A Disease-Specific Biobank for Advancing Clinical Research Into Myalgic Encephalomyelitis/Chronic Fatigue Syndrome. *Front. Neurol.* **2018**, *9*, 1026. [CrossRef] [PubMed]

16. NHS England. Compassion in Practice strategy and the 6Cs values. 2012. Available online: https://www.england.nhs.uk/6cs/wp-content/uploads/sites/25/2015/03/cip-6cs.pdf (accessed on 14 May 2019).

17. Cliff, J.M.; King, E.C.; Lee, J.-S.; Sepúlveda, N.; Wolf, A.-S.; Kingdon, C.; Bowman, E.; Dockrell, H.M.; Nacul, L.; Lacerda, E.; et al. Cellular Immune Function in Myalgic Encephalomyelitis/Chronic Fatigue Syndrome (ME/CFS). *Front. Immunol.* **2019**, *1*, 796. [CrossRef] [PubMed]

18. Grue, J. *Disability and Discourse Analysis*; Ashgate Publishing Ltd.: Farnham, UK, 2015; pp. 1–140.

19. Kingdon, C.C.; Bowman, E.W.; Curran, H.; Nacul, L.; Lacerda, E.M. Functional Status and Well-Being in People with Myalgic Encephalomyelitis/Chronic Fatigue Syndrome Compared with People with Multiple Sclerosis and Healthy Controls. *PharmacoEconomics Open* **2018**, *2*, 381–392. [CrossRef] [PubMed]

20. Arroll, M.A.; Howard, A. The letting go, the building up, [and] the gradual process of rebuilding: Identity change and post-traumatic growth in myalgic encephalomyelitis/chronic fatigue syndrome. *Psychol. Health* **2013**, *28*, 302–318. [CrossRef] [PubMed]

21. McManimen, S.; Devendorf, A.; Brown, A. Mortality in Patients with CFS/ME. *Fatigue* **2016**, *4*, 195–207. [PubMed]

22. Colby, J. ME—The Illness and Common Misconceptions: Abuse, Neglect, Mental Incapacity. *Tymes Trust: Professional Guide.* Available online: https://www.tymestrust.org/pdfs/metheillness.pdf (accessed on 1 April 2020).

23. Nacul, L.C.; Mudie, K.; Kingdon, C.C.; Clark, T.G.; Lacerda, E.M. Hand Grip Strength as a Clinical Biomarker for ME/CFS and Disease Severity. *Front. Neurol.* **2018**, *9*, 992. [CrossRef] [PubMed]

24. Kapur, N.; Webb, R. Suicide risk in people with chronic fatigue syndrome. *Lancet* **2016**, *387*, 1596–1597. [CrossRef]

 *healthcare*

Review

Elements of Suffering in Myalgic Encephalomyelitis/Chronic Fatigue Syndrome: The Experience of Loss, Grief, Stigma, and Trauma in the Severely and Very Severely Affected

Patricia A. Fennell [1,*], Nancy Dorr [2] and Shane S. George [2]

1   Albany Health Management Associates, Inc., Albany, NY 12203, USA
2   Department of Psychology, The College of Saint Rose, Albany, NY 12203, USA; dorrn@strose.edu (N.D.); georges947@strose.edu (S.S.G.)
*   Correspondence: communications@albanyhealthmanagement.com

Abstract: People who are severely and very severely affected by Myalgic Encephalomyelitis/Chronic Fatigue Syndrome (ME/CFS) experience profound suffering. This suffering comes from the myriad of losses these patients experience, the grief that comes from these losses, the ongoing stigma that is often experienced as a person with a poorly understood, controversial chronic illness, and the trauma that can result from how other people and the health care community respond to this illness. This review article examines the suffering of patients with ME/CFS through the lens of the Fennell Four-Phase Model of chronic illness. Using a systems approach, this phase framework illustrates the effects of suffering on the patient and can be utilized to help the clinician, patient, family, and caregivers understand and respond to the patient's experiences. We highlight the constructs of severity, uncertainty, ambiguity, and chronicity and their role in the suffering endured by patients with ME/CFS. A composite case example is used to illustrate the lives of severely and very severely affected patients. Recommendations for health care providers treating patients with ME/CFS are given and underscore the importance of providers understanding the intense suffering that the severely and very severely affected patients experience.

Keywords: ME/CFS; Chronic Fatigue Syndrome; severely and very severely ill; trauma; grief; chronic illness; fatigue; suicide; stress; uncertainty

Citation: Fennell, P.A.; Dorr, N.; George, S.S. Elements of Suffering in Myalgic Encephalomyelitis/Chronic Fatigue Syndrome: The Experience of Loss, Grief, Stigma, and Trauma in the Severely and Very Severely Affected. *Healthcare* **2021**, *9*, 553. https://doi.org/10.3390/healthcare 9050553

Academic Editors: Kenneth J. Friedman, Lucinda Bateman and Kenny Leo De Meirleir

Received: 31 March 2021
Accepted: 4 May 2021
Published: 9 May 2021

Publisher's Note: MDPI stays neutral with regard to jurisdictional claims in published maps and institutional affiliations.

1. ME/CFS: The Severely and Very Severely Affected

Understanding the depth and breadth of the suffering experienced by people who are severely and very severely affected by Myalgic Encephalomyelitis/Chronic Fatigue Syndrome (ME/CFS) is necessary in order to adequately care for them. These patients exhibit heterogeneous symptoms in multiple body systems and life domains. Their symptoms typically include fatigue, post-exertional malaise (PEM), sleep disturbance, cognitive impairment, orthostatic intolerance, muscle and/or joint pain, as well as flu-like symptoms including headaches, sore throat, swollen lymph glands, chills, and night sweats [1]. Commonly, patients also have bowel disorders, allergies, chemical sensitivities, light and sound disturbance, shortness of breath, and irregular heartbeat, etc. [1]. These symptoms fluctuate in kind, duration, frequency, and intensity over the lifetime of those afflicted [2–7]. In common with other chronic illnesses, these symptom impairments can affect multiple domains in patients' lives—the physical, the psychological, and the social-interactive [2,3,8,9]. Given the profound ongoing uncertainty, ambiguity and chronicity of ME/CFS, patients subsequently experience significant suffering. This suffering stems not only from the devastating physical symptoms, but also from ongoing stigma, loss, grief, and trauma. Through the lens of a phase model of chronic illness, the current paper examines these varieties of suffering. This phase framework illustrates the effects of suffering on the patient and can be utilized to help the clinician, patient, family, and caregivers understand and respond to the patient's experiences. A composite case example is used to illustrate the lives of severely and very severely affected patients.

The International Consensus Criteria categorize patient functionality, according to reductions from their premorbid levels, as satisfying the criteria for either a severe subgroup when they are rendered mostly bedbound or a very severe subgroup when they are rendered totally bedbound and require assistance with basic functions and tasks of daily living [10]. Symptom severity in patients with ME/CFS can be difficult to ascertain. Commonly, severity is assessed through the patient's ability to function with their condition. While functionality can vary between patients depending on the symptoms that are most predominant, research supports the persistence of physical and psychological fatigue severity, and primary symptoms, such as post-exertional malaise, cognitive dysfunctions, gastrointestinal issues, and impaired sleep, in maintaining reduced functionality [6,11]. Given that patients who have experienced severe reductions in functioning that render them homebound or bedbound account for approximately 25–29% of ME/CFS patients [12], there is a need for increased attention toward this subgroup.

2. Fennell Four-Phase Model: A General Description

Several models have been developed to describe the process of living with chronic illness that are applicable to patients with ME/CFS [2,13–16]. Fennell's Four-Phase Model [2] is useful for defining and describing four phases that occur in ME/CFS. This model has been thoroughly developed over the last several decades [2,3,17–22] through clinical encounters, patient testimonies, and empirical research [5,7,23–25]. The model explicitly captures the changing experience of patients over time in all domains of their lives—their physical, psychological, and social-interactive worlds.

Phase 1 of Fennell's Four-Phase Model is a state of chaos and crisis. It lasts from the onset of the disease through an emergency state when patients typically seek medical help. The phase usually concludes when patients receive a diagnosis or when their symptoms become stable and recognizable to them. Phase 2 is a time of stabilization. Initially, patients attain a plateau of symptoms. They become familiar with the symptoms and achieve some success in coping with them. Patients begin to feel a degree of control returning to their lives. In Phase 3, characterized by resolution, patients acknowledge that they have a chronic illness, and that their lives have thus been changed forever. With this recognition, however, can come existential despair. If patients are able to develop meaning in their lives, they begin to construct a new self that deals with the practical aspects of their illness; yet, they are not overwhelmed or totally preoccupied with it. Phase 4 is one of integration. Patients solidify their newly constructed self and reach out into the world again to participate as fully as they can in as complete a life as their physical condition permits.

Patients ordinarily proceed through the phases in sequence, but as in non-hierarchical stage models, they usually slip back into a prior phase or recycle, sometimes several times [26,27]. Attainment of integration (Phase 4) is not a permanent condition. Crises, especially those brought about by non-ME/CFS illnesses, accidents, or personal tragedies, can throw patients back into the chaos of Phase 1.

Patients present differing symptomatology during each phase and thus respond differently during specific instances of clinical interviews and data collection over time. If these responses are collapsed across phases, important distinctions about experience of the illness can be lost, which in turn may distort or obscure understanding of the illness [5].

The following section describes the course of ME/CFS using the Four-Phase Model to help illustrate how patients present in the physical-behavioral, the psychological, and the social-interactive domains during each phase of the model. The general description of the three domains in each phase includes a table outlining its characteristics. A composite case history illustrates how these generalizations manifest in actual patients, focusing on those who are severely and very severely affected.

3. ME/CFS through the Lens of the Four-Phase Model

3.1. Phase 1: Crisis

3.1.1. Physical-Behavioral Domain

The physical functioning of a person with ME/CFS in Phase 1 occurs in a linear progression through three periods (see Table 1). The patient copes or contends with the symptoms in the first period. Patients experience varying levels of severity and combinations of the aforementioned symptoms, including fatigue, flu-like symptoms, increasing cognitive confusion, muscle and/or joint pain, and non-restorative sleep. ME/CFS can have a gradual or a sudden onset. Since the onset of ME/CFS can frequently occur following an illness such as flu or a respiratory infection, patients may attribute their ME/CFS symptoms to a very slow recovery. They may sometimes try to ignore them.

Table 1. Phase 1: Crisis.

Physical-Behavioral

- Coping period
- Onset period
- Acute emergency period

Psychological

- Loss of identity and/or loss of psychological control
- Intrusive shame, poor self-esteem, despair
- Shock, disorientation, dissociation
- Fear of others, isolation, emotional lability

Social-Interactive

- Others experience shock, disbelief, and/or revulsion
- Vicarious traumatization
- Family and organizational coping
- Others fall within a continuum from suspicion to support

Next, patients enter a more intense onset period, where their symptoms more insistently demand their attention. It becomes more and more difficult for the patients to believe that (a) the symptoms are unrelated to the prior illness, or (b) they may be suffering a new or unresolved infection, or (c) they may be experiencing the symptoms of a different illness entirely. After a varying period of time, depending on the severity of symptoms, patients enter an acute emergency period and almost always seek medical help, even if they have not done so previously.

With most chronic illnesses, patients receive a reasonably firm diagnosis at the end of Phase 1. ME/CFS can be more problematic. It is poorly understood, and less familiar to many primary care physicians than illnesses such as rheumatoid arthritis or multiple sclerosis. The diagnostic process requires the presence of several symptoms occurring simultaneously as well as the exclusion of other illnesses that share some of the same symptoms. One of the defining symptoms—profound fatigue—needs to be ongoing for at least 6 months for patients to satisfy some ME/CFS case definitions. This particular diagnostic parameter in itself can mean that patients may live for a considerable time in crisis with debilitating—sometimes very debilitating—symptoms before they receive a diagnosis.

3.1.2. Psychological Domain

Severity of presentation of symptoms may impact the length of time between the first period of coping and clinically observed onset. Patients who have a lengthy onset—and ME/CFS patients often wait a long time for diagnosis [13]—may use denial as a coping mechanism [28]. Since they often receive little recognition and support from either the health care community or society at large, ME/CFS patients, especially those with a lengthy onset, may act with others in their life to deny their symptoms in order to hopefully maintain their daily lives. Stormorken and colleagues [13] suggest improvement can be delayed when patients engage in denial and do not accept that they are ill.

Typically in Phase 1, ME/CFS patients regularly report receiving conflicting advice and sometimes disbelief from health care professionals [29]. The patients do not know how to describe their condition to themselves or others [15]. They may feel emotional isolation, fear of others, mood swings, and intense confusion. They may also feel shame from the act of reporting such difficult symptoms and being disbelieved. They can also feel embarrassed if symptoms such as fatigue, cognitive impairment, and lack of energy are displayed in front of other people [13].

If their condition continues to deteriorate, denial can give way to intrusive feelings of fear, self-hatred, despair, and disorientation. As is the case with other diseases [30,31], some patients with ME/CFS report that they feel they are somehow responsible for what is happening to them, but they do not believe they can do anything about it.

By the time they contact their doctor, many ME/CFS patients present with urgency. They usually hold the locus of treatment and cure to be totally outside themselves. Similar to those with other chronic illnesses [32,33], patients with ME/CFS suffer from fear and shame, experience intrusive ideations about dying, but can sometimes exhibit a high degree of denial about their psychological suffering.

Homebound and bedbound ME/CFS patients experience an abundance of losses of their former selves due to their reduced functionality. Wiborg and colleagues [34] found that patients who were homebound had more impairment in home management, mobility, and functionality, as well as reporting an increased amount of physical problems as compared to other patients with ME/CFS. Homebound patients further struggled to hold paying jobs and thus experienced financial losses as well. The cumulative effect of all these physical losses frequently produces intense psychological suffering in patients with severe deficits in functionality [34–36].

Patients with severe or very severe ME/CFS disabilities experience a greater exacerbation of symptoms as losses of their physical capacities accumulate, and they can subsequently develop psychological symptoms as a result. Exacerbation of symptoms may compound when their psychological symptoms produce losses in their social life. These psychological symptoms, often appearing depression- or anxiety-like, may not be separate clinical diagnoses, but distinct emotional responses to the physical losses associated with a debilitating chronic illness [37]. Dancey and Friend [38] posit the importance of recognizing the intrusiveness of ME/CFS when assessing its psychological impact on patients. Their results suggest ME/CFS is an extremely intrusive illness that disrupts more aspects of patients' lives to a greater extent than does multiple sclerosis, laryngeal cancer, end-stage renal disease, irritable bowel syndrome, rheumatoid arthritis, and insomnia. As such, while depressive and anxious symptoms can result from the severity of the patient's condition, they are at least partially mediated by the intrusiveness of that condition. ME/CFS patients who are severely and very severely affected can thus experience greater psychological suffering as their reduced mobility, fatigue, impaired sleep, and cognitive dysfunctions intrude on more aspects of their life.

ME/CFS patients can also experience trauma. Research supports the existence of trauma symptoms in individuals who experience illness processes [39,40], and these can result from a variety of factors. Patients can experience stigmatization from a variety of sources, such as from their community, health care professionals, the media, and the public, and this can be traumatic. Their illnesses, often lacking in observable or measurable symptomatology, may be regarded as character failings rather than legitimate threats to well-being [41–43].

When these difficult to observe symptoms initially appear, patients who often cannot identify what they are experiencing have little or no tolerance for the uncertainty and ambiguity of their condition. Uncertainty is a primary element in illness experiences [44–47] and has been described as having a negative impact on how patients cope and their disease outcomes. The evidence suggests uncertainty creates suffering in a variety of chronic illnesses [48–52]. ME/CFS patients can experience uncertainty on three levels. Firstly, they can experience the uncertainty of the present and the unknowable future regarding all the symptoms they endure. Secondly, patients can experience uncertainty from their providers of care who may also be struggling with the unknowns and the unknowable regarding their patients' health outcomes and future [53]. Lastly, they suffer uncertainty from their culture and communities regarding the legitimacy, acceptability, and even the reality of their condition [53].

Not only does uncertainty play a role in the crisis phase, but ambiguity does as well [22]. The provider and the patient may be ruling out competing diagnoses and be beginning to suspect a tentative diagnosis of ME/CFS might be appropriate. However, what to do about an ME/CFS diagnosis is also ambiguous: determining the best course of action for this is always problematic because of the range of possible causes, assessments, and feasible treatments.

3.1.3. Social-Interactive Domain

When ME/CFS patients can no longer hide their symptoms from others, they find that some of their friends, family members, acquaintances, coworkers, and health care providers regard them as malingering or mentally ill. These others may think the ME/CFS patients are trying to avoid work [54], which increases the burden on everyone around them. The disbelief and suspicion that ME/CFS patients sometimes encounter can make them very cautious. These patients may be afraid to express how they actually feel or may misrepresent how they feel. Simultaneously, they may withdraw emotionally from others to avoid further rejection and negative stereotyping.

Some ME/CFS patients continue to try to work during Phase 1, which often creates difficulties with coworkers and supervisors. Increasingly, they may be late to work and unable to complete tasks in a timely manner. Eventually they are usually absent a great deal. Initial concern on the part of fellow workers can, therefore, give way to irritation and resentment as others have to take up the patients' workload. Some patients use their sick leave so quickly that they have to take unpaid leave or attempt to get disability. Disability may be difficult to obtain at this phase of the illness.

Three critical issues become evident during Phase 1. First, ME/CFS patients can be traumatized by the physical, psychological, and social impact of the acute emergency period [55–60]. Second, their friends, family, coworkers, and clinicians can be vicariously traumatized by what is happening [8,61–69]. Third, these significant others begin to queue up on a continuum that extends from suspicion to support in response to the ME/CFS patient's observable, decreased participation in activities. These social responses are often negative, if not in fact, stigmatizing, and can cause ME/CFS patients further secondary traumatization [2,3,70]. It is possible to propagate or mitigate ongoing traumatization. For example, the level of supervision and support in the health care organizations employing the clinicians can affect traumatization [71]. An effective health care organization provides support and supervision to providers who experience the daily grind of chronic suffering so they can avoid vicarious traumatization, subsequent burnout, and inflicting unintentional iatrogenic traumatization. The maturation, premorbidity, and comorbidity of the patients' social network also affect traumatization [55,72–74]. A supportive social network can tolerate the disruption of work and social exchange that the emergency period brings while containing and buffering it for the patients.

3.2. Elizabeth's Story: Phase 1

3.2.1. Physical-Behavioral Domain

Elizabeth is a married white woman in her late 30s. She has two children, Eva, age 13, and Michael, age 11. Jim, her husband, is a sales consultant, which often requires him to work out of town for one or two weeks at a time. Elizabeth works at a medical practice in the suburb where the family lives.

Over the past several months, Elizabeth has been increasingly distracted by a number of physical symptoms that are beginning to frighten her because they are interfering with her life and her work. She is exhausted most of the time and is not sleeping well. She thinks, at first, that she is just not completely recovered from a recent respiratory infection, but months are passing, and she still feels completely drained. Elizabeth has never been sick with more than occasional colds, so she rarely sees the doctor. At first, she does not think her situation is very different from that of a lot of her friends, because all the working mothers she knows are always tired, too. She just keeps trying to carry on with her regular activities, snatching whatever moments she can to nap or at least rest. However, as time is passing, she is finding it more and more difficult to get dressed, drive or work for any length of time when she arrives at her job.

Elizabeth is just entering Phase 1. Physically and behaviorally, Elizabeth is attempting to cope with her symptoms. Even though she feels very unwell, she tries to ignore or "push through" her symptoms, to push them out of her consciousness, and to continue her regular activities.

Eventually, however, Elizabeth's exhaustion, increased muscle pain, and headaches make it impossible for her to ignore her symptoms; she has trouble climbing the few steps into her front door. Now she is entering the acute onset period. Elizabeth decides to go see her primary care physician.

The doctor listens to Elizabeth describe her symptoms and gives her a physical examination. They talk a bit about her new limitations at work and at home. The doctor tells Elizabeth that test results do not reveal anything physically wrong with her. He suggests her exhaustion may be due to her poor sleep. He wonders if the many demands of her job and family life may be causing her difficulties. He also thinks that she may be mildly depressed, but not enough to require medication. He wants to follow up with her in six weeks and recommends that Elizabeth relax, try to get to bed earlier, cut back at work, and perhaps join an exercise class to help relieve her symptoms.

Elizabeth wants to follow the doctor's suggestions, but she does not dare cut back at work any more than she already has because the family needs the income from her job. Additionally, she cannot imagine how she could take an exercise class given her increasing difficulty ambulating from the bedroom to the bathroom. Additionally, it is not so much that she gets to bed late as that she wakes frequently at night and cannot get back to sleep again because her body aches and she feels as if she is oddly moving or vibrating. Her symptoms progressively worsen. Not only is she extremely fatigued, she experiences more and more trouble thinking. She is having memory problems, word finding, and concentration difficulty. She became tearful when she could not remember her home address when asked.

Elizabeth is now entering the acute emergency period. After six weeks, Elizabeth returns to her primary care physician who orders additional blood tests and refers her to a psychiatrist to assess her stress level and rule out clinical depression. The psychiatrist reports that Elizabeth appears to be suffering from reactive depression in response to her physical condition. The blood work results raise questions that cause the primary care doctor to refer Elizabeth to a rheumatologist. For months, she is examined, tested, and enters the diagnostic limbo of uncertainty and ambiguity. It is not until almost a year later that the rheumatologist is able to give Elizabeth a tentative diagnosis of Myalgic Encephalomyelitis or Chronic Fatigue Syndrome. Even now, some of the doctors are not really sure it is ME/CFS.

However, having a diagnosis, even a tentative one, makes an enormous difference to Elizabeth, for it finally gives her a way to at least partially understand and describe her experiences to herself and others.

3.2.2. Psychological Domain

During Elizabeth's lengthy coping and subsequent onset periods, she uses denial as a coping mechanism. Denial comes into play after her initial visit to her primary care physician, who tells her that he suspects she is mildly depressed and suffering from stress. She wants to believe that is an accurate diagnosis and so she agrees with her doctor, her husband, and the people at work that it is possible for her to return to her daily life.

However, as Elizabeth's symptoms worsen, other feelings begin to intrude. Like many people, Elizabeth has constructed two selves—a private persona and a public persona. Additionally, like others, Elizabeth reveals more or less of her private persona to individuals in her life depending on how intimate she is with them and what particular situation she is in. As Elizabeth's condition continues to deteriorate, she finds that her private persona is beginning to intrude on her public persona in ways that she cannot control. One day at a virtual staff meeting with the medical practice (where she now works part-time in billing from home, as meetings with her have to be virtual with her out of the office), Elizabeth suddenly bursts into tears. She feels embarrassed that she may have made her superior and coworkers feel uncomfortable, and as they comfort her, she feels like a burden to them. This behavior is not the self that Elizabeth recognizes. She cannot identify why she feels the way she does, and because of her uncertainty regarding her health, she feels as if she is losing control. She feels shame about her loss of control, and increased fear and despair about the uncertainty and ambiguity of her condition. She knows she feels terrible physically and is getting worse, and she worries she could actually be dying; at the same time, she feels dissociated, and she wonders if she could be losing her mind. It is important to remember that at this point no one has yet given Elizabeth's situation a definitive label.

Elizabeth has no effective way to express how she is feeling and when she tries, the people she talks to can only make up explanations and suggestions for improvement based on their own personal experiences, not on an understanding of what is happening to her. Elizabeth feels increasingly isolated because she fears what is happening to her and what other people will think of her. She is particularly afraid to talk to the person who used to be closest to her—her husband, Jim. Elizabeth and Jim have already been having marital difficulties due to conflicts over money and the amount of time Jim needs to be away from home. Elizabeth feels that Jim, by default, gets out of his fair share of home and child care duties. She struggles with basic activities of daily living and she increasingly relies upon her children's help. She fears she cannot quit her job, because the family cannot make ends meet without her part-time salary. In any case, she likes her job, which she used to do very well. She has received a lot of praise at work, especially from her supervisor, and she and Jim had both been hoping that she would get a promotion. The new position would have brought in more money, but Elizabeth would have to work full time and now that would be impossible.

In part, as a consequence of her pain, her fears, the lack of useful information about her condition, and her growing isolation, Elizabeth now begins to suffer emotionally with grief and anger from her suffering and losses. Half the time she is in tears, she says, and the other half she is furious. Jim has tried to be sympathetic, but now he is getting frustrated. The children act scared of her and disappear whenever possible. Even Elizabeth's coworkers find her muddled and distracted, whereas she used to be focused and attentive. Elizabeth is losing her life as she has known it, and she is frightened she will never get it back.

3.2.3. Social-Interactive Domain

Many of Elizabeth's psychological mechanisms and reactions result directly from what is happening in her social-interactive life. During the coping and onset periods, Elizabeth's family, friends, and coworkers respond in various ways to her experiences. They notice only that she is tired a lot of the time and missing work and for a while, they are sympathetic. She is a hard worker, and her female coworkers, particularly, have a strong personal understanding of how hard it is to juggle a job, home, and children. As Elizabeth accomplishes less and misses more and more work, however, they become critical. They have difficult lives too, but they manage to come to work, and they accomplish their assigned tasks.

Elizabeth's children think that she is acting strange. She does not behave like the mother they are used to. Jim finds her unpredictable and emotionally extreme. He is used to hearing Elizabeth say she is tired, but her complaints seem so serious that Jim is genuinely worried and urges her to go to the doctor. The doctor's diagnosis of suspected depression and stress seems reasonable to Jim, and he makes an effort to help more by staying in town or taking only short jobs away from home. However, that cannot go on forever, and Elizabeth does not seem to change. He thinks she could do a lot more if she tried, but she seems just to complain or sleep.

During the acute emergency period, while Elizabeth is being examined and tested extensively, Jim sometimes wonders whether anything is actually wrong with Elizabeth—maybe it is "all in her head". Some of her coworkers feel that way, too. Maybe Elizabeth is having some kind of emotional breakdown.

Finally, Elizabeth gets her diagnosis, however tentative, of ME/CFS. Although this gives her the relief of a name and an explanation, she now finds that the illness has put her squarely on the forefront of a cultural debate. Caught in a mesh of divergent popular beliefs regarding her partially understood disease, Elizabeth finds that some of her friends and coworkers—even some of the medical personnel she sees—view her negatively. For the first time in her life, Elizabeth begins to experience rejection by the society at large. She becomes very cautious about expressing her fears or revealing her pain because she does not want others to withdraw from her. Not only is she afraid of other people now, but her physical condition itself interferes with her reaching out socially.

Elizabeth's home life was already stressed prior to the onset of her illness, and after enduring the coping, onset, and acute emergency periods, her illness has greatly exacerbated the situation. At the medical practice, Elizabeth's immediate supervisor is sympathetic because the supervisor's sister has fibromyalgia, can no longer work, and is on disability. While lacking an understanding of the complexities of Elizabeth's condition, her supervisor is sympathetic to Elizabeth's problems and wants to help. Upper levels of management, however, think that Elizabeth should probably be replaced, and it is unclear how effective the supervisor's advocacy will be. Fortunately, on the medical front, Elizabeth's primary physician has become very involved with her case but cannot spend as much time as he would like to talk with Elizabeth and must focus on assessment and treatment of her acute physical symptoms. Time does not allow the doctor to discover in depth how Elizabeth is feeling or how she is coping with the whole illness experience. A few family members and friends are wondering whether she should consider filing for disability.

3.3. Phase 2: Stabilization

3.3.1. Physical-Behavioral Domain

As ME/CFS patients enter Phase 2, they usually proceed physiologically to a plateau. Their specific symptoms and their severity determine the functional level of their plateau. Their fatigue intensity follows predictable patterns, their muscle-joint pain remains stable or increases, they begin to recognize when they will have greater or lesser cognitive function, and so forth. During this period, symptoms stabilize and assume a somewhat familiar pattern. This in turn can help orient patients cognitively and psychologically. Table 2 lists the periods and characteristics of Phase 2.

3.3.2. Psychological Domain

When ME/CFS patients finally receive a diagnosis, they often feel an initial sense of profound relief [16,75]. Getting a name for their experience demystifies some of the disturbing uncertainties about their symptoms. Patients strongly desire to understand their illness experience [76,77], and even when ME/CFS patients do not get a diagnosis, they can begin to feel some increasing control when they discern a pattern in their symptoms and discover relationships between activities and symptoms. At the same time, their self-pathologizing and intrusive ideations usually decrease.

Table 2. Phase 2: Stabilization

Physical-Behavioral

- Plateau
- Stabilization

Psychological

- Increased caution and fear of secondary traumatization and wounding
- Social withdrawal from prior social circles; social searching for others with ME/CFS
- Medical/clinical service confusion; searching for appropriate and compassionate care
- Boundary and role confusion

Social-Interactive

- Interactive conflict or cooperation from others
- Vicarious secondary wounding
- Vicarious traumatic manifestation
- Social normalization failure

During both Phase 1 and Phase 2, ME/CFS patients can experience stigmatization, rejection, and iatrogenic traumatization [43,55,60,78–82]. As a result, they and their families may censor what they say and to whom. Patients often pretend to be well; they may attempt to "pass" for normal, but this presents a double-edged sword. While patients that pass may appear "normal," their symptoms persist and they face increased social pressure to conform in spite of their disabling symptoms [59]. When patients do present visible symptoms, they are more apt to experience internalized stigma, and may feel pressured to hide their condition, fearing that their condition may be exposed [41,43]. Thus, they frequently withdraw from hurtful social contacts, and when possible, they seek others with ME/CFS [2,3].

The initial relief occasioned by diagnosis usually fades quickly. As ME/CFS patients discover that their condition is not widely understood and that no treatment options promise a cure, they often begin searching for clinicians who can help them. This seeking behavior is natural to Phase 2 and, in many ways, can be a sign of emotional health. The patients are attempting to exert control over their suffering and to reject the disempowerment they experienced in Phase 1. However, confusion, urgency, and desperation can intensify as they consult with medical providers and encounter the ambiguity of conflicting opinions, assuming they are even well enough to attend medical visits. Some ME/CFS patients report a general lack of support, guidance, and knowledge from health care providers and clinicians when they try to find out more about their illness [75]. Indeed, without biomedical tests for diagnosis, patients can feel they are fighting to persuade their physician (and others) they are physically ill [76]. In addition, patients and their families often encounter disbelief of their reported symptoms by their physicians [83]. This results in the patients feeling socially disqualified and devalued. Not being believed is a form of stigma, in which they experience a subsequent loss of identity as well [84–86].

Patients may be traumatized by their clinicians when their condition is dismissed or trivialized because they perceive their patient-physician relationship is being "betrayed" [87]. The prevalence of uncertainty and ambiguity can also impact patients' propensity to continue to seek treatment [88], and can engender distrust toward health care professionals [89].

ME/CFS patients are also struggling against their new physical and cognitive limitations in Phase 2. They do not always know the limits of what they can do on a given day and can be confused about how and where to set boundaries. They can no longer perform as they used to, but familial and community pressures, to say nothing of their own internal desires, make them attempt to maintain their former roles and schedules. As the severely and very severely affected may repeatedly fail to live up to their former roles and expectations, their feelings of guilt and shame heighten, together with an increasing sense of purposelessness, worthlessness, and anomie.

ME/CFS patients gradually learn that they can no longer do "normal" tasks such as shopping or cleaning. The severely affected and very severely affected patients learn they cannot walk upstairs, drive, sit up in bed for very long, or have a normal social conversation without physical consequences. They sometimes cannot perform essential tasks such as paying bills. Similar to a stroke survivor, they are not completely confident about how their body, brain, or emotions will behave in any given situation.

As ME/CFS patients progress through Phase 2, they start to understand the relationship between their activities and their symptoms. With appropriate guidance, they also develop insight into their own attitudes and those of the people around them. Maintaining insight is notoriously difficult. For ME/CFS patients without strong clinical and social support, it can be easy to turn to destructive anodynes such as alcohol or drugs, which will usually spiral the patient back into a Phase 1 crisis. This can also happen when patients work beyond their capacity in an attempt to behave as they did before they became ill. Some patients never really leave Phase 1, while others endlessly cycle between Phase 1 and Phase 2 [2,3]. They are not able to achieve the acknowledgment of chronicity that comes during Phase 3 because they cannot tolerate the implications of permanent illness.

Physical disruption and the possibility of permanent illness can act as the catalyst for the loss of a normal life [84,86,90,91]. Losses rapidly accumulate when these physical disruptions bleed into the patients' social and emotional lives [35,36]. Patients' identities are called into question as they repeatedly face uncertainty and ambiguity regarding their condition and its prognosis. As they experience chronic sorrow about their condition, and as they are repeatedly stigmatized as a result of their disability, the patient grieves the losses in their identity associated with and resulting from these factors [36,84–86].

Patients may experience grief from the loss of their previous life, including their previous roles and relationships. Indeed, patients with ME/CFS (as well as those with fibromyalgia) reported more loss due to their illness than did patients with other chronically fatiguing diseases: more loss of social support by family and friends, more loss of recreational activities, and more loss of material possessions [92]. The subsequent grief from this variety of losses is often prolonged due to the chronicity of their condition. However, it is important to note that because grief is usually associated with death-related losses, patients are more apt to have their bereavement disenfranchised by societal reactions [93]. As a result, resolution is a difficult goal for patients experiencing grief. The cyclical nature of ME/CFS means that there is no way to simply "move past" the loss, and because it is repeatedly experienced, patients' normal grief responses are often pathologized and disenfranchised [2,93].

3.3.3. Social-Interactive Domain

In Phase 2, ME/CFS patients encounter increased conflict as friends, family members, coworkers, and some care providers can lose patience. By and large, society's model for illness is that of an acute condition. People are usually tolerant of ME/CFS patients when they first become sick, but this is with the expectation that they will eventually be cured and return to their normal functioning. The persistence of symptoms can predictably frustrate the patients' support networks.

As a result, patients may face several limitations in the roles they once occupied in their pre-crisis lives, as parents, employees, community leaders, and so forth. Role confusion can occur in patients as their position within their family changes, such as with the case of a mother, for example, who is too fatigued to care for her son, a husband who is unable to find joy in his relationship due to the intrusiveness of his condition, or a parent whose financial and functional losses have resulted in an inability to provide a sufficient income for their family [2,92,94].

Coupled with a chronic illness experience is the loss of various social aspects of the patient's old life that can no longer be recovered. For example, physical symptomatology, such as post-exertional malaise and pain-related sequelae, may significantly impact certain patients so that they experience a loss in their ability to maintain romantic and sexual relationships. Meanwhile, sleep-related disturbances and fatigue may significantly intrude in certain patients' daily energy expenditure, resulting in a loss in their ability to maintain platonic relationships [6,38]. Inevitably, patients' inability to fill roles they did prior to their illness experience can initiate or exacerbate trauma symptoms and cumulatively increase the burden put on them by their illness [92,95].

Oftentimes, the friends, family, and romantic partners—a group that makes up the support system of an individual—may share the patient's grief as well. Caregivers experience their own set of fears, anxieties, and even grief over the various aspects of the patient that have changed due to ME/CFS. Several lines of research support the vulnerability, frustration, guilt, uncertainty, loss of identity/role confusion, functional and financial losses, and lack of professional support in caregivers of chronically ill patients [83,96,97]. There is an unfortunate feedback loop here as patients perceive they may be a burden to their support system and this may worsen the patients' psychological symptoms.

The response to ME/CFS can dramatically alter the lifestyles, finances and work habits of spouses and parents. The life they had or had been working toward may not be possible now. Vacations, further education, marriage, or children may be indefinitely postponed. Families frequently divorce under the strain. Part of their frustration stems from their own experience of vicarious secondary wounding. Research suggests that populations that frequently and repeatedly engage with the traumatized individual, a group that can vary from family and romantic partners to the medical professionals and clinicians that treat them, can experience vicarious traumatization [61,65]. Oftentimes, even when the patient is able to receive help with their trauma, those vicariously traumatized are at a loss [83].

Friends and family can also suffer "guilt by association," that is, the family is stigmatized for the patient's illness. Even clinicians can find themselves stigmatized when they treat the ME/CFS population. As a result, it is not at all uncommon for significant others to depart sometime during Phase 2, which can create feelings of betrayal and trauma in the patient [87]. As ME/CFS patients face increasing difficulties with their support network, they begin to actively seek out a new network of friends and more information about their illness [98–101]. It is often among these others of "like kind" that patients begin to establish a nucleus for a new community of supporters who will accept them as they are, disabled with ME/CFS [2,3].

Patients who are severely or very severely affected are usually not employed when they are in Phase 2 and this can add to the financial stress that may have begun in Phase 1. If they are, they usually find their functioning at work stabilizes at the same time as their physical symptoms. Typically, these are individuals who can work from home on a very part time basis with no set schedule. In many cases, if possible, they ask for a leave of absence or take sick leave because they hope that a cure will allow them to return to full-time employment. Others ask for part-time work, quit their jobs, are asked to resign, or are fired. This adds a serious financial concern to all the other problems of ME/CFS. Adding to the financial strain is the cost of the increased medical care itself; research shows medical costs to be 50% higher for patients with ME/CFS as compared to patients with lupus or multiple sclerosis and three to four times higher than the average insured person [102].

3.4. Elizabeth's Story: Phase 2

3.4.1. Physical-Behavioral Domain

During Phase 2, Elizabeth attempts to carve order out of chaos. Her physical situation is very limited but seems to have stabilized. Only a few new symptoms have suddenly surprised her. None of her present symptoms seem to be getting progressively worse, at least for now. Her symptoms do not disappear, but they usually do not exceed patterns that she is beginning to decipher. If she has an especially bad night sleeping, she knows that she will probably have an especially bad day. She will have more pain and more cognitive confusion. If she does two hours of steady activity, she knows that her body aches are likely to increase and her glands will probably swell. Life is very difficult, but Elizabeth has identified a set of parameters around which she can function. Her health care professionals discuss a few of these parameters with her, but for the most part Elizabeth discovers them on her own. To some extent, her newfound knowledge also orients the people around her.

While in Phase 2, Elizabeth suffers two significant physical relapses beyond her usual baseline of functioning. Each time, she suddenly becomes far more exhausted than usual. She cannot even lift herself out of bed without physical assistance but feels as though she is being pulled down through her bed toward the center of the earth. Her head aches, her glands are very swollen, she has nausea, and she becomes intolerant of light and sound. She cannot organize her thinking at all. However, both times she relapses, she eventually returns to a plateau of stabilized symptoms that she recognizes and can negotiate.

3.4.2. Psychological Domain

Over time, Elizabeth's physician becomes increasingly convinced that the tentative diagnosis of ME/CFS does indeed fit her symptoms, and thus, her tentative diagnosis is confirmed. Initially, she feels enormous relief. Finally, she has an explanation for why she is so exhausted, why she cannot sleep, why she has muscle pain and headaches, and why she sometimes becomes cognitively confused. Her uncertainties also lessen as she begins to recognize her symptom pattern. Furthermore, diagnosis gives her a framework to learn about her condition so that she can exert a semblance of control over her life again. When she can read, she reads everything she can about ME/CFS and seeks out others with ME/CFS so that she can discuss her situation in a supportive setting.

However, Elizabeth quickly learns that the diagnosis does not explain how her illness started or what is going to happen in the future, so painful uncertainty and ambiguity return. No one seems to know what to do to cure her. No one can make her symptoms stop, and no one seems able to tell her how she is supposed to live her life under these conditions.

Elizabeth grew up believing that if she worked hard and told the truth, everything would eventually come out all right in her life. However, here she is, working hard to get better, telling the truth when she talks to family members and friends and clinicians, and yet a significant percentage of the time she finds not acceptance, but rejection, from many people. In fact, she is sometimes being blamed. So Elizabeth has become extremely cautious. She carefully censors what she says and to whom, and whenever she possibly can and for limited periods of time, Elizabeth carries on as though she is well and nothing is wrong with her. Elizabeth decides to avoid such secondary wounding by withdrawing from any social contacts that may evoke negative judgments. Instead, she tries to get in touch with other ME/CFS patients and ME/CFS advocates because they are likely to be helpful and will understand her situation. She continues to read about her condition and seeks sources of emotional sustenance to try and make up for the multiple losses she has suffered.

Because her medical outcome is uncertain, and Elizabeth still believes that a cure must be a possibility, she suspects that her health care professionals are not adequate to deal with her problem. She collects the names of other doctors from friends and new ME/CFS acquaintances and attempts to find a professional who will offer her better treatment and, she hopes, a cure.

Unfortunately, Elizabeth finds limited guidance and meets with confusing responses and even outright hostility as she consults other doctors, so she attempts alternative treatments. A practitioner of shiatsu massage listens to her with enormous empathetic patience, but she cannot continue to afford the sessions, and it is unclear if they are helpful. Elizabeth's cousin urges her to try acupuncture. A former coworker swears that a complicated vitamin and supplement regime returned her bedridden niece to full functioning and would do the same for Elizabeth.

Elizabeth has lost a sense of her boundaries. To others, she seems to have given up and they are confused, not grasping how exhausting it is for her to remain in ongoing contact. Elizabeth's family and employer are encouraging, even urging, her to return to her former roles and schedules. However, Elizabeth's efforts to return to these roles can have dangerous repercussions. She has trouble getting up in the morning and doing her basic activities of daily living. She showers weekly and washes her hair bi-weekly. Her husband has to make the children's school lunches. She can no longer serve on committees at her children's school because she can just barely keep up with basic self-care. In fact, she rarely leaves the house and only when her husband or a friend is driving. Her last attempt behind the wheel left her terrified and exhausted after a near miss at a stop sign she had not realized she had run through. Nothing about her body or her emotions or her mind acts the way it did in the past, and yet Elizabeth keeps trying to behave as though she were the person she used to be. Despite her efforts, Elizabeth fails daily at what she attempts, and daily she feels guilty and ashamed. Increasingly, she feels worthless.

3.4.3. Social-Interactive Domain

Elizabeth experiences growing conflicts with family, friends, and some of her medical care providers as they lose patience with her failure to become symptom-free or to adjust to her illness in a way that allows her to return to her former functioning. Although she has a diagnosis, such treatment as she receives does not produce rapid, let alone any significant, improvement. At one point when she was barely sleeping, Elizabeth's doctor put her on a course of medication. She somewhat improved, but she still has consistent difficulties.

Jim has told Elizabeth that she is no longer the person he married and this is not the life he signed up for. She has got to change if their marriage is to continue. Sex has stopped and she barely has the energy to regularly watch a movie with him. Elizabeth can tell that her coworkers are annoyed and believe that she could function a lot better if she just pulled herself together and put her mind on the job. One of them knows a ME/CFS patient of moderate functioning and tells others at work that she cannot understand why Elizabeth does not manage as well as her friend does. Elizabeth is not imperceptive. She knows that people think she is not trying hard enough. To make matters worse, a close friend with deep religious convictions has urged Elizabeth to pray, saying that if Elizabeth has a sincere desire to get better and asks for God's help, God will cure her. Elizabeth does not share her friend's convictions, but deep inside she fears that maybe she is sick because she is somehow unsatisfactory in God's eyes.

As Elizabeth goes through relapses, all the people in her life experience them as well. They become as exhausted by the process as Elizabeth does, and they are traumatized just as she is. Jim has lost the wife he married and the life he had, and his new life is not at all what he wants. Their son, Michael, has always liked school and has done well, but now his grades are beginning to suffer and it is difficult to participate in sports since his mom cannot drive anymore. Eva is behaving badly at home, and she has been acting out, and repeatedly disciplined at school. Both kids complain that their mom never comes to their soccer or baseball games and are embarrassed to rely on friends' parents to drive them. Elizabeth does not know how much of this is just a part of adolescence, or whether Eva and Michael are reacting to her health problems and her arguing with Jim over money, the division of labor at home, and her condition. Elizabeth's husband and children are not mean spirited. They are sad and scared to see this person who is very important to them suffer pain, confusion, and unhappiness. Outside the house, they suffer a kind of guilt by association. Eva's friends sometimes treat her as though she is as weird as her mother, and Eva overheard one of them say that Eva's mom was an alcoholic who was always hungover. Jim's boss is clearly concerned about whether Jim will be able to fulfill his job obligations, given the demands of Elizabeth's illness. Some secretly wonder whether ME/CFS might be contagious and just for safety's sake, many keep their distance.

Even Elizabeth's doctor is affected. Some clinicians who treat ME/CFS patients report that colleagues are skeptical when making referrals. Specialists in ME/CFS sometimes even meet skepticism in social settings. A neurologist to whom Elizabeth's doctor was going to refer her said that he did not think ME/CFS was a valid diagnosis and that he believed Elizabeth was suffering from a psychological condition.

As normalization failure is so common in Phase 2 [2,3], like the chronically ill person, those around them may turn to alcohol or drugs. People in the social network may avoid or even abuse the chronically ill individual. Any of these factors, or a totally new factor, can produce another crisis in the chronically ill, returning them to Phase 1. Jim's mother dies and the entire family must deal with that loss. Later, Elizabeth has a high fever during a bout with the flu that triggers a severe relapse.

Elizabeth is a fortunate person with chronic illness in that she has some warm and loyal friends. Her supervisor persuades management to let Elizabeth try working remotely with just a few hours a week doing billing as she is able. Additionally, a social worker newly affiliated with Elizabeth's doctor's office becomes involved in helping Elizabeth cope with ME/CFS.

3.5. Phase 3: Resolution

Without informed clinical guidance, many chronically ill people become caught in a repeating cycle of Phase 1 and Phase 2 [2,3]. Each new crisis produces new wounding and secondary wounding [55,72,74]. With luck, following each crisis the patient manages to arrive at a plateau of recognizable symptoms, until the next crisis sends the whole system into chaos again. Some people, particularly those on the margins of society who have almost no sources of support, never escape Phase 1, but are buffeted from crisis to crisis.

The personal, familial, and societal pressures heaped upon the patient to return to the pre-crisis life are enormous and can help maintain a cycle between Phase 1 and Phase 2. Most patients present to clinicians in Phase 1 or Phase 2. Repeatedly, they are urged by family and friends to try the next "cure", and in their growing desperation frequently do so. The clinicians also feel the omnipresent pressures to repeatedly treat the chronic patient. This approach will ultimately fail, frustrating all parties involved. Because they are all still struggling in the chaos of the crisis phase, or they are in the stabilization phase and have achieved some equanimity, however brief, they are not ready to grasp that it is highly improbable that they will return to the pre-crisis life and that they have to transition to a new life—a new way of being in the world.

This is the universal clinical treatment problem between Phase 2 and Phases 3 and 4: how to facilitate this painful transition for what is actually possible with a severe disabling chronic condition; otherwise patients never escape the futility of a Phases 1 and 2 cycle. Without an understanding of the nature, breadth, and depth of ME/CFS suffering and without a grasp of the longitudinal, cyclical experience of the phases, health care professionals will find it difficult to compassionately assess and treat ME/CFS patients. Important opportunities for a better, albeit a different life, can be missed.

3.5.1. Physical-Behavioral Domain

Most Phase 3 ME/CFS patients who are severely or very severely affected maintain a continued plateau, but relapses occur. Sometimes old symptoms worsen or new symptoms appear. Some ME/CFS patients experience modest periods of improvement and some have learned to balance activities to keep from relapsing [13,94]. If a relapse takes place, it is sometimes in response to the typical cycling of ME/CFS symptoms, but it may also be triggered by persistent attempts to engage in pre-crisis tasks, roles, and pursuits. True entry into Phase 3 comes, however, when ME/CFS patients recognize that they cannot perform as they used to in the past. Table 3 lists the stages and characteristics of Phase 3.

3.5.2. Psychological Domain

ME/CFS patients in Phase 3 suffer a secondary emotional crisis or grief reaction when they acknowledge the chronic nature of their condition. They finally realize that their lives have changed forever, and they begin the process of mourning their pre-crisis self. They typically feel demoralized and devalued, for they see that they can no longer carry out their previous roles in life—as parent, worker, lover, friend—in the way they had always thought they would. They may question what good they are, who they are, and why they should continue to exist at all. This appropriate, necessary grief reaction—their "dark night of the soul"—is a tenuous time. Individuals can be lost in their own understandable withdrawal, fall victim to predatory providers, or succumb to despair and thoughts of suicide [2]. Research suggests a relationship between suicidal ideation and illness identity [103,104], and as patients grieve losses that have resulted from their illness, the negative identity associated with the illness can be worsened. Yanos and colleagues [104] propose that patients' conceptualization of their illness identity can affect their hope and self-esteem, which further increases the likelihood of suicide. Given the susceptibility of ME/CFS patients to experience threats to their identity and role confusion, afflicted patients who feel stigmatized may turn to suicide due to a reduction in their self-esteem. Indeed, evidence supports this conclusion, as suicide has been identified as a cause of death for a significant number of ME/CFS patients [105,106]. Additionally, patients who commit suicide appear to lack sufficient depressive symptoms to qualify for a depression diagnosis, as their depressive symptoms may better be associated with grief [107]. Evidence suggests that losses associated with ME/CFS and perceived stigma may be likely risk factors for suicidal ideation [42]. Individuals may be traumatized at various points during this often cyclical process.

Table 3. Phase 3: Resolution

Physical-Behavioral

- Patients differ in their experiences, ranging from (a) continued plateau/stabilization/improvement, (b) emergency period or diminished functioning, or (c) relapse

Psychological

- Grief reaction and/or compassionate response to self
- Identifies with pre-crisis self
- Role and identity experimentation
- Returning internal locus of control
- Awareness of societal effects
- Spiritual development

Social-Interactive

- Breaking silence regarding disbelief of ME/CFS and stigmatization
- Confrontation regarding care and social roles
- Social and vocational role experimentation
- Potential integration or separation or loss of supporters

If, however, the patient with ME/CFS can work through this existential angst and establish meaning in their lives, they can then take marked steps toward constructing an authentic new self and a new life [84]. Indeed, individuals with a chronic illness or disability report higher post-traumatic growth than do those without a chronic illness or disability [108]. Many patients report new insight due to their illness experience [90] and some reported having a more confident and assertive personality as a result of having ME/CFS [16]. However, post-traumatic growth was found to be lower among women with ME/CFS than among women with rheumatoid arthritis, osteoarthritis, and multiple sclerosis, perhaps because of the greater levels of stigma women with ME/CFS felt due to having an unexplained illness [109].

Meaning is established over the phase process through three transformational steps: (1) the allowance of suffering as opposed to its rejection and the subsequent rejection of the suffering self; (2) the development of a compassionate response to the suffering of the rejected, sick, stigmatized self; and (3) the development of respect for their suffering and their ability to live with it and despite it. Creative activity is a successful path leading to the creation of meaning [108]. So, too, is a sustaining faith on the part of the care provider that the ME/CFS patient can construct an authentic new self.

As ME/CFS patients move through Phase 3, they develop a strongly internalized locus of control and increased tolerance for the uncertainty, ambiguity and chronicity of ME/CFS. They openly express compassion for themselves, and they begin to reconstruct an illness narrative that eliminates the harmful social messages they have endured until now. Their work to achieve meaning involves them in philosophical or spiritual development that offers an ongoing framework for adapting to new experiences, whether good or bad.

3.5.3. Social-Interactive Domain

Despite improvement in the psychological domain in Phase 3, ME/CFS patients may undergo even greater losses than they did in Phase 2 as significant others depart, clinicians give up, and friends disappear. They may continue to experience abandonment, isolation, and stigmatization. Associated with these personal losses are the subsequent losses in romantic partners, friendships, and even close family that deepens the patient's negative evaluation of their condition. A combination of the uncertainty and ambiguity they encounter [88,89,110] and the losses they experience chip away at the patient until they slip into another form of chronicity in their sorrow [111]. These losses are cumulative and compounding in nature; each new loss is another weight that produces and maintains grief.

At the same time, as they are able, they continue to pursue new friendships and the support of others with ME/CFS. Some Phase 3 patients can still feel engulfed in stigma, particularly when accompanied with the fear that their condition could be "exposed" to the public, further exacerbating their depressive symptoms and engagement with treatment [31,41,43,112]. Others may not, but neither group wants to remain silent. They are more willing to speak up against stigmatization and confront it instead of avoiding persons who behave inappropriately toward them. As they work through their existential issues and begin constructing a new authentic self, they begin experimenting with new social roles and sometimes, if able, with new part-time jobs or vocations. Some may engage in social or political activism related to ME/CFS [2,3].

3.6. Elizabeth's Story: Phase 3
3.6.1. Physical-Behavioral Domain

In Phase 3, Elizabeth experiences periods of stabilized symptoms, sometimes even minor improvement, but she still has relapses. Most of these are simply in the nature of ME/CFS. As Elizabeth comes to comprehend the chronicity, ambiguity, and uncertainty inherent in her condition, she lets go of her search for an elusive cure and works instead to integrate her illness into a new life.

3.6.2. Psychological Domain

Twice in Phase 2, Elizabeth suffered severe relapses brought about, in part, by her repeated attempts to do some of the things she did before her illness. Throughout that time, she wanted to be her former self, and everyone around her wanted her to be that person, too. However, repeated relapses have taught her that she cannot sustain the roles that she had always thought she would fulfill as spouse, parent, worker, or friend, or at least not in the way she used to imagine. Elizabeth comes to understand that her life has changed entirely and forever.

With the help and encouragement of her new ME/CFS friends and the social worker, she explores and expresses the grief she feels for the loss of her old self and she mourns the end of that life.

At this point, all the major existential questions come into play. Elizabeth wonders, "Who am I?" "What good am I?" "Why did this happen to me?" "Why should I live?" "Is there any value to my life?" During this painful period, she struggles to locate a meaning for her existence and her suffering. Elizabeth is very vulnerable. She could be lost because of her considerable social withdrawal. She could fall victim to cynical and predatory providers. She could give way to despair and attempt to kill herself.

However, again, Elizabeth is fortunate. Her new friends and the social worker help Elizabeth navigate the difficult course between necessary grieving for her past self and floundering in grief and reactive depression. She learns not to reject but to allow her new suffering self and to have compassion for herself and her suffering as well. This is not an easy task because Elizabeth is constantly receiving messages from people she knows telling her that if she stays the way she is, if she remains ill, then they no longer want her among them.

To move forward from grief and mourning, Elizabeth tries to discover meaning for what has happened to her and locate a way to live in the future. She begins to engage in philosophical or spiritual thinking in order to come to a new place. Elizabeth starts by learning to respect the person she is now—not the person she might become, but who she is right now.

As is typically the case, Elizabeth does this through a creative act that becomes an act of meaning development. She decides to write a journal describing her experiences. Other people she knows have done things as various as taking up painting, becoming ME/CFS advocates, even conducting online support groups. In composing her journal, Elizabeth re-creates herself—she integrates herself and begins to discover meaning in her experience.

Elizabeth draws heavily on her social worker's clinical skills, personal support, and encouragement and on a variety of wisdom traditions that seem to speak to her. Elizabeth is not religious in the traditional sense. However, since she has consciously begun thinking about the basic issues of life and meaning, she has discovered aspects of Buddhism and Celtic philosophy that resonate strongly with her personal vision of what is significant in life.

3.6.3. Social-Interactive Domain

The strides Elizabeth makes in her psychological evolution during Phase 3 do not occur in a benignly static social environment. She endures a considerable blow when Jim tells her that he wants a divorce and he has a new relationship. At first, Elizabeth worries terribly about how she will manage the loneliness and the bills. She still has a very limited job at the medical practice and she has always been on Jim's health plan. As part of the divorce settlement, he agrees to keep her and the children covered. He is also good about having the children visit regularly, which gives Elizabeth needed quiet time and reduces her anxiety that the divorce will cause harmful distance between the children and their father. Elizabeth knows that in this divorce experience, she is much more fortunate than one of her friends with ME/CFS who lost custody of her children and lost her home when she subsequently lost her job.

Encouraged by a ME/CFS friend, Elizabeth explicitly asks Eva and Michael for help at home. To her surprise, Eva responds enthusiastically, especially to cooking. Michael is good about drying dishes and starting a load of dirty clothes in the washer if reminded, but he has recently expressed wanting to live with his dad. Elizabeth, who could not have endured letting him go two years ago, now feels confident about his basic attachment to her and is planning to let him spend his next school year with his father.

At the medical practice, Elizabeth usually feels able to deal with her present limited job requirements. Her coworkers, to the extent they are aware, have gotten used to her condition and have more or less forgotten it. Her confidence was further bolstered when the social worker offered to conduct a workplace consultation for disability accommodation on her behalf. Elizabeth decided that it was not necessary at the present time, but she felt she had someone on her side if she should need it.

Elizabeth knows, however, that her job security depends almost entirely on her supervisor, and she has begun investigating other part-time work she might do from home. The social worker has also reminded Elizabeth that she is eligible for disability, and she has been inquiring among her ME/CFS friends about this as well.

In any social arena now, Elizabeth is less likely to keep silent about her illness. When people react badly or seek to label or stigmatize her, she may confront them about their bias. She has been surprised at how empowered such behavior makes her feel, and she is even thinking about becoming formally involved in advocacy work. The end of her marriage and the inevitable loss of some old friends and acquaintances have forced Elizabeth to consider new roles and to seek new friends. Although this experience has been intensely painful, Elizabeth continues to survive it. She is surprised to find how the process has turned out and how she is adjusting to her new self.

As Elizabeth freely acknowledges, it would have been very hard for her to navigate this passage without the informed help of her care providers—her primary care physician, the specialists, her friends, and the social worker. For example, the social worker suggested a number of books that she thought might help Elizabeth think about the philosophical questions involved and put her in touch with a social work expert in issues of major loss and trauma who helped Elizabeth discover what problems had bedrock significance for her.

3.7. Phase 4: Integration

3.7.1. Physical-Behavioral Domain

ME/CFS patients in Phase 4 may experience physiological plateau, improvement, or relapse. By this time, ME/CFS patients recognize the cyclic nature of their illness and no longer regard relapse as failure. Relapse is simply the beginning of another cycle that they must integrate. This understanding manifests the true nature of acceptance for ME/CFS patients, which is the integration of their ME/CFS into an ongoing, full life. When patients experience new symptoms or familiar ones worsen, they contact their health care providers immediately. Otherwise, they maintain a sensible monitoring of their condition and see clinicians on whatever schedules their clinicians have advised. Table 4 lists the characteristics of Phase 4.

Table 4. Phase 4: Integration

Physical-Behavioral
• Possible continued plateau or improvement or relapse
• Recovery period
Psychological
• Role and identity integration
• New personal best
• Continued spiritual and emotional development
Social-Interactive
• New and reintegrated supporters
• Alternative vocation and activities

3.7.2. Psychological Domain

Phase 4 ME/CFS patients have achieved a true integration of the pre-crisis self with the newly claimed respected self who has suffered and endured, similar to the "recovery in" framework discussed by Devendorf and colleagues [42]. They maintain this achievement through a daily commitment to allowing their suffering, meeting it with compassion, and treating it with respect. Life for them will necessarily include small daily acts of bravery in the presence of stigmatization, rejection, or the pains of ME/CFS itself. Patients do this in an exercise of free will, not because they are forced to. They formulate and then live up to a new "personal best." They continue to work on meaning development in conjunction with their continuing pursuit of philosophical or spiritual development.

3.7.3. Social-Interactive Domain

In Phase 4, ME/CFS patients, as they are able, continue to be involved in creative and social action. They expand their circle of supporters, but their increased self-assurance and self-confidence also permits them to attempt reintegrating old supporters who fell away in the past. They are often willing and able to help these people learn, if these old friends wish to be reintegrated. Although most severely and very severely affected ME/CFS patients have difficulty maintaining any work schedule, some of them, especially in Phase 4, find vocational activities that allow them to participate to the limit of their abilities in activities that they value.

3.8. Elizabeth's Story: Phase 4

3.8.1. Physical-Behavioral Domain

For the most part, Elizabeth experiences continuous plateau, and she enjoys occasional, limited improvement. However, she has also had three relapses—one severe and two lesser ones. Elizabeth now realizes that relapses happen, and she no longer regards them as some failure on her part. However, short of death, which will happen to everyone eventually, she intends to try to reintegrate herself after each relapse experience. Elizabeth comprehends that integration is the "philosophy of life" she should strive to maintain.

3.8.2. Psychological Domain

Elizabeth maintains her new self by consciously recognizing who she is now and by standing with herself. This does not mean that life has become easy. Frequently, Elizabeth cannot climb stairs at all. Sometimes she is so debilitated she must use a wheelchair at home, which she hates. She can still become cognitively confused, especially if she overextends. She still experiences some difficult moments of stigmatization and rejection, and even her own pain. However, she often speaks out against bias and has learned to endure the symptoms of her illness. She has created a new ideal self and takes pleasure in seeing how well she can live up to it.

Elizabeth finds that a constant, active, conscious consideration of meaning and purpose enriches her life and places her experiences, both positive and negative, in a context larger than herself. She still finds great solace in Buddhist conceptions of suffering, but she has also discovered a new trove of wisdom in the material discussed in an online Great Books virtual discussion group that she has joined.

3.8.3. Social-Interactive Domain

Elizabeth continues to nurture the new friendships that she began establishing in Phase 3. She also, through social media, sought out her younger sister, from whom she had been estranged as an adult, and the two have found they enjoy the openness and honesty of their new relationship as much as they like reminiscing about their childhood. Elizabeth's frankness about her condition and her refusal to accept derogatory estimations make it perfectly clear to people who she is now, and some admire her for it and see the truth of her self-assessment.

Elizabeth has changed her job situation. With her social worker's assistance, Elizabeth is receiving disability and is being accommodated at her very part time, remote job. Although Jim has remarried—an event that threw Elizabeth into an emotional crisis—they have remained civil, though distant, friends. Relations between the two are better than they have been for many years.

Elizabeth even dares to contemplate entering a meaningful sexual relationship again. One of her ME/CFS friends began online dating recently, which gives her hope, and she has met a man she likes very much in an online writing class. Because of his encouragement, she sent part of her journal to a ME/CFS patient advocacy website.

Elizabeth knows that crises and disasters happen all the time in life. She worries a lot about her children. One of her ME/CFS friends took a terrible turn for the worse and has been completely bedridden for 3 months. This scares Elizabeth terribly, for she knows the same could happen to her. However, Elizabeth is learning to separate those things she can control from those she cannot. Although it is a continuous effort, she endeavors to exert herself for the things she can affect and to endure with grace those she cannot.

4. Intersectionality

Examining Elizabeth's life with severe ME/CFS as a white woman does not capture the experiences of men, people of color, people of lower socioeconomic status (SES), or people who are in the lesbian, gay, bisexual, transgender, and queer community. For example, men with ME/CFS and other chronic illnesses may have additional issues to work through given the stereotype that men are supposed to be strong [86,113]. Additionally, it is important that appropriate attention is paid toward communities that are often overlooked in discussions of ME/CFS. Increasingly, evidence suggests that ME/CFS has a higher prevalence in communities of color [114,115]. Health disparities between white and minority American women are well documented in the literature [116], with poorer self-perceived health statuses in African American and Latina women, limitations in activity levels due to health, and the inability to work as a result of those limitations [117]. In their qualitative study, Bayliss and colleagues [118] highlight some profound aspects of what it is like being a minority navigating health care. Several participants shared concerns over a lack of communication, for example, that may be more relevant for communities of color. Indeed, evidence suggests that the patient's language preference plays a role in their access to health care [119,120]. Furthermore, participants in Baylis and colleague's [118] study also discussed discrimination and stereotyping when seeking treatment for ME/CFS. As patients of color face multiple hurdles in their attempts to seek treatment, their lack of treatment compounds with the coexisting hurdles they face as minorities.

Along with race, SES has been shown to be a risk factor for more severe presentations of ME/CFS [121]. This can be exacerbated when low SES patients experience difficulties in employment due to their condition. For example, patients who are homebound with ME/CFS are more likely to be unemployed as a result of their health status [34] and thus can suffer from decreases in SES as a result of their illness.

The multiple, compounding adversities that plague people of color can have a profound impact on their symptomatology and well-being. Turner and colleagues [122] posit the concept of cumulative stress burden in an attempt to operationalize the burden that affects an individual's health. It is therefore likely that patients who experience greater stress, such as patients of color and patients of low SES, may be at higher risk of stress-related health problems. Research suggests that there is an association between cumulative stress and health in various communities. Discrimination in African American and Latin American communities acts as a contributor to poor mental health. Both of these communities have been found to have high levels of depression along with moderate levels of anxiety and post-traumatic stress disorder [123]. Concurrent life events such as natural disasters also impact health and exacerbate health care disparities when there is a lack of protection of groups with compromised health [124], and as communities of color lack health care protections at a disproportionate scale [125], they too are at risk for cumulative stress-related concerns. Significant stressful disruptions can cause patients to re-enter the crisis phase before resuming the chronicity of their condition; such an occurrence may heighten the probability of traumatization and disenfranchise their grief. Thus, the debilitating nature of ME/CFS can have a particularly intrusive effect within populations that lack the protective factors that higher SES and racial majorities can have. Being severely and very severely affected with ME/CFS can be yet another barrier for those people of color and those with lower SES.

5. Utilizing the Four Phase Model: Managing Suffering, Chronicity, Uncertainty, Ambiguity, and Severity

The Four-Phase Model attempts to shift the focus of the standard care paradigm for chronically ill individuals. Most obviously, it moves from a unidirectional, acute model of illness to a conception that encompasses the cyclic nature of chronic conditions. The Four-Phase Model maps a process that most individuals do not enter into willingly. It recognizes that for patients with any chronic condition, the situation in their lives is *imposed*. The Four-Phase Model is also characterized by a broad inclusiveness. It addresses from a systems perspective the total environment of a patient's life, including the physical, psychological, and social domains. Its approach is distinctive in that it considers the physical and psychological domains as having equal importance with each other and with family and work-related issues. In actual experience, once patients have acquired methods for maintaining the best level of physical well-being they can, most chronic illness patients cite social and economic issues as causing them the greatest difficulties.

The model incorporates uncertainty and ambiguity as essential, irreducible aspects of chronic illness. It posits an integrated, meaningful life as the end sought in patient treatment, rather than a cure, which, by the definition of chronic illness, does not exist. Most people with chronic conditions will not experience cure; hence, they are better served by a paradigm that organizes itself around the uncertainty, chronicity, ambiguity and severity that reflects their actual situation. To this end, it is vitally important to recognize that the suffering engendered by the severely and very severely affected is dramatically increased, frequently to the extreme.

The model identifies four distinct phases in the illness experience of patients, though these phases do not follow an irreversible progression forward. Rather, the model acknowledges the cyclic pattern of the chronic illness experience and incorporates the concept that severe relapses, other illnesses, or non-illness-related crises may return patients temporarily to earlier phases. The Four-Phase Model asserts that patients who make a thoughtful, reflective progression through all four phases will have the knowledge, understanding, and techniques for moving through the phases more quickly a subsequent time, although the degree of severity that the patient experiences impacts the process. They will also have deeply assimilated the concept that it is possible for them to achieve a better life even within the limitations of their situation, and even if that situation is severe or worsens.

The Four-Phase Model presents what may be expected over time and indicates appropriate times and ways to intervene to improve the patient's quality of life by matching intervention to phase. Conversely, attempting interventions at the wrong time may prove less effective and may undermine the possibility of the same interventions being effective at a later phase when the patient would ordinarily be receptive to such interventions. The Four-Phase Model also suggests comprehensive interventions that take into account interactions among the multiple domains in which patients experience ME/CFS.

5.1. Use of Standard Interventions and Current Therapies

Adopting this model does not require clinicians to learn an entirely new battery of interventions, but to supplement those already in use with new techniques. The Four-Phase Model does not present an exclusive form of assessment and treatment. It complements standard assessment tools and current therapeutic methods with additional investigatory tools and interventions. It also covers the entire range of the patient's life experience, not only the patient's medicalized body or psyche. The model also frequently incorporates existing interventions that clinicians currently use. The techniques of stress management and of post-traumatic stress disorder (PTSD) reduction and management, for example, are particularly useful with Phase 1 patients who need to reduce the chaos of their experience before they can move forward. Many aspects of occupational therapy work well, particularly with Phase 2 patients as they try to gain insight into their activities and change habits that undermine the level of health that they can maintain. A host of grief therapies, meaning development, and art therapy techniques are especially useful during Phase 3, and the well-known benefits of journaling contributes to the writing and rewriting of the illness narrative, which is a distinctive feature of therapy throughout the four phases.

By serving as a narrative or cognitive map, the phase description helps to lessen the intense fear and anxiety frequently experienced by the severe and very severely affected ME/CFS patients and their families. It will also help them to know they are not alone, their experience is shared by others, and they are understood. They now have a method of validating their experiences and making them known to others. The narrative helps them develop a sense of what is happening to them and provides a degree of order and coherency about their illness experience. In addition, the mapping aspect of the phase process helps promote understanding and adjustment to the cognitive impairments in concentration, memory, and decision making that often affect those with ME/CFS.

5.2. How Health Care Professional Can Help

While pilot assessment and treatment programs, in all domains, are underway worldwide, patients, their families and friends live and suffer with ME/CFS. They struggle to have whole lives and clinicians and caregivers struggle to help them manage. How can health care professionals help?

During the brief time available in a patient visit, here are some important things health care professionals can do, in addition to the medical protocols:

- Demonstrate to patients an appreciation and understanding of the ME/CFS experience;
- Convey to patients the compassion that comes from an appreciation of what the severely and very severely affected patient is experiencing;
- Communicate to patients that you *believe* what they are saying about their experiences and symptoms;
- As time and opportunity permits, be open to learning more about this poorly understood syndrome;
- Have available a short list of therapists and specialists, including those who do trauma work, grief work, family therapy, couples therapy, sleep hygiene and occupational therapy, for referral, or to be part of the treatment team;
- Become familiar with the suicide hotline;
- Have available a list of patient support groups for information, education and support;
- Have a list of available support groups and assistance for caregivers;
- Remember that it is very difficult to care for those who do not recover in any conventional sense and clinician resources for support are always a good consideration as well.

Severely and very severely affected patients suffer profoundly. In order for health care professionals to adequately treat their patients, they need to understand all that composes and creates their suffering: struggling with uncertainty, ambiguity, chronicity, stigmatization, trauma, and rejection. These elements create losses for the patient and they subsequently grieve these many and varied losses, including lost friends, family, career, and life as they knew it (or imagined it). Not only do the patients grieve their losses and traumas, but so do the loved ones around them—spouses, parents, and children. Thus, in order to assess and treat, the suffering must first be described, understood, witnessed, and, most importantly, abided.

Author Contributions: Conceptualization, P.A.F.; writing—original draft preparation, P.A.F., S.S.G. and N.D.; writing—review and editing, P.A.F., N.D. and S.S.G. All authors have read and agreed to the published version of the manuscript.

Funding: This research received no external funding.

Institutional Review Board Statement: Not applicable.

Informed Consent Statement: Not applicable.

Data Availability Statement: Not applicable.

Acknowledgments: We express our gratitude to the patients and mentors of P.A.F., who are often one and the same.

Conflicts of Interest: The authors declare no conflict of interest.

References

1. Centers for Disease Control. Symptoms of ME/CFS. Available online: https://www.cdc.gov/me-cfs/symptoms-diagnosis/index.html (accessed on 15 February 2021).
2. Fennell, P.A. *Managing Chronic Illness Using the Four-Phase Treatment Approach: A Mental Health Professional's Guide to Helping Chronically Ill People*; John Wiley & Sons: Hoboken, NJ, USA, 2003.
3. Fennell, P.A. *The Chronic Illness Workbook: Strategies and Solutions for Taking Back Your Life*, 3rd ed.; Albany Health Management Publishing: Albany, NY, USA, 2012.

4. Friedberg, F.; Jason, L.A. *Understanding Chronic Fatigue Syndrome: An Empirical Guide to Assessment and Treatment*; American Psychological Association: Washington, DC, USA, 1998.
5. Jason, L.; Fennell, P.; Taylor, R.; Fricano, G.; Halpert, J. An Empirical Verification of the Fennell Phases of the CFS Illness. *J. Chronic Fatigue Syndr.* **2000**, *6*, 47–56. [CrossRef]
6. Jasona, L.A.; Zinn, M.L.; Zinn, M.A. Myalgic Encephalomyelitis: Symptoms and Biomarkers. *Curr. Neuropharmacol.* **2015**, *13*, 701–734. [CrossRef]
7. Reynolds, N.L.; Brown, M.M.; Jason, L.A. The Relationship of Fennell Phases to Symptoms Among Patients with Chronic Fatigue Syndrome. *Eval. Health Prof.* **2009**, *32*, 264–280. [CrossRef]
8. Engel, G. The need for a new medical model: A challenge for biomedicine. *Science* **1977**, *196*, 129–136. [CrossRef]
9. Moss-Morris, R.; Deary, V.; Castell, B. Chronic fatigue syndrome. In *Handbook of Clinical Neurology*; Elsevier: Amsterdam, The Netherlands, 2013; Volume 110, pp. 303–314.
10. Carruthers, B.M.; van de Sande, M.I.; De Meirleir, K.L.; Klimas, N.G.; Broderick, G.; Mitchell, T.; Staines, D.; Powles, A.C.P.; Speight, N.; Vallings, R.; et al. Myalgic encephalomyelitis: International Consensus Criteria. *J. Intern. Med.* **2011**, *270*, 327–338. [CrossRef] [PubMed]
11. Jason, L.A.; Corradi, K.; Torres-Harding, S.; Taylor, R.R.; King, C. Chronic Fatigue Syndrome: The Need for Subtypes. *Neuropsychol. Rev.* **2005**, *15*, 29–58. [CrossRef] [PubMed]
12. Committee on the Diagnostic Criteria for Myalgic Encephalomyelitis/Chronic Fatigue Syndrome; Board on the Health of Select Populations; Institute of Medicine. *Beyond Myalgic Encephalomyelitis/Chronic Fatigue Syndrome: Redefining an Illness*; National Academies Press: Washington, DC, USA, 2015. Available online: https://books.nap.edu/catalog/19012/beyond-myalgic-encephalomyelitischronic-fatigue-syndrome-redefining-an-illness (accessed on 2 December 2020).
13. Stormorken, E.; Jason, L.A.; Kirkevold, M. Factors impacting the illness trajectory of post-infectious fatigue syndrome: A qualitative study of adults' experiences. *BMC Public Health* **2017**, *17*, 952. [CrossRef]
14. Stormorken, E.; Jason, L.A.; Kirkevold, M. From good health to illness with post-infectious fatigue syndrome: A qualitative study of adults' experiences of the illness trajectory. *BMC Fam. Pract.* **2017**, *18*, 49. [CrossRef] [PubMed]
15. Ware, N.C. Toward A Model of Social Course in Chronic Illness: The Example of Chronic Fatigue Syndrome. *Cult. Med. Psychiatry* **1999**, *23*, 303–331. [CrossRef]
16. Whitehead, L.C. Quest, chaos and restitution: Living with chronic fatigue syndrome/myalgic encephalomyelitis. *Soc. Sci. Med.* **2006**, *62*, 2236–2245. [CrossRef] [PubMed]
17. Fennell, P.A. A systematic, four-stage progressive model for mapping the CFID experience. *CFIDS Chron.* **1993**, *1*, 40–46.
18. Fennell, P.A. CFS sociocultural influences and trauma: Clinical considerations. *J. Chron. Fatigue Syndr.* **1995**, *1*, 159–173. [CrossRef]
19. Fennell, P.A. The four progressive stages of the CFS experience: A coping tool for patients. *J. Chron. Fatigue Syndr.* **1995**, *1*, 69–79. [CrossRef]
20. Fennell, P.A. Capturing the different phases of the CFS illness. *CFIDS Chron.* **1998**, *11*, 3–16.
21. Fennell, P.A. A four-phase approach to understanding chronic fatigue syndrome. In *Handbook of Chronic Fatigue Syndrome*; Jason, L.A., Fennell, P.A., Taylor, R.R., Eds.; Wiley: Hoboken, NJ, USA, 2003; pp. 155–175.
22. Fennell, P.A. Phase-based interventions. In *Handbook of Chronic Fatigue Syndrome*; Jason, L.A., Fennell, P.A., Taylor, R.R., Eds.; Wiley: Hoboken, NJ, USA, 2003; pp. 455–492.
23. Jason, L.A.; Fricano, G.; Taylor, R.R.; Halpert, J.; Fennell, P.A.; Klein, S.; Levine, S. Chronic fatigue syndrome: An examination of the phases. *J. Clin. Psychol.* **2000**, *56*, 1497–1508. [CrossRef]
24. Van Hoof, E.; Coomans, D.; Cluydts, R.; De Meirleir, K. Association Between Fennell Phase Inventory Scores and Immune and RNase-L Parameters in Chronic Fatigue Syndrome. *J. Chron. Fatigue Syndr.* **2004**, *12*, 19–34. [CrossRef]
25. Hoof, E.; Coomans, D.; Cluydts, R.; Meirleir, K. The Fennell Phase Inventory in a Belgian Sample. *J. Chron. Fatigue Syndr.* **2004**, *12*, 53–69. [CrossRef]
26. Prochaska, J.O.; Redding, C.A.; Harlow, L.L.; Rossi, J.S.; Velicer, W.F. The Transtheoretical Model of Change and HIV Prevention: A Review. *Health Educ. Q* **1994**, *21*, 471–486. [CrossRef] [PubMed]
27. Sutton, S.R. Transtheoretical model of behavior change. In *Cambridge Handbook of Psychology, Health and Medicine*; Baum, A., Newman, S., Weinman, J., West, R., McManus, C., Eds.; Cambridge University Press: Cambridge, UK, 1997; pp. 180–183.
28. Pinxsterhuis, I.; Strand, E.B.; Stormorken, E.; Sveen, U. From chaos and insecurity to understanding and coping: Experienced benefits of a group-based education programme for people with chronic fatigue syndrome. *Br. J. Guid. Couns.* **2014**, *43*, 1–13. [CrossRef]
29. Guise, J.; McVittie, C.; McKinlay, A. A discourse analytic study of ME/CFS (Chronic Fatigue Syndrome) sufferers' experiences of interactions with doctors. *J. Health Psychol.* **2010**, *15*, 426–435. [CrossRef]
30. Callebaut, L.; Molyneux, P.; Alexander, T. The Relationship Between Self-Blame for the Onset of a Chronic Physical Health Condition and Emotional Distress: A Systematic Literature Review. *Clin. Psychol. Psychother.* **2017**, *24*, 965–986. [CrossRef] [PubMed]
31. Phelan, S.M.; Griffin, J.M.; Jackson, G.L.; Zafar, S.Y.; Hellerstedt, W.; Stahre, M.; Nelson, D.; Zullig, L.L.; Burgess, D.J.; Van Ryn, M. Stigma, perceived blame, self-blame, and depressive symptoms in men with colorectal cancer. *Psycho Oncol.* **2013**, *22*, 65–73. [CrossRef]

32. Trindade, I.A.; Ferreira, C.; Pinto-Gouveia, J. Chronic Illness-Related Shame: Development of a New Scale and Novel Approach for IBD Patients' Depressive Symptomatology. *Clin. Psychol. Psychother.* **2016**, *24*, 255–263. [CrossRef] [PubMed]
33. Trindade, I.A.; Duarte, J.; Ferreira, C.; Coutinho, M.; Pinto-Gouveia, J. The impact of illness-related shame on psychological health and social relationships: Testing a mediational model in students with chronic illness. *Clin. Psychol. Psychother.* **2018**, *25*, 408–414. [CrossRef] [PubMed]
34. Wiborg, J.F.; van der Werf, S.; Prins, J.B.; Bleijenberg, G. Being homebound with chronic fatigue syndrome: A multidimensional comparison with outpatients. *Psychiatry Res.* **2010**, *177*, 246–249. [CrossRef] [PubMed]
35. Golub, S.A.; Gamarel, K.E.; Rendina, H.J. Loss and growth: Identity processes with distinct and complementary impacts on well-being among those living with chronic illness. *Psychol. Health Med.* **2014**, *19*, 572–579. [CrossRef]
36. Ahlström, G. Experiences of loss and chronic sorrow in persons with severe chronic illness. *J. Clin. Nurs.* **2007**, *16*, 76–83. [CrossRef] [PubMed]
37. Rowe, P.C.; Underhill, R.A.; Friedman, K.J.; Gurwitt, A.; Medow, M.S.; Schwartz, M.S.; Speight, N.; Stewart, J.M.; Vallings, R.; Rowe, K.S. Myalgic Encephalomyelitis/Chronic Fatigue Syndrome Diagnosis and Management in Young People: A Primer. *Front. Pediatr.* **2017**, *5*, 121. [CrossRef]
38. Dancey, C.P.; Friend, J. Symptoms, impairment and illness intrusiveness–their relationship with depression in women with CFS/ME. *Psychol. Health* **2008**, *23*, 983–999. [CrossRef]
39. Burnham, K.E.; Cruess, D.G.; Kalichman, M.O.; Grebler, T.; Cherry, C.; Kalichman, S.C. Trauma symptoms, internalized stigma, social support, and sexual risk behavior among HIV-positive gay and bisexual MSM who have sought sex partners online. *AIDS Care* **2016**, *28*, 347–353. [CrossRef]
40. Cordova, M.J.; Giese-Davis, J.; Golant, M.; Kronenwetter, C.; Chang, V.; Spiegel, D. Breast Cancer as Trauma: Posttraumatic Stress and Posttraumatic Growth. *J. Clin. Psychol. Med. Settings* **2007**, *14*, 308–319. [CrossRef]
41. Brener, L.; Callander, D.; Slavin, S.; De Wit, J. Experiences of HIV stigma: The role of visible symptoms, HIV centrality and community attachment for people living with HIV. *AIDS Care* **2013**, *25*, 1166–1173. [CrossRef] [PubMed]
42. Devendorf, A.R.; A Rown, A.; A Jason, L. Patients' hopes for recovery from myalgic encephalomyelitis and chronic fatigue syndrome: Toward a "recovery in" framework. *Chronic Illn.* **2018**, *16*, 307–321. [CrossRef]
43. Horter, S.; Bernays, S.; Thabede, Z.; Dlamini, V.; Kerschberger, B.; Pasipamire, M.; Rusch, B.; Wringe, A. "I don't want them to know": How stigma creates dilemmas for engagement with Treat-all HIV care for people living with HIV in Eswatini. *Afr. J. AIDS Res.* **2019**, *18*, 27–37. [CrossRef]
44. Babrow, A.S.; Kasch, C.R.; Ford, L.A. The Many Meanings of Uncertainty in Illness: Toward a Systematic Accounting. *Health Commun.* **1998**, *10*, 1–23. [CrossRef]
45. Mishel, M.H. Uncertainty in Illness. *Image J. Nurs. Sch.* **1988**, *20*, 225–232. [CrossRef]
46. Mishel, M.H. Reconceptualization of the Uncertainty in Illness Theory. *Image J. Nurs. Sch.* **1990**, *22*, 256–262. [CrossRef] [PubMed]
47. Neville, K.L. Uncertainty in Illness. *Orthop. Nurs.* **2003**, *22*, 206–214. [CrossRef] [PubMed]
48. Choi, M.; Lee, J.; Oh, E.G.; Chu, S.H.; Sohn, Y.H.; Park, C.G. Factors Associated With Uncertainty in Illness Among People With Parkinson's Disease. *Clin. Nurs. Res.* **2018**, *29*, 469–478. [CrossRef]
49. Dudas, K.; Olsson, L.-E.; Wolf, A.; Swedberg, K.; Taft, C.; Schaufelberger, M.; Ekman, I. Uncertainty in illness among patients with chronic heart failure is less in person-centred care than in usual care. *Eur. J. Cardiovasc. Nurs.* **2013**, *12*, 521–528. [CrossRef] [PubMed]
50. Lee, I.; Park, C. The mediating effect of social support on uncertainty in illness and quality of life of female cancer survivors: A cross-sectional study. *Health Qual. Life Outcomes* **2020**, *18*, 1–11. [CrossRef] [PubMed]
51. Mullins, L.L.; Cote, M.P.; Fuemmeler, B.F.; Jean, V.M.; Beatty, W.W.; Paul, R.H. Illness intrusiveness, uncertainty, and distress in individuals with multiple sclerosis. *Rehabil. Psychol.* **2001**, *46*, 139–153. [CrossRef]
52. Reich, J.W.; Johnson, L.M.; Zautra, A.J.; Davis, M.C. Uncertainty of Illness Relationships with Mental Health and Coping Processes in Fibromyalgia Patients. *J. Behav. Med.* **2006**, *29*, 307–316. [CrossRef]
53. Mc Horton, S.; Poland, F.; Kale, S.; Drachler, M.D.L.; Leite, J.C.D.C.; A McArthur, M.; Campion, P.D.; Pheby, D.; Nacul, L. Chronic fatigue syndrome/myalgic encephalomyelitis (CFS/ME) in adults: A qualitative study of perspectives from professional practice. *BMC Fam. Pract.* **2010**, *11*, 89. [CrossRef]
54. Dickson, A.; Knussen, C.; Flowers, P. 'That was my old life; it's almost like a past-life now': Identity crisis, loss and adjustment amongst people living with Chronic Fatigue Syndrome. *Psychol. Health* **2008**, *23*, 459–476. [CrossRef]
55. A Alonzo, A. The experience of chronic illness and post-traumatic stress disorder: The consequences of cumulative adversity. *Soc. Sci. Med.* **2000**, *50*, 1475–1484. [CrossRef]
56. Baldwin, B.A. A paradigm for the classification of emotional crises: Implications for crisis intervention. *Am. J. Orthopsychiatry* **1978**, *48*, 538–551. [CrossRef]
57. Botha, K.F.H. Posttraumatic stress disorder and illness behavior in HIV+ patients. *Psychol. Rep.* **1996**, *79*, 843–845. [CrossRef] [PubMed]
58. Lindy, J.D.; Green, B.L.; Grace, M.C. Commentary. *J. Nerv. Ment. Dis.* **1987**, *175*, 269–272. [CrossRef] [PubMed]
59. Lonardi, C. The passing dilemma in socially invisible diseases: Narratives on chronic headache. *Soc. Sci. Med.* **2007**, *65*, 1619–1629. [CrossRef]

60. Weiss, M.G.; Ramakrishna, J.; Somma, D. Health-related stigma: Rethinking concepts and interventions 1. *Psychol. Health Med.* **2006**, *11*, 277–287. [CrossRef] [PubMed]
61. Doorn, K.A.-V.; Békés, V.; Prout, T.A.; Hoffman, L. Psychotherapists' vicarious traumatization during the COVID-19 pandemic. *Psychol. Trauma Theory Res. Pract. Policy* **2020**, *12*, S148–S150. [CrossRef] [PubMed]
62. Clark, M.L.; Gioro, S. Nurses, indirect trauma, and prevention. *Image J. Nurs. Sch.* **1998**, *30*, 85–87. [CrossRef]
63. Galovski, T.; A Lyons, J. Psychological sequelae of combat violence: A review of the impact of PTSD on the veteran's family and possible interventions. *Aggress. Violent Behav.* **2004**, *9*, 477–501. [CrossRef]
64. E Hayes, V. Families and children's chronic conditions: Knowledge development and methodological considerations. *Sch. Inq. Nurs. Pract. Int. J.* **1997**, *11*, 259–290.
65. Li, Z.; Ge, J.; Yang, M.; Feng, J.; Qiao, M.; Jiang, R.; Bi, J.; Zhan, G.; Xu, X.; Wang, L.; et al. Vicarious traumatization in the general public, members, and non-members of medical teams aiding in COVID-19 control. *Brain Behav. Immun.* **2020**, *88*, 916–919. [CrossRef]
66. Rolland, J.S. *Families, Illness, and Disability: An Interactive Treatment Model*; Basic Books: New York, NY, USA, 1994.
67. Lee, C.A.; Kessler, C.M.; Varon, D.; Martinowitz, U.; Heim, M.; Rosenberg, M.; Molho, P. Nonviolent (empathic) communication for health care providers. *Haemophilia* **1998**, *4*, 335–340. [CrossRef] [PubMed]
68. Scott, M.J.; Stradling, S.G. Post-traumatic stress disorder without the trauma. *Br. J. Clin. Psychol.* **1994**, *33*, 71–74. [CrossRef]
69. Yager, T.J.; Gerszberg, N.; Dohrenwend, B.P. Secondary Traumatization in Vietnam Veterans' Families. *J. Trauma. Stress* **2016**, *29*, 349–355. [CrossRef]
70. Fennell, P.A. Sociocultural context and trauma. In *Handbook of Chronic Fatigue Syndrome*; Jason, L.A., Fennell, P.A., Taylor, R.R., Eds.; Wiley: Hoboken, NJ, USA, 2003; pp. 73–88.
71. Fennell, P.A.; Levine, P.; Uslan, D.; Furst, G. Applications for the practicing physician: A multidisciplinary approach. In Proceedings of the American Association of Chronic Fatigue Syndrome, Seattle, WA, USA, 24–26 January 2001.
72. Fullilove, M.T.; Lown, E.A.; Fullilove, R.E. Crack 'Hos and Skeezers: Traumatic experiences of women crack users. *J. Sex Res.* **1992**, *29*, 275–287. [CrossRef]
73. Tingey, J.L.; Bentley, J.A.; Hosey, M.M. COVID-19: Understanding and mitigating trauma in ICU survivors. *Psychol. Trauma Theory Res. Pract. Policy* **2020**, *12*, 100. [CrossRef]
74. Turner, R.J.; Lloyd, D.A. Lifetime Traumas and Mental Health: The Significance of Cumulative Adversity. *J. Health Soc. Behav.* **1995**, *36*, 360. [CrossRef]
75. Ward, T.; Hogan, K.; Stuart, V.; Singleton, E. The experiences of counselling for persons with ME. *Couns. Psychother. Res.* **2008**, *8*, 73–79. [CrossRef]
76. Lacerda, E.M.; McDermott, C.; Msc, C.C.K.; Ba, J.B.; Cliff, J.M.; Nacul, L. Hope, disappointment and perseverance: Reflections of people with Myalgic encephalomyelitis/Chronic Fatigue Syndrome (ME/CFS) and Multiple Sclerosis participating in biomedical research. A qualitative focus group study. *Health Expect.* **2019**, *22*, 373–384. [CrossRef]
77. Drachler, M.D.L.; Leite, J.C.D.C.; Hooper, L.; Hong, C.S.; Pheby, D.; Nacul, L.; Lacerda, E.; Campion, P.; Killett, A.; McArthur, M.; et al. The expressed needs of people with Chronic Fatigue Syndrome/Myalgic Encephalomyelitis: A systematic review. *BMC Public Health* **2009**, *9*, 458. [CrossRef] [PubMed]
78. Alonzo, A.A.; Reynolds, N.R. Emotions and Care-Seeking during Acute Myocardial Infarction: A Model for Intervention. *Int. J. Sociol. Soc. Policy* **1996**, *16*, 97–122. [CrossRef]
79. Hampton, M.R.; Frombach, J. Women's experience of traumatic stress in cancer treatment. *Health Care Women Int.* **2000**, *21*, 67–76. [CrossRef]
80. Kwoh, C.K.; O'Connor, G.T.; Regan-Smith, M.G.; Olmstead, E.M.; A Brown, L.; Burnett, J.B.; Hochman, R.F.; King, K.; Morgan, G.J. Concordance between clinician and patient assessment of physical and mental health status. *J. Rheumatol.* **1992**, *19*, 1031–1037.
81. Salmon, P.; Peters, S.; Stanley, I. Patients' perceptions of medical explanations for somatisation disorders: Qualitative analysis. *BMJ* **1999**, *318*, 372–376. [CrossRef]
82. Starfield, B.; Wray, C.; Hess, K.; Gross, R.; Birk, P.S.; D'Lugoff, B.C. The influence of patient-practitioner agreement on outcome of care. *Am. J. Public Health* **1981**, *71*, 127–131. [CrossRef]
83. Catchpole, S.; Garip, G. Acceptance and identity change: An interpretative phenomenological analysis of carers' experiences in myalgic encephalopathy/chronic fatigue syndrome. *J. Health Psychol.* **2019**, 135910531983467. [CrossRef] [PubMed]
84. Arroll, M.A.; Howard, A. 'The letting go, the building up, [and] the gradual process of rebuilding': Identity change and post-traumatic growth in myalgic encephalomyelitis/chronic fatigue syndrome. *Psychol. Health* **2013**, *28*, 302–318. [CrossRef] [PubMed]
85. Larun, L.; Malterud, K. Identity and coping experiences in Chronic Fatigue Syndrome: A synthesis of qualitative studies. *Patient Educ. Couns.* **2007**, *69*, 20–28. [CrossRef]
86. Wilde, L.; Quincey, K.; Williamson, I. "The real me shining through M.E.": Visualizing masculinity and identity threat in men with myalgic encephalomyelitis/chronic fatigue syndrome using photovoice and IPA. *Psychol. Men Masc.* **2020**, *21*, 309–320. [CrossRef]
87. Klest, B.; Tamaian, A.; Mutschler, C. Betrayal Trauma, Health Care Relationships, and Health in Patients with a Chronic Neurovascular Condition. *J. Aggress. Maltreat. Trauma* **2016**, *26*, 1–16. [CrossRef]

88. Bs, N.S.; Taber, J.M.; Klein, W.M.P.; Ferrer, R.A. Evidence that perceptions of and tolerance for medical ambiguity are distinct constructs: An analysis of nationally representative US data. *Health Expect.* **2020**, *23*, 603–613. [CrossRef]

89. Neville, A.; Jordan, A.; Beveridge, J.K.; Pincus, T.; Noel, M. Diagnostic Uncertainty in Youth With Chronic Pain and Their Parents. *J. Pain* **2019**, *20*, 1080–1090. [CrossRef] [PubMed]

90. Asbring, P. Chronic illness—A disruption in life: Identity-transformation among women with chronic fatigue syndrome and fibromyalgia. *J. Adv. Nurs.* **2001**, *34*, 312–319. [CrossRef]

91. Ghezeljeh, T.N.; Nikravesh, M.Y.; Emami, A. Coronary heart disease patients transitioning to a normal life: Perspectives and stages identified through a grounded theory approach. *J. Clin. Nurs.* **2013**, *23*, 571–585. [CrossRef] [PubMed]

92. Assefi, N.P.; Coy, T.V.; Uslan, D.; Smith, W.R.; Buchwald, D. Financial, occupational, and personal consequences of disability in patients with chronic fatigue syndrome and fibromyalgia compared to other fatiguing conditions. *J. Rheumatol.* **2003**, *30*, 804–808.

93. Knight, C.; Gitterman, A. Ambiguous Loss and Its Disenfranchisement: The Need for Social Work Intervention. *Fam. Soc. J. Contemp. Soc. Serv.* **2018**, *100*, 164–173. [CrossRef]

94. Anderson, V.R.; Jason, L.A.; Hlavaty, L.E. A Qualitative Natural History Study of ME/CFS in the Community. *Health Care Women Int.* **2013**, *35*, 3–26. [CrossRef]

95. Rodrigo, L. A Rare Kind of Light at the End of a Rare Kind of Tunnel: Trauma of a Life-Altering, Chronic, Degenerative Disease and Its Redemption. *J. Loss Trauma* **2014**, *19*, 584–591. [CrossRef]

96. Curry, L.C.; Walker, C.A.; Moore, P.; Hogstel, M.O. Validating a model of chronic illness and family caregiving. *J. Theory Constr. Test.* **2010**, *14*, 10–16.

97. Mihelicova, M.; Siegel, Z.; Evans, M.; Brown, A.; A Jason, L. Caring for people with severe myalgic encephalomyelitis: An interpretative phenomenological analysis of parents' experiences. *J. Health Psychol.* **2016**, *21*, 2824–2837. [CrossRef] [PubMed]

98. Gilden, J.L.; Hendryx, M.S.; Clar, S.; Casia, C.; Singh, S.P. Diabetes Support Groups Improve Health Care of Older Diabetic Patients. *J. Am. Geriatr. Soc.* **1992**, *40*, 147–150. [CrossRef] [PubMed]

99. Hinrichsen, G.A.; Revenson, T.A.; Shinn, M. Does Self-Help Help? *J. Soc. Issues* **1985**, *41*, 65–87. [CrossRef]

100. Humphreys, K. Individual and social benefits of mutual aid self-help groups. *Soc. Policy* **1997**, *27*, 12–19.

101. Nash, K.B.; Kramer, K.D. Self-Help for Sickle Cell Disease in African American Communities. *J. Appl. Behav. Sci.* **1993**, *29*, 202–215. [CrossRef]

102. Valdez, A.R.; Hancock, E.E.; Adebayo, S.; Kiernicki, D.J.; Proskauer, D.; Attewell, J.R.; Bateman, L.; DeMaria, A.J.; Lapp, C.W.; Rowe, P.C.; et al. Estimating Prevalence, Demographics, and Costs of ME/CFS Using Large Scale Medical Claims Data and Machine Learning. *Front. Pediatr.* **2019**, *6*, 412. [CrossRef]

103. Johnson, M.L.; Cotler, J.; Terman, J.M.; Jason, L.A. Risk factors for suicide in chronic fatigue syndrome. *Death Stud.* **2020**, 1–7. [CrossRef] [PubMed]

104. Yanos, P.T.; Roe, D.; Lysaker, P.H. The Impact of Illness Identity on Recovery from Severe Mental Illness. *Am. J. Psychiatr. Rehabil.* **2010**, *13*, 73–93. [CrossRef]

105. Jason, L.A.; Corradi, K.; Gress, S.; Williams, S.; Torres-Harding, S. Causes of Death Among Patients With Chronic Fatigue Syndrome. *Health Care Women Int.* **2006**, *27*, 615–626. [CrossRef]

106. McManimen, S.L.; Devendorf, A.R.; Brown, A.A.; Moore, B.C.; Moore, J.H.; Jason, L.A. Mortality in patients with myalgic encephalomyelitis and chronic fatigue syndrome. *Fatigue Biomed. Health Behav.* **2016**, *4*, 195–207. [CrossRef] [PubMed]

107. Jacobsen, J.C.; Zhang, B.; Block, S.D.; Maciejewski, P.K.; Prigerson, H.G. Distinguishing Symptoms of Grief and Depression in a Cohort of Advanced Cancer Patients. *Death Stud.* **2010**, *34*, 257–273. [CrossRef] [PubMed]

108. Tolleson, A.; Zeligman, M. Creativity and Posttraumatic Growth in Those Impacted by a Chronic Illness/Disability. *J. Creat. Ment. Health* **2019**, *14*, 499–509. [CrossRef]

109. McInnis, O.A.; McQuaid, R.J.; Bombay, A.; Matheson, K.; Anisman, H. Finding benefit in stressful uncertain circumstances: Relations to social support and stigma among women with unexplained illnesses. *Stress* **2015**, *18*, 169–177. [CrossRef] [PubMed]

110. Han, P.K.; Zikmund-Fisher, B.J.; Duarte, C.W.; Knaus, M.; Black, A.; Scherer, A.M.; Fagerlin, A. Communication of Scientific Uncertainty about a Novel Pandemic Health Threat: Ambiguity Aversion and Its Mechanisms. *J. Health Commun.* **2018**, *23*, 435–444. [CrossRef]

111. Weingarten, K. Sorrow: A Therapist's Reflection on the Inevitable and the Unknowable. *Fam. Process.* **2012**, *51*, 440–455. [CrossRef]

112. Zeligman, M.; Varney, M.; Grad, R.I.; Huffstead, M. Posttraumatic Growth in Individuals With Chronic Illness: The Role of Social Support and Meaning Making. *J. Couns. Dev.* **2018**, *96*, 53–63. [CrossRef]

113. Flurey, C.; White, A.; Rodham, K.; Kirwan, J.; Noddings, R.; Hewlett, S. 'Everyone assumes a man to be quite strong': Men, masculinity and rheumatoid arthritis: A case-study approach. *Sociol. Health Illn.* **2017**, *40*, 115–129. [CrossRef]

114. Jason, L.A.; Richman, J.A.; Rademaker, A.W.; Jordan, K.M.; Plioplys, A.V.; Taylor, R.R.; McCready, W.; Huang, C.-F.; Plioplys, S. A Community-Based Study of Chronic Fatigue Syndrome. *Arch. Intern. Med.* **1999**, *159*, 2129–2137. [CrossRef]

115. Jason, L.A.; Katz, B.Z.; Sunnquist, M.; Torres, C.; Cotler, J.; Bhatia, S. The Prevalence of Pediatric Myalgic Encephalomyelitis/Chronic Fatigue Syndrome in a Community-Based Sample. *Child Youth Care Forum* **2020**, *49*, 563–579. [CrossRef]

116. Soto, G.J.; Martin, G.S.; Gong, M.N. Healthcare Disparities in Critical Illness. *Crit. Care Med.* **2013**, *41*, 2784–2793. [CrossRef] [PubMed]

117. Lillie-Blanton, M.; Martinez, R.M.; Taylor, A.K.; Robinson, B.G. Latina and African American Women: Continuing Disparities in Health. *Int. J. Health Serv.* **1993**, *23*, 555–584. [CrossRef] [PubMed]
118. Bayliss, K.; Riste, L.; Fisher, L.; Wearden, A.; Peters, S.; Lovell, K.; Chew-Graham, C. Diagnosis and management of chronic fatigue syndrome/myalgic encephalitis in black and minority ethnic people: A qualitative study. *Prim. Health Care Res. Dev.* **2013**, *15*, 143–155. [CrossRef]
119. Kim, W.; Keefe, R.H. Barriers to Healthcare Among Asian Americans. *Soc. Work. Public Health* **2010**, *25*, 286–295. [CrossRef] [PubMed]
120. Pearson, W.S.; Ahluwalia, I.B.; Ford, E.S.; Mokdad, A.H. Language preference as a predictor of access to and use of healthcare services among Hispanics in the United States. *Ethn. Dis.* **2008**, *18*, 93–97.
121. Pheby, D.; Saffron, L. Risk factors for severe ME/CFS. *Biol. Med.* **2009**, *1*, 50–74.
122. Turner, R.J.; Wheaton, B.; Lloyd, D.A. The Epidemiology of Social Stress. *Am. Sociol. Rev.* **1995**, *60*, 104. [CrossRef]
123. Myers, H.F.; Wyatt, G.E.; Ullman, J.B.; Loeb, T.B.; Chin, D.; Prause, N.; Zhang, M.; Williams, J.K.; Slavich, G.M.; Liu, H. Cumulative burden of lifetime adversities: Trauma and mental health in low-SES African Americans and Latino/as. *Psychol. Trauma Theory Res. Pract. Policy* **2015**, *7*, 243–251. [CrossRef] [PubMed]
124. Mokdad, A.H.; A Mensah, G.; Posner, S.F.; Reed, E.; Simoes, E.J.; Engelgau, M.M. Chronic Diseases and Vulnerable Populations in Natural Disasters Working Group when Chronic Conditions Become Acute: Prevention and Control of Chronic Diseases and Adverse Health Outcomes during Natural Disasters. *Prev. Chronic Dis.* **2005**, *2*, A04. [PubMed]
125. Brown, E.R.; Ojeda, V.D.; Wyn, R.; Levan, R. Racial and Ethnic Disparities in Access to Health Insurance and Health Care. 2000. Available online: https://www.kff.org/wp-content/uploads/2013/01/racial-and-ethnic-disparities-in-access-to-health-insurance-and-health-care-report.pdf. (accessed on 28 December 2020).

 *healthcare*

Opinion

ME/CFS: Past, Present and Future

William Weir [1,*] **and Nigel Speight** [2]

[1] Royal Free Hospital, London W1G 9PF, UK
[2] University Hospital of North Durham, Durham DH1 1QN, UK; speight@doctors.org.uk
*   Correspondence: wrcweir@hotmail.com

Abstract: This review raises a number of compelling issues related to the condition of Myalgic Encephalomyelitis/Chronic Fatigue Syndrome (ME/CFS). Some historical perspective is necessary in order to highlight the nature of the controversy concerning its causation. Throughout history, a pattern tends to repeat itself when natural phenomena require explanation. Dogma usually arrives first, then it is eventually replaced by scientific understanding. The same pattern is unfolding in relation to ME/CFS, but supporters of the psychological dogma surrounding its causation remain stubbornly resistant, even in the face of compelling scientific evidence to the contrary. Acceptance of the latter is not just an academic issue; the route to proper understanding and treatment of ME/CFS is through further scientific research rather than psychological theorisation. Only then will a long-suffering patient group benefit.

Keywords: myalgic encephalomyelitis/chronic fatigue syndrome (ME/CFS) controversy, dogma; psychological causation

Citation: Weir, W.; Speight, N. ME/CFS: Past, Present and Future. *Healthcare* **2021**, *9*, 984. https://doi.org/10.3390/healthcare9080984

Academic Editors: Kenneth J. Friedman, Lucinda Bateman and Kenny Leo De Meirleir

Received: 10 June 2021
Accepted: 25 July 2021
Published: 3 August 2021

Publisher's Note: MDPI stays neutral with regard to jurisdictional claims in published maps and institutional affiliations.

The history of human civilisation is littered with examples of natural phenomena, including human disease, initially explained by dogma. The dogma is initially created to fill a void in comprehension, but it is eventually replaced by rational scientific understanding. The creators of such dogma are often authoritarian, hierarchical figures who then ferociously defend their own creation. The classic manifestation of this was what Galileo encountered when he proposed, on the basis of careful observations through the new technology of the telescope, that the planet was not the centre of the solar system. Despite the scientific validity of his observations, he was threatened with torture by the Catholic inquisition if he did not recant. He was speaking truth to a powerful establishment, and the conflict came to a head in 1633, when, under severe duress, he was forced to withdraw his heretical ideas. The Catholic church reprieved him eventually, but not until 1992. Galileo's difficulties with the Catholic church are a good early example; he spoke scientific truth to authoritarian power and suffered the consequences.

There are similar examples of this pattern in medical history. Ignaz Semmelweis, working in an obstetric ward at the General Hospital in Vienna in 1846 noticed the large difference in mortality from puerperal fever between a ward where birthing women were attended by midwives, and another where the women were attended by doctors and medical students. The latter divided their time between the autopsy room and the ward. Semmelweis observed that the midwives washed their hands between deliveries, whereas the doctors and medical students did not, even after performing autopsies on the victims of puerperal fever. The patients attended by the doctors and medical students died, embarrassingly, more frequently from puerperal fever, and Semmelweis recognised correctly that a noxious agent was being transmitted from the autopsy room by unwashed hands. The precise cause, in the light of current scientific understanding, is now blindingly obvious. Unfortunately, for Semmelweis, this was before the discovery of disease-causing microbes. His medical colleagues, lacking Semmelweis' insights, were greatly offended by the implication that by not washing their hands, they were somehow responsible for the excess deaths. Consequently, he was hounded out of his post. He was ahead of his time, but nowadays he can be rightly regarded as a hero, having spoken truth to power, as Galileo did, at great personal expense.

John Snow's scientifically correct perceptions of cholera transmission were also up against a contemporary dogma. In 1855, he published his treatise, which found that the cause of cholera was spread through drinking water. What he wrote has stood the test of time and is now regarded as a model of scientific validity. He recognised that drinking water from known sources was the cause, and he was persistent enough to collect the data to prove this by undertaking painstaking house-to-house visits through the streets of London. Nonetheless, nearly 30 years elapsed before Robert Koch demonstrated the presence of *Vibrio cholerae* on the gut-lining of cholera victims at autopsy, having also been able to demonstrate the presence of this microbe in drinking water. At the time of Snow's publication, however, the prevailing dogma was that cholera was spread by rotten smells— "miasmata"—from dead bodies and rotting vegetable matter. Predictably, prominent members of the Royal College of Physicians at the time, declared Snow's work "untenable" because their dogma was being challenged.

This collective mindset meant that contemporary management of cholera was equally wide of the mark. Bloodletting and rectal infusions of mutton puree were among the conventional mainstream treatments. The latter was probably challenging given the profuse diarrhoea of cholera. Furthermore, contemporary survival from cholera was seen to be rather better at the London Homeopathic Hospital where patients were spared the lethally inappropriate practice of bloodletting. One of the main pathological characteristics of cholera is reduction of circulating blood volume due to the diarrhoea, causing massive salt and water depletion. Further reduction of blood volume by bloodletting certainly hastened the death of such patients.

Additional examples of the medical profession "getting it wrong" have continued since this time. General Paralysis of the Insane—a manifestation of tertiary syphilis— and multiple sclerosis were both considered to have a psychological basis [1] until the true physical basis was discovered. There are other examples: the tremor of Parkinson's disease had been attributed to "the expression of the moralistic man's suppressed desire to masturbate" [2] but we now know this to be untrue. More recently, the proposition that Helicobacter pylori infection could be the cause of peptic ulcers was up against the dogma that psychological stress was the major contributor. Hitherto, the presence of this strange organism in the stomach lining was regarded as insignificant, but trials with antibiotic therapy effectively refuted this idea, and the treatment of peptic ulcers was dramatically improved [3]. All of these examples illustrate a tendency to assume that, if no pathological mechanism can be demonstrated, then, by default, psychological disorder must be the problem. Inherent in such an assumption is the arrogant belief that routine laboratory tests infallibly exclude physical disorder.

The story of ME/CFS is a prime example of such dogma. Due to the fact that routine laboratory tests for the diagnosis of this condition usually produce "normal" results, the problem must be with the psyche. One of the foundation stones of this dogma was a paper published in the BMJ in 1970 in which the cause of the famous Royal Free outbreak of ME/CFS in 1955 was attributed to "mass hysteria". The authors did not interview any of the patients, nor any of the doctors involved; nonetheless, it seemed clear to them that the outbreak was due to mass hysteria because the majority of victims were women [4]. The background to this piece of sophistry was, and remains, the fashionable medical culture of linking physical symptoms to a psychological disorder. Mind and body are highly interactive, and certainly there are conditions in which psychological distress expresses itself with physical symptoms. Even so, there are many human diseases and infirmities in which the primary driver is physical pathology, with psychology playing a minor secondary role, if at all. Nevertheless, the psychological cognoscenti have not let this principle inhibit their wide-ranging suppositions about the role of psychology in the human condition. The result has been some very wild adventures in psychological theory. For example, a famous 20th century French psychologist once suggested in all seriousness that an erect penis could be expressed algebraically as the square root of minus one [5]. To his disciples, this was "boldly transgressive thinking", but most sane mathematicians of either sex will have been baffled, not least because in conventional mathematics, minus one does not have a square root. Nonetheless, a culture of similar nonsense has set the scene for equally fantastic theorisation concerning other manifestations of the human condition, including the cause(s) of ME/CFS.

As with previous examples of medical dogma, the belief that ME/CFS is "psychological" will eventually be consigned to the dustbin of medical history, alongside miasma theory and suchlike. Compelling evidence of physical causation is now accumulating but the authoritarian cabal who promoted the psychological dogma are even now trying to defend it in the face of irrefutable scientific evidence to the contrary. History repeats itself, to coin a phrase, given the stories of Galileo, Semmelweis and Snow, and the cabal referred to, do not yet recognise how badly placed they are in the historical narrative of ME/CFS. In some circumstances, the tendency of exponents to hold on to their dogma is reminiscent of the tenacious way conspiracy theorists are wedded to their particular false narrative. Sadly, the argument over the cause of ME/CFS would probably have remained academic but for one grim reality: treatment based on psychological dogma has damaged patients, some very severely.

Due to the fact that ME/CFS was due, amongst other things, to "abnormal illness beliefs, buried guilt and negative thoughts", the psychological advocates have always advised treatment intended to correct disordered psychology and its presumed consequences. The muscular weakness of ME/CFS was seen as simply due to "deconditioning" because of inactivity secondary to exercise phobia. Graded Exercise Therapy (GET) was, therefore, the answer, and abnormal illness belief and exercise phobia could be managed with Cognitive Behavioural Therapy (CBT). Both of these techniques have been widely promoted, supported in particular by the PACE trial [6], an egregious and expensive exercise in scientific sophistry whose methodology was so seriously flawed that it is now used as an example of how not to conduct scientific studies [7].

The damage caused by GET, in particular, has unfortunate historical precedents. As previously stated, bloodletting was particularly dangerous for cholera; likewise GET has caused significant harm for many ME/CFS patients, frequently consigning modestly mobile patients, adults and children alike, to a prolonged, bedbound, nasogastric-tube-fed existence. If GET were a drug, it would have been banned rapidly by the appropriate regulatory body, but in the UK, there is no such regulatory body for non-pharmacological treatments. This should be within the remit of the General Medical Council, but despite one of their stated functions being to "protect patients", many patients have been harmed in the way described.

In respect of children in the UK with ME/CFS, the psychological dogma has been particularly harmful. The UK paediatric establishment has not recognised the physical nature of the incapacity caused by ME/CFS. It has become increasingly fashionable in British paediatrics to apply the terms "Medically Unexplained Symptoms" (MUS) and "Perplexing Presentations" (PP) under the much wider umbrella of "potential Factitious Illness (FII)", on the specious grounds that if the doctor concerned cannot make a diagnosis, it is likely that the mother is "colluding" with her child's symptoms. Families of children with ME/CFS are particularly at risk of being trapped in such accusations, due to the dogma-led belief in psychological disorder when all routine tests are normal. As a result of this, children with ME/CFS have sometimes been removed by social services from the security of their own home. This can then be followed by grotesquely inappropriate treatment, one extreme example of which involved a severely impaired 12-year-old boy being left unsupported, deliberately, in a hydrotherapy pool. The intention being to force him to swim, thus revealing that he was physically unimpaired and had to overcome his abnormal illness beliefs and negative thoughts about his true physical capabilities. In reality he was so physically weak that he nearly drowned, unwittingly re-enacting the medieval test for witchcraft.

There are other examples in which non-existent psychological disorder was suspected: a teenage girl with severe ME/CFS was once visited at home by her GP. He said, "Now we are going to get to the bottom of the secret phobias that are causing your illness". The girl answered, "but I don't have any secret phobias", to which the doctor replied, "that's the thing about secret phobias, you don't know you've got them until we dig deep enough." In some egregious instances, children with ME/CFS whose condition predictably worsens with GET become bedbound. They then have an alternative diagnostic label applied, such as "Pervasive Refusal Syndrome." The skewed logic being that GET always helps ME/CFS; if it does not, the initial diagnosis of ME/CFS must have been wrong.

Mention has already been made of recognisable pathological abnormalities in ME/CFS, effectively rebutting the dogma of psychological causation. Even now the aforementioned authoritarian cabal continue to ignore or possibly regard such abnormalities as "downstream" of primary psychological disorder. Abnormalities of muscle metabolism in ME/CFS patients have now been clearly recognised, providing scientific insight into the characteristic intolerance of exercise [8,9]. The ME/CFS dogma attributes this to psychological causes, particularly "exercise phobia". It is now evident that calibrated exercise on a bicycle ergometer on two consecutive days indicates clear differences in muscle metabolism between ME/CFS patients and healthy but sedentary, i.e., deconditioned, controls. In the ME/CFS patients, the "anaerobic threshold" decreases on the second exercise day, whereas it increases in the controls as part of the process leading to increasing physical fitness [8,9].

In lay terms, the anaerobic threshold is the point at which muscles, exercising at maximum, switch to a metabolic pathway that does not use oxygen. This allows for a final burst of energy, followed within a few seconds by a sensation of exhaustion. High anaerobic thresholds are characteristic of athletes, particularly those undertaking endurance events that enable them to run long distances without hitting their anaerobic threshold. In non-athletic, but healthy people, repeated daily exercise causes the anaerobic threshold to rise, the result being increasing physical fitness. This does not happen in ME/CFS, and misguided attempts to force exercise on the patient has exactly the opposite effect for the reasons stated above. It is highly likely that such exercise on consecutive days will lower the anaerobic threshold even further. In badly affected patients, the effect of an extremely low anaerobic threshold is severe exercise intolerance, which manifests as profound exhaustion, even with the minimal effort of getting out of bed, or such activities as eating and swallowing. Such cases often arise as a consequence of enforced exercise, unwittingly and progressively lowering the anaerobic threshold, rendering a moderately affected and previously mobile patient even more exhausted. The result is a bedbound existence for prolonged periods, some even requiring tube-feeding because the level of exhaustion is such that chewing and swallowing a normal diet becomes physically impossible.

Studies in vitro of biopsied muscle from ME/CFS patients have shown metabolic defects that underpin the findings described above. Repeated electrical stimulation of isolated muscle fibres from ME/CFS patients reveals impairments of metabolism that are not seen in healthy controls [10]. Biopsied muscle is self-evidently separate from the owner's psyche, safely excluding any influence from this source. There are other studies that further demonstrate the physical basis of ME/CFS. Disorder of the hypothalamic/pituitary/adrenal axis (HPAA) has been recognised for at least 30 years [11–16] and may well be due to autoimmunity [17]. Reduced circulating cortisol levels are the result, with a similar reduction of HPAA responses to stresses, both physical and psychological [14]. As a consequence of this, long-standing ME/CFS patients, due to impaired ACTH output, have been shown to have significantly smaller adrenals compared to normal controls [18], and also a low circulating blood volume [19]. The latter is very likely to contribute to Postural Orthostatic Tachycardia Syndrome (POTS), a common complication of ME/CFS [19].

Immunological dysfunction is also a universal feature. Many patients, previously healthy, experience an acute infection at the onset of their ME/CFS. This can either be viral, bacterial or protozoan. The common denominator is clearly an immunological stimulus, a principle supported by the recognition that vaccination can play the same role for some. In healthy people an immune response is stimulated by the infection/vaccine, the response then shutting down when the infection/vaccine is cleared. The shutdown is due to a series of progressive checks and balances that operate efficiently in normal health. In ME/CFS, this does not happen, and immunological activity continues for reasons that are yet to be fully understood. The simplest analogy is that of a revolving door continuing to revolve with the exit blocked. Chronic inflammation is the sequel [20,21], with some researchers describing the immune system as "derailed" [22]. The resulting inflammatory process includes the brain, giving pathological validity to the term myalgic encephalomyelitis [23,24].

In conclusion, proper scientific research into the physical cause(s) of ME/CFS will eventually replace the damaging influence of pseudoscientific, psychological dogma. A reliable biomarker currently in development [25] is a big step in this direction. Also, the current Covid19 pandemic may be a cloud with a silver lining. "LongCovid", a devastating aftermath of Covid19 infection, is currently attracting research funding. The clinical presentations of "LongCovid" are strikingly similar to those of ME/CFS, and the underlying pathology may well be the same [26]. Hopefully, the funds referred to will be used for properly directed scientific searches for the precise cause of this pathology, rather than for a PACE mark 2. To paraphrase Albert Einstein: "the definition of insanity is to do the same thing again, expecting a different result". If sanity prevails, properly focussed scientific research will eventually bring much needed relief to a population of patients who have hitherto been very poorly served by the medical profession.

Author Contributions: W.W. and N.S. have 60 years of combined experience with ME/CFS and this review is the product of that experience. W.W. was a consultant in infectious disease at the Royal Free Hospital London and N.S. a consultant paediatrician at the University Hospital of North Durham. All authors have read and agreed to the published version of the manuscript.

Funding: This review did not require funding support.

Institutional Review Board Statement: Not applicable.

Informed Consent Statement: Not applicable.

Acknowledgments: To the ME/CFS patient community for whom effective treatment is long overdue.

Conflicts of Interest: The authors declare no conflict of interest.

References

1. Ghaemi, S.N. The rise and fall of the biopsychosocial model. *BJP* **2009**, *195*, 3–4. [CrossRef]
2. Booth, G. Psychodynamics in Parkinsonism. *Psychosom. Med.* **1948**, *10*, 1–14. [CrossRef]
3. Marshall, B.J.; Warren, J.R. Unidentified curved bacilli in the stomach of patients with gastritis and peptic ulceration. *Lancet* **1984**, *1*, 1311–1315. [CrossRef]
4. McEvedy, C.P.; Beard, A.W. Royal Free Epidemic of 1955: A reconsideration. *Br. Med. J.* **1970**, *1*, 7–11. [CrossRef] [PubMed]
5. Sokal, A.; Bricmont, J. *Fashionable Nonsense: Postmodern Intellectuals Abuse of Science*; Picador: New York, NY, USA, 1998; p. 27.
6. White, P.D.; Goldsmith, K.A.; Johnson, A.L.; Potts, L.; Walwyn, R.; DeCesare, J.C.; Baber, H.L.; Burgess, M.; Clark, L.V.; Cox, D.L.; et al. Comparison of adaptive pacing therapy, cognitive behavioural therapy, graded exercise therapy and specialist medical care for chronic fatigue syndrome (PACE): A randomised trial. *Lancet* **2011**, *377*, 823–836. [CrossRef]
7. Tuller, D. Trial by Error. The Troubling Case of the PACE Chronic Fatigue Syndrome Study. Available online: www.virology.ws/2015/10/21/trial-by-error (accessed on 10 June 2021).
8. Snell, C.R.; Stevens, S.R.; Davenport, T.E.; Van Ness, J.M. Discriminative value of metabolic and workload measurements for identifying people with chronic fatigue syndrome. *Phys. Ther.* **2013**, *93*, 1484–1492. [CrossRef] [PubMed]
9. Keller, B.A.; Pryor, J.L.; Giloteaux, L. Inability of ME/CFS patients to reproduce VO$_2$ peak indicates functional impairment. *J. Transl. Med.* **2014**, *12*, 104. [CrossRef]
10. Brown, A.E.; Jones, D.E.; Walker, M.; Newton, J.L. Abnormalities of AMPK activation and glucose uptake in cultured skeletal muscle cells from individuals with Chronic Fatigue Syndrome. *PLoS ONE* **2015**, *10*, e0122982. [CrossRef] [PubMed]
11. Demitrack, M.A.; Dale, J.K.; Straus, S.E.; Laue, L.; Listwak, S.J.; Kruesi, M.J.P.; Chrousos, G.P.; Gold, P.W. Evidence for impaired activation of the hypothalamic-pituitary-adrenal axis in patients with chronic fatigue syndrome. *J. Clin. Endocrinol. Metab.* **1991**, *73*, 1224–1234. [CrossRef]

12. Scott, L.V.; Medbak, S.; Dinan, T.G. Blunted adrenocorticotropin and cortisol responses to corticotropin-releasing hormone stimulation in chronic fatigue syndrome. *Acta Psychiatr. Scand.* **1998**, *97*, 450–457. [CrossRef]

13. De Becker, P.; De Meirleir, K.; Joos, E.; Campine, I.; Van Steenberge, E.; Smitz, J.; Velkeniers, B. Dehydroepiandrosterone (DHEA) response to i.v. ACTH in patients with chronic fatigue syndrome. *Horm. Metab. Res.* **1999**, *31*, 18–21. [CrossRef]

14. Gaab, J.; Hüster, D.; Peisen, R.; Engert, V.; Heitz, V.; Schad, T.; Schürmeyer, T.H.; Ehlert, U. Hypothalamic-pituitary-adrenal axis reactivity in chronic fatigue syndrome and health under psychological, physiological, and pharmacological stimulation. *Psychosom. Med.* **2002**, *64*, 951–962. [CrossRef]

15. Segal, T.Y.; Hindmarsh, P.C.; Viner, R.M. Disturbed adrenal function in adolescents with chronic fatigue syndrome. *J. Pediatr. Endocrinol. Metab.* **2005**, *18*, 295–301. [CrossRef]

16. Van Den Eede, F.; Moorkens, G.; Van Houdenhove, B.; Cosyns, P.; Claes, S.J. Hypothalamic-pituitary-adrenal axis function in chronic fatigue syndrome. *Neuropsychobiology* **2007**, *55*, 112–120. [CrossRef]

17. De Bellis, A.; Bellastella, G.; Pernice, V.; Cirillo, P.; Longo, M.; Maio, A.; Scappaticcio, L.; Maiorino, M.I.; Bellastella, A.; Esposito, K.; et al. Hypothalamic-Pituitary autoimmunity and related impairment of hormone secretions in chronic fatigue syndrome. *J. Clin. Endocrinol. Metab.* **2021**, dgab429. [CrossRef] [PubMed]

18. Scott, L.V.; Teh, J.; Reznek, R.; Martin, A.; Sohaib, A.; Dinan, T.G. Small adrenal glands in chronic fatigue syndrome: A preliminary computer tomography study. *Psychoneuroendocrinology* **1999**, *24*, 759–768. [CrossRef]

19. Van Campen, C.L.; Rowe, P.C.; Visser, F.C. Blood Volume Status in ME/CFS correlates with the Presence or Absence of Orthostatic Symptoms: Preliminary Results. *Front. Pediatr.* **2018**, *6*, 352. [CrossRef] [PubMed]

20. Montoya, J.G.; Holmes, T.H.; Anderson, J.N.; Maecker, H.T.; Rosenberg-Hasson, Y.; Valencia, I.J.; Chu, L.; Younger, J.W.; Tato, C.M.; Davis, M.M. Cytokine signature associated with disease severity in chronic fatigue syndrome patients. *Proc. Natl. Acad. Sci. USA* **2017**, *114*, E7150–E7158. [CrossRef]

21. Hornig, M.; Montoya, J.G.; Klimas, N.G.; Levine, S.; Felsenstein, D.; Bateman, L.; Peterson, D.L.; Gottschalk, C.G.; Schultz, A.F.; Che, X.; et al. Distinct plasma immune signatures in ME/CFS are present early in the course of illness. *Sci. Adv.* **2015**, *1*, e1400121. [CrossRef]

22. Metselaar, P.I.; Mendoza-Maldonado, L.; Yim AY, F.L.; Abarkan, I.; Henneman, P.; Te Velde, A.A.; Schönhuth, A.; Bosch, J.A.; Kraneveld, A.D.; Lopez-Rincon, A. Recursive ensemble feature selection provides a robust mRNA expression signature for myalgic encephalomyelitis/chronic fatigue syndrome. *Sci. Rep.* **2021**, *11*, 4541. [CrossRef]

23. Glassford, J.A. The neuroinflammatory etiopathology of Myalgic Encephalomyelitis/Chronic Fatigue Syndrome (ME/CFS). *Front. Physiol.* **2017**, *8*, 88. [CrossRef] [PubMed]

24. Tomas, C.; Newton, J. Metabolic abnormalities in chronic fatigue syndrome/myalgic encephalomyelitis: A mini-review. *Biochem. Soc. Trans.* **2018**, *46*, 547–553. [CrossRef] [PubMed]

25. Esfandyarpour, R.; Kashi, A.; Nemat-Gorgani, M.; Wilhelmy, J.; Davis, R.W. A nanoelectronics-blood-based diagnostic biomarker for myalgic encephalomyelitis/chronic fatigue syndrome (ME/CFS). *Proc. Natl. Acad. Sci. USA* **2019**, *116*, 10250–10257. [CrossRef]

26. Komaroff, A.L.; Bateman, L. Will Covid19 lead to Myalgic Encephalomyelitis/Chronic Fatigue Syndrome? *Front. Med.* **2020**, *7*, 606824. [CrossRef] [PubMed]

 *healthcare*

Review

The Lonely, Isolating, and Alienating Implications of Myalgic Encephalomyelitis/Chronic Fatigue Syndrome

Samir Boulazreg [1,*] **and Ami Rokach** [2]

1   Faculty of Education, University of Western Ontario, London, ON N6A 3K7, Canada
2   Department of Psychology, York University, Toronto, ON M3J 1P3, Canada; arokach@yorku.ca
*   Correspondence: Boulazreg@live.ca

Received: 17 July 2020; Accepted: 11 October 2020; Published: 20 October 2020

Abstract: This article provides a narrative review on myalgic encephalomyelitis/chronic fatigue syndrome (ME/CFS) through a psychosocial lens and examines how this impairment affects its sufferers during adolescence and adulthood, as well as how it impacts family caregivers and healthcare professionals' mental health. Since there has been a lack of investigation in the literature, the primary psychosocial stressor that this review focuses on is loneliness. As such, and in an attempt to help establish a theoretical framework regarding how loneliness may impact ME/CFS, loneliness is comprehensively reviewed, and its relation to chronic illness is described. We conclude by discussing a variety of coping strategies that may be employed by ME/CFS individuals to address their loneliness. Future directions and ways with which the literature may investigate loneliness and ME/CFS are discussed.

Keywords: myalgic encephalomyelitis; chronic fatigue syndrome; loneliness; psychosocial

---

1. The Lonely, Isolating, and Alienating Implications of ME/CFS

Myalgic encephalomyelitis/chronic fatigue syndrome (ME/CFS) is a debilitating neurological disorder known to produce a wide range of devastating symptoms best known to include extreme fatigue, pain, and post-exertional discomfort. Though it is thought to originate from a genetic predisposition and/or an interaction with a host of environmental factors (e.g., frequent injury), the exact precursors of this disorder are still not well-understood [1]. Researchers have, however, observed some uniformity in attempting to distinguish the properties of this illness; for example, patients with ME/CFS have been observed to have a reduced blood perfusion rate in the brain stem [2]. Additionally, abnormality in multiple brain structures that regulate pain has been observed [3], leading recent research studies to theorize that the ME/CFS brain's homeostatic processes that react to pain are aberrant in nature [1]. This unusual brain activity, viewed plausibly due to a viral infection that impacts the central nervous system [4] such as, for example, glandular fever shortly before a diagnosis [5], has overarching effects that can cause cognitive difficulties, sleep dysfunction, and immune system irregularity, amongst other debilitating outcomes [6].

From the start, the conceptualization of ME/CFS as an illness has been riddled with controversies and dismissiveness from the medical community. Initially deemed to be psychosomatic in nature [7] and as epidemic hysteria [8,9], this mislabeling has persisted to this day, leading researchers to make observation that the medical community continues to harbor "prejudiced opinions that it is not a real illness" [10] (p. 309). One example of this was seen in Canada; in 2016, the federal government's scientific panel rejected a grant application for ME/CFS research, implying that "it was not a disease" [11]. Though funding towards biomedical research devoted to researching ME/CFS was eventually accepted less than a year ago [11], this example provides an illustration of the ignorance faced by those living with the illness. Consequently, institutes such as the National Academies of Sciences Engineering and Medicine presently classify ME/CFS as a stigmatized illness [12].

While the ignorance directed towards this illness may be in large part be due to a holistic lack of scientific understanding surrounding the exact antecedents of this disorder [1] and the lack of a standardized procedure for determining its presence in patients [13], ME/CFS is known to cause irritation between the brain, the spinal cord, and the musculoskeletal system [14–17]. It is thus why the term myalgia encephalomyelitis was coined as the Latin word "myalgia" translates to muscle pain, "encephalo" to brain, and "myel" to spinal cord [18]. However, this understanding, as well as the incapacitating physical symptoms presented earlier, only tells one side of an ME/CFS sufferer's story. That is, sufferers of this disorder typically also carry with them a myriad of negative non-physical consequences affecting anxiety, depression, and overall well-being [14]. Put another way, an adjustment period filled with many new expected and unexpected vulnerabilities occurs following a diagnosis of ME/CFS. Additionally, while variability exists in the way this syndrome affects individuals [8], for the most part, the major non-physical changes that ME/CFS patients commonly must endure include learning how to cope with psychosocial impairment related to the family structure, a loss of self, and a reduced social network [10,19–22].

As such, this review aims to explore these psychosocial implications as they relate to ME/CFS in adolescence and adulthood, as well as to highlight its impact on the caregivers and parents of patients/children with ME/CFS. In doing so, a predominant focus on the lonely, alienating, and isolating features of this illness is explored. It is important to note that though some researchers have found themes of loneliness when investigating persons with ME/CFS (e.g., [10,21–24]), the consequences of loneliness as it pertains exclusively to ME/CFS have not been studied. Additionally, while some researchers have found themes of loneliness in their research (e.g., [10]), the authors of these studies were not interested in the variable of loneliness from the outset. Given that loneliness has been observed to influence the expression of symptoms in chronic illness (e.g., having the ability to exacerbate symptoms in chronic illnesses—a topic which is later explored in further detail), we wish to attend to this omission in the literature.

Thus, through a narrative review, this paper attempts to address multiple issues. It first aims to highlight the relevant psychosocial implications of ME/CFS; secondly, an overview of what loneliness is, how it relates to chronic illness (and illness in general), and its stigmatized connotations is presented; thirdly, this paper offers suggestions as to how to cope with loneliness stemming from and enhanced by chronic illness. This is done in hopes of constructing a theoretical framework for future research that wishes to bridge the gap in the literature between ME/CFS and loneliness.

2. Method

In making the assertion that no articles have directly investigated the impact of loneliness and ME/CFS, we conducted search queries that looked for ME/CFS keywords (i.e., "ME/CFS," "CFS/ME," "chronic fatigue syndrome," and "myalgic encephalomyelitis") and paired them with loneliness-related keywords (we used the terms "loneliness," "lonely," "isolation," "isolated," "alone," "alienating," and "alienation" as a possible pair to each of the ME/CFS keywords). Our search criteria, spanning 28 possible search queries, found no results on APA PsycNet, Google Scholar, and PubMed when requiring that at least one of each keyword be in the title of an article (e.g., a search query through google scholar was typed as the following: "allintitle: ME/CFS loneliness"). Next, we redid the search queries but, instead of requiring that a loneliness-related keyword be in the title, we changed this criterion to allow these words to appear anywhere in the abstract or in the body of an article.

Over 400 articles were researched. Of this, 137 studies were appropriate for our topics and were thus utilized. These articles mainly consisted of primary studies and textbooks; however, they also included literature reviews, meta-analyses, systematic-analyses, research instruments, and annual reviews. The search criteria were filtered between the years 1967 and 2020; however, a predominant emphasis was placed on articles within the last 10 years (of the total 137 studies, 18 dated between

1967 and 1999; 20 dated between 2000 and 2005; 23 dated between 2005 and 2009; 32 dated between 2010 and 2014, and 40 dated between 2015 and 2020).

3. Results

3.1. Psychosocial Factors of ME/CFS

The presence of chronic pain can have devastating effects on one's psychosocial functioning. In fact, when it comes to chronic pain, some researchers consider the accompanying psychosocial distress to be so severe in magnitude that they advocate for a dual-diagnosis—one that includes a component on pain severity while also emphasizing the debilitating function of one's social environment [25]. As Skelly and Walker [25] indicated, such a diagnosis would allow for the healthcare field to truly acknowledge "the way pain affects people's lives, how they adjust to this, and how it affects their behavior" (p. 253).

For example, depression and anxiety are almost always comorbid in the presence of ME/CFS [26,27]—so much so that some researchers postulate that the term "comorbid" does provide an accurate picture. For instance, Maes' [28] review of the neural pathways affected by ME/CFS and depression found that these two conditions do not exist independently of each other; instead, they are manifestations of similar damaged neural pathways. Thus, rather than using the term "comorbid," Maes [28] advocates for the identification of ME/CFS and depression as "co-associated" disorders. This linguistic distinction, though small, may benefit the ME/CFS community through earlier mental health intervention, as healthcare professionals would be vigilant of depression from the outset of diagnosis.

3.1.1. How ME/CFS Affects Sufferer's Mental Health

A common reason why mental health issues are high with patients with ME/CFS is the frequent observance of kinesiophobia (i.e., a fear of movement, [29,30], which may cause withdrawal and isolation from one's social circle due to the strenuous efforts of displacing oneself. Another common observance in ME/CFS individuals, which is devastating in tandem with kinesiophobia, is the act of catastrophizing [31,32].

Catastrophizing, defined as the general tendency to assume that the worst-case scenario will happen, presents challenges for the overly cautious ME/CFS sufferer. Pessimism, a fear of movement, and an intense irrational fear of expecting the worst to occur leads these individuals to isolate in an attempt to protect themselves from potentially negative exposure. However, this approach to confine and stay away from social and recreational pursuits is counterproductive; in fact, studies investigating the implications of social deprivation have shown that it instead increases pain perception [33,34]. Thus, kinesiophobia and catastrophizing may create a sort of negative feedback loop where an individual, wishing to mitigate symptoms, stays at home to "protect" themselves, only to have significant distress and an increase in pain.

While these issues may universally impact individuals with ME/CFS, the age-onset produces additional distinct circumstances that can affect psychosocial well-being in common ways. This is also true for caregivers of family members and patients of ME/CFS. As such, we devote the following section to explore further the different variations of ME/CFS' psychosocial impairments and stressors.

3.1.2. Psychosocial Implications of ME/CFS during Adolescence

While likely underestimated due to a lack of understanding and testing for ME/CFS, the prevalence of this diagnosis amongst teenagers varies between 0.11 and 1.9% [35,36]. Youth diagnosed with a chronic illness that damages the central nervous system (as ME/CFS has been observed to do [4]) are 40% at risk of a psychosocial impairment of some sort [37]. Additionally, compared to non-ME/CFS adolescents, teenagers battling this illness have shown disturbing social and emotional development, higher rates of depression, and a host of other negative implications related to school absenteeism [35].

Regarding school absenteeism, one study that included a sample of 81 adolescents with ME/CFS found that 45% percent of participants had missed a minimum of 50% of school during the previous six months. Such significant absenteeism lends itself to missed opportunities to develop peer relationships and social competence, future work-relevant skills (e.g., resilience and persistence), and academic and language development [38]. To make matters worse, the time missed from school does not even allow for optimal therapy to be had for these adolescents. For example, a study from the University of Bristol [39] found that only 10% of adolescents report having access to a specialist, and as much 94% of students report being disbelieved when disclosing their symptoms to health care professionals and school staff members.

Prolonged absence from school can also enhance a child's feelings of isolation due to reductions in networking and participation in sports and other social events [40]. Adolescents are also especially prone to body distortion, and chronic illness can heavily disrupt body image. As Vitulano [40] states, " . . . bodily changes, and treatment requirements are nagging reminders that they are 'different and damaged' in some way" (p. 587), further impairing social involvement, acceptance, and self-esteem.

It is not surprising then that ME/CFS during adolescence is associated with a substantial loss of self. To illustrate, we would like to draw attention to Parslow et al.'s [21] study, one of the first major systematic reviews and meta-ethnographies of the ME/CFS qualitative literature during adolescence. Their findings indicated (after translating and coding 10 studies and 82 participants' quotes on feelings of the syndrome) that the disruption and loss of self, alongside pain and social disturbances, was the most frequently emergent construct in the investigations reviewed. To further illustrate this, adolescents with ME/CFS were found to place their identities on their bodies; upon learning that their newfangled bodies limited their ability to behave as they used to, self-esteem and confidence, along with a loss of self, were negatively affected [10].

A loss in self may also be in part explained by the results of Winger et al.'s [22] qualitative study that found adolescents with ME/CFS attach significance to attending school and hang-outs with friends. When deprived of these events, the adolescent questions the meaning of life. Results from this study also found overwhelming feelings of loneliness and feelings of "sadness and guilt related to being a burden on one's family" [22] (p. 2652).

Lastly, sibling rivalry was also observed to be an issue for adolescents with ME/CFS. More specifically, enviousness and abandonment are felt when siblings of ME/CFS sufferers go outside the home and live the otherwise "normal" adolescent experiences. Wiliams-Wilson's [10] wrote about the effects of the illness, as experienced by Lisa, one of the participants: "Lisa believes that her having CFS has had a detrimental effect on her relationship with her sister; they argue frequently and are jealous of each other, they do not spend a lot of time together as her sister is either at school or the stables. Lisa mentions that she and her sister are jealous of each other, which is a point raised by other participants in the study" [10] (p. 200).

3.1.3. Psychosocial Implications of ME/CFS during Adulthood

Chronic illnesses, in general, significantly affect one's psyche. As Kiliçkaya and Karakaş [41] stated, "it is known that medical and psychosocial problems in chronic diseases cause negative emotions such as anger, distress, and unhappiness and that patients who have chronic illness feel loneliness. Hospital stays, taking medication, physical and social loss of function, economic setbacks, a changing body, and uneasiness in social relationships are factors that affect loneliness" (p. 486).

According to the World Health Organization's investigation in 15 centers across the Americas, Europe, Asia, and Africa, chronic and persistent pain affects 22% of the population globally [42,43]. Of the affected population dealing with severe chronic illnesses such as ME/CFS, independence and autonomy are greatly impacted causing individuals, well into adulthood, to have an increased dependence on older siblings, parents, and/or caregivers for everyday functioning [40,44]. Subsequently, many adult sufferers are left with many unpleasant uncertainties when envisioning the future.

For instance, patient testimonials reveal that a common concern for these adults includes fears of starting a family [25] (p. 189), which may develop into severe psychological trauma.

Economic setbacks and financial strain are further distressing side-effects for adults with ME/CFS. Many sufferers of the disorder are left too debilitated to travel to their office or work full-time, leaving them to seek fewer demanding jobs (with less pay) as a result [6]. Furthermore, in countries where patients have to pay for their medical bills (e.g., the United States), ME/CFS sufferers are expected to pay significant annual expenses related to treatment [45], leaving some with the illness to constantly worry and ruminate over how to clear any financial deficits. Thus, it is no wonder why a large body of literature reflects on the overall quality of life being significantly impacted as a result of ME/CFS [46].

This well-being is also affected by the disruptions caused to the overall routine of the ME/CFS sufferers. For instance, besides work and social habits, individual habits and the ability to partake in hobbies and recreational activities are also greatly impaired. To illustrate, Schweitzer et al. [47] found through a sample of 47 ME/CFS patients that 100% of respondents reported reduced recreational activities and 70% were forced to discontinue all physically active pastimes.

3.1.4. Caring for Someone with ME/CFS

Due to the disabling nature of ME/CFS, there is often a need to heavily rely on someone to look after and help individuals dealing with this disorder [19]. The duties of caregivers and caretakers are often numerous in scope; as [48] Girgis et al. describe it, caring for people with a chronic illness includes, but is not limited to, assistance with "mobility, transportation, communication, housework, management and coordination of medical care, . . . emotional support, assisting with personal care, organizing appointments, social services, assistance with social activities, . . . and managing finances" (p. 197).

As a result, and independent of whether this role is fulfilled primarily by a loved one or by a health care professional, looking after and caring for individuals with ME/CFS can also produce significant psychosocial maladjustment for the caregivers/caretakers themselves. One study that focused on tracking family caregivers found that, on average, caregivers spend 13 hours per day assisting family members with ME/CFS, and this profoundly affected sleep, work and study time, leisure time, and mood [49]. Furthermore, this same study found that 63% of family caregivers described feeling depressed as the main stressor endured as a result of looking after someone with ME/CFS [49].

This emotional toll can also be expressed in different ways. For example, Catchpole and Garip [19] found that parents often face skepticism and disbelief by their social circles (i.e., individuals outside the immediate family, such as friends and colleagues), causing them to withdraw from these networks and experience increased isolation as a result. This behavior was rationalized to be a common coping mechanism for caregivers as their isolation effectively shielded them from the possibility of this type of criticism.

Additionally, both parental caregivers and professional caretakers reportedly endured significant mental health problems as a result of caring for ME/CFS suffers, and, even worse, this mental health burden was also related to a reduced ability to effectively care for the persons with ME/CFS [20,50].

Parents also experience role ambiguity when tending to children with ME/CFS, as the illness is said to create some confusion on how to approach the parent-child dynamic, which can lead to a strained relationship. This was reasoned to be due to the illness creating reduced opportunities to enjoy shared activities that otherwise promote bonding experiences and allow the parent and child to relate to each other [19]. Another example of this occurs when the amount of missed school starts to accumulate—since parents often homeschool their children in these instances and assume the role of teacher, the parent-child relationship may be affected due to the frustrations inherent in coursework and homeschooling [10].

Related to household tensions, Missen et al. [20], tracking 28 mothers of children with ME/CFS, also found marital tensions and problems within the marriage related to the disorder as the theme that was the "broadest and most widely discussed" (p. 5). Perhaps contributing to marital tensions, feelings of grief are also felt in ME/CFS households. Put powerfully, and somewhat pessimistically, parental caregivers "mourned the perceived loss of their child's future" [19] (p. 5).

Another source of stress related to a child's diagnosis with ME/CFS that we previously briefly alluded to, and is otherwise under-discussed in the literature, is sibling rivalry. Earlier, we explained this from the point of view of the adolescent sufferers. However, the siblings of sufferers also experience significant angst as a result of a diagnosis of ME/CFS in the household, and parents have been said to often ruminate about the impact this has on the siblings of the child sufferer [20]. For example, as one parent testimonial revealed in Missen et al.'s [20] study, "The main thing I worry about is [little sister] because she can't now do the things that [child with ME/CFS] used to go out and do ... unless I leave [child with ME/CFS] at home. So, I feel guilt about leaving [child with ME/CFS] and going out having a good time with [little sister]" (pp. 509–510). Additionally, the siblings themselves experience a wide range of stressors. Houtzager et al. [51] observed that the siblings of chronically ill children sometimes display even more emotional and behavioral turbulence than parents do during the initial adjustment period (i.e., when a chronic illness diagnosis is first given), as they find themselves having to rapidly cope with losses of attention and companionship from both the parents and the affected sibling.

In some situations, parents, after making sacrifices (e.g., reducing work hours and time spent with friends) can also develop forms of anxiety as a result of long-term homebound treatment administered to their child with ME/CFS. For example, Williams-Wilson [10] recounted, "one woman told me, outside of an interview, that she had been confined to home for such a long time that she actually felt slightly panicky on the rare occasions she visited a public place such as a supermarket, feeling overwhelmed by the number of people and amount of noise, she described it as having become institutionalised." (pp. 252–253).

3.2. Loneliness

While the aforementioned psychosocial factors are numerous and greatly distressing, another equally troubling psychosocial impairment that individuals with ME/CFS must deal with exists—loneliness. Having so far provided a brief introduction to the ways in which this disorder may manifest itself detrimentally, we now do the same with loneliness by providing a brief background as to what it is, how it is expressed, and the stigma which is associated with it, before finally providing information on how it can affect chronic illness.

Loneliness, unlike solitude, which will soon be described, involves excruciating physical and mental suffering. Interestingly, we can find that the first thing that the biblical God named was loneliness, which is found to be associated with numerous somatic, psychosomatic, and emotional phenomena. Loneliness can be a reactive experience, that is aroused in response to a significant life change or loss, or it could be an essential experience which stems from one's infancy and is intertwined in the individual's personality [52]. Apparently, it was found that loneliness may have a significant or even profound impact on the brain and can affect reasoning, memory, hormone homeostasis, blood glucose levels, and one's manner of addressing of physical and mental stresses and illnesses [53].

While various theories regarding loneliness have been advanced, several characteristics are unmistakably part of that experience: while in solitude, we choose to be alone in order to do what can be done only alone, e.g., reflecting, creating, sculpting, writing, taking a walk in the woods, or communing with nature. Loneliness, in contrast, is painful, unwanted, and difficult to tolerate. It thus motivates humans to seek meaning and connection. If we explore it from an evolutionary perspective, we can notice the manner in which animals survive and thrive. They can do so only when they are part of the herd, for the deer who lags behind will become lunch for the waiting lions. Thus, like physical pain, loneliness has an important survival function even though it is unpleasant (see also [54]).

Loneliness is an integral part of being human and is experienced in order to encourage us to connect and remain part of the community. Loneliness is an experience that includes cognitions, emotions, and behaviors that are mostly negative, turbulent, and unpleasant [55,56]. Loneliness is a universal experience; as a uniquely subjective experience, it results from a combination of the individual's personality, social changes, and one's history. That history includes, of course, the various experiences and illnesses with which one may have been afflicted [57].

We do not just require others for our survival and growth; we also particularly need the presence of those who support us, whom we trust, and with whom we can interact, work together, and prosper [54]. Thus, the mere physical presence of others is insufficient.

As humans, we need to feel connected to significant others. In general, the prevailing view is that being alone and perceiving oneself as unloved and uncared for will result in loneliness [58]. Research is heightening our awareness that in the Western Hemisphere, today's fast-paced and constantly changing world where virtual reality can be seen as being on the brink of replacing the real one, people have little time and no energy to invest the effort required for establishing a connection with anyone beyond the narrow frame of their own hurried lives, living in and conforming to a culture that rewards nothing but the individual acquisition of power and money [59]. Cacioppo et al. [58] observed that "people are increasingly connected digitally, but the prevalence of loneliness (perceived social isolation) also appears to be rising. From a prevalence estimated to be 11–17% in the 1970s … loneliness has increased to over 40% in middle-aged and older adults … Over the past 40 years, loneliness has also become more widespread overseas" (p. 238) and is linked to poor physical and mental health outcomes.

At the time of this writing, COVID-19 has affected the entire globe, significantly changing the way we live. Dunham [60] indicated that loneliness could negatively affect the health of the brain as well as the immune system. This has also seemed to have been exacerbated by the confinement and lockdown. For instance, Fallon et al. [61] found through studying a sample of 431 individuals with chronic pain that they reported an increase in their pain perception and severity during the pandemic, which makes reviewing the effects of loneliness on those afflicted with ME/CFS even more poignant at this time.

3.2.1. The Stigma of Loneliness

Most people are reluctant to admit, even to themselves, that they are lonely. Though we may geographically alone, feel unimportant, and unloved, people seem ashamed to acknowledge, let alone admit, that they are lonely. That is a consequence of the Western culture's dictate that loneliness is a sign of weakness that should not affect "normal," "healthy," and "strong" people [62]. This denial does not eliminate loneliness; it simply conceals it from the world while we still hurt and feel alienated at times [63]. The increased use of drugs and alcohol, the purchasing and consumption of pornographic material, the very many calls to distress hotlines, and the rise in the number of suicides were found by research to be a consequence of the pain of loneliness that is not talked about and is not addressed. We can also see the footprints of loneliness in the increased number of divorces and religious fads. There is, clearly, a stigma to being lonely [64].

3.2.2. Loneliness and Illness

Boehm [65] pointed to the connection of our emotional well-being, our thoughts, emotions, and behaviors to our well-being (see also [66]):

> *"Individuals who are satisfied with their lives and who experience frequent positive emotions—that is, individuals with high levels of subjective well-being … —not only feel good but may also have reduced risk for developing coronary heart disease … subjective well-being may buffer against the harmful health consequences of stress and exert direct influence on bodily systems or may motivate healthy behavior"* (p. 1).

Modern medical science has been obsessed with death, which is a clear "enemy" in medical eyes, and, thus, medical research aims to eradicate the diseases that cause it [67]. Primary-care physicians consequently focus on the care of the patient and many times are not fully aware of the person who has the disease.

Illness is stressful and often frightening—even more to those who are physically disabled, immobile, or are close to the end of their journey on earth. Illness, in general, is a major stressor in one's life [68]. Fatigue, pain, or, in severe cases, immobility results in the body being in a continuous state of stress. Such a situation leads to hospitalization and may cause a wide range of short-term and long-term negative effects experienced by the patient [69]. In general, physical suffering and distress plunge the body into a state of continuous stress that may be exacerbated by the patients' negative psychological states. Such negative psychological states compound the patient's stress and may result in a perception that one's life is under threat, further suggesting that the illness is an uncontrollable or even unpredictable part of one's life [70].

Stress, including separation, loss, and feelings of hopelessness are known to compromise the immune system and can reduce the body's efficacy in fighting illness. Health deterioration is, thus, most probable in persons with already compromised immune functioning [71]. Loneliness, which is associated with a wide range of health problems, is linked to heightened morbidity and mortality. A positive correlation was found between social isolation and mortality [72]; a report in Australia by The National Heart Foundation reported strong evidence that social isolation contributes to coronary heart disease [73]. Loneliness has also been implicated in a lower level of quality of life [74].

The chronically lonely display negative mood, tend to withdraw socially, lack trust in others, and often are dissatisfied with their relationships [75]. Those with high loneliness tend to have poorer T-lymphocyte responses and show potentially harmful changes in natural killer cell activity [76]. Natural killer cells have a role in some cancers and inflammatory responses that have been observed in vascular disease [77].

3.2.3. Illness Conceptualization

This section provides a brief overview of how people conceptualize illness as it may clarify our understanding of why loneliness is such an influential experience in the progression of illness. As Leventhal et al. [78] found, the following components relate to illness conceptualization.

The disease's identity and label significantly influence patient behavior. For example, chest pain may be labeled "heartburn" and that will cause a very different behavioral reaction than the one labeled "heart attack." Similarly, when the illness indicates a minor physical problem, we can expect that less emotional arousal will be experienced than if it is of a more serious nature.

Diagnosis may not always concur with timelines. For instance, people diagnosed with hypertension may view it as acute (although it is a chronic condition), and that has a direct effect on how much they adhere to treatment because it significantly differs from the way they may address chronic illness.

After diagnosis, we search for the cause of the problem. The cause may intimate that we need to seek treatment for it and, moreover, may influence the degree of our compliance with instructions given by a healthcare professional. For instance, pain in our leg that resulted from a fall would generate a completely different reaction than if it was found to indicate bone cancer.

The consequences of the disease form the next component. For instance, cancer may be viewed as a death sentence and result in the patient feeling hopeless and consequently failing to seek active and lifesaving treatment.

The degree of controlling the disease is the final component. If patients perceive that the situation is beyond hope, they may not seek treatment. However, if they believe that the treatment can help or even cure them, they will actively and even aggressively seek to achieve healing.

Research has also shown that stronger immune systems are positively associated with stronger social support systems [79,80], whereas people who have fewer social ties are more susceptible to illnesses [71]. Those with a solid social support network commonly cope better with stress and chronic pain [81], have better health, and have lower rates of mortality [82]. The nature of people's connection to the community, and their perceptions of those relationships, significantly affect their physical and mental health [58].

What is even more staggering is that social isolation and loneliness rivaled cigarette smoking, high blood pressure, obesity, and a sedentary lifestyle as related to illnesses. Research has found a positive correlation between social support and health. Conversely, the opposite is also true. That is, those with the fewest social ties are up to four times more likely to die from illness and disease than those who had a good support system [54,81].

Segrin et al. [83] suggested that the lonely are less prone to behave in a health-promoting way, partly since they are not supported by others to adhere to a healthier lifestyle, which end up further increasing their chances to suffer from health problems [84]. Stress lowers the efficiency of the immune system, and loneliness can be a major stressor and contribute to ill health [85].

In sum, social relations seem able to protect us against the ill effects of stress, and those who lack social support end up with a greater allostatic load [83]. Loneliness may not only bring about illness, but it is known to corrupt the recuperative process [83,85].

3.2.4. Loneliness, Chronic Illness and Pain

Chronic pain is quite pervasive and is estimated to affect 36%, or about 120 million individuals, in the United States [43,86]. Morrissey [87], focusing on the illness and suffering of older adults, highlighted the negative effect chronic pain has on the quality of life of all sufferers and highlighting how this causes a focus in attention on the losses that come as a result of chronic illness or pain. Pain also affects our psychological well-being by making us focus on persistent thoughts and irrational beliefs (such as the earlier mentioned kinesiophobia) related to individual reactions to the experience of pain or illness [88,89]. Recent research has indicated that expectations, mood, and behavioral factors also affect chronic pain; this by itself can significantly affect a person's close relationships and social life [90]. Social isolation is also a major issue confronting chronic pain patients (Newton et al., 2013 [91]) and, thus, a strong association between chronic illness and pain and loneliness, as well as other emotions, has been found (e.g., [92,93])

Kool and Geenen [94] found, by comparing patients with fibromyalgia and rheumatic diseases, that patients with fibromyalgia were lonelier than those afflicted with rheumatoid arthritis. The same was found in a study on those with sickle cell disease [95]. High levels of social withdrawal and isolation were found in patients with neuropathic pain as they reported much social withdrawal and consequent isolation, and this, naturally, had an effect on both the patients and their spouses [96]. Loneliness was a major risk factor for the development of fatigue and depression in those patients [97], and social support and involvement have been found to be positively related to coping with pain [98,99].

Intrafamilial relationships can be a major source of personal resourcefulness for patients with chronic illnesses or pain. Family constellation can significantly impact the trajectory of chronic illness [100]. Research has repeatedly demonstrated a robust directional effect of loneliness on physical health across the lifespan [101].

Chronic illness may cause a loss of friends or family members and may thus intensify the loneliness that the ill person already experiences [102]. Loneliness is reported by patients who are forced to focus on their illness while the rest of the world, their family, and friends continue with their daily living [103]. "Individuals with high versus low chronic interpersonal stress were especially vulnerable to the negative effects of episodes of loneliness, showing greater loneliness-induced increases in cortisol … Beyond its physiological effects, one day's increase in loneliness has been associated with increases in the next day's symptoms, including exhaustion and fatigue, over and above the influence of the prior day's depressed affect and sleep duration" [104] (pp. 929–930). Further, Wolf and Davis [100] asserted that physical pain and perceived social exclusion (which we term loneliness) activate brain circuits in the central nervous system where there may be a "pain signature" that is interestingly activated by either a physical or social stimuli [105].

ME/CFS sufferers experience profound fatigue, exhaustion, the loss of muscle power, pain, joint tenderness, and cognitive dysfunction. In addition, these stressful symptoms cause headaches, sore throats, a loss of concentration, and short-term memory loss [106]. It is quite clear that being riddled with such symptoms for a lengthy period of time would make socializing, interpersonal connection, and remaining connected to others problematic and would most often require a termination of those relationships. Everyday activities become burdensome for people with ME/CFS. They often lose the ability to keep up with a conversation since they experience extreme trouble focusing on what the other person is saying and, moreover, processing the meaning of the words. ME/CFS patients have trouble not only processing information but also retaining it. Memory loss, particularly short-term memory loss, is another common cognitive complication of ME/CFS. Sufferers forget people's names, and that makes relating to them that much more difficult [107].

3.3. Coping with ME/CFS Induced Loneliness

As Biordi [108] pointed out, social isolation is a major aspect of chronic illness due to its significant impact on the patient and his or her support network. A variety of interventions, from high-touch and no-technology to low-touch and high technology use, have been suggested and tried. Here, we review a number of the main ones.

3.3.1. The Power of Empathy

Bharadvaj [109] rightfully observed that any chronic disabling condition (and especially ME/CFS) can make a person feel challenged, anxious, or even hopeless; all of which are closely related to energy levels and healthy functioning of the immune and neuroendocrine systems. Low energy levels may cause the suppression of the immune system and the imbalance of the hormonal pathways. Consequently, Bharadvaj [19] observed a strong association between ME/CFS and mental health issues. He concluded by suggesting that "perhaps the best support a healthcare provider can offer is empathy and understanding to an individual suffering from ME/CFS. From a place of trust and rapport between doctor and patient, communication can begin about diagnostic testing, therapeutic options, and follow-up care. Just as important is the individual's desire and hope in achieving wellness through lifestyle changes, psychological support, natural medicines, and anything else needed for the evolution in their health" (p. 92).

3.3.2. Keeping in Contact with the Outside World

Feeling isolated can be a very common problem for people with ME/CFS. Being socially connected is a basic human need. Being ill may not allow one to get together with friends, but, as suggested by Campling and Sharpe [26], one can socialize in a different way. For instance, instead of going out, you may wish to invite some friends to your home for a meal you ordered or prepared yourself. When one is struggling with illness, some friends drop away as they cannot cope with that person's illness. Making new social contacts and replacing those friends can address that problem, and leave the patient connected with the outside community.

3.3.3. Peer Counseling

Whether it is informal in structure or more formal, initiated, and supported by the professional healthcare worker, Riegel and Carlson [110] suggested that peer counseling can be quite effective and helpful. For instance, a telephone hotline can be set up at a clinic that helped peers befriend each other and enable them to provide emotional support and active listening over the phone. A peer counselor volunteer may also be able to visit clients at home or in institutions and offer more social contact. Peers can also provide a wealth of information on ways to connect with resources such as assisted transportation, volunteers, friendly or financial aid [111]. This may also take form in the presence of support groups.

3.3.4. Support Groups

A multitude of support and self-help groups exist for people in the general population, as well for those struggling with a debilitating illness. Research has indicated that support groups are very effective in meeting patients' social needs by allowing them to exchange information, offer mutual support, learn of ways to cope with what they are going through, and ease their physical and emotional pain [108,112,113]. The internet can direct the person to sites related to the chronic illness that he is battling, and some associations have both national and local support groups according to not only to the patient's illness but also their locations and ability to be mobile.

In addition, there may be ME/CFS support groups that are within reach of the patient where someone can get support and share their personalized experience on what it is to live with ME/CFS to the differently abled crowd. As we learn now, socializing does not have to happen face to face. Pen pals, friends, and telephone contacts can be very uplifting, as can emails, Facebook messages, etc. [114]. Brigden et al. [115] studied how adolescents coped with ME/CFS and noted that since the social connection was so important to them, they relied on the internet to connect to others and created a "community" of those suffering from the same illness, easing their isolation. They stated that in general, "the online world was less demanding and more flexible than offline relationships, especially in the context of a disabling and fluctuating illness" (p. 4) Additionally, one of their participants remarked that "It's just the support knowing that at any time during the day if I'm having a bad day I can literally go on and I know immediately I'll have support" (p. 4). Thus, it is suggested that whoever can gain access to the internet (and not just adolescents) may benefit similarly.

Some support groups may also be beneficial by offering physical activity. For example, Broadbent et al. [116] studied the use of aquatic exercise classes with ME/CFS patients for five weeks, offered in a biweekly manner. Besides the aerobic benefits, the results indicated that participants felt reduced social isolation and felt supported by their ME/CFS peers and exercise instructors, resulting in a reduction of pain, fatigue, and anxiety levels during post-treatment interviews [116].

3.3.5. Solitude

A person's choice to seek solitude is healthy. When it is chosen voluntarily, solitude is used for reflecting, centering, feeling spiritually connected, and finding inner peace and strength [117]. Solitude allows us to take a respite from everyday stresses, stimuli, and demands, and it also affords a better understanding of who we are, what do we want, and possibly how to get it [52]. ME/CFS sufferers may therefore benefit from a perspective shift on the extra time they have to themselves when being home-bound or otherwise away from their regular duties. Emphasizing personal growth may make these long bouts of aloneness more tolerable.

3.3.6. A Cognitive-Behavioral Approach to Illness

Campling and Sharpe [26] opined that ME/CFS patients who take control of their situation and actively attempt to help themselves are better able to overcome the emotional toll of their illness. On the other hand, those who believe that their symptoms are very severe, are caused by factors outside of themselves, and that they are "helpless" seem to be associated with greater disability. People relate to and are influenced by the people about them. While the research relating to this approach to coping is limited, we know that these social factors influence the degree to which an ill person struggles with ME/CFS.

Campling and Sharpe [26] indicated that pain which ME/CFS patients suffer from is made worse by muscle tension and, at times, is even caused by it. Consequently, they suggested that learning deep muscle relaxation may help reduce tension and ease the pain. They added that changing the way patients think and adopting a positive outlook will affect not only their mood and behavior but also their physiological state. They observed that "persistently inaccurate thinking can lead to poor coping and to bad effects on emotion, behavior, and bodily state. For instance, if a person constantly worries that things will go wrong, they will be chronically anxious, tend to avoid doing things, and be in a physiological state of tension and arousal" (p. 166). That will, in turn, affect the illness trajectory and their chances to recover. That may also influence them when they contemplate seeking treatment and their choice of preferred intervention, and it may further affect their nervous and immune systems.

To wit, we would like to re-emphasize the study by Fallon et al., [61] that was conducted in the midst of the COVID-19 lockdown in the UK (mid-April to early May 2020), where it was found that people with chronic pain reported self-perceived increases in levels of pain severity compared to the period before lockdown. The lockdown affected them more adversely than it affected the general population. They also reported greater increases in anxiety and depression, increased loneliness, and reduced levels of physical exercise. Evidently, the way the mind perceives one's illness is a key contributor to that individual's phenomenological experiences.

As such, Campling and Sharpe [26]) recommend cognitive behavioral therapy (CBT) to examine one's thoughts, enhance rational thinking, and encourage positive and proactive cognitions which will usher similarly proactive behaviors. "Does CBT work for people with ME/CFS? [they asked]. Yes, it does seem to help. It is not a cure, but research including a number of clinical trials in different centers has shown that about two-thirds of patients who take part in such a program are able to do more and feel better" (p. 169).

3.3.7. Religion

Research suggests that people rely on religion, spirituality, and faith to cope with illness and loneliness [118]. A study by Han and Richardson [119] identified spirituality as a coping strategy used to lessen loneliness in their sample of homebound elders. Religion was also found by Rokach [52] to assist in coping with loneliness in his research of both ill and non-ill samples. Being part of a religious community, as well as relying on one's faith that a higher power is overlooking one's life and suffering, was shown to ease suffering and help cope with loneliness.

3.3.8. Spirituality

Spirituality is known to be a source of strength for many people. As we clarified previously, spirituality and religiosity are not the same. Spirituality can be experienced regardless of a person's religious beliefs. Spirituality can promote the client's feelings of control, self-esteem, meaning, and purpose in life. Nurses, with compassionate listening and sharing, when accompanying the ill can teach and enhance spirituality in the patient. Among the practices that may enhance spirituality are meditation, reading, yoga, tai chi, pet therapy, journaling, listening to relaxing and pleasant music, and repeating a mantra [117,120].

Spirituality achieved through mindfulness and meditation-based practices may also prove beneficial. For example, Boellingus et al. [121] found that mindfulness-based interventions and loving-kindness meditation can produce increases in self-compassion. In turn, self-compassion was seen to correlate to better day-to-day functioning in chronically ill patients. More specifically, its presence decreased pain perception and depression symptoms and increased work and social adjustment [122].

While spirituality can both mean and be achieved in different ways from individual to individual (e.g., meditation, mindfulness, and acts of gratitude), the main core aiding agent from these acts, besides the ability to relax, is the ensuing feeling of control. As Friedberg [6] stated, "studies on coping in ME/CFS and FM have found that a sense of control over symptoms consistently predicts better functioning, regardless of limitations or disabilities." (p. 38) (see [123,124]).

3.3.9. Health Care Providers Therapeutic Use of Self

In an article directed at healthcare professionals, Holley [117] suggested that nurses can be a major source of social support in their patients' lives. Nurses are commonly perceived as trustworthy, compassionate, and knowledgeable, and they may even serve as confidants in many cases since they may be proficient in active listening. Even if the nurse cannot increase the patient's social circle, he or she can provide caring, genuineness, and high-quality contact. Biordi [108] highlighted the authentic intimacy a patient and nurse can share and pointed out that their relationship may be a powerful one. Validating a patient's importance as a human being, Biordi [108] suggested, can be as simple as stopping, making eye contact, and gently squeezing his or her hand.

As a way of summarizing the literature discussed thus far, Figure 1 provides a graphical recapitulation of the way ME/CFS and loneliness may interact through the different coping mechanisms and through the previously mentioned exacerbating effects.

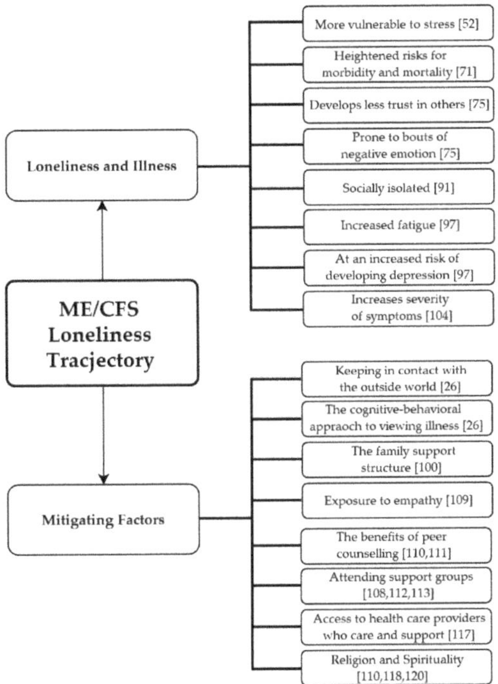

Figure 1. The interaction between ME/CFS and loneliness. [26,52,71,75,91,97,100,104,109–113,117,118,120].

4. Discussion

4.1. Summary

To reiterate, chronic illnesses that affect the central nervous system such as ME/CFS does produce psychosocial impairment in 40% of individuals [37]. Though this impairment varies from individual to individual, the age at which sufferers deal with the illness dictates broader and commonly reported themes. For example, adolescents plagued with this disorder miss significant amounts of schooling, which impedes social functioning and future career development skills and can lead to a loss of identity, all of which make young ME/CFS individuals question the meaning of life.

Additionally, a family who receives a diagnosis of ME/CFS for one of its members may experience disruptions of the family dynamic including sibling jealousies and rivalries, guilt, and, strained parent-child relationships resulting from parents and children needing to step into differing roles when assisting (e.g., a parent taking on the role of teacher when homeschooling or a child taking on the role of a parental figure when advising recommendations on what not to do). ME/CFS suffers who are single with no guardians and no dependents also have their own shares of concerns that they must deal with. This includes rumination and stress related to the financial impact of the disorder (e.g., the loss of work and the cost of treatment), fear about being unable to live a normal life and start a family, and decreased autonomy and an increase in reliance on a caregiver. Furthermore, the stigmatization of this illness results in dismissiveness and skepticism from peers, from authority figures (e.g., teachers and employers), and sometimes even from family members.

Concerning loneliness, the main focus of this article, we have provided a brief explanation of what loneliness is, how it may result in distress, unhealthy coping behaviors, and how it relates to chronic illness. In doing so, we have highlighted Leventhal et al.'s [78] study, which showed how one's conceptualization of illness—e.g., the labeling of the illness and the perception of it, control over how one feels about it, expected consequences, and level of hopefulness—can greatly aid or vastly worsen one's experience with their illness. Several coping strategies that caregivers and sufferers of ME/CFS may benefit from were also mentioned, including empathetic behaviors, the attempt to stay in touch with the outside world, peer counseling, support groups, solitude, and the cognitive-behavior approach to how to think about the illness. Additionally, we emphasized the important role healthcare professionals can have with their patients and spoke about the power of spirituality and religiousness as a buffer to ME/CFS-induced loneliness.

4.2. Future Directions

As we previously mentioned at the beginning of this article, there is a lack of investigation surrounding loneliness and how it affects individuals with ME/CFS. As such, we would like to raise some questions that would be of interest and offer insights into conducting research studies with this population.

Questions that glaringly present themselves are: can adequately managed and prolonged exposure to social support networks mitigate symptoms of pain in ME/CFS patients? Additionally, would being in a support group amongst other ME/CFS individuals offer the same buffers to loneliness non-ME/CFS groups? Might these effects be observable via online support groups (e.g., Zoom, Skype, etc.) and would they produce similar outcomes as in-person groups?

The length of illness and how it relates to loneliness are also of interest. For example, since ME/CFS symptoms are present for a minimum of six months and up to, in some cases, more than two years [21], a longitudinal study that tracks loneliness and how one perceives their diagnosis of ME/CFS (including pain, irritability, feelings of control) would be of great interest and could afford insight on whether or not lesser amounts of loneliness translate to a shorter length of pronounced distress faced by the illness. A specific look at personality traits, such as extraversion and introversion, and questionnaires related to perceptions of joy derived from outings, past job experiences/hobby enjoyment (e.g., Quality of Life Enjoyment and Satisfaction Questionnaire [125], The Minnesota Satisfaction Questionnaire [126], etc.) should also be noted and looked at for further perspectives on illness perception. For example, Davey et al. [122] found that individuals who ranked higher on openness to experience were more accepting of their own inner experiences dealing with chronic illness, resulting in significantly lower pain perception.

Additionally, while difficult, it would be fruitful to sample a comprehensive sample that includes many different cultures and/or backgrounds. Since different cultures are affected and tend to view loneliness differently [127–129], it would be interesting to observe if and how these cultural differences fare with respect to coping with ME/CFS. Answers to these questions would undoubtedly result in better treatment protocols and healthcare expectations.

5. Conclusions

Considering that loneliness, its accompanying stigma, and illness conceptualization have a devastating impact in exacerbating chronic illness, we deem the current lack of investigation between loneliness and ME/CFS a major omission in the ME/CFS literature. In closing, we wish to end this article on a quote from Williams-Wilson [10], a researcher who suffers from ME/CFS herself and who investigated the qualitative experiences of adolescents with ME/CFS; drawing from one of the emergent themes of her study, and her personal experiences, she remarked, "finding other people in the same situation as you, with the same struggles and daily trials makes one feel less alone and different from the rest of the world; it provides a sense of affinity and justification and helps alleviate feelings of isolation and loneliness." (p. 317). It is thus a healthcare imperative that we take the necessary steps to study and demystify the illness' alienating and isolating aspects so that those suffering with ME/CFS can feel empowered and compassion from the medical community when dealing with the disorder. Future research may explore the assistance that others, family members, friends, and the community at large can offer those who are struggling with ME/CFS loneliness-related stress and emotional pain.

Author Contributions: Conceptualization, A.R. and S.B.; methodology, S.B. and A.R.; investigation A.R. and S.B.; resources, S.B. and A.R.; data curation, S.B. and A.R.; writing—original draft preparation, A.R. and S.B.; writing—review and editing, S.B. and A.R.; All authors have read and agreed to the published version of the manuscript.

Funding: This research received no external funding.

Conflicts of Interest: The authors declare no conflict of interest.

References

1. Nacul, L.; O'Boyle, S.; Palla, L.; Nacul, F.E.; Mudie, K.; Kingdon, C.C.; Cliff, J.M.; Clark, T.G.; Dockrell, H.M.; Lacerda, E.M. How Myalgic Encephalomyelitis/Chronic Fatigue Syndrome (ME/CFS) Progresses: The Natural History of ME/CFS. *Front. Neurol.* **2020**, *11*, 826. [CrossRef] [PubMed]

2. Costa, D.C.; Tannock, C.; Brostoff, J. Brainstem perfusion is impaired in chronic fatigue syndrome. *QJM Int. J. Med.* **1995**, *88*, 767–773. [CrossRef]

3. Bradley, L.A.; McKendree-Smith, N.L.; Alberts, K.R.; Alarcón, G.S.; Mountz, J.M.; Deutsch, G. Use of neuroimaging to understand abnormal pain sensitivity in fibromyalgia. *Curr. Rheumatol. Rep.* **2000**, *2*, 141–148. [CrossRef] [PubMed]

4. Nijs, J.; Meeus, M.; Van Oosterwijck, J.; Ickmans, K.; Moorkens, G.; Hans, G.; De Clerck, L.S. In the mind or in the brain? Scientific evidence for central sensitisation in chronic fatigue syndrome. *Eur. J. Clin. Investig.* **2011**, *42*, 203–212. [CrossRef]

5. Feltham, C.; Hanley, T.; Winter, L.A. (Eds.) *The SAGE Handbook of Counselling and Psychotherapy*; Sage: London, UK, 2017.

6. Friedberg, F.; Bateman, L.; Bested, A.C.; Davenport, T.; Friedman, K.J.; Gurwitt, A.; Jason, L.A.; Lapp, C.W.; Stevens, S.R.; Underhill, R.A.; et al. *ME/CFS: A Primer for Clinical Practitioners*; International Association for Chronic Fatigue Syndrome/Myalgic Encephalomyelitis: Chicago, IL, USA, 2012.

7. Neu, D.; Mairesse, O.; Montaña, X.; Gilson, M.; Corazza, F.; Lefèvre, N.; Linkowski, P.; Le Bon, O.; Verbanck, P. Dimensions of pure chronic fatigue: Psychophysical, cognitive and biological correlates in the chronic fatigue syndrome. *Graefe's Arch. Clin. Exp. Ophthalmol.* **2014**, *114*, 1841–1851. [CrossRef]

8. Friedman, K.J.; Bateman, L.; Bested, A.; Nahle, Z. Editorial: Advances in ME/CFS Research and Clinical Care. *Front. Pediatr.* **2019**, *7*, 370. [CrossRef]

9. McEvedy, C.P.; Beard, A.W. Royal Free Epidemic of 1955: A Reconsideration. *BMJ* **1970**, *1*, 7–11. [CrossRef]

10. Williams-Wilson, M. "I Had to Give Up So, So Much": A Narrative Study to Investigate the Impact of Chronic Fatigue Syndrome (CFS) on the Lives of Young People. Ph.D. Thesis, Bournemouth University, Poole, UK, 2009.

11. Canadian Institutes of Health Research. Government of Canada Invests $1.4M in Biomedical Research to Improve the Quality of Life of People Living with Myalgic Encephalomyelitis. 2019. Available online: https://www.canada.ca/en/institutes-health-research/news/2019/08/government-of-canada-invests-14m-in-biomedical-research-to-improve-the-quality-of-life-of-people-living-with-myalgic-encephalomyelitis.html (accessed on 14 July 2020).

12. Committee on the Science of Changing Behavioral Health Social Norms; Division of Behavioral and Social Sciences and Education. *Ending Discrimination Against People with Mental and Substance Use Disorders*; The National Academies Press: Washington, DC, USA, 2016.

13. Jason, L.A.; Sunnquist, M. The Development of the DePaul Symptom Questionnaire: Original, Expanded, Brief, and Pediatric Versions. *Front. Pediatr.* **2018**, *6*, 330. [CrossRef]
14. Chu, L.; Valencia, I.J.; Garvert, D.W.; Montoya, J.G. Onset Patterns and Course of Myalgic Encephalomyelitis/Chronic Fatigue Syndrome. *Front. Pediatr.* **2019**, *7*, 12. [CrossRef]
15. Kerr, J.R. Epstein-Barr Virus Induced Gene-2 Upregulation Identifies a Particular Subtype of Chronic Fatigue Syndrome/Myalgic Encephalomyelitis. *Front. Pediatr.* **2019**, *7*, 59. [CrossRef]
16. Proal, A.D.; Marshall, T. Myalgic Encephalomyelitis/Chronic Fatigue Syndrome in the Era of the Human Microbiome: Persistent Pathogens Drive Chronic Symptoms by Interfering with Host Metabolism, Gene Expression, and Immunity. *Front. Pediatr.* **2018**, *6*, 373. [CrossRef] [PubMed]
17. Gandevia, S.C. Spinal and Supraspinal Factors in Human Muscle Fatigue. *Physiol. Rev.* **2001**, *81*, 1725–1789. [CrossRef] [PubMed]
18. The Terminology of ME & CFS. Invest in ME Research. 2005. Available online: http://www.investinme.org/Article%20010-Encephalopathy%20Hooper.shtml (accessed on 14 July 2020).
19. Catchpole, S.; Garip, G. Acceptance and identity change: An interpretative phenomenological analysis of carers' experiences in myalgic encephalopathy/chronic fatigue syndrome. *J. Health Psychol.* **2019**. [CrossRef] [PubMed]
20. Missen, A.; Hollingworth, W.; Eaton, N.; Crawley, E. The financial and psychological impacts on mothers of children with chronic fatigue syndrome (CFS/ME). *Child Care Health Dev.* **2011**, *38*, 505–512. [CrossRef]
21. Parslow, R.M.; Harris, S.; Broughton, J.; Alattas, A.; Crawley, E.; Haywood, K.; Shaw, A. Children's experiences of chronic fatigue syndrome/myalgic encephalomyelitis (CFS/ME): A systematic review and meta-ethnography of qualitative studies. *BMJ Open* **2017**, *7*, e01263. [CrossRef]
22. Winger, A.; Ekstedt, M.; Wyller, V.B.; Helseth, S. 'Sometimes it feels as if the world goes on without me': Adolescents' experiences of living with chronic fatigue syndrome. *J. Clin. Nurs.* **2013**, *23*, 2649–2657. [CrossRef]
23. Åsbring, P.; Närvänen, A.-L. Women's Experiences of Stigma in Relation to Chronic Fatigue Syndrome and Fibromyalgia. *Qual. Health Res.* **2002**, *12*, 148–160. [CrossRef]
24. Fisher, H.; Crawley, E. Why do young people with CFS/ME feel anxious? A qualitative study. *Clin. Child Psychol. Psychiatry* **2012**, *18*, 556–573. [CrossRef]
25. Skelly, M.; Walker, H.; Carson, R.; Schoen, D. *Alternative Treatments for Fibromyalgia and Chronic Fatigue Syndrome*; Hunter House Publishers: Nashville, TN, USA, 2006.
26. Campling, F.; Sharpe, M. *Chronic Fatigue Syndrome (ME/CFS)*; Oxford University Press: Oxford, UK, 2000.
27. Morriss, R.K.; Ahmed, M.; Wearden, A.J.; Mullis, R.; Strickland, P.; Appleby, L.; Campbell, I.T.; Pearson, D. The role of depression in pain, psychophysiological syndromes and medically unexplained symptoms associated with chronic fatigue syndrome. *J. Affect. Disord.* **1999**, *55*, 143–148. [CrossRef]
28. Maes, M. An intriguing and hitherto unexplained co-occurrence: Depression and chronic fatigue syndrome are manifestations of shared inflammatory, oxidative and nitrosative (IO&NS) pathways. *Prog. Neuro-Psychopharmacol. Biol. Psychiatry* **2011**, *35*, 784–794. [CrossRef]
29. Nijs, J.; De Meirleir, K.; Duquet, W. Kinesiophobia in chronic fatigue syndrome: Assessment and associations with disability. *Arch. Phys. Med. Rehabil.* **2004**, *85*, 1586–1592. [CrossRef] [PubMed]
30. Nijs, J.; Vanherberghen, K.; Duquet, W.; De Meirleir, K. Chronic Fatigue Syndrome: Lack of Association Between Pain-Related Fear of Movement and Exercise Capacity and Disability. *Phys. Ther.* **2004**, *84*, 696–705. [CrossRef] [PubMed]
31. Meeus, M.; Nijs, J. Central sensitization: A biopsychosocial explanation for chronic widespread pain in patients with fibromyalgia and chronic fatigue syndrome. *Clin. Rheumatol.* **2006**, *26*, 465–473. [CrossRef] [PubMed]
32. Petrie, K.; Moss-Morris, R.; Weinman, J. The impact of catastrophic beliefs on functioning in chronic fatigue syndrome. *J. Psychosom. Res.* **1995**, *39*, 31–37. [CrossRef]
33. Karayannis, N.V.; Baumann, I.; Sturgeon, J.A.; Melloh, M.; Mackey, S.C. The Impact of Social Isolation on Pain Interference: A Longitudinal Study. *Ann. Behav. Med.* **2018**, *53*, 65–74. [CrossRef]
34. Zhang, M.; Zhang, Y.; Kong, Y. Interaction between social pain and physical pain. *Brain Sci. Adv.* **2019**, *5*, 265–273. [CrossRef]
35. Collin, S.M.; Norris, T.; Nuevo, R.; Tilling, K.; Joinson, C.; Sterne, J.A.C.; Crawley, E. Chronic Fatigue Syndrome at Age 16 Years. *Pediatrics* **2016**, *137*, e20153434. [CrossRef]

36. Nijhof, S.L.; Maijer, K.; Bleijenberg, G.; Uiterwaal, C.S.P.M.; Kimpen, J.L.L.; Van De Putte, E.M. Adolescent Chronic Fatigue Syndrome: Prevalence, Incidence, and Morbidity. *Pediatrics* **2011**, *127*, e1169–e1175. [CrossRef]

37. Northam, E.A. Psychosocial impact of chronic illness in children. *J. Paediatr. Child Health* **1997**, *33*, 369–372. [CrossRef]

38. Kearney, C.A.; Graczyk, P. A Response to Intervention Model to Promote School Attendance and Decrease School Absenteeism. *Child Youth Care Forum* **2013**, *43*, 1–25. [CrossRef]

39. University of Bristol. 1 in 50 16-Year-Olds Affected by Chronic Fatigue Syndrome. 2015. Available online: https://www.sciencedaily.com/releases/2016/01/160125090619.htm (accessed on 27 May 2020).

40. Vitulano, L.A. Psychosocial issues for children and adolescents with chronic illness: Self-esteem, school functioning and sports participation. *Child Adolesc. Psychiatr. Clin. N. Am.* **2003**, *12*, 585–592. [CrossRef]

41. Kiliçkaya, C.; Asi Karakaş, S. The effect of illness perception on loneliness and coping with stress in patients with Chronic Obstructive Pulmonary Disease (COPD). *Int. J. Caring Sci.* **2016**, *9*, 481.

42. Gureje, O.; Von Korff, M.; Simon, G.E.; Gater, R. Persistent Pain and Well-being. *JAMA* **1998**, *280*, 147–151. [CrossRef]

43. Johannes, C.B.; Le, T.K.; Zhou, X.; Johnston, J.A.; Dworkin, R.H. The Prevalence of Chronic Pain in United States Adults: Results of an Internet-Based Survey. *J. Pain* **2010**, *11*, 1230–1239. [CrossRef] [PubMed]

44. Williams, K.L.; Morrison, V.; Robinson, C.A. Exploring caregiving experiences: Caregiver coping and making sense of illness. *Aging Ment. Health* **2013**, *18*, 600–609. [CrossRef]

45. Jason, L.A.; Benton, M.C.; Valentine, L.M.; Johnson, A.; Torres-Harding, S.R. The Economic impact of ME/CFS: Individual and societal costs. *Dyn. Med.* **2008**, *7*, 6. [CrossRef] [PubMed]

46. Densham, S.; Williams, D.; Johnson, A.; Turner-Cobb, J.M. Enhanced psychological flexibility and improved quality of life in chronic fatigue syndrome/myalgic encephalomyelitis. *J. Psychosom. Res.* **2016**, *88*, 42–47. [CrossRef] [PubMed]

47. Schweitzer, R.; Kelly, B.J.; Foran, A.; Terry, D.; Whiting, J. Quality of life in chronic fatigue syndrome. *Soc. Sci. Med.* **1995**, *41*, 1367–1372. [CrossRef]

48. Girgis, A.; Lambert, S.; Johnson, C.; Waller, A.; Currow, D. Physical, Psychosocial, Relationship, and Economic Burden of Caring for People with Cancer: A Review. *J. Oncol. Pract.* **2013**, *9*, 197–202. [CrossRef]

49. Cruz, D.D.A.L.M.D.; Pimenta, C.A.M.; Kurita, G.P.; Oliveira, A.C. Caregivers of Patients with Chronic Pain: Responses to Care. *Int. J. Nurs. Terminol. Classif.* **2004**, *15*, 5–14. [CrossRef]

50. Nacul, L.; Lacerda, E.; Pheby, D.; Campion, P.; Molokhia, M.; Fayyaz, S.; Leite, J.C.D.C.; Poland, F.; Howe, A.; Drachler, M.D.L. Prevalence of myalgic encephalomyelitis/chronic fatigue syndrome (ME/CFS) in three regions of England: A repeated cross-sectional study in primary care. *BMC Med.* **2011**, *9*, 91. [CrossRef] [PubMed]

51. Houtzager, B.A.; Grootenhuis, M.A.; Caron, H.N.; Last, B.F. Sibling Self-Report, Parental Proxies, and Quality of Life: The Importance of Multiple Informants for Siblings of a Critically Ill Child. *Pediatr. Hematol. Oncol.* **2005**, *22*, 25–40. [CrossRef] [PubMed]

52. Rokach, A. *The Psychological Journey to and from Loneliness*; Elsevier: Amsterdam, The Netherlands, 2019.

53. Hawkley, L.C.; Cacioppo, J.T. Loneliness and pathways to disease. *Brain Behav. Immun.* **2003**, *17*, 98–105. [CrossRef]

54. Cacioppo, J.T.; Cacioppo, S.; Capitanio, J.P.; Cole, S.W. The Neuroendocrinology of Social Isolation. *Annu. Rev. Psychol.* **2015**, *66*, 733–767. [CrossRef] [PubMed]

55. Danneel, S.; Maes, M.; Vanhalst, J.; Bijttebier, P.; Goossens, L. Developmental Change in Loneliness and Attitudes Toward Aloneness in Adolescence. *J. Youth Adolesc.* **2017**, *42*, 815. [CrossRef]

56. Rokach, A. Three of humankind's universal experiences: Loneliness, illness and death. In *Lonelness in Life: Education, Business, Society*; Kowalski, C., Rokach, A., Cangemi, J.P., Eds.; McGraw-Hill: New York, NY, USA, 2015; pp. 119–132.

57. Rokach, A.; Sha'ked, A. *Together and Lonely: Loneliness in Intimate Relationships—Causes and Coping*; Nova Publishers: Hauppauge, NY, USA, 2013.

58. Cacioppo, S.; Grippo, A.J.; London, S.; Goossens, L.; Cacioppo, J.T. Loneliness. *Perspect. Psychol. Sci.* **2015**, *10*, 238–249. [CrossRef]

59. Friedman, R.L. Widening the therapeutic lens: Sense of belonging as an integral dimension of the human experience. (Doctoral dissertation). *Diss. Abstr. Int. Sect. B Sci. Eng.* **2007**, *68*, 3394.

60. Loneliness can Directly Impair Immune System, Increase Risk of Death: Study. CTV News. 16 June 2020. Available online: https://www.ctvnews.ca/health/loneliness-can-directly-impair-immune-system-increase-risk-of-death-study-1.4986159 (accessed on 5 July 2020).

61. Fallon, N.; Brown, C.; Twiddy, H.; Brian, E.; Frank, B.; Nurmikko, T.; Andrej Stancak, A. Adverse effects of COVID-19 related lockdown on pain, physical activity and psychological wellbeing in people with chronic pain. *medRxiv* **2020**. [CrossRef]

62. Perlman, D.; Joshi, P. The revelation of loneliness. *J. Soc. Behav. Personal.* **1987**, *2*, 63–76.

63. Rokach, A. Loneliness and the Life Cycle. *Psychol. Rep.* **2000**, *86*, 629–642. [CrossRef]

64. Lasgaard, M.; Friis, K.; Shevlin, M. "Where are all the lonely people?" A population-based study of high-risk groups across the life span. *Soc. Psychiatry Psychiatr. Epidemiol.* **2016**, *51*, 1373–1384. [CrossRef] [PubMed]

65. Boehm, J.K.; Chen, Y.; Williams, D.R.; Ryff, C.D.; Kubzansky, L.D. Subjective well-being and cardiometabolic health: An 8-11year study of midlife adults. *J. Psychosom. Res.* **2016**, *85*, 1–8. [CrossRef] [PubMed]

66. Cameron, L.D.; Leventhal, H. Self-regulation, health and illness: An overview. In *The Self-Regulation of Health and Illness Behaviour*; Cameron, L.D., Leventhal, H., Eds.; Routledge: London, UK, 2003; pp. 1–13.

67. Sullivan, M.D. The new subjective medicine: Taking the patient's point of view on health care and health. *Soc. Sci. Med.* **2003**, *56*, 1595–1604. [CrossRef]

68. Sellick, S.M.; Edwardson, A.D. Screening new cancer patients for psychological distress using the hospital anxiety and depression scale. *Psycho Oncol.* **2007**, *16*, 534–542. [CrossRef]

69. Rokach, A. Health, Illness, and the Psychological Factors Affecting Them. *J. Psychol.* **2019**, *153*, 1–5. [CrossRef]

70. Rattray, J.; Johnston, M.; Wildsmith, J.A.W. Predictors of emotional outcomes of intensive care. *Anesthesia* **2005**, *60*, 1085–1092. [CrossRef] [PubMed]

71. Kiecolt-Glaser, J.K.; McGuire, L.; Robles, T.F.; Glaser, R. Psychoneuroimmunology: Psychological influences on immune function and health. *J. Consult. Clin. Psychol.* **2002**, *70*, 537–547. [CrossRef]

72. Lauder, W.; Mummery, K.; Jones, M.; Caperchione, C.M. A comparison of health behaviours in lonely and non-lonely populations. *Psychol. Health Med.* **2006**, *11*, 233–245. [CrossRef]

73. Bunker, S.J.; Colquhoun, D.M.; Esler, M.D.; Hickie, I.B.; Hunt, D.; Jelinek, V.M.; Oldenburg, B.F.; Peach, H.G.; Ruth, D.; Tennant, C.C.; et al. "Stress" and coronary heart disease: Psychosocial risk factors. *Med. J. Aust.* **2003**, *178*, 272–276. [CrossRef]

74. Bramston, P.; Pretty, G.; Chipuer, H. Unravelling Subjective Quality of Life: An Investigation of Individual and Community Determinants. *Soc. Indic. Res.* **2002**, *59*, 261–274. [CrossRef]

75. Ernst, J.M.; Cacioppo, J.T. Lonely hearts: Psychological perspectives on loneliness. *Appl. Prev. Psychol.* **1999**, *8*, 1–22. [CrossRef]

76. Steptoe, A.; Owen, N.; Kunz-Ebrecht, S.R.; Brydon, L. Loneliness and neuroendocrine, cardiovascular, and inflammatory stress responses in middle-aged men and women. *Psychoneuroendocrinology* **2004**, *29*, 593–611. [CrossRef]

77. Seeman, T.E.; Singer, B.H.; Ryff, C.D.; Love, G.D.; Levy-Storms, L. Social Relationships, Gender, and Allostatic Load Across Two Age Cohorts. *Psychosom. Med.* **2002**, *64*, 395–406. [CrossRef] [PubMed]

78. Leventhal, H.; Leventhal, E.A.; Cameron, L. Representations, procedures, and affect in illness self-regulation: A perceptual-cognitive model. In *Handbook of Health Psychology*; Baum, A., Revenson, T.A., Singer, J.E., Eds.; Earlbaum: Mahwah, NJ, USA, 2001; pp. 133–151.

79. Glaser, R.; Kiecolt-Glaser, J.K.; Bonneau, R.H.; Malarkey, W.; Kennedy, S.; Hughes, J. Stress-induced modulation of the immune response to recombinant hepatitis B vaccine. *Psychosom. Med.* **1992**, *54*, 22–29. [CrossRef]

80. Hagerty, B.M.; Williams, R.A.; Coyne, J.C.; Early, M.R. Sense of belonging and indicators of social and psychological functioning. *Arch. Psychiatr. Nurs.* **1996**, *10*, 235–244. [CrossRef]

81. Brannon, L.; Feist, J. *Health Psychology: An Introduction to Behavior and Health*; Thomson Wadsworth: Belmont, CA, USA, 2004.

82. REPRINT: Social Networks, Host Resistance, and Mortality: A Nine-Year Follow-up Study of Alameda County Residents. *Am. J. Epidemiol.* **2017**, *185*, 1070–1088. [CrossRef]

83. Segrin, C.; Domschke, T. Social Support, Loneliness, Recuperative Processes, and Their Direct and Indirect Effects on Health. *Health Commun.* **2011**, *26*, 221–232. [CrossRef]

84. Hawkley, L.C.; Cacioppo, J.T. Aging and Loneliness. *Curr. Dir. Psychol. Sci.* **2007**, *16*, 187–191. [CrossRef]

85. Segrin, C.; Passalacqua, S.A. Functions of Loneliness, Social Support, Health Behaviors, and Stress in Association With Poor Health. *Health Commun.* **2010**, *25*, 312–322. [CrossRef]

86. Piotrowski, C. Chronic pain patients and loneliness: A systematic review of the literature. In *Loneliness in Life: Education, Business, Society*; Kowalski, C., Rokach, A., Cangemi, J.P., Eds.; McGraw-Hill: New York, NY, USA, 2015; pp. 189–202.

87. Morrissey, M.B.Q. Phenomenology of Pain and Suffering at the End of Life: A Humanistic Perspective in Gerontological Health and Social Work. *J. Soc. Work End-Of-Life Palliat. Care* **2011**, *7*, 14–38. [CrossRef]

88. Piotrowski, C. Assessment of Pain: A Survey of Practicing Clinicians. *Percept. Mot. Skills* **1998**, *86*, 181–182. [CrossRef] [PubMed]

89. Turk, D.C.; Okifuji, A. Psychological factors in chronic pain: Evolution and revolution. *J. Consult. Clin. Psychol.* **2002**, *70*, 678–690. [CrossRef] [PubMed]

90. Dansie, E.J.; Turk, D.C. Assessment of patients with chronic pain. *Br. J. Anaesth.* **2013**, *111*, 19–25. [CrossRef] [PubMed]

91. Newton, B.J.; Southall, J.L.; Raphael, J.H.; Ashford, R.L.; LeMarchand, K. A Narrative Review of the Impact of Disbelief in Chronic Pain. *Pain Manag. Nurs.* **2013**, *14*, 161–171. [CrossRef]

92. Dewar, A. Assessment and management of chronic pain in the older person living in the community. *Aust. J. Adv. Nurs.* **2006**, *24*, 33–38.

93. Tse, M.M.Y.; Wan, V.T.; Vong, S.K. Health-Related Profile and Quality of Life Among Nursing Home Residents: Does Pain Matter? *Pain Manag. Nurs.* **2013**, *14*, e173–e184. [CrossRef]

94. Kool, M.B.; Geenen, R. Loneliness in Patients with Rheumatic Diseases: The Significance of Invalidation and Lack of Social Support. *J. Psychol.* **2012**, *146*, 229–241. [CrossRef]

95. Asnani, M.; Fraser, R.; Lewis, N.A.; Reid, M. Depression and loneliness in Jamaicans with sickle cell disease. *BMC Psychiatry* **2010**, *10*, 40. [CrossRef]

96. Sofaer, B.; Moore, A.P.; Holloway, I.; Lamberty, J.M.; Thorp, T.A.S.; Dwyer, J.O. Chronic pain as perceived by older people: A qualitative study. *Age Ageing* **2005**, *34*, 462–466. [CrossRef]

97. Jaremka, L.M.; Fagundes, C.P.; Glaser, R.; Bennett, J.M.; Malarkey, W.B.; Kiecolt-Glaser, J.K. Loneliness predicts pain, depression, and fatigue: Understanding the role of immune dysregulation. *Psychoneuroendocrinology* **2013**, *38*, 1310–1317. [CrossRef]

98. Benka, J.; Nagyova, I.; Rosenberger, J.; Čalfová, A.; Macejova, Z.; Middel, B.; Lazúrová, I.; Van Dijk, J.P.; Groothoff, J.W. Social support and psychological distress in rheumatoid arthritis: A 4-year prospective study. *Disabil. Rehabil.* **2011**, *34*, 754–761. [CrossRef]

99. Zhou, X.; Gao, D.-G. Social Support and Money as Pain Management Mechanisms. *Psychol. Inq.* **2008**, *19*, 127–144. [CrossRef]

100. Rosland, A.-M.; Heisler, M.; Piette, J.D. The impact of family behaviors and communication patterns on chronic illness outcomes: A systematic review. *J. Behav. Med.* **2011**, *35*, 221–239. [CrossRef] [PubMed]

101. Hawkley, L.C.; Masi, C.M.; Berry, J.D.; Cacioppo, J.T. Loneliness is a unique predictor of age-related differences in systolic blood pressure. *Psychol. Aging* **2006**, *21*, 152–164. [CrossRef]

102. Barlow, M.A.; Liu, S.Y.; Wrosch, C. Chronic illness and loneliness in older adulthood: The role of self-protective control strategies. *Health Psychol.* **2015**, *34*, 870–879. [CrossRef]

103. Stringer, H. Unlocking the emotions of cancer. *PsycEXTRA Dataset* **2015**, *45*, 34–38. [CrossRef]

104. Wolf, L.D.; Davis, M.C. Loneliness, daily pain, and perceptions of interpersonal events in adults with fibromyalgia. *Health Psychol.* **2014**, *33*, 929–937. [CrossRef] [PubMed]

105. Wager, T.D.; Atlas, L.Y.; Lindquist, M.A.; Roy, M.; Woo, C.-W.; Kross, E. An fMRI-Based Neurologic Signature of Physical Pain. *N. Engl. J. Med.* **2013**, *368*, 1388–1397. [CrossRef]

106. Vella-Baldacchino, M.D.; Schembri, M.; Vella-Baldacchino, M. Myalgic encephalomyelitis/chronic fatigue syndrome (ME/CFS). *Malta Med. J.* **2014**, *26*, 17–22.

107. Abrams, L. *Chronic Fatigue Syndrome*; Thomson Gale: Farmington Hills, MI, USA, 2003.

108. Biordi, D.L. Social isolation. In *Chronic Illness: Impact and Interventions*; Lubkin, I., Larsen, P., Eds.; Jones and Bartlett: Burlington, MA, USA, 2002; pp. 119–147.

109. Bharadvaj, D. *Natural Treatments for Chronic Fatigue Syndrome*; Greenwood Publishing Group: Westport, CT, USA, 2008.

110. Riegel, B.; Carlson, B. Is Individual Peer Support a Promising Intervention for Persons With Heart Failure? *J. Cardiovasc. Nurs.* **2004**, *19*, 174–183. [CrossRef]

111. Perese, E.F.; Wolf, M. Combating Loneliness Among Persons with Severe Mental Illness: Social Network Interventions' Characteristics, Effectiveness, And Applicability. *Issues Ment. Health Nurs.* **2005**, *26*, 591–609. [CrossRef] [PubMed]

112. Purk, J.K. Support Groups: Why Do People Attend? *Rehabil. Nurs.* **2004**, *29*, 62–67. [CrossRef] [PubMed]

113. Stewart, M.J.; Craig, D.; MacPherson, K.; Alexander, S. Promoting Positive Affect and Diminishing Loneliness of Widowed Seniors Through a Support Intervention. *Public Health Nurs.* **2001**, *18*, 54–63. [CrossRef] [PubMed]

114. Ali, N. *Understanding Chronic Fatigue Syndrome: An Introduction for Patients and Caregivers*; Rowman & Littlefield: Lanham, MD, USA, 2015.

115. Brigden, A.; Barnett, J.; Parslow, R.M.; Beasant, L.; Crawley, E. Using the internet to cope with chronic fatigue syndrome/myalgic encephalomyelitis in adolescence: A qualitative study. *BMJ Paediatr. Open* **2018**, *2*, e000299. [CrossRef]

116. Broadbent, S.; Coetzee, S.; Beavers, R.; Horstmanshof, L. Patient experiences and the psychosocial benefits of group aquatic exercise to reduce symptoms of Myalgic Encephalomyelitis/Chronic Fatigue Syndrome: A pilot study. *Fatigue Biomed. Health Behav.* **2020**, *8*, 84–96. [CrossRef]

117. Holley, U.A. Social Isolation: A Practical Guide for Nurses Assisting Clients with Chronic Illness. *Rehabil. Nurs.* **2007**, *32*, 51–58. [CrossRef]

118. Warner, C.B.; Roberts, A.R.; Jeanblanc, A.B.; Adams, K.B. Coping resources, loneliness, and depressive symptoms of older women with chronic illness. *J. Appl. Gerontol.* **2007**, *38*, 1–29. [CrossRef]

119. Han, J.; Richardson, V.E. The Relationship Between Depression and Loneliness Among Homebound Older Persons: Does Spirituality Moderate This Relationship? *J. Relig. Spiritual. Soc. Work Soc. Thought* **2010**, *29*, 218–236. [CrossRef]

120. Walton, J.; Craig, C.; Derwinski-Robinson, B.; Weinert, C. I Am Not Alone: Spirituality of Chronically Ill Rural Dwellers. *Rehabil. Nurs.* **2004**, *29*, 164–168. [CrossRef]

121. Boellinghaus, I.; Jones, F.W.; Hutton, J. The Role of Mindfulness and Loving-Kindness Meditation in Cultivating Self-Compassion and Other-Focused Concern in Health Care Professionals. *Mindfulness* **2012**, *5*, 129–138. [CrossRef]

122. Davey, A.; Chilcot, J.; Driscoll, E.; McCracken, L.M. Psychological flexibility, self-compassion and daily functioning in chronic pain. *J. Context. Behav. Sci.* **2020**, *17*, 79–85. [CrossRef]

123. Ray, C.; Jefferies, S.; Weir, W.R.C. Life-events and the course of chronic fatigue syndrome. *Br. J. Med. Psychol.* **1995**, *68*, 323–331. [CrossRef] [PubMed]

124. Vercoulen, J.H.; Swanink, C.M.; Fennis, J.F.; Galama, J.M.; Van Der Meer, J.W.; Bleijenberg, G. Prognosis in chronic fatigue syndrome: A prospective study on the natural course. *J. Neurol. Neurosurg. Psychiatry* **1996**, *60*, 489–494. [CrossRef] [PubMed]

125. Endicott, J.; Nee, J.; Harrison, W.; Blumenthal, R. Quality of Life Enjoyment and Satisfaction Questionnaire: A new measure. *Psychopharmacol. Bull.* **1993**, *29*, 321–326.

126. Weiss, D.J.; Dawis, R.V.; England, G.W. Manual for the Minnesota Satisfaction Questionnaire. *Minn. Stud. Vocat. Rehabil.* **1967**, *22*, 120.

127. Hansen, T.; Slagsvold, B. Late-Life Loneliness in 11 European Countries: Results from the Generations and Gender Survey. *Soc. Indic. Res.* **2015**, *129*, 445–464. [CrossRef]

128. Rokach, A. The Effect of Gender and Culture on Loneliness: A Mini Review. *Emerg. Sci. J.* **2018**, *2*, 59–64. [CrossRef]

129. Triandis, H.C. The psychological measurement of cultural syndromes. *Am. Psychol.* **1996**, *51*, 407–415. [CrossRef]

Publisher's Note: MDPI stays neutral with regard to jurisdictional claims in published maps and institutional affiliations.

 *healthcare*

Review

The Impact of Severe ME/CFS on Student Learning and K–12 Educational Limitations

Faith R. Newton

Department of Education, Delaware State University, Dover, DE 19901, USA; fnewton@desu.edu

Abstract: Children with ME/CFS who are severely ill are bedbound and homebound, and oftentimes also wheelchair-dependent. Very seriously affected children are often too sick for doctor's office visits, let alone school attendance. The most recent data estimate that 2–5% of children may be severely affected or bedridden. However, there is no recent research that confirms these numbers. The severely ill receive little help from their schools, and are socially isolated. This article outlines several suggestions for the type of education that students with ME/CFS should be receiving and develops a preliminary sketch of the web of resources and emergent techniques necessary to achieve these outcomes.

Keywords: chronic illness; housebound; chronic fatigue syndrome; myalgic encephalomyelitis; severely ill

Citation: Newton, F.R. The Impact of Severe ME/CFS on Student Learning and K–12 Educational Limitations. *Healthcare* **2021**, *9*, 627. https://doi.org/10.3390/healthcare9060627

Academic Editors: Kenneth J. Friedman, Lucinda Bateman and Kenny Leo De Meirleir

Received: 30 March 2021
Accepted: 1 May 2021
Published: 25 May 2021

Publisher's Note: MDPI stays neutral with regard to jurisdictional claims in published maps and institutional affiliations.

1. Article

Educators and clinicians focused on children suffering from mild-to-moderate ME/CFS have made significant strides in the past decade in supporting chronically ill students in achieving academic success. Classroom teachers and other education professionals are increasingly designing effective Individual Educational Plans (IEPs), creating supportive classroom learning environments, and modifying curricula to emphasize mastery over completion [1]. Many treating physicians effectively collaborate with schools to unify clinical and instructional best practices, becoming effective allies for students and families [2,3]. Work remains to refine successful techniques and spread their implementation, but a foundational template for educating children with ME/CFS has been achieved [4,5].

The same cannot be said regarding the education of children whose ME/CFS symptoms are so severe as to render them housebound and/or bedbound. Despite sometimes heroic measures by family members, caregivers, and visiting teachers, children with ME/CFS often face spending years in social and intellectual isolation, devoid of challenging cognitive stimulation or the opportunity to learn new skills and information. Though estimates remain little more than guesses because bedbound students are rarely disaggregated in reported statistics, it is likely that thousands of children, adolescents, and young adults suffering from ME/CFS are essentially being "warehoused" in their own bedrooms [6–8]. Pendergrast et al. [9] stated in a study of adult housebound patients with ME/CFS that "25% or more are confined to their homes (housebound) or completely bedbound." Although, there are no studies outlining the percentages of children, it would not surprise this author if the numbers were as high. This condition commands our attention because there is evidence that children with ME/CFS can learn and lead far more enriched lives whether or not they are physically capable of leaving their bedrooms. Innate human dignity requires us to explore, develop, and execute strategies for alleviating this situation.

This article seeks to accomplish three modest goals:

(1). delineate the impact of severe pediatric ME/CFS on student learning, and how that impact inherently limits the positive impact that even the most innovative K–12 educational programs can expect to achieve;

(2). establish an initial foundation—in philosophical and process terms—for the *type* of education that students with ME/CFS *should be* receiving, and the educational outcomes to be expected; and

(3). develop a preliminary sketch of the web of resources and emergent techniques necessary to achieve these outcomes.

2. The Impact of Severe ME/CFS on Student Learning and K–12 Educational Limitations

Myalgic Encephalomyelitis/Chronic Fatigue Syndrome (ME/CFS) is characterized by profound, medically unexplained fatigue including a range of sleep, pain, cognitive, neuroendocrine, and immune symptoms. Post-exertional malaise (the provocation of fatigue by physical, cognitive, or orthostatic stress resulting in symptoms including cognitive fogginess, headache, light-headedness, flu-like symptoms, and muscle aches) and persistent generalized fatigue make school attendance irregular even in mild-to-moderate cases; the most common feature of pediatric ME/CFS observable in the classroom is the array of attention, concentration, and memory deficits known collectively as "Brain Fog" [10]. Children with ME/CFS often view school as a frustrating, painful experience, a situation further complicated by symptom variability. Some students will have only mild symptoms requiring modest accommodations, while others will present with more severe and debilitating symptoms. Children with moderate-to-extreme cases will be homebound or bedbound, resulting in protracted or even permanent non-attendance at school [3,11]. Speight pointed out that the muscular, gastric, fatigue-based, neural, and cognitive symptoms in severe ME/CFS "can actually be worse than that suffered by patients with other chronic conditions such as multiple sclerosis and cancer," and that "abdominal pain may be so severe as to interfere with nutrition, and in some cases are due to an added complication, Mast Cell Activation Disorder" [12].

Mental health consequences are equally profound, as Boularzreg and Roach note, with catastrophism, pessimism, depression, and isolation, leading to "a substantial loss of self" because "adolescents with ME/CFS attach significance to attending school and hang-outs with friends, [and] when deprived of these events, the adolescent questions the meaning of life" [13]. As one bedbound patient noted:

"Imagine having the flu, being severely jet-lagged, and having not slept for two days but without the sinus and lung congestion—that's the closest I've found to a description of what this illness is like, but it understates it. I lack mental clarity; I mix up words, and I have memory problems and trouble focusing. I have an enormous need for sleep, which never refreshes. There is an overwhelming, permanent, and intense malaise" [11].

A mother of two sons in their mid-twenties who have been homebound or bedbound for a decade reports that they often cannot be upright in bed, and that severe light, sound, and touch sensitivities restrict their use of computers. Their attention spans are often limited to irregular intervals 20 min long, and while "their intellect is not impaired, the [physical] resources available to apply that intellect are severely limited."

American public schools are chronically ill-equipped to educate such children, adolescents, and young adults [14,15]. Educators lack training for dealing with chronically ill children in their classrooms, let alone those permanently isolated at home [16,17]. Homebound instruction is governed not by consistent national standards and best practices, but by state statutes and school district policies. These often have time-consuming bureaucratic requirements to qualify for services (sometimes mandating re-qualification every three months); severely limiting both the availability of tutors (who may have no formal training for teaching chronically ill children); and tie all education to the courses the student would be taking were s/he actually attending school [18–20]. Additional barriers are imposed by an unwillingness to extend deadlines; the expectation that services will be duration-limited and focused on eventual "re-entry" into the regular classroom; and no responsibility for providing educational services past age 21 [21]. Leaving aside nationally specific legal requirements, similar situations pertain internationally; there is significant literature pertaining to these issues in Australia, the Czech Republic, India, New Zealand, Japan, Turkey, across the European Union, and elsewhere [4,22–28].

3. Why Children with Severe ME/CFS Must Become Semi-Independent Lifelong Learners

Most innovations introduced to support the education of bedbound individuals suffering from severe ME/CFS have been technology-based. Robotics are intended to allow these students virtual access to live classrooms in a real-time, interactive three-dimensional setting [29]. New kinds of bed tables or trays have been developed to allow more comfortable access to computers. An increasingly broad array of fully online courses has become available, with their use expanding beyond the original purpose of credit recovery to what is often a default choice suggested for homebound students [30].

These innovations are intended as aides to making the existing patchwork system function, rather than rethinking the philosophy, organization, and objectives of educating bedbound students. They are sometimes perceived by school districts as comparatively inexpensive and not requiring significant investments of staff or resources, and often lump all bedbound students into a single category. A robotic telepresence is of extremely limited usefulness to a child with ME/CFS whose working attention span for active learning exists in irregularly spaced 20 min intervals, and who has enough issues with processing speed without having to operate a complex interface while trying to pay attention to a lecture or group activities. Many of the solutions to these issues are being pursued in hospitals, not schools [31].

In fact, for decades, the most innovative work in educating bedbound students (and older patients as well) has originated in hospitals providing long-term care rather than the educational system. In 1942, Dr. Elena D. Gall of Hunter College established a continuing education program at Goldwater Memorial Hospital in New York City, which had a population of 1850 patients permanently confined to a hospital setting. Designed in 1937 by architect Isadore Rosenfield, who "was as concerned with the careful design of a patient's bedside lamp as he was with the circulation patterns of the thousands of people who would use the facility each day," Goldwater Memorial embodied an "unwavering patient-centric design approach" [32].

Gall theorized that the patients "are thinking, intelligent individuals who still function on an intellectual level," but that "agencies which offer continued education to interested adults did not reach out to people who were to be hospital bound for many years." She created a continuing education program that was multigenerational, democratically run by the patients themselves, and focused on lifelong learning rather than specifically tied to credentialing within the public education system: "The patients, ranging from teenagers to elderly people, meet every Friday night in one of the recreation rooms of the hospital. Each year they themselves choose the area of interest for the term of study. One year it was English literature, another time it was world affairs. This year they decided on current events. They have their own roster of duly elected officers" [33,34]. This program appears to have operated successfully until at least 1970, when Goldwater Memorial ceased to be a research institution, but it had a long-term impact on hospital-based education. When Goldwater Memorial closed in 2013 the range of educational programs transferred to successor institutions included "community-based English as a Second Language programs, General Educational Development training, higher education programs and vocational services" [35].

Gall's Goldwater Memorial program is a comprehensive model for the education of bedbound individuals that provides a potential alternative to our current patchwork approach. First and foremost, the model emphasizes supported lifelong learning based on individual interest for personal growth as part of a total patient-care package, largely decoupled from the credential and course-based emphasis of public education. Gall also insisted on creating what we would currently call "a community of learners" as a critical element [34]. This represents at the foundation a new pedagogical approach to educating bedbound ME/CFS patients, especially when updated with new technological capacities;

lessons from home-schooling for chronically ill children; advances in "gamification" of curriculum and new language-learning strategies; and innovative approaches to designing learning spaces for bedbound students.

Additionally, diverse elements outside American public education seem almost to have been crafted toward Gall's model, or a close variant. There is a robust body of research on the "gamification of education," with a significant subset focused on meeting the needs of special needs children. Gamification, although it has been variously defined since the first use of the term in 2008, generally refers to the practice of using elements of structure and rewards from digital gaming to support students in solving problems and remaining motivated, and experts constantly reiterate that "in gamification applications, students need not have to have toys, electronic devices, etc., and not always play games in order to learn" [36,37]. Materials presented via gamification are more likely to seamlessly meld short-term and long-term rewards for mastery, and are significantly more adaptable to the kinds of learning differences that chronically ill children (including those suffering from ME/CFS) manifest than traditional online courses [38–40]. Some examples include Kahoot, Quizizz, Quizlet and Gimkit. Transforming a classroom environment using gamification elements can enhance student learning and increase student understanding of the subject matter in a way that is enjoyable for students. For our bedbound and housebound patients, this brings creativity, play and collaboration into their lives [41].

A second example addresses the criticality of establishing a learning community to provide long-term psychosocial support for bedbound students. There is an understandable tendency in American society—especially after a year spent dealing with the COVID-19 pandemic—to consider virtual support communities as the default option here. In the Kohzikode district of the State of Kerala in India, however, education professionals are exploring a different strategy, that of physically constructing "resource rooms" in the houses of chronically bedbound students "to help differently abled bedridden students develop social and other skills" [23]. Besides computers, these rooms may contain "an FM radio, daily living aids, reading and writing materials, education and play materials, therapy balls, therapeutic seats, coloring books, and crayons" [23].

What makes these home-learning resource centers particularly innovative is that they are specifically linked to building social ties to a friendship peer group "from the school where the differently abled child is enrolled. They interact with the child during weekends or holidays to help them improve their skills. Resource teachers plan activities for the children keeping in mind their physical and mental condition" [23]. This emphasis on learning communities integrates well with new findings in regard to the impact of severe ME/CFS on children's sense of self, and the emerging use of a "social," or constructivist, approach to the teaching, assessment, and support of special needs students [13,42].

4. Moving forward to Enrich the Lives of Children Bedbound with Severe ME/CFS

In terms of education, bedbound children, adolescents, and young adults suffering from severe ME/CFS currently exist in a lonely frontier at the borders of public education, long-term care, psychotherapy, technological innovation, and communities too often completely unaware of any obligation to citizens they never see. The problem is not that we do not know enough about the requisite elements to enrich these young people's lives through education, but that there is currently no framework to coordinate and deploy the knowledge, skills, supports, and technologies which already exist. Not only does the model of best practices have to be built, but we—as a society—have to figure out who will be responsible for enacting it.

There is reason to be hopeful, however. Slightly more than a decade ago we faced a similar state of affairs in supporting children with mild-to-moderate cases of ME/CFS to be educationally successful within the public schools. A broad coalition of researchers, clinicians, educators, parents, and volunteers coalesced to address that challenge, and they have made major progress. Students with ME/CFS increasingly find that schools are better prepared to teach them on their own terms. That process began with a few educators raising the issue, followed by a national dialogue among the concerned parties. Despite the loss of the Chronic Fatigue Syndrome Advisory Committee at the Department of Health and Human Services, which coordinated many of those early discussions, there are today a number of robust networks that cross disciplinary boundaries and are capable of initiating such conversations.

Funding: This research received no external funding.

Institutional Review Board Statement: Not applicable.

Informed Consent Statement: Not applicable.

Data Availability Statement: Not applicable.

Conflicts of Interest: The author declare no conflict of interest.

References

1. Tollit, M.; Politis, J.; Knight, S. Measuring School Functioning in Students with Chronic Fatigue Syndrome: A Systematic Review. *J. Sch. Health* **2017**, *88*, 74–89. [CrossRef] [PubMed]
2. Knight, S.; Politis, J.; Garnham, C.; Scheinberg, A.; Tollit, M. School Functioning in Adolescents with Chronic Fatigue Syndrome. *Front. Pediatr.* **2018**, *6*, 302. [CrossRef] [PubMed]
3. Newton, F. Improving academic success for students with myalgic encephalomyelitis/chronic fatigue. *Fatigue Biomed. Health Behav.* **2015**, *3*, 97–103. [CrossRef]
4. Rowe, K. Paediatric patients with myalgic encephalomyelitis/chronic fatigue syndrome value understanding and help to move on with their lives. *Acta Pediatr.* **2020**, *109*, 790–800. [CrossRef] [PubMed]
5. Newton, F. Meeting the Educational Needs of Young, ME/CFS Patients: Role of the Treating Physician. *Front. Pediatr.* **2019**, *7*, 104. [CrossRef] [PubMed]
6. Newhart, V.A.; Olson, J.; Warschauer, M.; Eccles, J. *Go home and Get Better: An Exploration of Inequitable Educational Services for Homebound Children*; UC Irvine: Irvine, CA, USA, 2017; Available online: https://escholarship.org/uc/item/8js3h9zc (accessed on 22 March 2021).
7. Sable, J.; Plotts, C.; Mitchell, L. *Characteristics of the 100 Largest Public Elementary and Secondary School Districts in the United States: 2008–2009*; Statistical Analysis Report; American Institutes for Research: Arlington, VA, USA, 2010.
8. Jason, L.; Benton, M.; Valentine, L.; Johnson, A.; Torres-Harding, S. The Economic impact of ME/CFS: Individual and societal costs. *Dyn. Med.* **2008**, *7*, 1–8. [CrossRef]
9. Pendergrast, T. Housebound versus nonhousebound patients with myalgic encephalomyelitis and chronic fatigue syndrome. *Tric. Chronic Illn.* **2016**, *12*, 292–307. [CrossRef]
10. Rowe, P.C.; Underhill, R.A.; Friedman, K.J.; Gurwitt, A.; Medow, M.S.; Schwartz, M.S.; Speight, N.; Stewart, J.M.; Vallings, R.; Rowe, K.S. Myalgic Encephalomyelitis/ Chronic Fatigue Syndrome Diagnosis and Management in Young People: A Primer. *Front. Pediatr.* **2017**, *5*, 121. [CrossRef]
11. Cornes, O. Commentary: Living with CFS/ME. *BMJ* **2011**, *342*, d3836. [CrossRef]
12. Speight, N. Severe ME in Children. *Healthcare* **2020**, *8*, 211. [CrossRef]
13. Boularzeg, S.; Rokach, A. The Lonely, Isolating, and Alienating Implications of Myalgic Encephalomyelitis/Chronic Fatigue Syndrome. *Healthcare* **2020**, *8*, 413.
14. Flanagan, M. Educating Chronically Ill Students through the Lens of the Classroom Teacher: An Interpretative Phenomenological Analysis. Ph.D. Thesis, Northeastern University, Boston, MA, USA, 2015.
15. Kanthor, H. Educating the chronically ill child. *Am. J. Disabl. Child* **1983**, *37*, 207–208. [CrossRef]
16. Clay, D.; Cortina, S.; Harper, D.; Cocco, C.; Drogar, D. Schoolteachers' Experiences with Childhood Chronic Illness. *Child. Health Care* **2004**, *33*, 227–239. [CrossRef]
17. Nabors, L. Teacher Knowledge of and Confidence in Meeting the Needs of Children with Chronic Medical Conditions: Pediatric Psychology's Contribution to Education. *Psychol. Sch.* **2008**, *45*, 217–226. [CrossRef]
18. Newhart, V.A. Are They Present? Homebound Children with Chronic Illness in Our Schools and the Use of Telepresence Robots to Reach Them. Ph. D. Thesis, University of California Irvine, Irvine, CA, USA, 2018.
19. An, Y.; Reigeluth, C. Creating Technology-Enhanced, Learner-Centered Classrooms: K–12 Teachers' Beliefs, Perceptions, Barriers, and Support Needs. *J. Digit. Learn. Teach. Educ.* **2011**, *28*, 54–62. [CrossRef]
20. Irwin, M.; Elam, M. Are We Leaving Children with Chronic Illness Behind? *Phys. Disabil. Educ. Relat. Serv.* **2011**, *30*, 67–80.
21. Kaffenberger, C. School Reentry for Students with a Chronic Illness: A Role for Professional School Counselors. *Prof. Sch. Couns.* **2006**, *9*, 223–230. [CrossRef]
22. Pliskova, B.; Snopek, P. Primary School Teachers' Awareness of Chronic Diseases of Children. *Acta Educ. Gen.* **2017**, *7*, 111–121. [CrossRef]
23. Staff Reporter. Bedridden children to get resource rooms at homes. *The Hindu Times*, 20 February 2019.
24. Monk, F. *The Children left Bedbound by Fatigue. RNZ*; Radio New Zealand: Wellington, New Zealand, 2018.

25. Osaki, H. Home/Hospital-Bound Education in Japan—From a Survey on Home/Hospital-Bound Education. *J. Spec. Educ. Asia Pac.* **2005**, *1*, 27–32.
26. Akkok, F. The Past, Present and Future of Special Education: The Turkish Perspective. *Mediterr. J. Educ. Stud.* **2001**, *6*, 15–22.
27. Moravcikova, D.; Snopek, P. Analysis of the Teacher's Work with Ill Children. In Proceedings of the European Proceedings of Social & Behavioural Sciences, Kyrenia, Cyprus, 9–12 July 2015.
28. Mangal, S.K. *Educating Exceptional Children*; PHI Learning: Dehli, India, 2007.
29. Newhart, V.A.; Olson, J. Going to School on a Robot: Robot and User Design Interfaces that Matter. *ACM Trans. Comput. Hum. Interact.* **2019**, *26*, 1–28. [CrossRef] [PubMed]
30. Glick, D.B. *The Evolution of K-12 Learning Policy from a Void to a Patchwork*; David A Glick & Associates: Marietta, GA, USA, 2009.
31. Soares, N.; Kay, J.C.; Craven, G. Mobile Robotic Telepresence Solutions for the Education of Hospitalized Children. *Perspect. Health Inf. Manag.* **2017**, *14*, 1e.
32. Giarudet, C. *Autopsy of a Hospital: A Photographic Record of Coler-Goldwater on Roosevelt Island*; Urban Omnibus: New York, NY, USA, 2014.
33. Berkowitz, B. A Program of Continued Education for the Chronically Ill. *J. Except. Child.* **1949**, *16*, 15–32. [CrossRef]
34. Gall, E. The Adult Education Program at Goldwater Memorial Hospital. Ph.D. Thesis, Columbia University, New York, NY, USA, 1949.
35. Coler-Goldwater Long-Term Acute Care Hospital. *Community Health Needs Assessment and Implementation Strategy*; City Health and Hospitals Corporation: New York, NY, USA, 2013.
36. Ceker, E.; Ozdamli, F. What "gamification" is, and what it's not. *Eur. J. Contemp. Educ.* **2017**, *6*, 221–228.
37. Brull, S.; Finlayson, S. Importance of Gamification in Increasing Learning. *J. Contin. Educ. Nurs.* **2016**, *47*, 372–375. [CrossRef]
38. Nacke, L.; Deterding, C. Editorial: The maturing of gamification research. *Comput. Hum. Behav.* **2017**, *71*, 450–454. [CrossRef]
39. Kirayovka, G.; Angelova, N.; Yordanova, L. Gamification in Education. In Proceedings of the 9th International Balkan Education and Science Conference, Edirne, Turkey, 16–18 October 2014.
40. Muntean, C. Raising engagement in e-learning through gamification. In Proceedings of the 6th International Conference on Virtual, Cluj-Napoca, Romania, 28–29 October 2011.
41. Haiken, M. (12 February 2021). 5 Ways to Gamify Your Classroom. Available online: https://www.iste.org/explore/In-the-classroom/5-ways-to-gamify-your-classroom#:~{}:text=Gamification%20is%20about%20transforming%20the,student%20 understanding%20of%20subject%20matter (accessed on 22 March 2021).
42. Anastasiou, D.; Kauffman, J.M. A Social Constructionist Approach to Disability: Implications for Special Education. *Except. Child.* **2011**, *77*, 367–384. [CrossRef]

Review

The Development of a Consistent Europe-Wide Approach to Investigating the Economic Impact of Myalgic Encephalomyelitis (ME/CFS): A Report from the European Network on ME/CFS (EUROMENE)

Derek F.H. Pheby [1,*], Diana Araja [2], Uldis Berkis [3], Elenka Brenna [4], John Cullinan [5], Jean-Dominique de Korwin [6,7], Lara Gitto [8], Dyfrig A Hughes [9], Rachael M Hunter [10], Dominic Trepel [11,12] and Xia Wang-Steverding [13]

[1] Society and Health, Buckinghamshire New University, High Wycombe HP11 2JZ, UK
[2] Department of Dosage Form Technology, Faculty of Pharmacy, Riga Stradins University, Dzirciema Street 16, LV-1007 Riga, Latvia; Diana.Araja@rsu.lv
[3] Institute of Microbiology and Virology, Riga Stradins University, Dzirciema Street 16, LV-1007 Riga, Latvia; Uldis.Berkis@rsu.lv
[4] Department of Economics and Finance, Università Cattolica del Sacro Cuore, Largo Agostino Gemelli 1, 20123 Milan, Italy; elenka.brenna@unicatt.it
[5] School of Business & Economics, National University of Ireland Galway, University Road, H91 TK33 Galway, Ireland; john.cullinan@nuigalway.ie
[6] Internal Medicine Department, University of Lorraine, 34, cours Léopold, CS 25233, F-54052 Nancy CEDEX, France; jean-dominique.dekorwin@univ-lorraine.fr
[7] University Hospital of Nancy, Rue du Morvan, 54511 Nancy, France; jd.dekorwin@chru-nancy
[8] Department of Economics, University of Messina, Piazza Pugliatti 1, 98122 Messina, Italy; lara.gitto@unime.it
[9] Centre for Health Economics & Medicines Evaluation, Bangor University, Bangor LL57 2PZ, UK; d.a.hughes@bangor.ac.uk
[10] Institute of Epidemiology & Health, Royal Free Medical School, University College London, London NW3 2PF, UK; r.hunter@ucl.ac.uk
[11] School of Medicine, Trinity College Dublin, College Green, D02 PN40 Dublin 2, Ireland; trepeld@tcd.ie
[12] Global Brain Health Institute, School of Medicine, Trinity College Dublin, College Green, D02 PN40 Dublin 2, Ireland
[13] Warwick Medical School, University of Warwick, Coventry CV4 7AL, UK; xiasteverding@gmail.com
* Correspondence: derekpheby@btinternet.com

Received: 27 February 2020; Accepted: 31 March 2020; Published: 7 April 2020

Abstract: We have developed a Europe-wide approach to investigating the economic impact of Myalgic Encephalomyelitis/Chronic Fatigue Syndrome (ME/CFS), facilitating acquisition of information on the economic burden of ME/CFS, and international comparisons of economic costs between countries. The economic burden of ME/CFS in Europe appears large, with productivity losses most significant, giving scope for substantial savings through effective prevention and treatment. However, economic studies of ME/CFS, including cost-of-illness analyses and economic evaluations of interventions, are problematic due to different, arbitrary case definitions, and unwillingness of doctors to diagnose it. We therefore lack accurate incidence and prevalence data, with no obvious way to estimate costs incurred by undiagnosed patients. Other problems include, as for other conditions, difficulties estimating direct and indirect costs incurred by healthcare systems, patients and families, and heterogeneous healthcare systems and patterns of economic development across countries. We have made recommendations, including use of the Fukuda (CDC-1994) case definition and Canadian Consensus Criteria (CCC), a pan-European common symptom checklist, and implementation of prevalence-based cost-of-illness studies in different countries using an agreed data list. We recommend using purchasing power parities (PPP) to facilitate international comparisons, and EuroQol-5D as a generic measure of health status and multi-attribute utility instrument to inform future economic evaluations in ME/CFS.

Healthcare **2020**, *8*, 8; doi:10.3390/healthcare8020088

Keywords: ME/CFS; economic impact; cost-of-illness studies; economic evaluation; healthcare systems

---

1. Introduction

Myalgic Encephalomyelitis/Chronic Fatigue Syndrome (ME/CFS) is a poorly understood, serious, complex, multi-system disorder, characterized by symptoms lasting at least six months, with severe incapacitating fatigue not alleviated by rest, and other symptoms, many autonomic or cognitive in nature, including profound fatigue, cognitive dysfunction, sleep disturbances, muscle pain, and post-exertional malaise, which lead to substantial reductions in functional activity and quality of life [1–3]. Symptomatology, severity and disease progression are extremely variable. ME/CFS most commonly occurs between the ages of 20 to 50 years, butan affect all age groups, while some three quarters of patients are female [4–6]. There are no Europe-wide prevalence data, but if the commonly held belief that there are some 250,000 sufferers in the UK is correct [7], then there may be some two million patients in Europe as a whole.

In terms of estimating the economic costs of ME/CFS, the current state of the art, and its historical development, have recently been described by Brenna and Gitto, who carried out a comprehensive literature review [8,9]. The literature was reviewed chronologically and detailed the evolution of economic studies of ME/CFS, including cost of illness studies and economic evaluations (e.g., cost effectiveness and cost utility analyses) of specific interventions. The authors also drew attention to the failure of many patients with the condition to be correctly diagnosed, which renders problematic attempts to determine the economic burden of the disease. Another problem they identified was that of determining direct, indirect and intangible costs in a context where there is lack of agreement over, and inconsistent use of, case definitions. This, in turn, is reflected in a lack of agreement regarding the incidence and prevalence of the condition. The prevalence in developed countries appears to be within the range of 0.2–1%, but this is highly dependent on case definition, while there is published research on ME/CFS in only a small number of countries, notably Australia, USA and the UK. Brenna and Gitto [8] concluded that a clearer definition of the population prevalence of ME/CFS would make it easier to reach a general consensus on its economic burden. This, they argued, would assist the development of appropriate guidelines to manage the disease.

In this context, the current classification of ME/CFS in the *International Statistical Classification of Diseases and Related Health Problems* (ICD), the ICD-10 (10th revision) and ICD-10-CM (Clinical Modification) is confusing and may contribute to this lack of clarity, since the code G93.3, within 'Diseases of the Nervous System', specifies 'post-viral fatigue syndrome' and is applicable to 'benign myalgic encephalomyelitis', while 'chronic fatigue syndrome' is an applicable term within 'chronic fatigue, unspecified' (R53.82), within the symptoms chapter [10]. The situation may be improved by the intended implementation in 2022 of ICD-11, in which it is proposed to list 'post-viral fatigue syndrome' in Chapter 08 (Diseases of the nervous system) under the code 8E49 [11]. While this may create a source of confusion by making an assumption about the aetiology of the condition, which cannot always be demonstrated, this is mitigated and clarified by the addition of 'benign myalgic encephalomyelitis' and 'chronic fatigue syndrome' as inclusion terms, and of 'fatigue' as an exclusion.

In relation to the economic burden of ME/CFS, Brenna and Gitto [8] highlighted that the most substantial component is the indirect costs that arise as a result of productivity losses. Indeed, attempts have been made in the last decade to evaluate the societal costs of ME/CFS, especially in terms of occupational outcomes, such as absenteeism, work incapacity, and early retirement. For example, estimates in the UK population suggest an average yearly productivity loss due to discontinuation of employment of £ 22,684 per patient in the period 2006–2010, with a significant gap between women (£ 16,130) and men (£ 44,515) [12]. To these figures should be added other direct non-medical costs, such as informal care provided by relatives, neighbours or friends, as well as intangible costs related to diminished quality of life. The study also highlighted the need to identify common diagnostic criteria and to alert policy makers to the overall economic burden of this disease (i.e., its impact on society as a whole) and other non-health impacts.

This article reports on work undertaken by Working Group 3 of the European Network on ME/CFS (EUROMENE) in the development of a consistent Europe-wide approach to investigating the economic impact of ME/CFS. It is structured as follows. It starts with an overview of EUROMENE and Working Group 3, followed by a detailed discussion of the challenges and issues involved in developing a consistent Europe-wide approach to measurement that have been identified by the group. After this, we set out a range of recommendations for addressing each of the challenges, while the final section concludes.

2. European Network on ME/CFS (EUROMENE)—A European Cost Action Project

The EUROMENE research network was formally established in 2016 as a collaborative, Europe-wide, consortium aiming to address serious gaps in knowledge of ME/CFS. The network now involves 22 countries, with four working groups focused on: epidemiology; biomarkers and diagnostic criteria; clinical research; and socio-economics. All working groups have the active involvement of researchers from across Europe [13].

The intended long-term impact of the EUROMENE collaboration is that of "preventing ME/CFS, determining suitable treatments or avoiding unnecessary treatment in order to improve patients' quality of life". In terms of the rationale for a socio-economic work package, we estimated that the annual burden of ME/CFS in Europe, on the basis of extrapolation from UK estimates [14], could be in the region of € 40 bn if the prevalence and cost burden associated with each case were similar to those found in the UK. Therefore, even a modest 1% reduction in the overall burden could deliver cost savings of around €400 million/year, while there may be scope for significantly reducing the economic cost of ME/CFS through effective prevention and treatment, though the costs of such initiatives also need to be recognised. The EUROMENE Action aims to "promote further research on ME/CFS with high economic impact" [15].

In this context, the terms of reference for Working Group 3 (Socio-economics) require it to coordinate efforts to determine the societal impact of ME/CFS, to appraise the economic implications from the disease, and to do so by enabling the estimation of the burden of ME/CFS to society and the provision of long-term trend estimates for societal impact [15]. In addition, specific objectives of the working group include: to survey the existing data from European countries pertaining to economic losses attributable to ME/CFS; to develop approaches to calculating the direct and indirect economic burdens due to ME/CFS; and to provide an integrated outcome assessment framework [15].

3. Challenges in Developing a Consistent Europe-Wide Approach to Measuring the Economic Impact of ME/CFS

In this section we describe the progress made by EUROMENE Working Group 3 in the development of a Europe-wide approach to investigating the economic impact of ME/CFS. We start by setting out an overview of the challenges faced, followed by more in-depth discussion of the most relevant issues. In particular, we focus on challenges relating to case definition and prevalence rates, case ascertainment, and the determination of costs, as well as Europe-wide comparisons.

3.1. Overview of Challenges

As noted above, there are major challenges rendering economic analyses of ME/CFS problematic, and these include the use of different case definitions and the unwillingness of many doctors to diagnose it. Both of these factors have major impacts on incidence and prevalence estimates, while, in addition, we have established that there is a lack of routinely collected data which could contribute to such estimates. In particular, since a high proportion of patients with ME/CFS remain undiagnosed, there is no obvious way to estimate the costs incurred by such patients. Finally, issues arising in respect of international comparisons of the economic impact of ME/CFS include the heterogeneous nature of patterns of organisation and delivery of health and social care across Europe, differences in the availability of social support and welfare benefits, the varying levels of economic development and wealth of different European countries, and the problem of comparing economic impacts when different currencies are in use in different countries, and are subject to variations in exchange rates.

Our review of the position regarding the economic impact of ME/CFS has led us to identify a number of challenges that need to be addressed if progress is to be made in this area. These issues are summarised in Table 1 and are addressed in more detail in the subsequent sub-sections.

Table 1. Challenges in assessing the economic impact of ME/CFS in Europe.

1.	Case definition	ME/CFS is a syndrome, defined in terms of its symptomatology rather than its underlying pathology. Work done in this area is therefore dependent on case definitions, which of their very nature are arbitrary. In addition, there are numerous case definitions in use, which vary substantially in sensitivity and specificity, and do not necessarily identify the same population.
2.	Incidence and prevalence	Little is known about the incidence or prevalence of ME/CFS. Very little work has been done in this area in Europe, except in the UK. Conclusions drawn from UK experience, or indeed from work done in other countries, in particular the USA and Australia, cannot be readily extrapolated to Europe as a whole, because the extent of natural variation between populations is unknown.
3.	Failure to diagnose	A high proportion of doctors, in particular GPs, refuse to recognise ME/CFS as a genuine clinical entity, and as a result do not diagnose it. Even in countries where ME/CFS is officially recognised, this proportion may be as high as 50%. It is not possible therefore to obtain accurate prevalence data through the use of service utilisation data.
4.	Determination of costs and losses	Any attempt to determine costs and economic losses attributable to ME/CFS must take into account direct and indirect costs incurred both by healthcare systems, patients and families, as well as productivity losses. This applies equally to patients who have been diagnosed as having ME/CFS and those who have not received a diagnosis, including for the reasons outlined in (iii) above. It is likely to be difficult to identify costs for this latter group, for obvious reasons.
5.	Variation within the ME/CFS population	It can be hypothesised that, for example, severely affected people (housebound or bedbound) may incur greater overall costs than mildly or moderately affected people. There is no information available which could shed light on this, which clearly requires further research.
6.	Heterogeneity of national economies and health care systems	Against such a background, it is clearly an uphill struggle to reach meaningful conclusions about the costs and losses attributable to ME/CFS across Europe, particularly given the variety of systems of healthcare delivery in Europe, and varying stages of economic development.

3.2. Case Definition and Prevalence Rates

The major problem in determining the overall economic burden of disease attributable to ME/CFS, i.e., little agreement over case definition, has been considered by a number of authors. For example, Brurberg et al. [16] reviewed the comparability of case definitions and identified papers in which different case definitions have been applied to the same patient populations, making possible direct comparisons of the impact differences in case definition have on apparent prevalence. They listed 20 case definitions developed from 1988 onwards, which tend to define different populations and which therefore impact significantly on the perceived prevalence of the disease, and also levels of severity and hence of need for care within the identified patient population.

The case definition most commonly used for research purposes has been that produced by the US Center for Disease Control in 1994, otherwise known as the Fukuda definition [17]. More recently, the Canadian Consensus Criteria (CCC), which identify a more severely affected group of patients than the Fukuda definition, have been widely accepted [3]. A UK study found that some 50% of those patients who conformed to the Fukuda definition conformed also to the CCC [18], while a parallel study in the UK concluded that there were advantages to using both definitions, in order to take advantage of the greater sensitivity of the Fukuda definition and the greater specificity of the CCC [5]. A new case definition has been proposed by the Institute of Medicine (IOM, 2015) [19] that received international recognition. Because of their relative simplicity, the IOM criteria seem useful for screening patients in clinical practice, the CCC criteria being used to confirm the diagnosis of ME/CFS. Working Group 1 of EUROMENE (Epidemiology) proposes that the Fukuda and CCC definitions should be used in all participating European countries [20]. Working Group 3 (Socio-economics) accepts this guidance and also recommends use of a symptom checklist, enabling data to be collected of such a nature that mapping algorithms can be applied to them, enabling conformity to both Fukuda and CCC to be determined. Such a symptom checklist was developed by Osoba et al. [21].

A study in three English regions [18] found a prevalence rate of 0.19% conforming to the CDC-1994 definition in the age group 18–64, but only 0.10% conforming to the more recent Canadian definition [3]. The comparative assessment previously reported by Jason et al. in 2004 [22] found that the Canadian criteria selected cases with less psychiatric co-morbidity, more physical functioning impairment, and more fatigue/weakness, neuropsychiatric and neurological symptoms than the CDC-1994 definition. Two papers comparing the CDC-1994 and Australian definitions are cited. Lindal et al., in Iceland, found population prevalences of 2.1% (CDC-1994) and 7.6% (Australian) respectively [23], while Wessely et al., in England, found that use of the CDC-1994 definition produced a population prevalence of 2.6%, while the equivalent figure using the Australian definition was 1.4% [24]. Brurberg et al. [16] attributed this variation in prevalence obtained using the Australian definition to differences in data collection methods; the CDC-1994 definition appeared more robust and less likely to be affected by variations in data collection methods.

In conclusion, it is clear that, if comparable data on the economic impact of ME/CFS are to be collected across Europe, there needs to be comparable data on the prevalence of the illness, and this in turn requires agreement on case definition. Most of the work done to date in this area has used the CDC-1994 definition [17], which cannot be ignored because of its widespread use in the past but is not ideal for epidemiology. This is because it was not designed for that purpose, but rather to enable well-characterised and relatively homogeneous groups of patients to be identified for clinical trials. As regards cost of illness, one can hypothesise that overall costs, at a national level, will appear greater using the more sensitive CDC-1994 definition, while costs per case will be greater using the Canadian definition [3]. In adopting this two-fold approach, we are acting in concert with the epidemiology and the diagnostic methods/biomarkers working groups of EUROMENE.

3.3. Case Ascertainment

As stated above, it is well known that many doctors do not diagnose ME/CFS, further complicating the estimation of accurate prevalence data. It also makes it much more difficult to determine the economic impact associated with the illness when using service utilisation data.

Among the countries participating in the EUROMENE network, the only published work on case ascertainment we have been able to identify comes from Ireland, Belgium, Norway and the UK. In Ireland, Fitzgibbon et al. found that 58% of GPs accepted CFS as a distinct entity in 1997 [25], while in Belgium, a survey of patients attending a fatigue clinic concluded that only 35% of GPs had experience of CFS, with only 23% having sufficient knowledge to treat the condition [26]. A Norwegian study found that the quality of primary care was rated poor by 60.6% of ME/CFS patients [27]. In a survey of 811 GPs in South-West England, with a response rate of 77%, 48% did not feel confident with making a diagnosis of ME/CFS and 41% did not feel confident in treatment, though 72% of GPs accepted ME/CFS as a recognisable clinical entity [28]. Bayliss et al. also found that many GPs lacked confidence and knowledge in diagnosing and managing people with ME/CFS [29]. They made available to GPs an online training module and an information pack for patients, but nearly half of all patients in their study (47%) failed to receive it. Finally, a study in South Wales concluded that the level of specialist knowledge of CFS in primary care was low, and only half the GP respondents in their survey believed that the condition actually existed [30].

As part of our work, we carried out a survey among EUROMENE participants regarding GP diagnosis of ME/CFS. Responses were received from Bulgaria, France, Germany, Ireland, Italy, Latvia, the Netherlands, Norway, Romania, Spain and the UK, and discussed at meetings of the Working Group. Only in Latvia, Norway and the UK was it reported that GPs have lists of all registered patients, which could provide denominator data for primary care-based prevalence studies. In many countries, the proportion of people with ME/CFS presenting to a GP was not known. Where estimates were made, these varied from 20% to 100% per annum. In turn, the proportion of those people with ME/CFS who, having consulted a GP, are referred to specialist care, was estimated at about 60% in Latvia and 80% in Spain. In France, it was thought that the majority were referred, while in the UK it likely varied according to region. The proportion of patients with ME/CFS who self-refer to specialist services was thought to be around 30–40% in Latvia and 80% in Spain. In the UK the figure was thought to be very low, and in most countries this was not known.

Specialist care is highly variable in nature, and different clinical specialties are involved in the different secondary care centres that offer services. In many countries, such services are non-existent. There is official guidance on treatment pathways for ME/CFS in Spain, Norway, the Netherlands and the UK. In Italy and Latvia, the majority of GPs do not recognize ME/CFS as a genuine entity. This is also true of Spain as a whole, though not of Catalonia. In France, it is generally regarded as psychological in nature. In both the UK and the Netherlands it is officially recognised, though many GPs still refuse to accept this. In Catalonia, GPs were said to be confident in diagnosing ME/CFS, but in Latvia, Norway, the Netherlands, France and the UK, there was considerable lack of confidence. The fact that this is a diagnosis made essentially through exclusion of other possible diagnoses undoubtedly contributes to this. The proportion of patients with ME/CFS who consult their GPs and are in fact diagnosed by them was generally said to be low or unknown. In those countries where a proportion was estimated (Spain, France, UK), it was thought to be around 20–50%.

Overall, it is clear that, in Europe, a high proportion of GPs, which is likely to be at least 50%, do not recognise ME/CFS as a genuine clinical entity and therefore do not diagnose it. Among those GPs who do recognise its existence, there is a marked lack of confidence in making the diagnosis and managing the condition. Therefore, estimates of the public health burden of the illness, even where these exist, are likely to underestimate substantially its true prevalence.

3.4. Determination of Costs

The overall economic burden of ME/CFS within participating European countries could be determined by the implementation of cost of illness studies. These would have to be prevalence rather than incidence based, since little is known about the prognosis of the disease. There have been Europe-wide cost of illness studies in other conditions, such as cancer [31,32], and the output from such studies can be invaluable, both in informing health and social care policy, and facilitating the management of health and care services. For example, Tarricone states: "COI is a descriptive study that can provide information to support the political process as well as the management functions at different levels of the healthcare organisations. To do that, the design of the study must be innovative, capable of measuring the true cost to society; to estimate the main cost components and their incidence over total costs; to envisage the different subjects who bear the costs; to identify the actual clinical management of illness; and to explain cost variability. In order to reach these goals, COI need to be designed as observational bottom-up studies" [33].

There have been relatively few cost of illness studies of ME/CFS. Those that exist were undertaken in the USA, Australia and the UK, the latter being the only European country where such studies have been carried out. Hunter et al. [34] compared three such studies, by Collin et al. [12], McCrone et al. [14], and Sabes-Figuera et al. [35], and two trials which contained cost data, by McCrone et al. [36], and Richardson et al. [37]. One potential problem in comparing the outcomes of such studies arises from the multiplicity of case definitions that exist for ME/CFS, as noted earlier. Whereas the cost of illness studies by both Collin et al. [12] and McCrone et al. [14] used the Fukuda definition [17], Sabes-Figuera et al. [35] undertook a primary care based study of chronic fatigue (not ME/CFS) and used a case definition not dissimilar to, but less stringent than, that of the National Institute for Health and Clare Excellence (NICE) [38], which is less restrictive than the Fukuda definition [39].

Other studies which did not meet the inclusion criteria for the 2020 Health report [37] include three American studies, by Jason et al. [40], Lin et al. [41] and Reynolds et al. [42]. The study by Jason et al. was an archive-based database study which used the CDC-1994 definition [43], as did the population-based telephone survey by Lin et al. [41]. The study by Reynolds et al. involved analysis of data from a population-based epidemiological study in Wichita, Kansas [43], which also used the CDC-1994 (Fukuda) definition. The final cost of illness study identified was an Australian population-based study by Lloyd and Pender [44], which predated the CDC-1994 definition and used the 1990 Australian definition [6].

A further cost of illness study undertaken in the UK by researchers at Sheffield Hallam University in 2007 for the charity Action for ME surveyed nearly 3000 people with ME/CFS, recruited through patient organisations. It concluded that, at that time, the total costs of ME/CFS could have been over £ 10,000 p.a. per patient, or £ 0.6 billion and £ 2.1 billion per year nationally, depending on the prevalence estimate used [45]. More than 90% of this was due to loss of income, with NHS healthcare costs quite small in comparison, though it was not made clear how cases were defined in the study.

As part of EUROMENE Working Group 3's activities, a study was undertaken in Latvia to explore to what extent the economic burden of ME/CFS could be determined from routinely collected process data. In order to do this, authors DA and UB made use of data from the Latvian Centre for Disease Prevention and Control (CDPC) and The National Health Service (NHS) of Latvia. Patient-related data were classified by ICD-10 code. ICD-10 codes of interest for this study were G93.3 Post-viral fatigue syndrome, R53 Malaise and fatigue, in particular R53.82 Chronic fatigue, unspecified (which is not identified separately in official statistics), and B94.8 (Sequelae of other specified infectious and parasitic diseases). CDCP data from primary care indicated that approximately 700 patients had ICD-10 code G93.3 assigned, while there were approximately 15,000 with ICD-10 code R53, and about 70 with code B94.8. In toto, these constitute about 0.8% of the Latvian population, which is considerably higher than the prevalence found in other comparable populations. Therefore, it is likely, though unconfirmed, that the category R53 includes a great many patients with illnesses other than ME/CFS. Category G93.3, by contrast, looks like a significant underestimate of the true population prevalence. The total of recorded health system costs for all these categories in 2017 was € 63,893,580, but the authors noted that a great deal of additional data would be required in order to make an accurate determination of the real costs to Latvian society of ME/CFS. These included numbers of confirmed ME/CFS diagnoses, costs of illness and out of pocket treatment costs per patient, patient reported outcomes, the benefits of management of ME/CFS, and return on investment [46].

Overall, on the basis of the activities of Working Group 3 on the determination of costs, we believe any attempt to measure societal economic losses attributable to ME/CFS must take into account direct and indirect costs incurred both by healthcare systems, patients and families, and this applies equally to patients who have been diagnosed as having ME/CFS and those who have not received a diagnosis. As noted, it is likely to be difficult to identify this latter group, for obvious reasons. Furthermore, we also believe that due to variation within the ME/CFS population, it can be hypothesised that, for example, severely affected people (housebound or bedbound) may incur greater costs than mildly or moderately affected people. There is no information available which could shed light on this, which clearly requires further research.

3.5. Europe-Wide Comparisons

A comprehensive review of the financing and organisation of health care in the European Union, conducted by WHO for the European Observatory on Health Systems and Policies in 2009, documented in detail the diversity of such arrangements throughout Europe [47]. Similar diversity is found in terms of health outcomes and general levels of health, but no correlation was found between accessibility of health care and funding levels in a somewhat crude analysis [48].

Against such a background, it is clearly a challenge to reach meaningful conclusions about differences in costs and losses attributable to ME/CFS across Europe, particularly given the variety of systems of healthcare delivery that exist, as well as varying stages of economic development. This is because there is a problem making valid comparisons of health care costs between countries which differ markedly in terms of wealth and levels of economic development. We propose that, for ME/CFS, purchasing power parity (PPP) adjustments should be used in any comparisons. This is a method for comparing the price of goods between countries. Using a 'basket of goods' of items commonly purchased by consumers, such as bread, milk, and shampoo, PPP is a ratio of the total cost of these goods between two countries. In this way, one can compare what 1 unit of currency can buy across different countries and convert the values back to a single reference currency [49]. For Europe, the obvious choice for the reference currency is the Euro.

In terms of specific data items required for conducting comparable cost of illness studies across countries, the need for more comprehensive data collection at the level of the individual patient is supported by other sources. Jo [50] has itemised the range and scope of the data required to support cost of illness studies, and there may be variation across countries in the availability of data items due to differences in the organisation and funding of health care. These include the data required to identify both system costs and costs to the individual with ME/CFS and those close to him or her. A recent study undertaken in Italy to determine costs to the individual assessed the direct and indirect costs of ME/CFS via a questionnaire distributed via Italian patient associations [51]. By estimating the cost of medical procedures and the cost of lost working time, the study arrived at an estimate for the total economic burden of the disease. The questionnaire was discussed in detail by Working Group 3 both in terms of its specificity and applicability to different countries. This study could, the Working Group felt, be repeated in other countries, in order to enable the acquisition of data capable of direct comparison between countries. The study also aimed to relate the cost impact on people with ME/CFS to their clinical condition and the severity of the disease through the use of the EuroQol-5D instrument to assess health status [52,53]. An alternative approach to achieving this objective could involve the use of instruments for resource use measurement [54,55].

4. Recommendations

Our review has led us to a number of conclusions and recommendations as to how research in this area could be advanced. For example, there are a number of serious omissions in the availability of data necessary for investigation of the economic impact of ME/CFS and these are summarised in Table 2. The main areas of concern relate to: case definition; case identification; prevalence and incidence rates; economic cost estimates; data items; and data audits.

In addition, a major problem we have identified is the lack of any information on the economic impact of severe ME/CFS. For this reason, these patients were necessarily excluded from this review, but, since it can be hypothesised that the economic impact of such illness is likely to be greater than that of mild or moderate disease, this is clearly a subject that requires further investigation. It is of considerable importance that this be confirmed empirically, and the scale of any such variation be determined, as a necessary prerequisite to the overall determination of the economic impact of ME/CFS in Europe.

Table 2. Summary of Recommendations.

	Area of Concern	Recommendation
1	Case definition	That there should be Europe-wide adoption of the Fukuda (CDC-1994) case definition alongside the Canadian Consensus Criteria (CCC).
2	Case identification	That a common symptom checklist should be used, capable of being mapped by algorithms onto both the Fukuda case definition and the CCC.
3	Prevalence and incidence	Better descriptive epidemiological information is required, as a basis for economic investigation. This should include information concerning the proportion of severely affected people, as there are likely to be different cost implications for such people, in comparison with those with mild or moderate illnesses.
4	Economic investigation	Prevalence based cost of illness studies, based on these case definitions, should be carried out in different countries, to determine the overall cost burden attributable to ME/CFS.
5	Data items	A list of data items required for cost of illness studies has been identified (though not reported here). Individual participating countries should examine this, to ensure that, insofar as these are derivable from routine data collection, systems are in place to ensure that they are collected.
6	Data audit	The availability in participating countries of the relevant data items referred to above which are required for cost-of illness studies should be examined, with a view to achieving convergence, and facilitating international comparisons.
7	Relationship between disease severity and economic impact	The EuroQol-5D instrument [52,53] should be used as a generic measure of health status and as a multi-attribute utility instrument to determine the relationship, if any, between disease severity and economic impacts, as in the Italian study reported in this document [51], and to inform future economic evaluations in ME/CFS. We further recommend that the Italian study be replicated in other countries, to enable international comparisons to be made.
8	International comparisons and compilation of Europe-wide statistics	Given the diversity of patterns of health care organisations and health funding, as well as of outcomes and general levels of health, and of national wealth and levels of economic development, we recommend the use of purchasing power parities (PPP) in order both to make valid international comparisons and to collate meaningful statistics at a European level.

5. Conclusions

Problems impeding research into the economic impacts of ME/CFS in Europe include those of arbitrary case definitions, incomplete and inadequate information on incidence and prevalence, failure to diagnose the condition on the part of a high proportion of doctors, and how to determine costs and losses, given variation within the ME/CFS population (e.g., between severely affected and mildly or moderately affected patients, with different cost implications for each of these categories). There are also important issues relating to the heterogeneity of national economies and health care systems within Europe. As a result, we have made a set of recommendations concerning how further progress might be made in this area, in particular in relation to case definition, case identification, descriptive epidemiology, methods of economic investigation, data items required to support such investigation, data audit of quality and completeness, and international comparisons and the compilation of Europe-wide statistics. The relationship between disease severity and economic impact should be a high priority for future research.

Author Contributions: All authors participated in the discussions which led to the development of the initial draft of this paper, and all contributed to its revision. The initial version was drafted by D.F.H.P. on the basis of the contributions of all the authors, and the final revised version was prepared by J.C. Conceptualisation, D.F.H.P., D.A., U.B., E.B., J.C., J.-D.d.K., L.G., D.A.H., R.M.H., D.T. and X.W.-S. Methodology, D.F.H.P., D.A., U.B., E.B., J.C., J.-D.d.K., L.G., D.A.H., R.M.H., D.T. and X.W.-S. Validation, D.F.H.P., D.A., U.B., E.B., J.C., J.-D.d.K., L.G., D.A.H., R.M.H., D.T. and X.W.-S. Investigation, D.F.H.P., D.A., U.B., E.B., J.C., J.-D.d.K., L.G., D.A.H., R.M.H., D.T. and X.W.-S. Writing—original draft preparation, D.F.H.P. Writing—review and editing, D.F.H.P., D.A., U.B., E.B., J.C., J.-D.d.K., L.G., D.A.H., R.M.H., D.T. and X.W.-S. (all authors contributed to the review, while J.C. edited the revised draft, and D.F.H.P. undertook the final formatting). Visualisation, D.F.H.P., D.A., U.B., E.B., J.C., J.-D.d.K., L.G., D.A.H., R.M.H., D.T. and X.W.-S. Project administration, D.F.H.P. All authors have read and agreed to the published version of the manuscript.

Funding: This research received no external funding.

Acknowledgments: The work of EUROMENE was made possible by funding from the European Union to support network activities, administered by the COST Association, Brussels, Belgium.

Conflicts of Interest: The authors declare no conflict of interest.

Abbreviations

ME/CFS	Myalgic Encephalomyelitys/Chronic Fatigue Syndrome
EUROMENE	European Network on ME/CFS
COI	Cost of illness
EuroQol-5D	An instrument for measuring quality of life
Bn	Billion
CCC	Canadian Consensus Criteria
PPP	Purchasing Power Parity
ICD-10 CM	The international diagnostic classification standard for reporting diseases, disorders, injuries and health conditions for all clinical and research purposes

References

1. Lindan, R. Benign Myalgic Encephalomyelitis. *Can. Med. Assoc. J.* **1956**, *75*, 596–597.
2. Acheson, E.D. The clinical syndrome variously called myalgic encephalomyelitis, Iceland disease and epidemic neuromyasthenis. *Am. Med.* **1959**, *26*, 589–595. [CrossRef]
3. Carruthers, B.M.; Jain, A.K.; De Meirleir, K.L.; Peterson, D.L.; Klimas, N.G.; Lerner, A.M.; Bested, A.C.; Flor-Henry, P.; Joshi, P.; Powles, A.P.; et al. Myalgic encephalomyelitis/chronic fatigue syndrome: Clinical working case definition, diagnostic and treatment protocols. *J. Chronic Fatigue Syndr.* **2003**, *11*, 7–116. [CrossRef]
4. Johnstone, S.C.; Staines, D.R.; Marshall-Gradisnik, S.M. Epidemiological characteristics of chronic fatigue syndrome/myalgic encephalomyelitis in Australian patients. *Clin. Epidemiol.* **2016**, *8*, 97–107. [CrossRef] [PubMed]
5. Pheby, D.; Lacerda, E.; Nacul, L.; de Lourdes Drachler, M.; Campion, P.; Howe, A.; Poland, F.; Curran, M.; Featherstone, V.; Fayyaz, S.; et al. A disease register for ME/CFS: Report of a pilot study. *BMC Res. Notes* **2011**, *4*, 139–146. [CrossRef] [PubMed]
6. Lloyd, A.R.; Hickie, I.; Boughton, C.R.; Wakefield, D.; Spencer, O. Prevalence of chronic fatigue syndrome in an Australian population. *Med. J. Aust.* **1990**, *153*, 522–528. [CrossRef] [PubMed]
7. Action for ME. Available online: https://actionforme.org.uk/what-is-me/introduction/ (accessed on 6 January 2020).
8. Brenna, E.; Gitto, L. The economic burden of Chronic Fatigue Syndrome/Myalgic Encephalomyelitis (CFS/ME): An initial summary of the existing evidence and recommendations for further research. *Eur. J. Pers. Cent. Healthc.* **2017**, *5*, 413–420. [CrossRef]
9. Brenna, E.; Gitto, L. I costi economici e sociali della syndrome da affaticamento cronico/encefalomielite mialgica: Una prima rassegna della letteratura sull'argomento. *Politiche Sanitarie* **2018**, *2*, 59–66.
10. ICD-codes.com. Clinical Code Database. Available online: https://icd-codes.com/icd10cm/ (accessed on 6 January 2020).
11. Dx Revision Watch. Available online: https://dxrevisionwatch.com/2019/04/30/icd-11-recently-processed-proposals-for-postviral-fatigue-syndrome-me-cfs-fatigue-and-bodily-distress-disorder/ (accessed on 6 January 2020).
12. Collin, S.M.; Crawley, E.; May, M.T.; Sterne, J.A.C.; Hollingworth, W. The impact of CFS/ME on employment and productivity in the UK: A cross-sectional study based on the CFS/ME national outcomes database. *BMC Health Serv. Res.* **2011**, *11*, 217. [CrossRef]
13. EUROMENE. Available online: http://www.euromene.eu (accessed on 20 October 2019).
14. McCrone, P.; Darbishire, L.; Ridsdale, L.; Seed, P. The economic cost of chronic fatigue and chronic fatigue syndrome in UK primary care. *Psychol. Med.* **2003**, *33*, 253–261. [CrossRef]
15. Memorandum of Understanding of COST Action 15111—EUROMENE. Available online: http://www.cost.eu/actions/CA15111/#tabs/Name;overview (accessed on 18 March 2020).
16. Brurberg, K.G.; Fønhus, M.S.; Larun, L.; Flottorp, S.; Malterud, K. Case definitions for chronic fatigue syndrome/myalgic encephalomyelitis (CFS/ME): A systematic review. *BMJ Open* **2014**, *4*, e003973. [CrossRef] [PubMed]

17. Fukuda, K.; Straus, S.E.; Hickie, I.; Sharpe, M.C.; Dobbins, J.G.; Komaroff, A.; International Chronic Fatigue Syndrome Study Group. The Chronic Fatigue Syndrome: A comprehensive approach to its definition and study. *Ann. Intern. Med.* **1994**, *121*, 953–959. [CrossRef] [PubMed]

18. Nacul, L.C.; Lacerda, E.M.; Pheby, D.; Campion, P.; Molokhia, M.; Fayyaz, S.; Drachler, M.; Poland, F.A.; Howe, A.; Leite, J.C. Prevalence and incidence of myalgic encephalomyelitis/ chronic fatigue syndrome (ME/CFS) in three regions of England: A repeated cross-sectional study in primary care. *BMC Med.* **2011**, *9*, 91. [CrossRef] [PubMed]

19. Institute of Medicine (IOM). *Beyond Myalgic Encephalomyelitis/Chronic Fatigue Syndrome: Redefining an Illness*; The National Academies Press: Washington, DC, USA, 2015.

20. EUROMENE Working Group 1. *European Network on Myalgic Encephalomyelitis/Chronic Fatigue Syndrome (EUROMENE)—Working Group 1 (Epidemiology) Deliverable 7: Joint Publication Manuscript on ME/CFS European Practices in Diagnosis and Treatment Based on Survey Data*; 2017; Available online: http://www.euromene.eu/ workinggroups/deliverables/deliverable.html (accessed on 18 March 2020).

21. Osoba, T.; Pheby, D.; Gray, S.; Nacul, L. The Development of an Epidemiological Definition for Myalgic Encephalomyelitis/Chronic Fatigue Syndrome. *J. Chronic Fatigue Syndr.* **2007**, *14*, 61–84. [CrossRef]

22. Jason, L.A.; Torres-Harding, S.R.; Jurgens, A.; Helgerson, J. Comparing the Fukuda criteria and the Canadian case definition for chronic fatigue syndrome. *J. Chronic Fatigue Syndr.* **2004**, *12*, 2004–2052. [CrossRef]

23. Lindal, E.; Stefansson, J.G.; Bergmann, S. The prevalence of chronic fatigue syndrome in Iceland—A national comparison by gender drawing on four different criteria. *Nord. J. Psychiatry* **2002**, *56*, 273–277. [CrossRef] [PubMed]

24. Wessely, S.; Chalder, T.; Hirsch, S.; Wallace, P.; Wright, D. The prevalence and morbidity of chronic fatigue and chronic fatigue syndrome: A prospective primary care study. *Am. J. Public Health* **1997**, *87*, 1449–1455. [CrossRef]

25. Fitzgibbon, E.J.; Murphy, D.; Shea, K.; Kelleher, C. Chronic debilitating fatigue in Irish general practice: A survey of general practitioners' experience. *Br. J. Gen. Pract.* **1997**, *47*, 618–622.

26. Van Hoof, E. The doctor-patient relationship in chronic fatigue syndrome: Survey of patient perspectives. *Qual. Prim. Care* **2009**, *17*, 263–270.

27. Hansen, A.H.; Lian, O.S. How do women with chronic fatigue syndrome/myalgic encephalomyelitis rate quality and coordination of healthcare services? A cross-sectional study. *BMJ Open* **2016**, *6*, e010277. [CrossRef]

28. Bowen, J.; Pheby, D.; Charlett, A.; McNulty, C. Chronic Fatigue Syndrome: A survey of GPs' attitudes and knowledge. *Family Pract.* **2005**, *22*, 389–393. [CrossRef] [PubMed]

29. Bayliss, K.; Riste, L.; Band, R.; Peters, S.; Wearden, A.; Lovell, K.; Fisher, L.; Chew-Graham, C.A. Implementing resources to support the diagnosis and management of Chronic Fatigue Syndrome/Myalgic Encephalomyelitis (CFS/ME) in primary care: A qualitative study. *BMC Fam. Pract.* **2016**, *17*, 66. [CrossRef] [PubMed]

30. Thomas, M.A.; Smith, A.P. Primary healthcare provision and Chronic Fatigue Syndrome: A survey of patients' and General Practitioners' beliefs. *BMC Fam. Pract.* **2005**, *6*, 49. [CrossRef]

31. Luengo-Fernandez, R.; Leal, J.; Gray, A.; Sullivan, R. Economic burden of cancer across the European Union: A population-based cost analysis. *Lancet Oncol.* **2013**, *14*, 1165–1174. [CrossRef]

32. Ó Céilleachair, A.J.; Hanly, P.; Skally, M.; O'Neill, C.; Fitzpatrick, P.; Kapur, K.; Staines, A.; Sharp, L. Cost comparisons and methodological heterogeneity in cost-of-illness studies: The example of colorectal cancer. *Med. Care* **2013**, *51*, 339–350. [CrossRef] [PubMed]

33. Tarricone, R. Cost-of-illness analysis: What room in health economics? *Health Policy* **2006**, *77*, 51–63. [CrossRef]

34. Hunter, R.M.; James, M.; Paxman, J. *Chronic Fatigue Syndrome/Myalgic Encephalomyelitis: Counting the Cost—Full Report*; 2020 Health: London, UK, 2018.

35. Sabes-Figuera, R.; McCrone, P.; Hurley, M.; King, M.; Donaldson, A.N.; RidsdaleS, L. The hidden cost of chronic fatigue to patients and their families. *BMC Health Serv. Res.* **2010**, *10*, 56. [CrossRef]

36. McCrone, P.; Sharpe, M.; Chalder, T.; Knapp, M.; Johnson, A.L.; Goldsmith, K.A.; White, P.D. Adaptive Pacing, Cognitive Behaviour Therapy, Graded Exercise, and Specialist Medical Care for Chronic Fatigue Syndrome: A Cost-Effectiveness Analysis. *PLoS ONE* **2012**, *7*, e40808. [CrossRef]

37. Richardson, G.; Epstein, D.; Chew-Graham, C.; Dowrick, C.; Bentall, R.P.; Morriss, R.K.; Wearden, A.J. Cost-effectiveness of supported self-management for CFS/ME patients in primary care. *BMC Fam. Pract.* **2013**, *14*, 12. [CrossRef]

38. Turnbull, N.; Shaw, E.J.; Baker, R.; Dunsdon, S.; Costin, N.; Britton, G.; Kuntze, S. *NICE Guideline—Chronic Fatigue Syndrome/Myalgic Encephalomyelitis (or Encephalopathy): Diagnosis and Management of Chronic Fatigue*

Syndrome/Myalgic Encephalomyelitis (or Encephalopathy) in Adults and Children; Royal College of General Practitioners: London, UK, 2007; pp. 65–66.

39. Reeves, W.C.; Lloyd, A.; Vernon, S.D.; Klimas, N.; Jason, L.A.; Bleijenberg, G.; Evengard, B.; White, P.D.; Nisenbaum, R.; Unger, E.R. Identification of ambiguities in the 1994 chronic fatigue syndrome research case definition and recommendations for resolution. *BMC Health Serv. Res.* **2003**, *3*, 25. [CrossRef]

40. Jason, L.A.; Benton, M.C.; Valentine, L.; Johnson, A.; Torres-Harding, S. The Economic impact of ME/CFS: Individual and societal costs. *Dyn. Med.* **2008**, *7*, 6. [CrossRef] [PubMed]

41. Lin, J.-M.S.; Resch, S.C.; Brimmer, D.J.; Johnson, A.; Kennedy, S.; Burstein, N.; Simon, C.J. The economic impact of chronic fatigue syndrome in Georgia: Direct and indirect costs. *Cost Eff. Resour. Alloc.* **2011**, *9*, 1. [CrossRef] [PubMed]

42. Reynolds, K.J.; Vernon, S.D.; Bouchery, E.; Reeves, W.C. The economic impact of chronic fatigue syndrome. *Cost Eff. Resour. Alloc.* **2004**, *2*, 4. [CrossRef] [PubMed]

43. Reyes, M.; Nisenbaum, R.; Hoaglin, D.; Unger, E.R.; Emmons, C.; Randall, B.; Stewart, J.A.; Abbey, S.; Jones, J.F.; Gantz, N.; et al. Prevalence and incidence of chronic fatigue syndrome in Wichita, Kansas. *Arch. Intern. Med.* **2003**, *163*, 1530–1536. [CrossRef]

44. Lloyd, A.R.; Pender, H. The economic impact of chronic fatigue syndrome. *Med. J. Aust.* **1992**, *157*, 599–601. [CrossRef] [PubMed]

45. Bibby, J.; Kershaw, A. *How Much is ME Costing the Country*; The Survey and Statistical Research Centre, Sheffield Hallam University: Sheffield, UK, 2007.

46. Araja, D.; Pheby, D.; Hunter, R.; Brenna, E.; Gitto, L.; Berkis, U.; Lunga, A.; Ivanovs, A.; Murovska, M. *Patient Reported Outcomes in Evaluation of Socio-Economic Impact of Myalgic Encephalomyelitis/Chronic Fatigue Syndrome (ME/CFS) to Society. Science Week*; Riga Stradins University: Riga, Latvia, 2019.

47. Thomson, S.; Foubister, T.; Mossialos, E. *Financing Health Care in the European Union: Challenges and Policy Responses*; WHO (for the European Observatory on Health Systems and Policies): Geneva, Switzerland, 2009.

48. Björnberg, A. *Euro Health Consumer Index 2017: Report*; Health Consumer Powerhouse: Marseillan, France, 2018.

49. Eurostat/OECD. *Eurostat/OECD Methodological Manual on Purchasing Power Parities*; Publications Office of the European Union: Luxembourg, 2012.

50. Jo, C. Cost of illness studies: Concepts, scopes and methods. *Clin. Mol. Hepatol.* **2014**, *20*, 327–337. [CrossRef]

51. Beretta Ardino, R.; Lorusso, L. La sindrome da affaticamento cronico/encefalomielite mialgica: Le caratteristiche della malattia e il ruolo dell'Associazione Malati CFS. *Politiche Sanitarie* **2018**, *2*, 91–95.

52. EQ-5D. Available online: http/euroqol.org (accessed on 20 October 2019).

53. New version of the EQ-5D-5L. Available online: https://euroqol.org/new-version-of-the-eq-5d-5l-user-guide-now-available-for-download/ (accessed on 8 January 2020).

54. DIRUM. Available online: www.DIRUM.org (accessed on 7 January 2002).

55. Ridyard, C.H.; Hughes, D.A.; DIRUM Team. Development of a database of instruments for resource-use measurement: Purpose, feasibility, and design. *Value Health* **2012**, *15*, 650–655. [CrossRef]

Appendix

Audio and Audiovisual Links

This monograph contains links to an audio recording and several videos.

1. We are providing a link to the audio recording of Whitney Dafoe's first-person description of life as a very severely ill ME/CFS patient. The intent is to provide patients who are too ill to read the article with the ability to listen to a recorded version. We gratefully acknowledge Michael Thurston for the sound editing and enhancement of the audio tracks of the narration of Whitney Dafoe's description of living life as a very severely affected ME/CFS patient. We thank Baruch Zeichner, host of Paradigms Podcast/Radio: inspiring people, music, life on Earth, a radio talk show/podcast, for hosting the audio file.

Link to audio file of Whitney Dafoe's article:

https://paradigms.life/Whitney_Dafoe/Dafoe-Full-Version.mp3

2. We also provide a QR code which links to the audio file for the convenience of some.

QR for Whitney Dafoe's article:

3. We are providing a link to pre-existing videos that serve as an audio–visual summary. Videos express what words alone cannot. We thank Natalie Boulton and her son Josh Biggs of the U.K. for permission to use a link to their films (http://www.dialogues-mecfs.co.uk/films/severeme/). Natalie became the caregiver to her daughter in 1990, relinquishing careers as an artist and teacher. Her son, Josh Biggs, served as the cameraman and editor of their films.

Link to summary videos: http://www.dialogues-mecfs.co.uk/films/severeme/

MDPI
St. Alban-Anlage 66
4052 Basel
Switzerland
www.mdpi.com

Healthcare Editorial Office
E-mail: healthcare@mdpi.com
www.mdpi.com/journal/healthcare